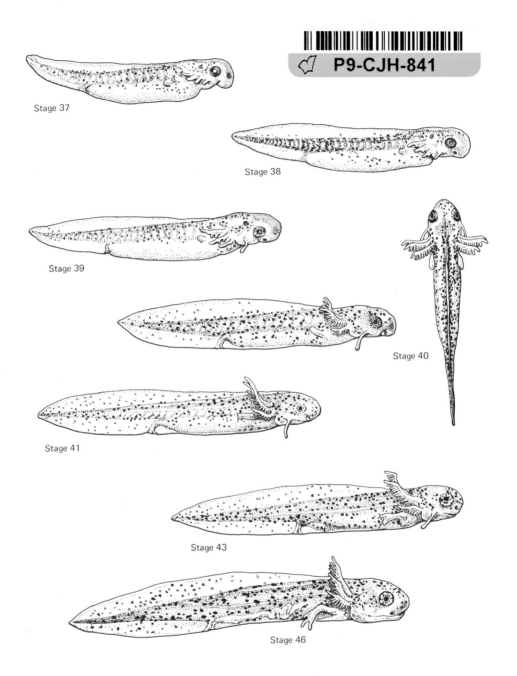

P9-CJH-841

Stage 37

Stage 38

Stage 39

Stage 40

Stage 41

Stage 43

Stage 46

Stages in Development of *Ambystoma punctatum* (Redrawn from photographs of the original drawings from the Ross G. Harrison Silliman Lectures ORGANIZATION AND DEVELOPMENT OF THE EMBRYO, ed. Sally Wilens. (1969) Yale University Press, New Haven.)

PATTEN'S FOUNDATIONS OF EMBRYOLOGY

Nirmal K. Das
Lennoxville
22 November 1994

PATTEN'S FOUNDATIONS OF EMBRYOLOGY

FIFTH EDITION

BRUCE M. CARLSON, M.D., PH.D.

Department of Anatomy and Biological Sciences
University of Michigan

McGraw-Hill, Inc.
New York St. Louis San Francisco Auckland Bogotá
Caracas Lisbon London Madrid Mexico City Milan
Montreal New Delhi San Juan Singapore
Sydney Tokyo Toronto

This book was set in Times Roman by the William Byrd Press.
The editors were Elizabeth Dollinger and Jack Maisel;
the cover designer was Fern Logan;
the production supervisor was Joe Campanella.
Arcata Graphics/Halliday was printer and binder.
Cover micrograph of the neural tube and somite in the
developing chick embroyo by K.W. Tosney.

PATTEN'S FOUNDATIONS OF EMBRYOLOGY

678910 HDHD 99876543

ISBN 0-07-009902-2

LIBRARY OF CONGRESS
Library of Congress Cataloging-in-Publication Data

Patten, Bradley Merrill, 1889–1971.
 [Foundations of embryology]
 Patten's foundations of embryology/Bruce M. Carlson. — 5th ed. p. cm.
 Bibliography: p.
 Includes index.
 ISBN 0-07-009902-2
 1. Embryology. I. Carlson, Bruce M. II. Title. III. Title: Foundations of embryology.
QL955.P23 1988
599.03'3—dc 19
                                                              87-28586
                                                              CIP

This book is printed on acid-free paper.

ABOUT
THE AUTHOR

Bruce M. Carlson is a professor in the anatomy and biology departments at the University of Michigan. Since receiving his M.D. and Ph.D. degrees from the University of Minnesota, he has taught courses in both developmental biology and medical embryology over the past 20 years. Dr. Carlson maintains an active research program on the development and regeneration of limbs and muscles and has written or edited 15 books and over 100 articles in his field. For his research on regeneration, he was awarded the Newcomb-Cleveland prize of the American Association for the Advancement of Science. Dr. Carlson has also written a number of articles for encyclopedias and fishing magazines.

TO JEAN AND THE KIDS

CONTENTS

PREFACE

In preparing the fifth edition of the *Foundations of Embryology*, I have made some major changes in and revisions of the previous edition. Approximately 40 percent of the text of this edition is new, and over 100 new illustrations have been added. Yet the fundamental objective of the book remains unchanged: to provide a coherent description of normal embryonic development so that the student will acquire an organizational frame of reference for understanding more advanced concepts of the mechanisms of both normal and abnormal development.

Presentation of the normal morphology of development, however, is not the sole goal of this text. In contemporary developmental biology the ideal goal is to integrate morphological, experimental, molecular, and conceptual approaches to studying development. If all these approaches were presented in any detail, the text would become unmanageably large for standard college courses. Therefore, it has been necessary to set up priorities for the depth in which material is presented. As before, the backbone of this introductory text is a coherent description of the way in which the fertilized egg develops into a free-living organism. However, in this edition the treatment of developmental anatomy per se has been streamlined so that students with only a single course in introductory biology should be able to follow the text. I believe that it is important for students to know what causes the structures they see in the laboratory to form. Thus I have placed considerable emphasis on the tissue interactions, migrations, substrate relations, etc., that are involved in this process. Treatment of such areas involves not only the facts but the techniques used to obtain new data. Areas where biochemical or molecular investigations have made major contributions to our understanding of development have also been included in this edition, but for economy of space the techniques used to obtain these data have not been described in great detail. A major goal of this edition has been to show where some of the well-investigated model systems currently popular in developmental biology relate to the systematic study of embryogenesis. This is designed to facilitate the student's ability to relate different ways of studying development to one another.

Major specific changes in the text are as follows:

(1) Addition of sea urchin and mouse fertilization and early development (Chaps. 3 through 5),

(2) Greatly increased emphasis on the role of the extracellular matrix and cell adhesion molecules in development (Chap. 1 and many others),

(3) Expansion of coverage of the neural crest and somitogenesis (Chap. 6),

(4) Addition of a general section of cytodifferentiation and major updating of muscle and cartilage differentiation (Chap. 9),

(5) Addition of a new chapter on the skin (Chap. 10),

(6) Greater emphasis on the use of genetic mutants in the study of development (several chapters),

(7) Complete reorganization of the chapter on the nervous system (Chap. 12),

(8) Inclusion of a number of summary tables throughout the text,

(9) Reduction in the length of the Appendix.

As always, a book of this type is the product of a joint effort of a number of people besides the author. Margaret Croup, who again prepared the new artwork for this edition, continued to work magic with the crude sketches and ideas that were presented to her. The sage advice of William Brudon, who did the artwork for the third edition, has been appreciated by both of us. Thanks to the expertise and hard work of Jackie Rodgers, the entire text was entered into a word processor in a manner that resulted in a great saving of time for me. A number of scientific colleagues were very gracious in allowing the use of their original illustrations to clarify the text. Special thanks are due to my teaching partner, Kathryn Tosney, for ideas and comments as well as for supplying me with a number of her beautiful scanning electron micrographs of embryos. The reviewers of the manuscript were very helpful with their comments and suggestions. Finally, I would like to thank the editorial and production staff at McGraw-Hill for their efforts.

Bruce M. Carlson

EMBRYOLOGY—ITS SCOPE, HISTORY, AND SPECIAL FIELDS

HISTORICAL BACKGROUND

How we develop before we are born has always been a matter of great interest. "Where did I come from?" is one of the first thoughtful questions a child asks. Many primitive societies showed an intense interest in our prenatal origins, but their curiosity often led to unfounded speculation and mysticism. Aristotle's work on embryos is significant for us, but not because of the information he secured, surprisingly accurate as some of it was. Rather, his work is for us a symbol of the beginning of the turning of the human mind away from superstition and conjecture and toward observation. Unfortunately, such an approach did not take firm root. Through much of the Middle Ages the spark that the better Greek and Roman scholars had been attempting to fan was smothered by bigotry and authoritarianism.

But the manner of approach was not the only reason for the lag in the growth of knowledge about embryology. The early phases of development involve exceedingly minute structures, and curiosity and a willingness to learn by observation were not enough. Galen, it is true, had learned much about the structure of relatively advanced fetuses, but it was not until the close of the seventeenth century, when the microscope was being developed into an efficient instrument, that the early stages of embryology could be studied effectively.

The human sperm was first seen by Hamm and Leeuwenhoek in 1677, shortly after ovarian follicles were described by de Graaf (1672). Even then the significance of the gametes in development was not understood. Two camps grew up, one contending that the sperm contained the new individual in miniature (Fig. 1-1), which was merely nourished in the ovum, and the other

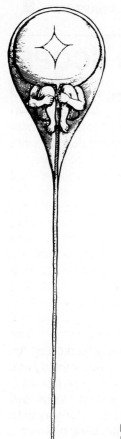

FIGURE 1-1
Reproduction of Hartsoeker's drawing of a spermatozoon showing a preformed individual (homunculus) in the sperm head. (From *Essai de Diop-
trique*, Paris, 1694.)

arguing that the ovum contained a minute body which was in some way
stimulated to growth by the seminal fluid. For a time, the ovists seemed to gain
the ascendency when Bonnet (1745) discovered that the eggs of some insects
can develop parthenogenetically. But preconceived ideas are persistent, and
the war between the spermists and the ovists continued to be bitter and
vituperative. Ardor for the spermist cause was not dampened even by the
inevitable absurdity of the encasement concept—the implication that each
miniature must in turn enclose the miniature of the next generation, and so on
for all future generations.

   This bootless controversy continued into the next century, until it was finally
laid to rest by the studies of Spallanzani (1729–1799) and Wolff (1733–1794).
The work of Lazzaro Spallanzani is of special interest to us, for it was an initial
step in bringing the experimental method to bear on embryological problems.
By means of an ingeniously planned series of experiments, he demonstrated

that both the female and the male sex products are necessary for the initiation of development.

Spallanzani's contemporary, Kaspar Friedrich Wolff, at the age of 26, wrote a brilliant thesis setting forth his conception of *epigenesis*. The hypothesis that embryological development occurs through progressive growth and differentiation rapidly replaced the old encasement theories. Although this was an important step forward, it rested too largely on purely theoretical grounds to give a lasting impetus to the subject. For more than half a century little was published to advance knowledge of the early stages of development, even though accurate observation and recording were becoming more common.

The important work of Karl Ernst von Baer (1828) first emphasized the fact that the more general basic features of any large group of animals appear earlier in development than do the special features that are peculiar to different members of the group. This concept is sometimes referred to as *von Baer's law*. It was von Baer also who demonstrated the existence of the germ layers in embryos. However, the real significance of these layers could not be grasped until the cellular basis of animal structure became known. With the formulation of the cell theory by Matthias Schleiden and Theodor Schwann (1839), the foundations of modern embryology and histology were simultaneously laid. The knowledge that the adult body is composed entirely of cells and cell products paved the way for a realization of the basic fact of embryology: The body of the new individual is developed from a single cell that is formed by the union, in fertilization, of a germ cell contributed by the male parent with a germ cell contributed by the female parent. Thus, although there had been curiosity since before the dawn of written history and although critical observation had begun, with Aristotle, to replace conjecture, it was not until the development of the microscope, the advent of the experimental method, and the discovery of the cellular structure of the body that embryology began to become a science.

EMBRYOLOGY

Essentially all higher animals start their lives from a single cell, the fertilized ovum (*zygote*). As its name implies, the zygote has a dual origin from two gametes—a spermatozoon from the male parent and an ovum from the female parent. The time of fertilization, when the spermatozoon meets the egg, represents the starting point in the life history, or *ontogeny*, of the individual. In its broadest sense, ontogeny refers to the individual's entire life span.

A century ago the great German biologist August Weismann (1834–1914) made the important distinction between the *soma* (body) and the germ-cell line (*gametes*). Weismann thought that the germ-cell line was all-important for perpetuation of the species, and that the soma was primarily a vehicle for protecting and perpetuating the germ plasm. In more contemporary biological thought, particularly with the ever-increasing emphasis on sociobiology, this viewpoint may seem somewhat restrictive, but it does provide a convenient

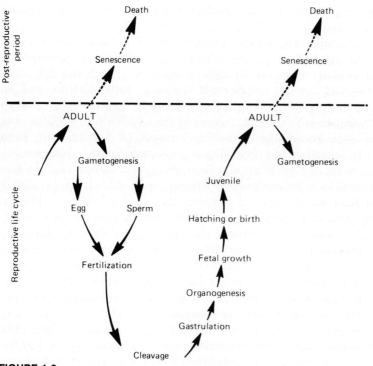

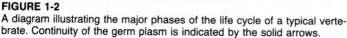

FIGURE 1-2
A diagram illustrating the major phases of the life cycle of a typical vertebrate. Continuity of the germ plasm is indicated by the solid arrows.

framework for looking at the perpetuation of life (Fig. 1-2). Once an
individual has passed the reproductive years, the remainder of its ontogeny
does not provide direct physical input into the generative process. Strictly
speaking, embryology is usually regarded as the period starting with fertilization and ending with metamorphosis in Amphibia, hatching in birds, and
birth in mammals. Books on vertebrate embryology, however, also deal with
the development and maturation of gametes (*gametogenesis*). Thus, this text
covers most of the phases of ontogeny shown below the dashed line in Fig.
1-2. It is important, however, to recognize that hatching, birth, and metamorphosis merely represent convenient landmarks in a continuing process
and that development in reality consists of an uninterrupted series of
correlated events.

The study of embryology now encompasses a bewildering array of approaches, facts, techniques, and concepts. It is impossible to cover them all in
a single course or a single book. Over the years various special fields have
developed within the general subject of embryology; this has been the logical
outcome of progress, both conceptual and technical, in the natural sciences as
a whole.

Earlier studies were chiefly concerned with learning the basic structural pattern of the embryonic body. Interest gradually shifted, however, from general body configuration to more detailed studies of the structure and arrangement of the minute internal organs of the embryo. Work of this type received a great impetus and gained much in accuracy from techniques that were developed between 1880 and 1890: the making of serial sections and the wax-plate method (His and Born) of making accurate reconstructions from such sections. Toward the turn of the century, attention began to shift to the finer cellular structure of embryos. However, there was still relatively little physiological or experimental work, and publications were concerned primarily with drawings and explanations of the structural features of embryos of various ages. Work of this type is generally characterized as *descriptive embryology.*

Having its roots in the same type of descriptive work, the field of *comparative embryology* arose late in the nineteenth century. A driving force behind the development of this field was a great interest in evolution, which was the dominating factor in biology during this period. Early comparative studies were carried out on the most readily available forms and on those which were easiest to deal with by simple techniques. The embryos of marine invertebrates were, and continue to be, very popular objects of study. With improving methodology, a wealth of detailed information has been gathered from the study of many different types of embryos. As human material has become more readily available, the relationships between the early developmental stages of humans and those of many other animals have been strikingly well established.

The gradual acquisition of detailed and accurate information on the structural stages embryos pass through in the course of their development paved the way for the growth of *experimental embryology.* This branch of the subject aims at ascertaining the factors which activate or regulate developmental processes. Descriptive embryology tells us when and how a process is carried out; experimental embryology seeks to find out why a process occurs at that specific time and in just that particular manner. One of the pioneers in this field was Wilhelm Roux (1850–1924), who performed a seminal experiment that ushered in the era of experimental embryology. As a test of the concepts of preformation and epigenesis, Roux (1888) destroyed one cell (blastomere) of a two-cell frog embryo with a hot needle. He wanted to learn whether the remaining cell would give rise to only half an embryo (as would be expected in accordance with preformation) or whether that cell could, in its subsequent development, restore the deficiency. Although his experimental results proved to be somewhat misleading, other investigators, stimulated by this work, soon showed that if the cells of a two-cell frog embryo are entirely separated, each cell is capable of giving rise to a complete individual. This procedure provided experimental proof of the untenability of the preformationist doctrine and laid the foundations for a new field. Roux coined the German word *Entwicklungsmechanik* for such experimental studies; its literal translation into English is *developmental mechanics.* Waddington (1956) felt that this term carries the unfortunate implication that the phenomena involved are largely physical and machinelike. He preferred the term *epigenetics*, which expresses the concept that "development is brought about through a series of causal interactions between the various parts; and also reminds one that genetic factors are among the most important determinants of development" (Waddington, 1956, p. 10).

Recent spectacular advances in molecular biology have given new impetus to the field of *chemical embryology*. Earlier in this century chemical studies involving embryos were largely descriptive (Needham, 1931). Biochemical investigations are currently playing a basic role in broadening our knowledge of the physiology of the embryo. These investigations are helping us understand how, through the activities of deoxyribonucleic acid (DNA) and ribonucleic acid (RNA), the information contained in the genetic material of the fertilized ovum presides over the fabrication of the specific chemical and structural components of the embryo.

Teratology is the branch of embryology concerned with the study of malformations. Drawings and images of abnormal individuals are among the oldest biological records. In ancient times the birth of a "monster" was supposed to portend events to come. As a matter of fact, the word *monster* is derived from the Latin verb meaning *to show*. This usage was based on the belief that the birth of a malformed infant was a supernatural way of showing what future happenings to expect. In the Middle Ages the writings on teratology seemed to degenerate into contests to discover who could assemble the most bizarre malformations. When an author was falling behind his competitors, he apparently had no compunctions about drawing on his imagination to fabricate weird monstrosities. Those interested in this phase of teratology can find fascinating illustrations in Gould and Pyle's *Anomalies and Curiosities of Medicine* (Fig. 1-3). Recently, work in teratology has taken on an entirely new aspect. With birth defects having moved well up among the top 10 causes of death in countries with advanced levels of medicine, great effort and much money are being spent to identify and eliminate factors causing congenital defects. Investigations into the mechanisms by which *teratogens* (substances that cause birth defects) interfere with normal development are becoming increasingly biochemical and pharmacological in orientation.

The extremely rapid growth in recent years of research related to problems of conception and contraception has led to the establishment of a discipline that is commonly called *reproductive biology*. In addition to more practically oriented problems involving techniques of fertilization and contraception, this field places heavy emphasis on normal gametogenesis, the endocrinology of reproduction, the transport of gametes and fertilization, early embryonic development, and the implantation of the mammalian embryo.

A currently popular way of looking at embryonic development is through the approach known as *developmental biology*. Broad in scope, this field includes not only embryonic development but also postnatal processes such as normal and neoplastic growth, metamorphosis, regeneration, and tissue repair at levels of complexity ranging from the molecular to the organismal. The focus of developmental biology is on processes and concepts rather than specific

FIGURE 1-3
Early drawings purporting to illustrate cases of human malformation. (A)
The bird-boy of Paré (about 1520). (B) Single monsters, part human and
part animal (Schwalbe, 1906–1909).

morphological structures, and both plant and animal systems are studied.
Ideally, developmental biology and the more classically oriented methods of
embryology should not be looked upon as competing methods of studying the
embryo but rather as complementary approaches, each offering exciting insight
into the way development is accomplished.

EMBRYOLOGY IN CONTEMPORARY SOCIETY

In recent years a virtual explosion of technology has turned into common
practice ideas that were the science fiction of only a few decades ago. Much of
the new technology is based on the results of laboratory research, which will be
described more fully in subsequent chapters.

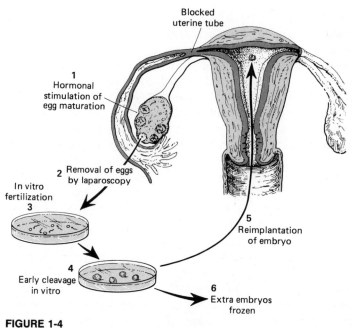

FIGURE 1-4
Schematic drawing of human in vitro fertilization and embryo transfer.

The "test-tube baby" is now not only a reality but a relatively commonplace event in many medical centers. Many different research and technological advances had to be combined to permit the application of this method. Interestingly, many of these advances were developed by animal breeders with purely economic goals in mind. In humans, the technique, called *in vitro fertilization and embryo transfer*, has allowed some childless couples to have children from their own genetic heritage. It is used in cases where both the mother and the father are capable of producing viable eggs and sperm cells, but because of a blockage in the women's uterine tubes the ovulated eggs are unable to be fertilized in her body and then become transported to her uterus (Fig. 1-4).

The first problem is obtaining fertile eggs from the mother. This is accomplished through two technical advances. One involves the administration of a fertility-enhancing drug to the mother. This results in her producing several eggs, rather than the usual single egg, at the time of ovulation. (Many women who in recent years have given birth to four to seven babies at one time had previously taken fertility-promoting drugs.) Just before the eggs would normally be shed from the ovary, a doctor, using a technique called *laparoscopic surgery*, inserts a tube into her pelvic cavity and under direct observation removes the ripe eggs from the ovary without the need for a major surgical procedure. The eggs are placed in a dish (hence the term *in vitro*, which means *in glass*) and mixed with the father's sperm. After many years of unsuccessful attempts, reproductive biologists learned what environmental conditions are required for fertilization outside the body. The fertilized egg is then allowed to develop for a few days in its artificial incubator.

Meanwhile, the body of the mother is hormonally conditioned so that her uterus can accept the embryo. While the embryo still consists of just a tiny ball of cells, it is sucked up into a tube and then released inside the cavity of the mother's uterus, where if all goes well it will attach and then complete a normal pregnancy. Some women who are able to produce fertile eggs but are unable to carry an embryo to term in their own uteri have made arrangements with other women to act as *surrogate mothers*. For about $10,000 (the standard rate in 1987), the surrogate mother agrees to have another couple's embryo implanted into her uterus and bring it to term. When the baby is delivered, the surrogate mother turns the baby over to its genetic parents. As of this writing, the first case in which a surrogate mother refused to give up her baby has been reported. It is still not possible to raise a mammalian embryo from conception to maturity entirely outside the body, but the major barriers appear to be technical rather than conceptual.

It is now common to fertilize more than one of the woman's eggs at the same time. After one embryo has been implanted in her uterus, the remainder are frozen. With the proper technique, a mammalian embryo can be frozen, stored, and later thawed as needed and then implanted into a uterus. This technique is routinely used with domestic animals and in humans; if the first implanted embryo fails to survive, frozen ones can be thawed and implanted until the supply of embryos has run out. Thus embryo banks are now a reality. In actual practice the extra human embryos are destroyed when a baby resulting from an artificial conception is born.

The ability to manipulate early mammalian embryos has increased dramatically in recent years, and now it is possible to produce chimeras between two or more individuals of the same species or even different species (see Chap. 4). It is not yet possible to clone vertebrates from single cells, as can be done in higher plants, but an approximation of cloning can be obtained by transplanting the nucleus from a somatic cell into an enucleated egg (Fig. 1-27). Although most commonly performed on amphibians, this has also been successfully done in mice (Illmensee and Hoppe, 1981). It is becoming increasingly common to transfer genes from one species into the egg of another species. This technique has been used mainly to study gene action, but it has been shown that transferred genes can exert significant effects on the host. For example, the transfer of a rat growth hormone gene into a fertilized mouse egg results in the development of a mouse that is much larger than normal (Fig. 1-5). Such techniques have the potential to be applied in the treatment of certain genetic diseases.

Other recent techniques make possible the diagnosis and/or treatment of genetic diseases and birth defects before a baby is born. Examination of a small amount of the amniotic fluid (see Chap. 7) that surrounds an embryo makes it possible to determine the sex of a baby before it is born and to detect the

FIGURE 1-5
A pair of 10-week old mice. The one on the left, which is a normal mouse, weighs 21.2 gm. The one on the right, a littermate of the normal mouse, carries a transferred rat gene coding for growth hormone. It weighs 41.2 gm. (Photomicrograph courtesy of R. Brinster. From Palmiter et al., 1982. *Nature 300*:611.)

presence of genetic conditions that could lead to a defective child. The application of ultrasound and new x-ray imaging techniques allows the diagnosis of many anatomical defects in fetuses. Some of these can be dealt with by means of intrauterine surgery. More and more the revolution in contemporary biology is permitting the application of techniques that allow one to manipulate human reproduction and embryonic development. A major challenge now is to cultivate the wisdom and foresight to deal with both the application and consequences of these techniques.

FUNDAMENTAL PROCESSES AND CONCEPTS IN DEVELOPMENT

Although a large part of this text is devoted to the morphology of embryonic development, it is important to recognize that the organs and structures which we see grossly or in microscopic preparations are the end products of dynamic processes that are as much an integral part of development as are the structures themselves. In this section we shall describe briefly some of the processes and concepts that are of vital importance in explaining why an embryo develops the way it does. Specific examples of many of these fundamental processes will be described later in the text as they relate to characteristic morphological events in embryogenesis.

Intracellular Synthesis and Its Regulation

From the moment of fertilization, embryonic development at all levels is a direct or indirect result of synthetic activities within cells. The DNA within the nucleus is the repository of much of the genetic information within the cell, and the transcription of this information from DNA to RNA and its subsequent translation into proteins are familiar subjects to all students of biology. One of the most important aspects of embryonic development is the nature of the regulatory mechanisms that restrict or permit the synthesis of specific proteins and other macromolecules. The intricacies of the mechanisms controlling nucleic acid and protein synthesis are beyond the scope of this text, but a review of some intracellular synthetic and regulatory pathways relating to development is in order. Figure 1-6 shows a generalized model of a cell; it stresses only intracellular structures and pathways that will be referred to later in this text.

In an *interphase cell* (one between mitotic divisions) certain portions of the nuclear DNA molecule are free of restricting proteins that bind to the DNA and can direct the synthesis of messenger RNA (mRNA). This step is known as *transcription* (Fig. 1-6, one). The newly formed RNA molecules commonly contain regions (*exons*) that code for specific segments of a protein molecule and other regions (*introns* or *intervening sequences*) that appear to contain no information directly involved in the amino acid sequence of the protein to be formed. In a series of steps commonly called *mRNA processing* (Fig. 1-6, two), the introns are enzymatically cut out and the remaining exons are spliced together to form the definitive mRNA molecule. After processing, the newly formed mRNA molecules migrate from the nucleus into the cytoplasm of the cell via pores in the nuclear membrane. Once in the cytoplasm, the mRNA molecules may follow either of two chief pathways, depending on the type of molecule and the type of cell. For the formation of protein molecules that are destined to function within the cell (structural proteins and most enzymes), the mRNA molecules link up with ribosomes to form polyribosomes, the length of which varies according to the size of the protein that is being made. If, however, the mRNAs are coding for proteins that will be secreted from the cell (e.g., collagen, immunoglobulins), the mRNA forms complexes with the rough endoplasmic reticulum. The polypeptide chains that are formed in the rough endoplasmic reticulum are then transported to the Golgi apparatus, where they are commonly linked with polysaccharide molecules. From the Golgi complex, the finished proteins are then brought to the cell membrane within vesicles and emptied into the medium surrounding the cell.

Regulatory mechanisms operate at almost every level of the protein synthetic pathway. Some are strictly intracellular, whereas others are extracellular influences that are effected through intracellular pathways. It is becoming increasingly apparent that many of the extracellular influences are mediated by receptor molecules located at the cell surface (Fig. 1-7).

An important enzyme found at the cell membrane is adenyl cyclase, which catalyzes the reaction of adenosine triphosphate into cyclic adenosine monophosphate (ATP→cAMP). Cyclic AMP acts as a general intracellular messenger and final common pathway for the effects of a number of compounds that act on the cell surface. The enzyme adenyl cyclase controls the rate of cAMP synthesis. According to one hypothesis, when a substance reacts with a specific

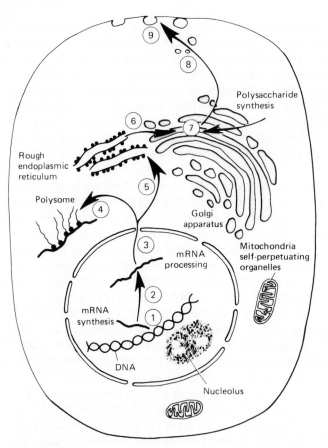

FIGURE 1-6
Generalized model of a cell, emphasizing the pathway of protein synthesis. In protein synthesis, messenger RNA is first synthesized on the DNA template (1). After processing within the nucleus (2), the mRNA leaves the nucleus through nuclear pores (3). The synthesis of intracellular proteins (4) is accomplished by polysomes, which consist of molecules of mRNA associated with ribosomes. Synthesis of proteins for export from the cell is accomplished on the rough endoplasmic reticulum (5). From there they are transported to the Golgi apparatus (6), where they may be complexed with newly synthesized polysaccharides (7). Small membrane-bound vesicles containing the proteins leave the Golgi apparatus (8) and, when they reach the cell membrane (9), fuse with it and release the protein molecules by a process called exocytosis.

cell-surface receptor, the complex migrates along the cell membrane until it binds to the adenyl cyclase molecule. This activates the adenyl cyclase, resulting in an increase in the synthesis of cAMP. The cAMP, acting as a "second messenger," can influence, usually by means of activation, a number of intracellular processes. Prominent among them is the activation of *protein*

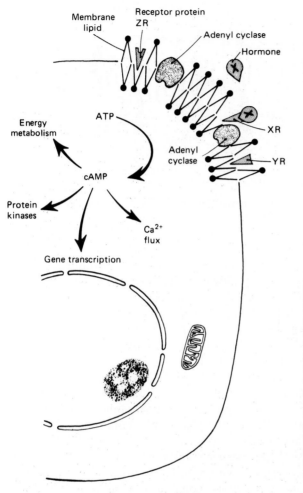

FIGURE 1-7
Diagram of the cell membrane and the cyclic AMP system within a generalized cell. The cell membrane is a lipid bilayer with protein molecules interspersed in different areas and at different levels. When a hormone or some other substance joins up with a special receptor (X-XR in this case), the receptor-hormone complex may then move to an adenyl cyclase molecule, which becomes activated and catalyzes the formation of cAMP from ATP. Several intracellular activities influenced by cAMP are shown here.

kinases (enzymes which phosphorylate proteins), which in turn may stimulate metabolic pathways, such as the breakdown of glycogen. Other activities stimulated by increasing cAMP levels are gene transcription, energy metabolism, and calcium flux (Fig. 1-7). Other molecules that interact with receptors on the cell surface apparently bypass the cAMP pathway. Their influence is directed to the nucleus by other intracellular mechanisms.

Cell Surface

In developmental processes the *cell surface* can be viewed in many ways. According to traditional cell physiology, the cell surface is represented by a membrane made up of a lipid bilayer through which molecules are transported into and out of the cell. However, the cell surface is much more than that. It is

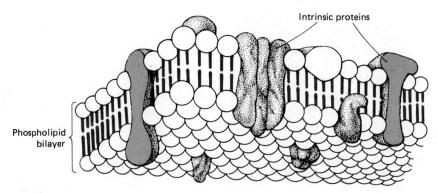

FIGURE 1-8
Schematic diagram of the plasma membrane, showing intrinsic proteins embedded in a lipid bilayer.

the means by which the cell tests or samples its external environment. To a large extent this is accomplished by specific protein *receptor molecules* which are embedded in the plasma membrane. The numerous proteins associated with the surface provide many cells with a unique molecular identity which plays an important role in cellular interactions during development. Differences in the cell surface constitute the basis on which cells of the immune system detect and deal with cells and substances foreign to the body.

To appreciate the developmental role of the cell surface, one must be aware of how it is put together. The bulk of the plasma membrane is a bilayer made up of phospholipid molecules, with the hydrophobic lipid components (hydrocarbon chains) meeting in the middle and the hydrophilic polar heads exposed along the outer and inner surfaces of the membrane (Fig. 1-8). The lipid bilayer is semifluid, and components of the plasma membrane can become concentrated or dispersed over the membrane within minutes. Embedded within the membrane are a variety of protein molecules. Some of these proteins penetrate through both layers of the plasma membrane; others are embedded in only the outer or inner leaflet of the membrane. Many of the membrane proteins are receptors for specific molecules (e.g., hormones, growth factors) that the cell encounters. Others mediate the attachment of specific molecules to the inner or outer surface of the cell. Many of the membrane proteins contain carbohydrate side chains which confer specific functional or antigenic properties to the cell.

An important but still poorly understood component of the cell surface is the class of *glycosphingolipid* molecules. These molecules, of which there are at least 130 varieties, constitute about 5 percent of the lipid molecules in the outer surface of the plasma membrane. They possess a fatty acid side chain that is embedded in the plasma membrane and a free carbohydrate chain that is composed of various combinations of simple sugar molecules and projects from the outer surface of the cell. The free carbohydrate chains are important in regulating many aspects of cell-surface activity and can also modify activities as fundamental as cell division (Hakomori, 1986). Many of the antigenic properties and recognition phenomena of individual cells can be attributed to their surface glycosphingolipids.

In addition to specific molecular components, the cell surface contains a number of junctional complexes which bind one cell to the next. Among the more prominent are *desmosomes*, which bind epithelial cells together in small spots, almost like rivets (Fig. 11-7). Desmosomes also serve as focal points for the attachment of fibrillar intracellular proteins. Another spotlike junction is the *gap junction* (Fig. 1-19), which mediates communication between two cells. Along the surface of many epithelia are *tight junctions*, which bind adjacent cells together, forming an impermeable barrier to the outside. Tight junctions also prevent the mingling of membrane proteins on either side of the junction.

Cell adhesion is an important property of most embryonic structures. A number of crucial experiments have shown that like cells tend to stick together and sort out from cells of a different sort. This phenomenon was first demonstrated early in the century by H. V. Wilson (1907), who squeezed a sponge through a silk mesh and dissociated it into individual cells. The dissociated cells later reaggregated and ultimately formed a new sponge. In later work, when two species of sponges were thus treated, the disaggregated cells sorted out according to species and the two original types of sponge re-formed. In a study by Holtfreter (Townes and Holtfreter, 1955) on amphibian embryos, cells of different germ layers or organ primordia displayed similar properties. Since these early experiments, much effort has been expended in trying to understand the nature of the cell-surface properties that would account for the phenomena described by the early embryologists. One of the early insights was that cells tend to separate from one another if Ca^{2+} and Mg^{2+} are removed from the surrounding medium. Explanations of why like cells aggregate more readily than do unlike cells include differential adhesion and a variety of molecular intercellular aggregation factors (*ligands*) which bind to specific types of cells.

Recently, a class of specific glycoprotein *cell adhesion molecules* (CAMs) has been identified (Edelman, 1983). Two major CAMs are dominant in early embryonic development: N-CAM (isolated from neurons) and L-CAM (originally isolated from embryonic liver cells). A third CAM (Ng-CAM) is associated with neurons and their surrounding glial cells (see Chap. 12). Despite their specific names, these CAMs are found on the precursor cells of many kinds of tissues, and they can be present or absent at different stages of a cell's life history (Fig. 6-18). There are also significant developmental changes in a given CAM. For example, the embryonic N-CAM molecule contains a large number of sialic acid molecules, whereas the adult form contains only a third as many. In their capacity as adhesion molecules, the CAMs appear to be involved in a number of morphogenetic events during embryogenesis. Specific examples will be given later in the text.

In addition to cell adhesion, the cell surface is involved in many important developmental processes. All humoral agents (e.g., hormones, growth factors, drugs) that affect cells must interact with the cell surface, usually by means of receptor molecules. The development of differences in surface properties is usually considered to be responsible for the major cellular displacements (e.g., the morphogenetic movements that occur during gastrulation; see Chap. 5) that are so prominent at certain stages of development. Many of the cellular interactions that are involved in the generation of pattern and form in complex structures are thought to rely heavily on surface interactions of the involved cells.

Properties of the cell membrane can be studied in many ways. The surface morphology of cells can be examined with scanning electron microscopy (Fig. 1-24) or, at a finer level, with a technique known as freeze fracture (Fig. 1-19B). Valuable tools for marking specific cell surface components are antibodies to specific cell surface molecules and *lectins*, a family of primarily plant glycoproteins which bind specifically to defined carbohydrate sequences from the cell surface (Etzler, 1985). Various electrophoretic techniques have allowed the identification of a large number of specific membrane proteins.

Extracellular Matrix

Cells do not exist or function in a vacuum; nor, in most circumstances, do they directly touch one another, even in the most densely packed tissues. Instead, they are embedded in or rest upon an *extracellular matrix*, a macromolecular meshwork that varies in composition from one tissue to the next and from one developmental period to the next (Hay, 1981). Epithelial-cell layers rest upon a basal lamina, a thin sheetlike form of extracellular matrix. Cartilage cells and bone cells are embedded in a massive extracellular matrix designed to support great weight. The spaces between different tissues are filled with fascia, an extracellular matrix that serves as both a biological packing material and a means of transmitting mechanical tension. A tendon represents an extreme example of an extracellular matrix designed to transmit powerful mechanical forces from a muscle to a bone.

Several major classes of macromolecules constitute the extracellular matrix. *Collagen* is the generic term for a family of glycoproteins which are characterized by having glycine as every third amino acid and also by possessing two amino acids, hydroxyproline and hydroxylysine, which are rarely found in other proteins. The basic unit of collagen, called *tropocollagen*, consists of three separate polypeptide chains (α chains) of about 1000 amino acids each, twisted in a left-handed helix. Tropocollagen molecules polymerize in a staggered fashion to form the familiar banded collagen fibers that can be seen with the electron microscope (Fig. 1-9). It is now clear that vertebrate tissues contain a number of different types of collagens—possibly as many as 20. The collagen types, determined largely by differences in the α chains, have different properties and are located in different places in the body. Those of greater

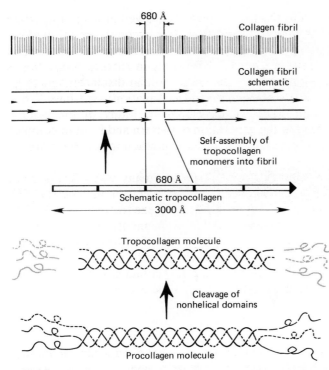

FIGURE 1-9
Schematic representation of the self-assembly of the collagen fibril
from procollagen molecules. Adapted from several sources.

relevance to embryonic development are summarized in Table 1-1. Most of the
collagens form distinct fibers, but type IV collagen, which is the collagen found
in basal laminae, is distributed as a loose meshwork among the other components of the basal lamina.

TABLE 1-1
MAJOR TYPES OF COLLAGEN AND THEIR DISTRIBUTION

Type	Attachment protein	Distribution in body
I	Fibronectin	Skin, bone, tendons, teeth, cornea, ligaments, interstitial connective tissue (about 90% of collagen is type I)
II	Chondronectin	Cartilage, notochord, vitreous body (eye), cornea (chick)
III	Fibronectin	Skin, blood vessels, sclera, many organs, skeletal muscle
IV	Laminin	Basal laminae
V	Fibronectin	Placenta, blood vessels, smooth muscle
X	Chondronectin(?)	Hypertrophying cartilage

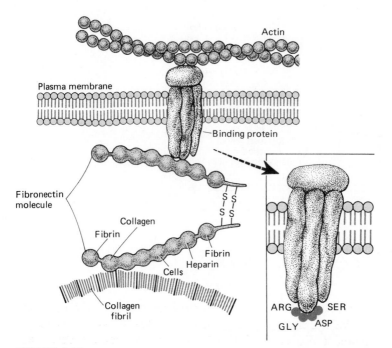

FIGURE 1-10
Diagram of the fibronectin molecule and its binding to cells and components of the extracellular matrix. The labels on the lower limb of the fibronectin molecule show the location of specific binding sites. Inset—cellular binding by fibronectin is accomplished by four amino acids, arginine (arg), glycine (gly), asparagine (asp), and serine (ser), attaching to a binding protein in the plasma membrane.

The *attachment glycoproteins* are involved in attaching cells to other components of the extracellular matrix. In developmental processes characterized by the migration or extension of cells, the attachment glycoproteins are an important feature of the substrates through which the cells move.

By far the best understood of the attachment glycoproteins is *fibronectin*, a dimer with similar polypeptide subunits of 220,000 to 250,000 daltons. At one end these subunits are joined by disulfide bonds. Each subunit is divided into distinct domains with specific functional characteristics (Fig. 1-10). Of particular relevance to embryology are the domains binding to cells and to collagen. Through these domains it is becoming possible to understand the binding of cells to their substrates. It has recently been shown (Pierschbacher and Rouslahti, 1984) that a sequence of only four amino acids (arginine–glycine–asparagine–serine) on the fibronectin molecule accounts for its cell-binding properties. It is now becoming apparent that these amino acids bind to a cluster of three glycoproteins (140,000 daltons) which are inserted in the plasma membrane of cells. Sites of fibronectin attachment to cells are also areas upon which bundles of actin, an important intracellular contractile protein, con-

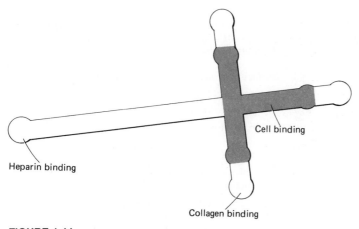

Cell binding

Heparin binding

Collagen binding

FIGURE 1-11
The laminin molecule, showing binding regions.

verge. Fibronectin is associated in particular with nonepithelial cells and with types I, III, and IV collagen (Table 1-1), and it confers stability on mature cells. Interestingly, malignant tumor cells bind poorly to fibronectin; this may explain, in part, their invasive properties.

A glycoprotein with an analogous function is *chondronectin*, which, as the name implies, mediates the attachment of *chondrocytes* (cartilage cells) to type II collagen in cartilage matrix. Chondronectin consists of several subunits linked by disulfide bonds, but its functional properties are poorly understood.

The other major attachment glycoprotein is *laminin* (Fig. 1-11), a cross-shaped molecule composed of three A chains of 200,000 daltons each and one B chain of 400,000 daltons. Laminin is a major component of basal laminae, where it binds cells to type IV collagen and other matrix molecules.

Glycosaminoglycans (GAGs), formerly called *mucopolysaccharides*, constitute another of the fundamental groups of extracellular matrix molecules. Although they are large molecules, most consist of repeated disaccharide units (Table 1-2). Glycosaminoglycans bind large amounts of water, which is important in maintaining the physical and mechanical properties of different types of extracellular matrix. The water-binding properties of hyaluronic acid make it particularly important in early developmental processes (Toole, 1982).

Proteoglycans are immense molecules of the extracellular matrix with molecular weights in the millions. A proteoglycan consists of a brushlike monomer, with a protein core and numerous glycosaminoglycan branches (Fig. 1-12). These monomers are in turn linked by a special protein to a backbone of hyaluronic acid. Proteoglycans are intertwined among the collagen fibers in the extracellular matrix. Most of the glycosaminoglycan molecules are integral components of proteoglycans, and their types and proportions vary from one tissue to the next.

TABLE 1-2
COMMON GLYCOSAMINOGLYCANS AND THEIR REPEATING DISACCHARIDE SUBUNITS

Glycosaminoglycan	Repeating subunit
Hyaluronic acid	D-Glucuronic acid + N-acetylglucosamine
Dermatan sulfate (chondroitin sulfate B)	L-Iduronic acid (or glucuronic acid)
Chondroitin 4- or 6-sulfate (chondroitin sulfate A or C)	D-Glucuronic acid + N-acetylglucosamine 4- or 6-sulfate
Keratan sulfate	Galactose + N-acetylglucosamine sulfate
Heparan sulfate	D-Glucuronic acid + iduronic acid + N-acetylglucosamine + glucosamine
Heparin sulfate	Glucuronic acid + glucosamine sulfate

With improved biochemical and immunological methodologies, it has become possible to make increasingly accurate descriptions of the distribution and types of extracellular matrix in tissues and organs (Fig. 1-13). Starting with the role of hyaluronic acid in raising the fertilization membrane of the egg (see p. 126), the extracellular matrix is a prominent component of virtually all developing systems. There is increasing evidence that the extracellular matrix mediates important interactions among cells during critical developmental periods. Cellular migrations are heavily dependent on the nature of the substrate through which the cells move. For example, neural crest cells migrate

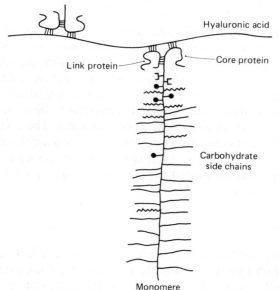

Hyaluronic acid

Link protein

Core protein

Carbohydrate side chains

Monomere

FIGURE 1-12
The structure of a proteoglycan molecule. Proteoglycan monomers, consisting of a core protein and numerous carbohydrate side chains, attach to a central filament of hyaluronic acid with the aid of a link protein (Adapted from Hascall and Hascall).

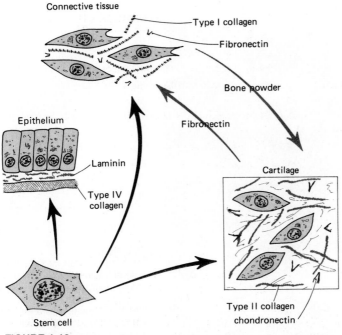

FIGURE 1-13
A model showing the influence of attachment proteins on the phenotypic expression of cells. Different sets of matrix proteins are associated with different cellular phenotypes. (Redrawn from Hewitt and
Martin after Kleinman et al., 1980 in *Current Research Trends in Prenatal Craniofacial Development*, Elsevier.)

through a meshwork of matrix fibrils (Fig. 1-14), and several studies (Fig. 6-18)
have emphasized the importance of fibronectin as a determinant of their
migratory behavior. In some systems (e.g., the cornea) high concentrations of
hyaluronic acid are associated with cell migration, and its removal with
hyaluronidase coincides with the end of the migratory stage. Even in postnatal
life, the evidence strongly points to fibronectin as being the important substrate
over which epidermal cells from a fresh skin wound must migrate. There is
increasing evidence that laminin is important in promoting the outgrowth of
nerve fibers both in the embryo and in regeneration after injury in the adult.
Other examples of the role of the extracellular matrix in embryonic development are scattered throughout the text.

Cell Division

Cell division is one of the fundamental properties of living systems, and it is of
vital importance in many developmental processes. An increase in cell number
is an obvious consequence of cell division. This is one of the basic mechanisms

FIGURE 1-14
Scanning electron micrograph of dorsal ectoderm and underlying extracellular matrix of an early chick embryo. The fine fibrillar material is a loose matrix of collagen fibrils, and the small round bodies are complexes of fibronectin and glycosaminoglycans. (Courtesy of K. Tosney.)

underlying growth in both embryonic and postembryonic systems. Less obvious, however, is the fact that a certain minimum number of cells is sometimes required for the development of certain structures in the embryo. In the embryonic limb bud, for instance, a deficiency of cells will typically result in the formation of a hand with less than the normal number of digits rather than a hand with the normal number of smaller digits.

For other developmental events the process of cell division itself seems to be crucial. It now seems that most, if not all, embryonic tissue interactions involving a qualitative change in the structural or functional state of groups of cells take place in populations that are characterized by a high rate of cell division. Mitosis may act as a destabilizing agent that permits the activation of certain groups of genes that previously had been tightly repressed.

Cell division is one component of the *cell cycle*. The life history of a cell can be conveniently divided into four periods (Fig. 4-5A). Immediately after mitosis and the separation of the dividing cell into daughter cells, the G_1 (gap 1) period, often called the *interphase*, commences. Its length is extremely variable. In rapidly cleaving embryos just after fertilization, the G_1 phase is very short and sometimes may not even exist. At the other extreme, the G_1 phase of mature neurons persists throughout the remainder of the life of the cell because further cell division does not occur. Cells of this type are called *postmitotic cells*. During the G_1 period the cell carries out its normal set of activities, such as specific synthesis, secretion, conduction, and contraction.

If a cell is in a dividing population, the synthesis of DNA, preliminary to mitosis, will occur. The period of DNA synthesis is called the S phase of the cycle. This is followed by a G_2 (gap 2) phase, which constitutes the period between the end of DNA synthesis and the beginning of mitosis itself (M phase). The process of mitosis is illustrated in Fig. 3-4.

Gene Activation

Genes are typically not active in the zygote, where they are tightly complexed with basic proteins called *histones*. The chromosomal DNA plus its enveloping histones is called *chromatin*, and the densely staining chromatin (*heterochromatin*) that can be seen within the nucleus at both the light and electron microscopic levels represents inactivated, or *repressed*, genetic material. As development begins, certain groups of genes become activated, or *derepressed*, by being freed from their associated histones. Derepressed DNA represents potentially functional genes. The first genes to become derepressed are those involved with the proliferative and general metabolic activity of the cell. As cleavage progresses and the embryo enters the stage of gastrulation, the first tissue-specific genes become activated. Later, during the period of organogenesis and histogenesis, other genes controlling more specific activity of differentiated cells come into play (Fig. 1-15).

It is estimated that at any given stage of development not more than 5 to 10 percent of the genes are active; the rest remain repressed. Studies on giant *polytene chromosomes* in insects have shown that at a given stage of development certain genes are activated, whereas at another stage the same genes are repressed and other genes are activated (Fig. 1-16).

Restriction and Determination

Within the fertilized ovum lies the capability to form an entire organism. In many vertebrates the individual cells resulting from the first few divisions after fertilization retain this capability. In the jargon of embryology, such cells are described as *totipotent*. As development continues, the cells gradually lose the ability to form all the types of cells that are found in the adult body. It is as if they were funneled into progressively narrower channels. The reduction of the developmental options permitted to a cell is called *restriction*. Very little is known about the mechanisms that bring about restriction, and the sequence and time course of restriction vary considerably from one species to another. Nevertheless, an example representing a general pattern of restriction during development may serve to clarify the concept (Fig. 1-17).

Shortly after fertilization the zygote undergoes a series of cell divisions, called *cleavage*, during the early phase of which the cells commonly remain

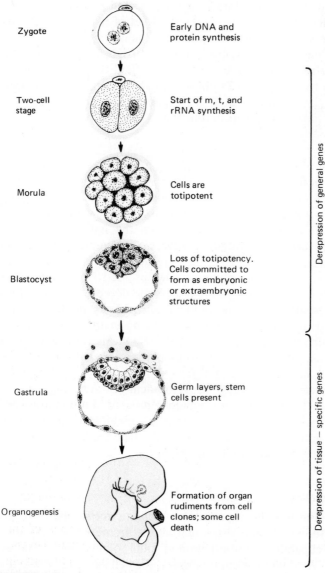

Zygote — Early DNA and protein synthesis

Two-cell stage — Start of m, t, and rRNA synthesis

Morula — Cells are totipotent

Blastocyst — Loss of totipotency. Cells committed to form as embryonic or extraembryonic structures

Derepression of general genes

Gastrula — Germ layers, stem cells present

Organogenesis — Formation of organ rudiments from cell clones; some cell death

Derepression of tissue — specific genes

FIGURE 1-15
Scheme of early mammalian development, stressing important
properties of the embryos and the varieties of genetic regulation.
[Adapted from B. Konyukhov, 1976, *The Genetic Control of the Development of Organisms* (Russian), Znanie, Moscow.]

totipotent. The period of cleavage comes to an end when certain cells in the
embryo undertake extensive migrations and rearrange themselves into three
primary germ layers during a process known as *gastrulation*. Named on the

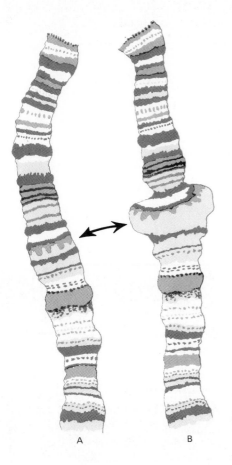

FIGURE 1-16
Drawing of a segment of a giant polytene chromosome in the fly, *Sarcophaga*, showing the banding pattern. (A) One of the bands (*arrow*) has just begun to puff. (B) Two days later the puffing is much larger, indicating activation of the genetic material in that part of the chromosome.

A   B

basis of their relative positions, the outermost layer is the *ectoderm*, the innermost is the *endoderm*, and between the two is the *mesoderm*. By this time at least one stage of restriction has usually occurred, so that the cells of the three germ layers are now locked into separate developmental channels and are no longer freely interchangeable. The potential options open to the cells of the ectodermal channel are shown in Fig. 1-17. In the next major developmental event, part of the ectoderm becomes thickened and is henceforth committed to forming the brain, the spinal cord, and other associated structures. This stage of development is commonly called *neurulation*. The remainder of the ectodermal cells can no longer form these structures and have thus undergone another phase of restriction. Soon, as a result of tissue interactions with the newly forming brain, groups of ectodermal cells become committed to forming the lens and inner ear, whereas the remainder of the ectoderm ultimately loses this capacity.

Subsequent developmental events see the ectoderm further subdivided into groups of cells destined to form cornea; hair, scales, or feathers; cutaneous glands; or simply epidermis. When restriction has proceeded to the point at

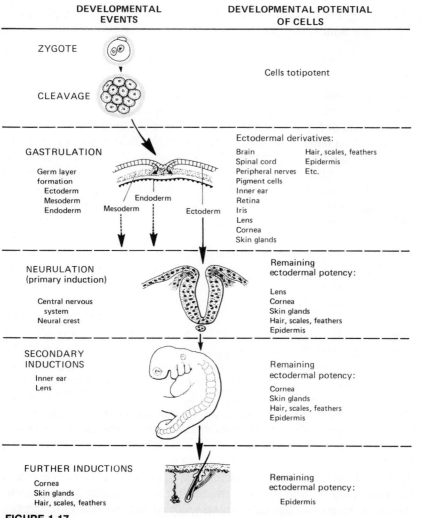

DEVELOPMENTAL EVENTS	DEVELOPMENTAL POTENTIAL OF CELLS

ZYGOTE

CLEAVAGE

Cells totipotent

GASTRULATION

Germ layer formation
Ectoderm
Mesoderm
Endoderm

Mesoderm   Endoderm   Ectoderm

Ectodermal derivatives:

Brain   Hair, scales, feathers
Spinal cord   Epidermis
Peripheral nerves Etc.
Pigment cells
Inner ear
Retina
Iris
Lens
Cornea
Skin glands

NEURULATION
(primary induction)

Central nervous
system
Neural crest

Remaining
ectodermal potency:

Lens
Cornea
Skin glands
Hair, scales, feathers
Epidermis

SECONDARY
INDUCTIONS

Inner ear
Lens

Remaining
ectodermal potency:

Cornea
Skin glands
Hair, scales, feathers
Epidermis

FURTHER INDUCTIONS

Cornea
Skin glands
Hair, scales, feathers

Remaining
ectodermal potency:

Epidermis

FIGURE 1-17
Diagram illustrating restriction during embryonic development. The column to the right of the figures demonstrates the progressive restriction of the developmental capacity of cells along one track, ultimately leading to the formation of epidermis. The column to the left of the figures describes major developmental events that remove groups of cells from the epidermal track.

which a group of cells becomes committed to a single developmental fate (for example, the formation of cornea), we say that *determination* of these cells has taken place. Thus, determination represents the final step in the process of restriction. The mechanisms that bring about determination of various groups of cells are receiving intensive study, but, as in the case of restriction, much remains to be learned. Usually, however, tissue interactions called *inductions* (see page 29) shortly precede the process of determination (and some phases of restriction) and are almost certainly involved in some manner.

Differentiation

Whereas restriction and determination signify the progressive limitation of the developmental capacities of cells in the embryo, *differentiation* refers to the actual morphological or functional expression of the portion of the genome that remains available to a particular cell or group of cells. Differentiation is really the process by which a cell becomes specialized, and the final product is called a *differentiated cell*. Although in many respects differentiation is a cellular event, a cell rarely undergoes differentiation in isolation. It is becoming increasingly apparent that differentiation of many tissues in the embryo will not occur unless a minimum critical number of cells are present. Typically, differentiation in vivo is a communal process that occurs within groups of similar cells. Nevertheless, much of the most incisive analysis of differentiation has been performed in vitro, and increasingly, the differentiation of cells descended from a single clonal precursor cell is being investigated.

There are many ways of looking at differentiation. From the biochemical standpoint, differentiation may be viewed as the process by which a cell chooses one or a few specialized synthetic pathways, for example, the synthesis of hemoglobin by erythrocytes or of specific crystallin proteins by the lens. Functional differentiation can be looked upon as the development of contractility by muscle fibers or as the development of conductivity along a nerve. From the morphological standpoint, final differentiation is represented by a myriad of specific cell shapes and structures. A comparison between the histological properties of morphologically differentiated and undifferentiated cells is given in Table 1-3. Although exceptions can be given for every category, this table should prove useful as a general guide.

Definitions of differentiation vary greatly, and it is beyond the scope of this section to treat them in detail. The most restrictive definition would limit differentiation to the maturation of a cell during a single cell cycle—often the terminal cycle. Other, broader definitions would include the maturation of a cell and its descendants over the span of several cell cycles. Irrespective of the working definition, differentiation can follow several general pathways. One type of differentiation pathway, which is without question a terminal one, results in a population of highly specialized cells which have lost their nuclei. Examples of this are the platelets and erythrocytes in the bloodstream of higher vertebrates and the cells of the outer layer of the epidermis. For other cells that retain their nuclei, differentiation may be expressed by the synthesis of highly specialized intracellular molecules, such as contractile proteins in muscle, or by the secretion of extracellular substances, such as hormones and collagen fibrils. A more detailed treatment of cellular differentiation is given in Chap. 9.

At the tissue level, differentiation can often be recognized as characteristic morphological changes occurring in groups of cells in certain locations and at

TABLE 1-3
CHARACTERISTICS OF MORPHOLOGICALLY UNDIFFERENTIATED VERSUS
DIFFERENTIATED CELLS

Characteristic	Undifferentiated cells	Differentiated cells
Nuclear size	Larger	Smaller
Nucleocytoplasmic ratio	High	Low
Nuclear chromatin	Dispersed	Condensed
Nucleolus	Prominent	Less prominent
Cytoplasmic staining	Basophilic	Acidophilic
Ribosomes	Numerous	Less numerous
RNA synthesis	Greater	Lesser
Mitotic activity	Great	Reduced
Metabolism	Generalized	Specialized

certain times. The process by which individual tissues take on a characteristic appearance through differentiation of their component cells is called *histogenesis*. At this level it is often difficult to separate histogenesis from morphogenesis.

Morphogenesis

The entire group of processes which mold the external and internal configuration of an embryo is included under the general term *morphogenesis*. Morphogenesis remains one of the major mysteries of biology, and our knowledge of this field is so slight that it can be compared with the state of genetics before the rediscovery of Mendel's laws.

A bewildering array of phenomena can be included under the overall umbrella of morphogenetic events in the vertebrate embryo. Consider, for example, the branching of the lungs, the form of the limbs, the shape of the eyeball, the pattern of blood vessels in the thorax, the intricate structure of a feather, and the complex loops and whorls on the fingertips. All are the result of morphogenetic processes.

How, then, does one approach the study of morphogenesis? In the absence of a general theory, it is first necessary to have some facts in hand. This has been most effectively done by identifying morphogenetic processes that are relatively easy to characterize and then subjecting these processes to genetic or experimental analysis. Classical models for morphogenetic analysis have included slime molds, bacteriophages, and the regenerating amphibian limb. Within the sphere of vertebrate embryology, the formation of the neural tube, limb development, and the branching of internal glands have been subjected to intensive analysis. With increasing information about morphogenesis in certain systems, attempts are now being made to categorize morphogenetic phenomena and construct hypotheses that will explain them.

Many types of processes can contribute to the morphogenesis of a structure. One of the most fundamental but least understood is *pattern formation* (Malacinski and Bryant, 1984). There is evidence that very early in the development of many structures, before the onset of cell differentiation, an invisible pattern or blueprint is laid down and that further development is guided by this pattern. The pattern is not always fixed and firm because at certain periods experimental manipulations or natural events can result in parts of the plan not being followed by the developing structure. Whether developing internal organs follow the same sorts of patterns as do external structures or the body as a whole remains to be seen. At present, intensive efforts are being directed toward uncovering a genetic basis for pattern formation, especially in *Drosophila*.

Assuming that some sort of pattern has been set, the next phase in morphogenesis consists of realization of the pattern. This is accomplished by employing familiar processes in special ways. Some of these processes are illustrated in Fig. 1-18. They may include cell proliferation, migration, aggregation, secretion of extracellular substances, change in cell shape, and even localized cell death. How the cells within the primordium of a structure communicate with one another to carry out the instructions inherent in a pattern is poorly understood. A currently popular way of looking at cell communication during morphogenesis is based on the concept of *positional information*. A simplified explanation of this concept is that a given cell is able to (1) recognize its position on a coordinate system that is set up within the primordium of an organ and (2) differentiate according to its position. For additional details as well as applications to specific developing systems, the reader is referred to two reviews by Wolpert (1969, 1971).

Induction

One of the most remarkable features of embryology is the precision with which developmental signals are generated and transmitted to the appropriate receptor. These signals may be of many types, and their effects may be made manifest in a variety of ways. One of the most important systems of embryonic signal calling is the process of *induction*. By induction we mean an effect of one embryonic tissue (the *inductor*) on another, so that the developmental course of the responding tissue is qualitatively changed from what it would have been in the absence of the inductor. One of the classic examples of embryonic induction is the formation of the lens of the eye as a result of the inductive action of the optic cup on the overlying ectoderm (Spemann, 1901, 1912). Details of this inductive system will be presented in Chap. 13. A fundamental inductive event in the embryo takes place as the germ layers are becoming established during gastrulation. Part of the mesodermal layer (specifically the chordamesoderm, see page 224) acts upon the overlying ectoderm, resulting in the appearance of a thickened plate of cells that will ultimately form the central nervous system. The specific interaction is called *primary embryonic induc-*

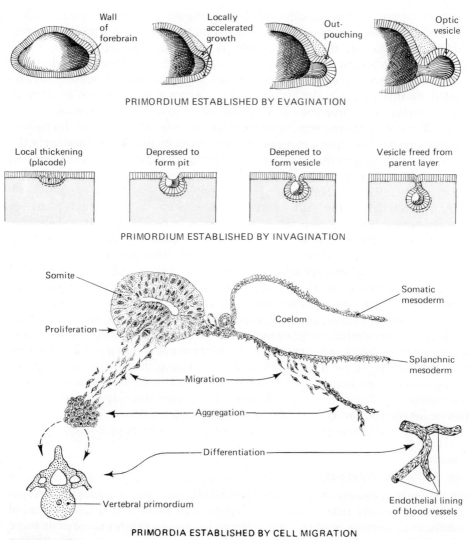

Wall of forebrain

Locally accelerated growth

Out-pouching

Optic vesicle

PRIMORDIUM ESTABLISHED BY EVAGINATION

Local thickening (placode)

Depressed to form pit

Deepened to form vesicle

Vesicle freed from parent layer

PRIMORDIUM ESTABLISHED BY INVAGINATION

Somite

Proliferation

Somatic mesoderm

Coelom

Splanchnic mesoderm

Migration

Aggregation

Differentiation

Vertebral primordium

Endothelial lining of blood vessels

PRIMORDIA ESTABLISHED BY CELL MIGRATION

FIGURE 1-18
Diagrams illustrating some of the different ways in which primordial cell groups may arise from parent cell layers.

tion. (For an excellent review, see Saxén and Toivonen, 1962.) Subsequent inductive events in the embryo are commonly called *secondary inductions*.

The nature of the inductive stimulus and its mode of transmission have been the subject of intensive research. Several varieties of inductive mechanisms have been suggested. One is the extracellular diffusion of inductive substances secreted by the inducing tissue. There is considerable evidence in favor of this mechanism in primary embryonic induction, resulting in the formation of the nervous tissues (Saxén and Toivonen, 1962; Toivonen et al., 1975). Other forms of induction appear more likely to be contact-mediated, either by direct cell-to-cell contact or through the extracellular matrix secreted by the cells involved in the interaction (Lehtonen, 1976; Hay, 1977). In 1956, Grobstein reported that direct contact between inducing and responding tissues in the kidney is not required; placing a porous filter between the two tissues did not halt induction. Inductive reactions do not occur through a nonporous membrane. More recent research, however, has shown that in many transfilter induction processes close cell contact does occur by means of small cellular processes growing into the pores of the filter from both sides of the membrane (Lehtonen and Saxén, 1975).

During the 1930s several groups of investigators attempted to define chemically the nature of the inductive effect evoked by the dorsal lip of the amphibian blastopore. It was soon found that a wide variety of killed tissues could duplicate the inductive effect of some of the natural inductors. Several classes of chemicals, ranging from proteins and nucleoproteins to steroids, elicited inductive effects similar to those produced by the cells of the dorsal lip of the blastopore. As more agents—including inorganic ions and even slight damage to the cells of the responding tissue—were found to produce inductive effects, embryologists turned their attention to the responding tissues.

There is some evidence that certain inductors may be specific, to a greater or lesser extent, in directing the fate of the responding tissues. This sort of induction is often called *instructive interaction*. It is also apparent that many inducing agents merely act as nonspecific triggers, or *evocators*, to release a response already encoded in the cells of the responding tissue. This type of induction is called a *permissive interaction*. Despite considerable research, little is known about how an inductive stimulus is received and processed by the responding tissues.

Intercellular Communication

One of the fundamental properties of living things—whether a flock of birds or a collection of organelles within a cell—is the ability of the components of a biological community to generate signals and to respond in turn to signals from other members of that community. The developing embryo can be looked upon as a community of cells whose integrity and activities depend on a well-developed system of intracellular communication. We have already seen how one type of communication, embryonic induction, can bring about profound qualitative changes in subsequent development. It has been recognized for many years that cellular communication must exist in induction and in the reaggregation of dissociated embryonic cells. Not until recently, however, have embryologists been able to approach this phenomenon with any degree of real understanding.

Steps are now being taken toward recognizing some of the means by which individual cells communicate with one another. In certain instances, for example, it has been shown that very small electric currents, inorganic ions,

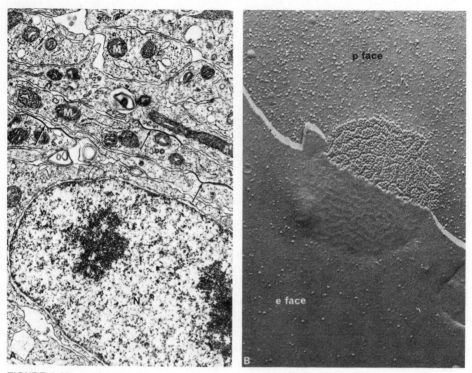

FIGURE 1-19
(A) Transmission electron micrograph through the apical ectodermal ridge (see Fig. 11-7) of the limb bud in a chick embryo. Gap junctions between adjacent cells are indicated by arrows. ×19,000. Abbreviations: N, nucleus, M, mitochondria. (B) Freeze-fracture replica of a gap junction within the apical ectodermal ridge of a quail embryo at a stage comparable to the chick in A. In making a freeze-fracture preparation the tissue is frozen at very low temperatures and then cleaved with a special apparatus. This procedure splits membranes into their inner and outer components along the interface between the hydrophobic ends of the lipid molecules that constitute the two sheets of the membrane. The fractured membranes are then examined with the electron microscope. The P face shows the surface of the inner portion of the fractured cell membrane whereas the E face refers to the outer portion of the membrane. The gap junction itself is the large aggregate of particles in the center of the photograph. (From J. F. Fallon, and R. O. Kelly, 1977. *J. Embryol. Exp. Morph. 41:223.* Courtesy of the authors and publisher.)

and even relatively large molecules can pass from one cell to its neighbors (Lowenstein, 1970). Such intracellular communication takes place in localized regions called gap junctions (Larsen, 1983), where the membrane of one cell is in intimate contact with that of another (Fig. 1-19).

Cell Movements

At numerous periods during embryonic life, cells or groups of cells move from one part of the embryo to another (Table 5-1). Some movements consist of short migrations of individual cells, whereas others involve the massive

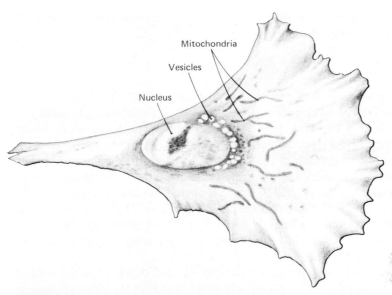

Mitochondria

Vesicles

Nucleus

FIGURE 1-20
Drawing of a mesenchymal cell moving in culture. The advancing edge (right)
is ruffled, whereas the trailing edge (left) is tapered.

dislocation of groups or sheets of cells over relatively great distances. Individual cells in embryos commonly migrate by means of ameboid movements. Although these cells are mesenchymal in appearance, they may originate from any of the three germ layers. In ameboid movement, the cell is continually testing its surroundings and its activity is characterized by the presence of a ruffled membrane along the leading surface (Fig. 1-20). A unique form of individual cell movement occurs in early avian embryos: The primordial germ cells move from the wall of the yolk sac into the bloodstream and are carried via the blood to the gonads (see Chap. 3). Examples of individual cells moving by ameboid movement are the migration of cells away from the neural crest (ectoderm), the spreading out of mesodermal cells during germ-layer formation, and the migration of primary germ cells (endoderm) from the yolk sac to the gonads in mammalian embryos. These processes will be dealt with in greater detail later in the text. Cell movements have increasingly been shown to be intimately tied to the relationships between the cells and the surrounding extracellular matrix (see Fig. 6-18).

Movement as a sheet is principally a property of epithelial cells, particularly those of the ectodermal germ layer. The migration of the cells during gastrulation in amphibians and the spreading of cells over the yolk in bird embryos are good examples of this phenomenon. Little is known about what causes sheets of cells in an embryo to move (rev. by Gustafson and Wolpert, 1963; Trinkhaus, 1969). The movement of cells as sheets is not confined to embryos. A simple cut in the skin of an adult vertebrate mobilizes the epidermis on either side of the defect, and within hours the wound is covered by a new layer of epidermal cells.

Cell Death

It may seem paradoxical that destructive processes, even the death of cells, play a vital role in the development of embryos. Nevertheless, cell death is a necessary component of many phases of development (Glücksmann, 1951). Although perhaps most spectacularly represented in some postembryonic events, such as the resorption of the tail, intestine, and opercular membrane of metamorphosing tadpoles and the liquefaction of most internal organs of a metamorphosing insect larva, cell death also occurs in many regions of avian and mammalian embryos. For example, separation of digits in the embryonic hand or foot is preceded by well-defined areas of cell death. Details of the way this process is involved in sculpturing the chick wing are given later in the text.

Although the exact mechanism responsible for cell death is poorly understood, the process appears to be genetically determined. In the chick (Saunders et al., 1962) the death of certain groups of cells becomes irreversibly fixed; if they are transplanted to another location, they still die according to a predetermined schedule.

Hormones sometimes play an important role in stimulating the death of cells. The primitive female (Müllerian) genital ducts in the embryo regress in the presence of the male gonad and its secretions, whereas the male ducts, which lie alongside them, are stimulated to further growth. In the case of the central nervous system, death is the fate of motor nerve cells that fail to make functional contact with a muscle fiber.

The Clonal Mode of Development

It has become increasingly clear that many structures in the embryo arise from the descendants of small numbers of cells (Mintz, 1971). A group of cells arising from a single precursor is called a *clone*. This concept arose from immunological studies in which it was shown that after the introduction of a foreign antigen into the body, a single immunologically competent cell undergoes a massive proliferative response and subsequently produces antibody against the antigen. This represents the basis for the "clonal selection" theory of Burnet (1969). Many tumors also arise as clones descended from a single malignant cell. Some examples of clonal development in the embryo are the formation of the body of the mammalian embryo from only 3 cells of the 64-cell embryo (page 179) and the origin of large portions of the central nervous system from well-defined cells of the early embryo (Fig. 12-1).

An important consequence of clonal selection in the embryo is that many cells in the early embryo are destined not to participate in subsequent development. Why these cells are not selected for further proliferation instead of the precursors of the clones is not known; presumably they ultimately die, but their fate remains obscure. Also unknown are the specific times when the clonal precursor cells of embryonic structures are selected and the mechanisms of selection.

Regulation and Regeneration

During early development of the entire organism or of specific systems, most vertebrate embryos have an uncanny ability to recognize whether the structure is intact. If part of a structure is lost by accident or through experimental manipulation, the loss is recognized and reparative processes are set in motion. If this occurs before differentiation of the structure has set in, the restoration of the missing material is called *regulation*.

Regulation is the basis for the development of identical twins. In mammals, including human beings, twinning usually results from the subdivision of embryos during early stages of cleavage (Fig. 1-21). Each half of the embryo is able to compensate for the lost tissues and develop into a perfectly normal individual. Occasionally separation of the two portions of the embryo is incomplete. This results in the formation of *conjoined twins* (Fig. 1-21). Commonly, when entire individuals or parts of organs are incompletely separated, one structure is a mirror image of the other (*Bateson's rule*). The reason for this reversal of symmetry is not known. In the normal development of the armadillo, the embryo breaks up at the four-cell stage, producing identical quadruplets.

Areas of the body that are able to reconstitute lost portions are sometimes called *morphogenetic fields* (Gurwitsch, 1944; Weiss, 1939). A morphogenetic field is a region of the body, such as that surrounding an appendage bud, in which the cells as a group are somehow cognizant of the overall nature of the structure to be formed. Thus, if cells are somehow removed from the field or extra cells are added to the field, the primordium as a whole adjusts to the change and the cells establish a harmonious relationship with one another, resulting in the formation of a normal structure. Morphogenetic fields have boundaries, which can be defined experimentally but not anatomically, and if all the cells within a field are removed, the structure does not form. Chapter 11 describes regulative properties within the limb field.

Sometimes in the late embryo or in postnatal life a missing structure can be replaced. If differentiation of recognizable structures has already occurred, the process of replacement is called *regeneration*. One of the main features of a regenerating system is the formation of a mass of primitive-appearing cells (the *regeneration blastema*) that demonstrate many of the properties of the embryonic primordium of the structure. One of the most difficult problems in both regulation and regeneration is how the cells remaining in the field are able to recognize that something is missing. Regulative activities within morphogenetic fields are now commonly interpreted on the basis of positional information of the component cells.

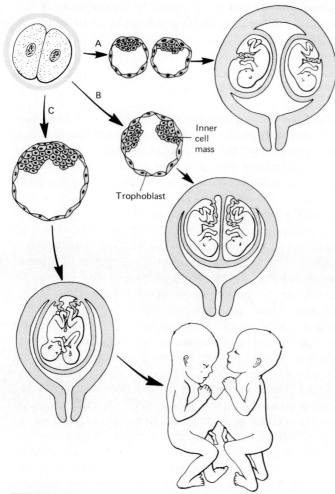

FIGURE 1-21
Modes of monozygotic twinning. (A) A cleaving embryo may split at
an early stage of cleavage, allowing the two portions to develop as
completely separate embryos. (B) At a later stage of development
the inner cell mass may split into two separate masses, both enclosed within the same shell of trophoblast. This is the most common mode of development of human twins. (C) If the inner cell
mass does not become completely subdivided, conjoined twins
may result.

Growth

When one compares the bulk of the human ovum, a spheroidal cell about 0.15
mm in diameter, with that of an adult human being, it is obvious that the
amount of growth involved is quantitatively astronomical. Even more striking
is the example of the whale, in which an ovum about the same size as the

human ovum produces an adult body weighing several tons. *Growth* can be defined in many ways, but perhaps the simplest definition is *an increase in mass*. This implies a concomitant increase in the number of cells, and in embryonic systems this is indeed the case. Normally accompanying the increase in mass is an increase in linear dimensions, but in some circumstances involving changes in form as well, the length of a structure may increase in the absence of an increase in its mass. Moreover, mass may increase in the absence of cell divisions if the cells undergo *hypertrophy*. One would expect from the varied characteristics of the tissues themselves and from the different ways in which they grow that they would not all increase at a constant rate. What is meant by *differential growth* goes far beyond this obvious situation. In embryology this term is employed to cover different rates of growth of the same kinds of tissue in different locations and at different times. One of the striking features of young embryos is the rapid growth of the cephalic region. This results in the formation of a disproportionately large head in the embryo and fetus. Later, when growth in the cephalic region becomes relatively less rapid, the rest of the body catches up and adult proportions are established (Fig. 1-22). This is a manifestation of the effect of differential growth on the proportions of the body as a whole. A specific term for the disproportionate growth of body parts during postembryonic life is *allometric growth*.

How is growth controlled? This is another of the major mysteries of biology. With respect to overall growth of the body, there are two major patterns of growth. In *determinate growth*, the body grows up to a certain point that is characteristic of the species and sex and then growth ceases. This is the characteristic pattern of growth in birds and mammals, but the enormous difference in growth potential between a pigmy shrew and a blue whale remains impossible to explain. The typical pattern of growth in the lower vertebrates is

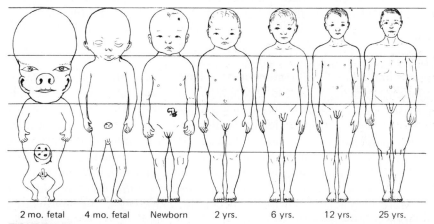

| 2 mo. fetal | 4 mo. fetal | Newborn | 2 yrs. | 6 yrs. | 12 yrs. | 25 yrs. |

FIGURE 1-22
Two fetal and five postnatal stages drawn to the same total height to show the characteristic age changes in the proportions of various parts of the body. (Redrawn from Scammon.)

indeterminate growth, in which growth continues throughout the lifetime of the individual, although at a reduced rate later in life. This characteristic makes it possible to determine the age of fish by examining the annual growth rings on their scales or in cross sections of certain skeletal elements. Despite the two patterns of growth, the problem of different degrees of growth potential remains. The difference in size between a guppy and a whale shark is as striking an example of the range of growth potential as that between a shrew and a whale. Some aspects of growth are obviously due to *growth hormone*, but all the growth hormone in the world would still fail to produce a shrew the size of a whale.

Some progress has been made in understanding certain components of growth at the tissue level. Several specific growth-stimulating factors have been isolated and chemically characterized (Papaconstantinou and Rutter, 1978). *Nerve growth factor* (Chap. 12) acts specifically on sensory and sympathetic nerves. Investigations of the nature of nerve growth factor brought to light the existence of an *epidermal growth factor* (Carpenter and Cohen, 1978), and later research uncovered a *fibroblast growth factor* (Gospodarowicz et al., 1978). The existence of a substance regulating hematopoiesis (*erythropoietin*) has been recognized for a number of years.

There is also evidence, although it is not universally accepted, for the existence of tissue-specific inhibitors of growth. These have been called *chalones* (Bullough et al., 1967). Chalones are thought to be glycoproteins and act by inhibiting or slowing the rate of mitosis in the tissues that produce them. Chalones are characterized by (1) being formed by the same tissues in which they act, (2) being cell-specific (e.g., epidermal chalone affects only epidermis), and (3) lacking species or even class specificity (e.g., epidermal chalone from the codfish acts on mammalian epidermis).

Whether each type of tissue and cell will have its own stimulators and inhibitors of growth remains to be seen. This area of biology is still in its infancy, but research to date has revealed the existence of some very interesting control systems that operate in both the embryo and the adult.

Recapitulation

The story of individual development sketches for us an approximate outline of the evolutionary changes passed through by our forebears. This concept is known as the *biogenetic law of Müller and Haeckel*. The general idea of recapitulation was first propounded by Müller (1864) on the basis of his studies of the development of invertebrates. Haeckel (1868) formulated its principles much more fully and called it the *biogenetic law*. In essence the law tells us that *an animal in its individual development passes through a series of constructive stages like those in the evolutionary development of the race to which it belongs*. More technically and more succinctly, *ontogeny is an abbreviated recapitulation of phylogeny*.

In recent years the biogenetic law has been subjected to considerable criticism. Most of the objections to it have been directed against attempts to apply it too rigidly to details. It is readily apparent that recapitulation does not consist simply of adding more recent phylogenetic traits onto old ones. In particular, one would not expect the embryo to pass through stages in which many of the specialized structural features of present-day lower chordates are emphasized, for many of these are specific adaptations that diverge from the mainstream of chordate evolution. Rather, ontogenetic recapitulation is a conservative process which retains the basic ontogenetic stages of more primitive forms. Thus, in a mammalian embryo only the most fundamental steps of early development and the establishment of major organ systems, such as the heart and large blood vessels, would resemble those of a fish embryo. There would be a greater similarity between mammalian and reptilian or avian embryos (Fig. 1-23), and this similarity would be apparent for a greater portion of embryonic life. In many cases ontogenetic processes leading to the formation of specialized structures in lower species are discarded or greatly reduced, and new ones are superimposed.

Often, however, a phylogenetically newer structure will make use of some component of the older one during the early phases of its development. This can be seen in the human placenta, the major organ of exchange between the embryo and the mother. The major blood vessels supplying and draining the placenta are homologous with those supplying the allantois, which subserves a similar exchange function in the embryos of birds and some lower mammals. In the human, the allantois itself remains vestigial but the allantoic blood vessels become incorporated into the phylogenetically newer circulatory system of the placenta.

Heredity and Environment

Both heredity and environment are of vital importance in development, but in quite different ways. *Heredity* establishes the inherent potentialities of a developmental system or an individual. *Environment* determines how far an individual can go toward a full realization of this inheritance.

At a purely biological level, differences in water temperature can result in the formation of one or two more or fewer vertebrae than the normal number in trout embryos. In some fishes and in most turtles and crocodiles, phenotypic sex is dependent on the temperature at which the eggs are raised. In many types of turtles, low incubation temperatures (below 28°C) cause most embryos to become males and higher temperatures (above 30°C) result in most or all individuals becoming females.

There are also conditions in which heredity and environment interact. An interesting example of this was brought to light by Frazer and his colleagues (1954, 1957). They found that when pregnant females of a particular genetic strain of mice were fed heavy doses of cortisone, cleft palates resulted in practically 100 percent of their offspring. With exactly the same treatment mice in a different genetic strain showed cleft palates in only 17 percent of their offspring. When the same experimental procedures were applied to animals

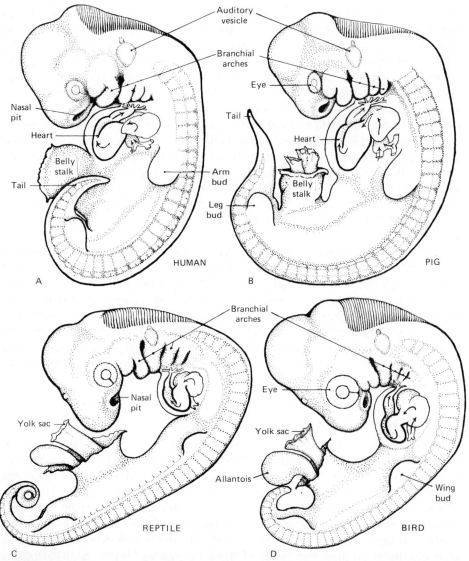

FIGURE 1-23
Embryos of (A) human, (B) pig, (C) reptile, and (D) bird at corresponding developmental stages. The striking resemblance of the embryos to one another is indicative of the fundamental similarity of the processes involved in their development. (From William Patten, 1922, *Evolution*, Dartmouth College Press, Hanover, N. H.)

resulting from the crossing of these two strains, approximately 40 percent of the offspring showed the defect.

Following these dramatic results, Frazer and his colleagues studied the rate of growth of the palatal shelves in the embryos of the strain in which the incidence of the defect was high as compared with the strain in which it was low. Those in the high-incidence strain showed a slow growth rate of the palatal shelves. They had barely enough growth energy to meet and fuse with each other if they were not disturbed in any way. In contrast, those in the low-incidence strain showed a high growth rate in the palatal shelves. In other words, there was enough reserve vigor in their growth so that in over 80 percent of the cases the shelves fused in spite of the same disturbing treatment that caused 100 percent defects in the other strain. Here, then, is a situation in which neither heredity nor environment can be said to be the sole cause of the defective development. The interaction between the two determines the degree of vulnerability.

METHODS USED IN THE STUDY OF EMBRYONIC DEVELOPMENT

Over the years many methods have been devised for studying various aspects of embryonic development. These range from the examination of entire embryos with the naked eye or simple lenses to extremely sophisticated molecular probes. All techniques have their uses, and it is important to recognize that the choice of technique is determined by the question that is being asked. This section will provide a brief survey of the major methods and techniques that are used in the study of vertebrate embryogenesis. These methods will be described to a greater or lesser extent depending on the emphasis placed on them throughout the text.

Direct Observation of Living Embryos

The earliest technique used in embryology was the direct observation of embryos, either with the naked eye or with simple lenses. After the introduction of the microscope, gametes and small embryos were examined in greater detail. Direct observation, particularly of a living embryo, provides one with a good overall view of the embryo and impresses the observer with the dynamic and often sweeping changes that constitute embryonic development. A major disadvantage of direct observation is that often resolution of finer details is sacrificed in favor of the total picture. With small structures these disadvantages can sometimes be overcome by the use of special optical techniques, such as phase contrast microscopy, or by the use of *vital dyes*, which permit the identification and tracing of specific cells or cell groups. *Microcinematography* is a powerful tool for investigating the development of entire embryos or groups of cells. This technique provides a moving picture of development, usually accelerated by a considerable extent, that can later be subjected to quantitative analysis. Anyone who has seen time-lapse films of developmental processes, such as cleavage or the outgrowth of a nerve fiber, cannot fail to be amazed at the number and precision of the changes that are taking place.

Examination of Fixed Material

It was recognized early that direct observation of living embryos is inadequate for the analysis of many aspects of development. At times one wishes to arrest a process at a critical phase so that the material can be examined at leisure. This is normally accomplished by means of *fixation*, in which the embryo is treated with various chemicals, such as formalin or glutaraldehyde, that will preserve structures as faithfully as possible without causing undue distortion or other artifacts in the tissue. About a century ago fixed material was used to prepare serial microscopic sections of entire embryos; this was one of its earliest uses. From these sections, the three-dimensional internal structure of an entire embryo can be reconstructed. This method of observation is still commonly used in student laboratories as well as in research. With the advent of electron microscopy, attention has shifted to the finer details of embryonic structure, but the basic principles, as well as the advantages and disadvantages, remain. A new dimension in the study of embryos was added with the application of *scanning electron microscopy*. This technique produces a three-dimensional view of entire embryos or parts of embryos with a clarity and resolution that had been previously unattainable (Fig. 1-24), and it has proved immensely valuable in sorting out structural relationships between cells and tissues in the embryo. With the ready availability of laboratory computers, techniques of *image analysis* are coming into more common use in the study of the embryonic structure. In some cases, computer reconstructions made from serial sections of embryos are replacing the old wax-plate methods of reconstruction. In other cases, the power of the computer can be used to provide information not readily apparent by observation alone, for example, the distribution of mitotic activity in a developing organ (Fig. 1-25).

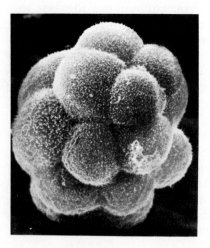

FIGURE 1-24
Scanning electron micrograph of a 4-day cleavage stage (blastocyst) of a hamster embryo. The cells bulging on the surface (trophoblast) will form extraembryonic membranes rather than the embryo proper. The small projections on the surface of the cells are microvilli. ×700. (From P. Grant, B. O. Nilsson, and S. Bergström, 1977. *Fert. and Steril.*, *28*:866. Courtesy of the authors and the publisher.)

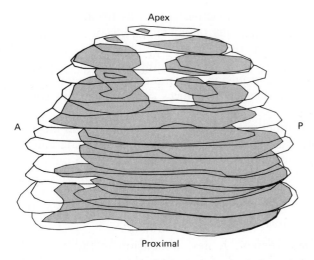

Apex

A   P

Proximal

FIGURE 1-25
Three-dimensional computer reconstruction of a regenerating salamander forelimb. From serial sections of the regenerate, it is possible to reconstruct a three-dimensional image of the skeleton (gray) within the gross contours of the regenerate (which in this case has not yet developed hand structures). With the cross-sectional data in the computer, one can display the reconstruction from a variety of angles. (Courtesy of T. Connelly).

Histochemical Methods

Histochemistry is a method of localizing specific chemical substances or sites of chemical activity on morphological structures that are disturbed as little as possible. Typically, the tissue or embryo is rapidly frozen in liquid nitrogen and sectioned with a special low-temperature microtome called a *cryostat*. The tissue is then placed on a glass slide and subjected to a specific chemical reaction that leads to the deposition of a colored product at the site of enzymatic activity or at a place where certain molecules are concentrated. In some cases it is possible to obtain very fine resolution of histochemical reactions at the electron microscopic level. One disadvantage of histochemistry at the light microscopic level is that the staining methods often do not show specific structural features of the embryo or organ with great clarity.

Autoradiography

One of the useful by-products of the atomic age has been the widespread availability of radioactive isotopes for use in biomedical research. Precursors of macromolecules containing radioactively tagged atoms can be introduced into a biological system and then followed through a metabolic cycle by various analytical means. One such means of analysis is called *autoradiography*, a method that allows the localization of a radioactive isotope within cells or tissues by employing methods similar to those used in photography (Baserga and Malamud, 1969).

Typically, an embryo or a part of an embryo is placed in a solution containing a radioactively labeled amino acid or a precursor of DNA or RNA. After a given period, the embryo is removed from the radioactive solution and sectioned for microscopic examination, but in addition to the usual histological procedures the tissue sections are covered with a photographic emulsion and kept in the dark for several weeks. Radioactive emissions from the isotope, which has been incorporated into proteins or nucleic acids of the embryo, impinge on the emulsion during the period of exposure. The emulsion is then developed in much the same manner as is used for photographic film, and tiny grains of silver are deposited in the emulsion over the cells containing labeled atoms. With this as a guide, the labeled structures can be localized with a microscope.

Autoradiography is now routinely employed at both the light and electron microscopic levels. With ordinary processing techniques, autoradiography is limited to localizing labeled macromolecules that are not dissolved out of the tissues. Newer, painstaking techniques have been developed for autoradiographic localization of soluble substances, such as steroid hormones. These require the tissues to be frozen immediately and then dried in such a way that displacement of the labeled soluble compounds does not occur.

Autoradiographic techniques have been of great help in localizing sites of nucleic acid and protein synthesis in embryos (Fig. 1-26) and in tracing the movements of cells. They do not, however, allow the direct analysis of their

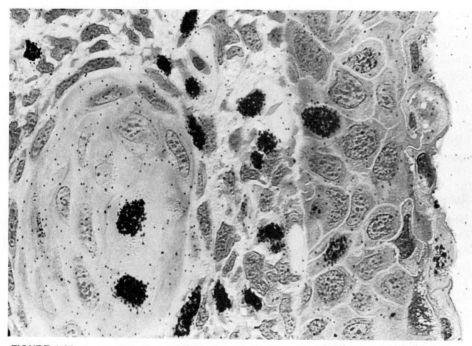

FIGURE 1-26
Example of an autoradiograph of a regenerating salamander limb. The animal was injected with ^{3}H-thymidine and the limb was fixed an hour later. Photographic emulsion is coated over the slide and developed after several weeks exposure. Dense accumulations of silver grains over a nucleus indicate that the cell was undergoing DNA synthesis when the isotope was administered. (Courtesy of T. Connelly.)

chemical form in the tissues. Thus, it is extremely important to employ the correct labeled precursor molecules and to recognize the chemical and technical pitfalls that can lead to misinterpretation of results.

An autoradiographic technique that promises to narrow the gap between morphology and molecular biology is *in situ hybridization*. Now that it is possible to construct laboratory-made complementary DNA (cDNA) to specific RNAs, radioactive cDNAs can be added to tissue sections suspected of containing the mRNA in question. If that mRNA is present, the labeled cDNA hybridizes with the corresponding nucleotides of the mRNA. With the labeled cDNA tightly bound to the RNA, an autoradiograph is prepared in the standard manner. The presence of silver grains tells where in the cell or the tissue the RNA molecules are located (Fig. 9-2).

Tracing Methods

Many types of markers have been used to trace cell movements in the growing embryo. Some of the classic studies of cell movements have involved the application of nontoxic markers to small groups of cells. Certain stains, such as Nile blue sulfate and neutral red, can be applied to living cells without harming them. These are known as *vital dyes*. Changes of position of cells treated with these dyes can be followed through an extensive period of growth before the dye becomes so diffused that identification is no longer possible. We shall have occasion to see the application of such techniques in connection with the cell movements in amphibian gastrulation (Fig. 5-6). Marking may also be carried out by placing finely divided, physiologically inert carbon particles, such as those of blood charcoal, on a small group of cells. The results of experiments involving carbon marking will be used in discussing cell movements in the neighborhood of the primitive streak in the chick.

A recent innovation in tracing methods consists of injecting tiny amounts of *horseradish peroxidase* (HRP) into cells. This enzyme is distributed throughout the cell, and its presence can be demonstrated by several reactions. When a cell containing injected HRP divides, the enzyme is distributed in the daughter cells. Because it persists in cell progeny, HRP has become a major tool in mapping cell lineages. Horseradish peroxidase can be injected into one cell of an early embryo. The embryo is allowed to develop for a time, and then it is sectioned. Cells descended from the cell originally labeled with HRP retain the label, whereas it is absent from other cells in the embryo (Fig. 12-1). Because it becomes distributed throughout a cell, HRP is also injected into developing nerve cells. It becomes distributed throughout the long processes of the cell, enabling the investigator to determine where the processes go and with what other cells they are connected.

Cells in which DNA has been heavily labeled with radioactive isotopes have been used for extremely precise localization of migrating cells. Such isotopic labeling of DNA has the disadvantage that the label is diluted by each cell division. This renders it unsuitable for long-term tracing of rapidly dividing cells.

An important category of marker, especially for long-term tracing of cells, consists of "natural" markers. By introducing into an embryo cells which differ from those of the host by virtue of size, pigmentation, isoenzyme types, chromosomal complement, or number of nucleoli, a stable marker is effected. Sex chromatin (Fig. 3-40) has been used with some success, but the most recent major advance in tracing methods has involved the use of Japanese quail cells as marker grafts in chick embryos (Fig. 9-9). Differences in nuclear size and morphology and the ease of grafting pieces of quail tissue into homologous sites in chick embryos have permitted investigators to solve a number of long-standing problems regarding the cellular origin and composition of certain tissues and organs.

Immunological Methods

Cells of different types, or even different developmental stages of the same cell type, contain different proteins and polysaccharides in the cytoplasm or on their surface membranes. These macromolecules are antigenic; i.e., when injected into another animal, such as a mouse or a rabbit, they provoke the immune cells of the animal to form antibodies against them. If prepared properly against specific *antigens*, antibodies can be valuable in studies of development because they can be used to probe for the specific presence or absence of the antigenic molecule in question in a tissue or organ. In embryological studies it is common to take a section of a tissue suspected of containing an antigen and cover it with a fluid containing an antibody against that antigen. If the antigen is present in the tissue, the antibody will combine with it. This antigen-antibody complex, however, cannot be detected as it stands. It is common to add next a second antibody directed against the first antibody. However, the second antibody is complexed to a marker molecule, commonly one with fluorescent properties. The marker can be detected with fluorescence microscopy, and its location on the tissue indicates the presence of the antigen in question. The name commonly given to the localization of specific molecules in tissues by means of antibodies is *immunocytochemistry*.

A powerful new immunological technique involves the production of *monoclonal antibodies*. This technique is designed to produce exceptionally pure and specific antibodies. Details of the technique are well summarized by Milstein (1980), but in brief it consists of first injecting the antigen into a mouse. Later, antibody-producing cells from the spleen are removed from the mouse and fused with a cell from a type of tumor called a myeloma. The fused cells (now called *hybridomas*) can be maintained in culture. When a hybridoma is shown to produce an antibody of interest, it is cultured as a single clone and allowed to multiply. Its cellular descendants all produce the same highly specific variety of monoclonal antibody, and with the proper culture technique they can be maintained almost indefinitely as factories for that specific antibody.

Immunocytochemical techniques directed at both cells and components of the extracellular matrix have provided valuable insight into both the localization of minute amounts of developmentally important antigens and the time of their appearance as development progresses. Specific examples of the use of these techniques will be given throughout the text.

Microsurgical Techniques

Much of the fundamental information about causative mechanisms in embryonic development, particularly those involving tissue interactions, has been obtained through the use of microsurgical techniques. Work with embryos often no more than a few millimeters in length has necessitated the development of special tools, such as glass or tungsten needles and loops of baby's hair, instead of the scalpels and forceps commonly associated with the usual types of surgery. For work with extremely small embryos or groups of cells, micromanipulators, instead of the hands, are used to hold the instruments.

Microsurgical techniques are used in many types of experiments. One of the simplest is *ablation*, or removal of part of an embryo to determine what effect the absence of that structure will have on the remainder of the embryo. *Transplantation* and *explantation* are commonly used surgical techniques that have found wide application in embryological studies.

Explantation consists of excising a small sample of embryonic tissue and growing it in an artificial environment. Explants may be handled in various ways. One method is to graft the excised tissue into a host organism in such a location that it is well supplied with nutritive materials but must grow and differentiate without the influence of the other tissues of its own body that normally surround it. In working with bird embryos it is common to explant a small group of primordial cells from a young individual to the *chorioallantoic membrane* in an older host. With mammalian embryos favorable locations are the anterior chamber of the eye and a vascular area of the peritoneum. Explantation experiments have provided much information on how the tissue can adapt to and differentiate in a new location. Some embryonic primordia show a striking capacity for *self-differentiation*, indicating that within the transplanted cells there is sufficient information to direct the development of the organ.

In embryological studies tissues are sometimes transplanted to other sites on the same embryo (*autografting*), but often tissues or organs from a donor embryo are grafted to hosts of a different species (*heterografting*) or even of a different order (*xenografting*). Some types of transplantation experiments involve relatively minor shifts or rotations of the tissues, but in other types of transplantation an embryonic structure is moved far from its normal location. It has been found that transplanted embryonic tissues and the tissues of the host often do not lie passively side by side but may exert profound influences on the course of development of their new neighbors. As we shall see, examples of this type of influence have been particularly striking in the field of embryonic induction.

A recent application of transplantation involved not tissues or organs but components of single cells. Using the technique of nuclear transplantation

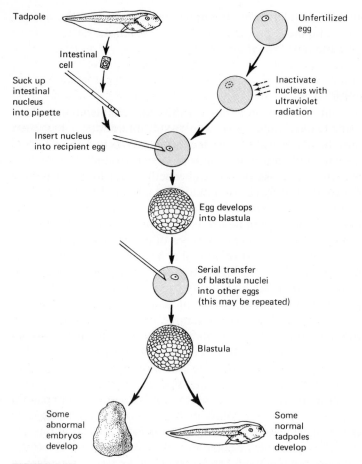

FIGURE 1-27
Outline of a nuclear transplantation procedure in Amphibia (*Xenopus*).
A nucleus from an intestinal epithelial cell is sucked into a micropi-
pette and then injected into a recipient egg, the nucleus of which has
been inactivated with ultraviolet light. The egg then develops into a
blastula. When mature nuclei are transplanted, it is often necessary to
make several serial transfers of daughter nuclei from blastulae into
other eggs in order to create appropriate conditions for full development. Only a small percentage of the embryos develop normally. In
the remainder, development is grossly abnormal and becomes arrested. (After Gurdon.)

developed by Briggs and King (1952), Gurdon (1962) transplanted the nucleus
from the intestinal epithelium of a postmetamorphic frog (*Xenopus*) into an egg
whose nucleus had been inactivated by *ultraviolet* (UV) *radiation* (Fig. 1-27).
An adult frog developed from the egg. This experiment demonstrated that even
the nucleus of an intestinal epithelial cell still contains a sufficient endowment

of genetic information to guide the development of an entire mature animal from the egg.

In at least one case the technique of transplantation has been employed to economic advantage. A group of South African sheep ranchers wanted to begin raising a special strain of sheep native to Scotland. To ship a sufficient number of adult sheep to South Africa by boat would have required a long and costly trip. This problem was met by removing newly fertilized eggs from the Scottish sheep and transplanting the early embryos into the uteri of rabbits. The rabbits were then flown to South Africa, where the sheep embryos were removed from the rabbits and transplanted into the uteri of local ewes. In due time normal lambs of the Scottish strain were delivered to the South African ewes and the long-distance transfer of the herd of sheep was completed (Hunter et al., 1962).

Clinical application of the technique of *embryo transfer* resulted in the first birth of a human conceived outside the uterus. A woman in England was unable to have a baby because of a blockage of her uterine tubes that prevented ovulated eggs from reaching the uterus. Two British reproductive biologists, Robert Edwards and Patrick Steptoe, obtained an ovum from the woman's ovary and fertilized the egg in vitro with her husband's sperm. The embryo was allowed to develop to the eight-cell stage and then was transplanted to the woman's uterus. The embryo became implanted in the uterine lining, and an uneventful pregnancy and the birth of a normal baby girl resulted.

Culture Techniques

One of the most interesting and instructive ways of studying embryonic development is to grow components of embryos or even whole embryos in an artificial environment. Depending on the nature of the explanted material, the techniques are known as cell, tissue, organ, and even whole embryo culture. Each type of culture requires slightly different methods, but the principle of culture is the same. The embryonic material is placed into dishes or tubes of glass or plastic and surrounded by an artificial culture medium designed to resemble as closely as possible the environment surrounding the material in its normal site in the embryo. The ideal culture medium is completely defined chemically, but commonly it is necessary to add undefined biological factors such as serum or even extracts from entire embryos to provide necessary growth factors. There is increasing awareness that the nature of the substrate beneath the cells, whether it is the plastic of the dish or added extracellular matrix material, is an important determinant of the success of the culture.

The culture method was developed in a revolutionary experiment by Ross G. Harrison (1907), who was looking for a method to demonstrate the growth of nerves (Fig. 12-18). In the time since his pioneer work, culture methods have contributed greatly to our understanding of developmental processes. Today, the culture of tissue or organ rudiments is often routine, and for a number of cell types, e.g., muscle (Konigsberg, 1963), it is possible to produce differentiated clones from single precursor cells. Recently there has been increasing interest in the culture of complex organs or even whole mammalian embryos, but progress in refining culture techniques is often slow and the work is frequently frustrating.

A major advantage of culture techniques is that the surrounding medium or the tissue itself can often be altered in defined ways that would never be possible in vivo. A disadvantage, particularly in cell and tissue culture, is that it is sometimes difficult to distinguish processes that operate only in culture conditions from those which occur naturally in the embryo.

Biochemical and Molecular Techniques

The biochemical analysis of embryonic tissues, particularly with the use of the newer techniques of molecular biology, is one of the most rapidly growing ways of studying development. The biochemical techniques used to study embryonic systems would include almost all of those now in use in biochemistry, so in the space allotted here only general categories will be mentioned. Among the older methods are purely chemical techniques designed to determine the presence or absence of specific compounds and their amounts. Needham's (1931) treatise summarizes much of this material. Analysis of enzyme activity is frequently used in studies on the metabolic properties of embryos. In these experiments, a chemical reaction involving the mediation of an enzyme (as do most biological reactions) is allowed to occur and the amount of some reaction product is commonly measured—usually by spectrophotometric means—and compared with a reference curve.

Separation methods are widely used in studies of embryos. The first separation techniques were paper chromatography and electrophoresis. These techniques make use of the physical properties of compounds, such as amino acids or proteins, which give them different migratory properties in solutions or in electrical fields. Among the molecules that can be separated by electrophoresis are *isoenzymes* (Markert, 1975). Isoenzymes (isozymes) are different forms of molecules with the same enzymatic activity but with slightly different structures that enable them to be separated from one another. Because different isozyme forms of a given enzyme are often formed by separate populations of cell types at various times, they often make good developmental markers (Fig. 9-11).

Another family of separation techniques involves the centrifugation of solutions or tissue homogenates containing macromolecules. After prolonged centrifugation at high speeds, subcellular fractions or different classes of macromolecules become stratified according to their size and density.

Column chromatographic methods have been developed to separate many classes of compounds according to various physical characteristics. In typical column techniques, a glass column is loaded with beads or other special materials that have been developed to allow the differential migration of molecules, and a solution containing a family of macromolecules—commonly nucleic acids—is added to the tube. The fluid in the column is allowed to drip from the bottom into a fraction collector, which is a container of some sort full of tubes. At periodic intervals the tubes are moved; their content of molecular material is a reflection of the rate of passage of the molecules through the column. The fluid in the tubes may then be examined for its content of the molecules in question, or if isotopic labeling is also used, as is commonly the case, the fluid in each tube is also analyzed for radioactivity.

A number of highly specialized techniques have been devised for demonstrating particular species of information-containing nucleic acids. These techniques take advantage of the unique sequence of bases that constitute a strand of DNA or RNA. They involve the *hybridization* of a simple strand of DNA with a strand of DNA or RNA so that there is a matching up of complementary sets of bases. Hybridization techniques have proved to be very valuable in demonstrating the presence or absence of specific sequences of bases in the nucleic acids of embryonic cells.

Some of the most spectacular recent advances in our knowledge of development have involved the application of the new gene technologies. These include the preparation of *recombinant DNA*, the construction of synthetic genes, and the ability to prepare specific *molecular probes*. Recombinant DNA technology allows one to produce large quantities of a given DNA sequence. This is accomplished with the help of *plasmids*, small circular molecules of DNA that replicate independently in bacteria. Both eukaryote DNA and the plasmid DNA are simultaneously treated with one of a family of restriction enzymes, which cleave DNA strands at specific combinations of base pairs. Both types of DNA are split in the same way, and if they are in a solution together, some fragments split off from the eukaryote DNA reattach to the plasmid DNA and re-form a new circle which includes a segment of the eukaryotic DNA. When these recombinant plasmids are introduced into bacteria, such as *Escherichia coli*, the plasma DNA replicates and the eukaryotic DNA is also transcribed. Because the bacteria have an almost unlimited process of multiplication, the eukaryotic DNA enzyme can be produced in large quantities. The genetic information in the recombinant DNA can then be used to produce gene products, such as protein hormones. An early application of recombinant DNA technology was the production of human insulin (Goeddel et al., 1979).

A very useful variety of molecular probe is the cDNA molecule that can be produced by incubating a mixture of defined mRNA, nucleotides and the enzyme *reverse transcriptase*, which permits the synthesis of a single strand of DNA from the mRNA. As was mentioned earlier in this chapter, radioactively labeled cDNAs can be added to sections of test tissues (in situ hybridization) to test for the presence of the corresponding mRNAs in the tissue. If the complementary RNA is present, the labeled cDNA hybridizes with the mRNA and the complex can be detected autoradiographically.

Irradiation Techniques

Various forms of irradiation have been used in embryological studies, mainly to inflict some form of damage on parts of the embryo. For some experiments

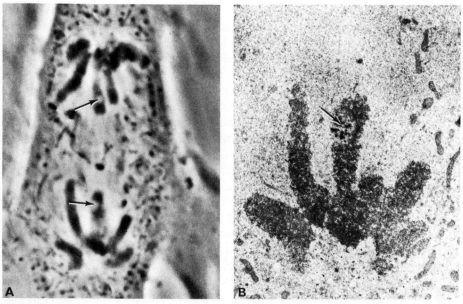

FIGURE 1-28
Effects of irradiating chromosomes of cultured cells [rat kangaroo line (PTK$_2$)] with an argon laser beam. (A) In a phase-contrast micrograph, the irradiated area appears as pale spots within the chromosome (arrows). ×1300. (B) In an electron micrograph of the lower set of chromosomes, the irradiated area contains aggregates of electron-dense material (arrow). ×3100. (From J. B. Rattner, and M. W. Berns, 1974. *J. Cell Biol. 62*:526. Courtesy of the authors and publisher.)

x-rays are brought to focus on small areas of an embryo to provide a circumscribed area of tissue injury or to inactivate a group of cells. UV rays are sometimes used for the same purpose, but they lack the deep penetrating power of x-rays. Laser beams are proving to be another valuable tool in producing sharply localized lesions in embryos (Berns and Saleti, 1972). Their precision is already such that small areas of individual chromosomes can be destroyed (Fig. 1-28).

Inhibitory Agents and Teratogens

In the analysis of embryonic development, many types of chemical agents have been used to inhibit normal embryonic processes. In some cases, the mechanism of action of a chemical inhibitor is quite well defined and an alteration in development can be attributed to a specific disturbance in a metabolic pathway. For example, the antibiotic actinomycin D is known to inhibit the synthesis of RNA. If actinomycin D is administered to a very early embryo, development proceeds normally for a short time but is soon markedly inhibited (Gross and Cousineau, 1964). One can infer that the stage which was inhibited in its development required the synthesis of new RNA molecules.

In most cases the exact mechanism of the disturbing action caused by a chemical or a radiation (such as an x-ray or UV ray) is not yet known. The effects of these agents can be interpreted only at a less specific level, sometimes involving the intermediary effect of one tissue on another rather than a direct disturbance in a chemical process.

In recent years a number of drugs have been shown to cause many types of abnormal development. Such drugs are said to have a teratogenic effect and are commonly referred to as *teratogens*. A striking example of a chemical teratogen acting on human development occurred in West Germany and several other countries during the early 1960s. A large number of children were born with an unusual type of defect of the limbs. In severe forms of this defect the proximal segments of both arms and legs are missing and the hands and feet appear to grow directly from the body. Because such limbs resemble the flippers of a seal, the condition was called *phocomelia* (seal appendage). It was soon discovered that these deformities were caused by a supposedly safe sedative called *thalidomide*, which was commonly used by pregnant women in those countries. This drug has been taken off the market, but because of its sometimes tragic effects the screening procedures for all new drugs now include rigorous tests to determine whether the drugs exert ill effects upon embryos.

Developmental Genetics and the Use of Mutants and Genetic Markers

One of the really powerful tools for the dissection of complex developmental processes is the approach of *developmental genetics*. Studies of genetic mutants that affect specific stages of development have provided information that in some species (*Drosophila* being an outstanding example) are allowing investigators to begin to link gene structure and complex morphogenetic phenomena. The use of *lethal mutants* has assumed increasing importance in the analysis of embryonic development. Identifying where and when development first goes wrong in a mutant strain often makes it possible to pinpoint the effects of certain specific genes on developmental processes. A very valuable series of lethal mutants is represented by the *T*-complex in mice. Genes of the *T*-complex are located on chromosome 17, and they play a role in controlling a variety of cellular interactions involving intercellular recognition events in the early embryo. The *T*-complex gene products are important both in spermatozoa and in early embryonic stages. Interestingly, another large gene family, the *H-2* complex, which is also found on chromosome 17, affects cellular recognition in the mouse's immune system during late embryonic stages and in adult life. Mutants of the different *T* alleles are lethal and cause development to cease at specific stages. The sensitive points of some of the major *T*-complex mutations are illustrated in Fig. 1-29. Specific details of the deleterious effects of mutants at the different *T* alleles are given in Chap. 5.

Not all mutants are lethal, of course. As a general rule, one can say that the later in development a gene begins to act, the less likelihood there is that a

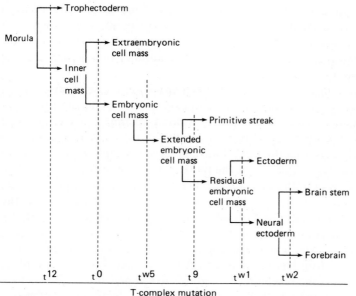

T-complex mutation

FIGURE 1-29
Schematic representation of the effects of different mutations of the *T* complex in early mouse development. Embryos homozygous for the mutations listed along the bottom of the diagram show impaired progress in the developmental dichotomies indicated by the diverging arrows the upper part of the diagram. (Adapted from J. Klein, 1975. *Biology of the Mouse Histocompatibility—2 Complex.* Springer Verlag, N. Y.)

mutant will be lethal. In albinism, for example, the lack of color is due not to the absence of pigment cells but rather to the absence of a specific enzyme (tyrosinase) that is required in the synthetic pathway of the black pigment, melanin. As a rule, the more an animal is studied, the more mutant genes are discovered. Among the vertebrates, the mouse is by far the best-studied species, with several hundred defined mutant genes, but mutant strains in axolotls, *Xenopus*, and the chicken have also proved quite useful in embryological studies.

Genetic markers are also useful in developmental studies, often as tracers. Particularly valuable have been strains of mice that have developed isoenzymatic differences in the form of certain enzymes. Identification of specific isoenzymes by electrophoretic methods has proved useful as a tracing method.

The methodological approaches listed in this section are being utilized in investigations that are placing us on the threshold of a new era in understanding the controlling and regulating factors in development. Their combination is tending to break down the old boundaries between various fields of scientific research and is resulting in a more unified approach to the study of embryological problems.

Unfortunately, some of the newer methods of research involve apparatus that is too expensive to be in every laboratory. Any lack of special equipment should merely indicate the desirability of directing research efforts into channels not requiring such equipment. One should not be unduly impressed either by the availability of apparatus or by the lack of it. Rather, it should be realized, as Ebert (1966, p. 50) has pointed out, that many of the most significant advances in our understanding of developmental mechanisms "were made with techniques of utter simplicity, the only requisite being glass needles and hair loops, physiological saline, and the ability to pose the right questions." It is important to remember also that "new methods can usually only be applied to old material; and new ideas do not suddenly emerge full-fashioned as Aphrodite was born from the chaotic sea; they are built up laboriously on the foundation of previous work" (Waddington, 1956, p. 5). A thorough and accurate knowledge of events during normal development is the foundation on which experimental embryology must rest. Without it there is no sound basis for interpreting the significance of experimental results.

CHAPTER 2

REPRODUCTIVE ORGANS AND THE SEXUAL CYCLE

REPRODUCTIVE ORGANS

As a prelude to the study of fertilization and embryonic development, it is important to understand how and where sex cells (*gametes*) are produced and, in animals that employ the reproductive strategy of internal fertilization and the bearing of live young, the anatomical and functional processes that are designed to bring the gametes together and later to nourish the developing embryo. This chapter will concentrate on the reproductive organs in mammals and the hormonal changes that are of crucial importance in the production of gametes and their meeting at fertilization and in the maintenance of the embryo in its mother's uterus.

Female Reproductive Organs

The reproductive organs in the human female and their relations to other structures in the body are shown in Figs. 2-1 and 2-2. The paired gonads, the *ovaries*, are located in the pelvic cavity. Each ovary lies close to a funnel-like opening (*ostium tubae*) at the end of a *uterine tube*. Around the abdominal orifice of the tube are characteristic fringelike processes called *fimbriae*. The lining of the uterine tube is thrown up into numerous complex folds and the epithelial surface contains many ciliated cells (Fig. 2-3) whose beat causes strong fluid currents to flow toward the uterine cavity. When an ovum is liberated from the surface of the ovary, it enters the fimbriated end of the uterine tube and passes slowly along the tube to the uterus. There, if it has been fertilized, it becomes attached and is nourished during prenatal development.

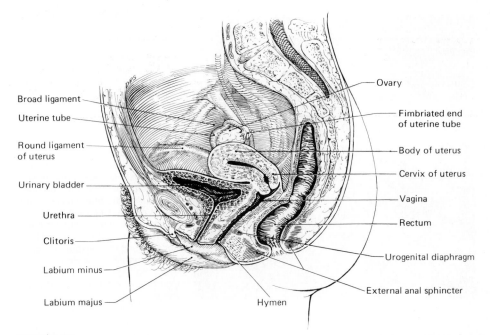

Broad ligament
Uterine tube
Round ligament of uterus
Urinary bladder
Urethra
Clitoris
Labium minus
Labium majus

Ovary
Fimbriated end of uterine tube
Body of uterus
Cervix of uterus
Vagina
Rectum
Urogenital diaphragm
External anal sphincter

Hymen

FIGURE 2-1
Sagittal section, adult female pelvis. (Redrawn, with slight modifications, from Sobotta, *Atlas of Human Anatomy*. Courtesy, G. L. Stechert & Company, New York.)

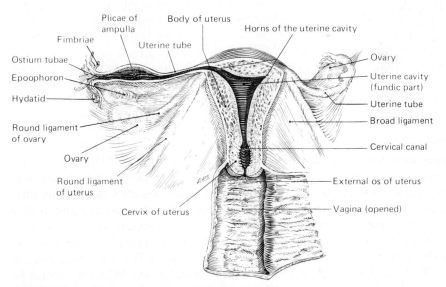

Plicae of ampulla
Body of uterus
Horns of the uterine cavity
Fimbriae
Ostium tubae
Uterine tube
Epoophoron
Hydatid
Round ligament of ovary
Ovary
Round ligament of uterus
Cervix of uterus

Ovary
Uterine cavity (fundic part)
Uterine tube
Broad ligament
Cervical canal
External os of uterus
Vagina (opened)

FIGURE 2-2
Internal reproductive organs of the female, spread out and viewed in ventral aspect. The vagina, uterus, and right uterine tube have been opened to show their internal configuration. (Redrawn, with slight modifications, from Rauber-Kopsch, *Lehrbuch und Atlas der Anatomie des Menschen*. Courtesy, Georg Thieme Verlag KG, Stuttgart.)

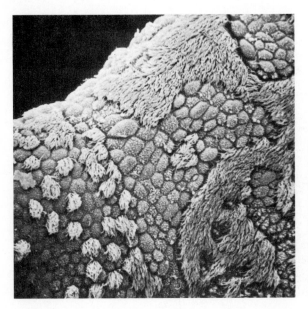

FIGURE 2-3
Scanning electron micrograph of the mucosal surface of the ampullary portion of the human uterine tube during the premenstrual (late luteal) phase. Cells with long tufts of cilia are scattered among nonciliated cells. ×1000. (From H. Ludwig, and H. Metzger, 1976. *The Human Female Reproductive Tract*, Springer-Verlag, Berlin. Courtesy of the authors and publisher.)

The human *uterus* is a pear-shaped organ which in the nonpregnant condition has thick walls, is richly vascular, and is well supplied with smooth muscle. The body of the uterus is continuous caudally with the neck or *cervix*, a region characterized by an attenuated lumen, thick walls, and glands of a different type from those occurring in the body of the uterus. The cervix of the uterus projects into the upper part of the vagina, which has the double function of an organ of copulation and a birth canal.

The external genitalia of the female are a complex of structures grouped about the vaginal orifice. Collectively they constitute the *vulva*. The outermost structures are a pair of fat-containing folds of skin known as the *labia majora* (Fig. 2-1). Within the cleft between the labia majora is a second, smaller pair of skin folds that are highly vascular and devoid of fat, the *labia minora*. Partially enwrapped by the labia minora where they meet anteriorly is the *clitoris*, a small erectile organ which is the homologue of the penis of the male. In the vulva, about midway between the clitoris and the vaginal orifice, is the opening of the *urethra*. The vaginal orifice is located in the posterior part of the vulva (Fig. 2-1). In virgins the entrance into the vagina is narrowed by a thin fold of tissue known as the *hymen*.

Male Reproductive Organs

The general arrangement and relationships of the male reproductive system are shown in Figs. 2-4 and 2-5. The *testes*, unlike the ovaries, do not lie in the abdominal cavity; instead, they are suspended in a pouchlike sac called the *scrotum*. Because of their location in the scrotum and the specialized arrange-

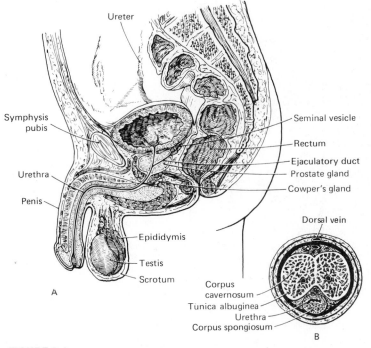

Ureter

Symphysis pubis

Urethra

Penis

A

Seminal vesicle

Rectum

Ejaculatory duct

Prostate gland

Cowper's gland

Dorsal vein

Epididymis

Testis

Scrotum

Corpus cavernosum

Tunica albuginea

Urethra

Corpus spongiosum

B

FIGURE 2-4
(A) Lateral view of the male reproductive organs. (B) Cross-section
through the penis to show the arrangement of its masses of erectile tissue
(the paired corpora cavernosa) and the unpaired corpus spongiosum.

ment of the vascular supply to the testes (a countercurrent heat-exchange
system), the temperature of the testes is several degrees lower than that of the
abdominal cavity. This is a requirement for the normal production of spermatozoa. The spermatozoa are produced in a large number of highly convoluted
seminiferous tubules. The total length of the seminiferous tubules is astonishing. Bascom and Osterud (1925) estimated that the seminiferous tubules from
one testis of a mature boar would extend 3200 meters if laid end to end. The 360
meters of seminiferous tubules in the human testes account for the production
of approximately 95 million spermatozoa per day.

The spermatozoa must pass over a long and elaborate series of ducts before
reaching the outside. From the seminiferous tubules they find their way through
short, straight ducts, the *tubuli recti*, into an irregular network of slender
anastomosing ducts known as the *rete testis*. From the rete testis the spermatozoa are collected by the *ductuli efferentes*, which in turn pass them on by way
of the much coiled duct of the *epididymis* into the *ductus deferens*. At the distal
end of the ductus deferens is a glandular dilation known as the *seminal vesicle*.
It was once believed that, as the name implies, the seminal vesicles serve as a
reservoir in which the spermatozoa are stored pending their ejaculation. We

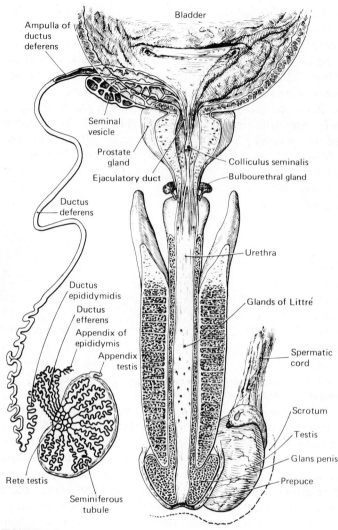

FIGURE 2-5
Schematic plan of the male reproductive organs spread out in frontal aspect.

now know that the spermatozoa are stored in the epididymis and ductus deferens and that the seminal vesicles are glandular organs which produce a secretion that serves as a vehicle for the spermatozoa and contributes to their nutrition.

The external genitalia in the male consist of the scrotum, containing the testes, and the penis. The *penis* contains three rodlike masses of erectile tissue held together by dense connective tissue and covered by freely movable skin. The paired dorsal structures are the *corpora cavernosa*. The single erectile mass located beneath the corpora cavernosa is the *corpus spongiosum* (Fig. 2-4B). In the shaft of the penis it is smaller than the corpora cavernosa, but distally it expands to form the *glans*. It is traversed throughout its length by the urethra. Opening into the urethra are numerous small mucus-producing glands, the *glands of Littré*, (Fig. 2-5). These glands become active in sexual excitement and produce a lubricating fluid which facilitates intromission.

When, during the coital climax (*orgasm*), the spermatozoa are discharged, they enter the urethra by way of the *ejaculatory ducts* (Fig. 2-5). At the same time, the contents of the seminal vesicles, the *prostate gland*, and the *bulbourethral glands* (*Cowper's glands*) are forcibly evacuated into the urethra, providing a fluid medium in which the spermatozoa become actively motile. This mixture of secretions with spermatozoa suspended in it (*semen*) is swept out along the urethra by rhythmic muscular contractions culminating in *ejaculation*.

SEXUAL CYCLE IN MAMMALS

Reproduction in mammals is a closely orchestrated process which requires the coordinated preparation of many tissues in the body of the female. Not only must an ovum be liberated from the ovary, the tissues of the female reproductive tract must be ready to transport both eggs and sperm to a common site where fertilization can occur. In the event of fertilization, the early embryos must be carried to a portion of the uterus that is prepared both to receive an embryo and to meet its nutritional requirements throughout the duration of pregnancy. In the behavioral realm, the female must signal to the male her readiness for copulation, and the male in turn must be ready to respond.

Most of the preparations for reproduction are of a cyclic nature. The changes in structural and functional characteristics of both male and female reproductive tissues are mediated by hormones, often interacting in tightly controlled feedback loops. It is now recognized that there is a substantial neural influence on reproduction and that environmental and psychic influences can exert profound effects on reproductive patterns.

Estrous Cycle in Mammals

Sexual periodicity is as a rule much less strongly developed in the male than in the female. In some animals, such as those of the deer family, there is a brief period of intense sexual activity at one particular season of the year and then a long period during which there is sexual impotence and cessation of spermatogenesis. More commonly, especially among the primates, the male is sexually potent throughout adult life. A brief period of pronounced sexual activity, when it does occur in males, is known to animal breeders as the "rutting season." It always corresponds in time with the females' period of strong mating impulse, which breeders call the "period of heat" and biologists speak of as the *estrus*.

Originally the term *estrus* referred merely to the existence of a period of strong sexual desire made evident through behavior. As more information has been acquired about the concomitant changes going on within the body, it has become evident that estrus occurs close to the time of ovulation and that the characteristic behavior is simply an external indication that all the complicated internal mechanisms of reproduction are ready to become functional. If pregnancy does not occur at this time, regressive changes follow and another period of preparation must ensue before conditions are again favorable for reproduction. This repeated series of changes is known as the *estrous* or *sexual cycle* (Hansel and Convey, 1983). In the absence of pregnancy, its phases are (1) a short time of complete preparedness for reproduction accompanied by sexual desire (*estrus*), (2) a period during which the fruitless preparations for pregnancy undergo regression (*post-* or *metestrum*), and (3) a period of rest (*diestrum*), followed by (4) a period of active preparatory changes (*proestrum*) leading up to the next estrus, when everything is again in readiness for reproduction (Fig. 2-6).

There is wide variation among animals in regard to the length of time occupied by this cycle. In some it occurs only once in an entire year, with the estrus being so placed seasonally that when the young are born, conditions are

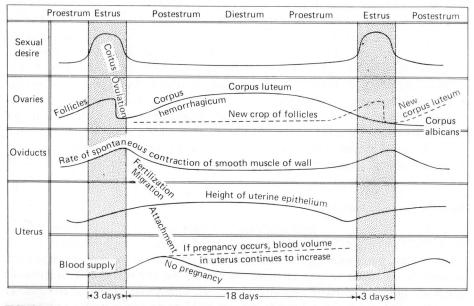

FIGURE 2-6
Graph showing correlation of changes which occur during the estrous cycle in the sow. Note the correlation of the important events leading toward pregnancy (coitus, ovulation, fertilization, and the migration of the ovum through the oviduct to the uterus, and finally its attachment to the uterine mucosa) with the height of local activity as indicated by the curves. (Compiled from the work of Corner, Seckinger, and Keye.)

favorable for their rearing. Species having only one breeding season in a year are said to be *monestrous*. Other animals exhibit several breeding periods in a year; they are said to be *polyestrous*. On the basis of our current knowledge, it would appear that a polyestrous rhythm is the underlying condition in mammals generally. Many factors mask or modify it in different cases, but in the forms which have been most fully studied it is unmistakenly present. It is not unreasonable to suppose that an annual estrus such as that exhibited by the deer family has become established through the suppression of other periods, primarily because of the regular recurrence of pregnancies of long duration following what was originally merely the most favorable of several estrous periods. It is well known, furthermore, that the estrous cycle may be interrupted by many things other than pregnancy. Thus starvation, extreme exposure, or severe sickness may cause the suppression of an estrus. A contributing cause in reducing a polyestrous rhythm to a monestrous one, operative in females failing to become pregnant, might well be the severity of the conditions in which many wild animals live during the winter or during a dry season.

Many animals (sheep, for example) that have only one breeding season a year when living in the wild state develop a polyestrous rhythm when living under domestication. An underlying polyestrous condition may thus be obscured when a pregnancy of long duration follows each estrus; this occurs normally among many wild animals. It becomes apparent, however, when such an animal, under conditions of domestication or under experimental conditions in the laboratory, is not permitted to become pregnant. Then, after an unfruitful estrus, there is a brief interval occupied by regression, rest, and preparation, followed shortly by another estrus. Since living conditions under domestication are relatively uniform, suppression of an estrus through starvation or exposure does not occur, and the estrous periods keep recurring at fairly regular intervals until one of them is consummated by pregnancy.

Other mammals with short periods of gestation, such as the rabbit, may be polyestrous except in the winter. In these animals light is apparently a critical initiating factor. Only when the average daily amount of light gets above a certain threshold does the hypophysis become active in the production of enough *follicle-stimulating hormone* (FSH) to set the whole reproductive cycle into operation (Fig. 2-7A).

Primate Menstrual Cycle

In primates the sexual cycle in females is characterized by *menstruation*, the discharge from the uterus of blood, mucus, and cellular debris at periodic intervals (approximately 4 weeks in humans). In humans menstruation usually commences (*menarche*) when the female is 12 to 14 years old and continues until the time of *menopause*, which ordinarily occurs during the late forties. The usual duration of the menstrual discharge is from 4 to 5 days, but there is considerable individual variability in both the length of period and the interval at which it occurs.

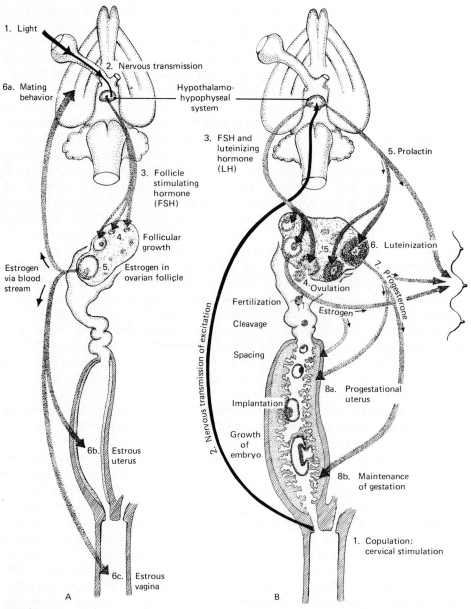

FIGURE 2-7

Diagrams showing the sequence of events in the reproductive cycle of the rabbit. (A) Reactions leading to the estrous state. (B) Sequence of events following mating. (Redrawn with slight modifications from Witschi, 1956. *Development of Vertebrates*. Courtesy of the author and W. B. Saunders Company, Philadelphia.)

The menstrual cycle may be divided into three major phases: (1) the *menses*, (2) the *proliferative (follicular) phase*, and (3) the *secretory (luteal) phase*. It will be easiest to follow the changes if we commence with conditions just after

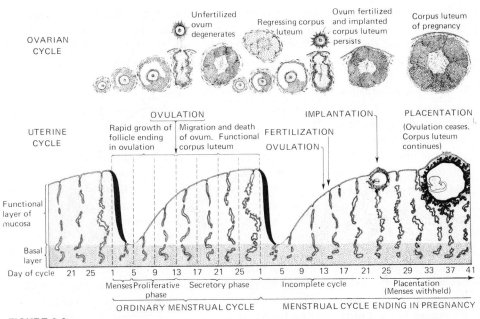

FIGURE 2-8
Graphic summary of changes in the endometrium during an ordinary menstrual cycle and a subsequent cycle in which pregnancy occurs. The correlated changes in the ovary are suggested above, in their proper relation to the same time scale. (Modified from Schroder.)

a menstrual period has ended. Figure 2-8 represents very schematically the changes occurring in the mucosal lining of the uterus during the cycle. The black descending bands signify the abrupt decrease in thickness which results from the menstrual sloughing. The tubular structures shown within the mucosa represent uterine glands. The more darkly shaded lower part of the mucosa is the so-called basal layer, which is below the level involved in sloughing. It is from this layer that the repair and growth processes of the proliferative phase are initiated. Restoration of the epithelial lining is accomplished with surprising rapidity by proliferation of the cells of the deep part of the glands, which have remained undisturbed in the basal layer. The glands increase in length as the mucosa increases in thickness, but throughout the proliferative phase they remain slender and relatively straight, and their lumina are small and devoid of any conspicuous amount of secretion.

After ovulation the proliferative phase gradually changes over into the secretory phase. The walls of the glands become irregular and the size of their lumen increases, and a considerable amount of secretion can be seen within the glands. There is also a striking increase in the conspicuousness of the small arteries supplying the superficial portion of the mucosa, and they extend nearer to the surface. These arteries tend to follow a spiral course, and their coiling becomes much more marked at this stage. A week after ovulation has occurred the whole histological picture reflects heightened activity. The glands are greatly distended, the small blood vessels are engorged, and the thickness of the mucosa increases from the 1 mm or less that was there immediately after the last period to perhaps 4 or 5 mm. At this stage the uterus is fully prepared to implant and nourish a young embryo. (See the cycle represented in the right side of Fig. 2-8.)

If the implantation of a fertilized ovum does not occur, the activities of the secretory phase end in the brief ischemic phase which immediately precedes menstruation. There is reduced blood flow to the superficial zone of the uterine mucosa, although the blood flow in the vessels supplying the deeper layers remains uninterrupted. In the superficial zone white blood corpuscles begin to migrate into the stroma, and the tissues, deprived of an active circulation, begin to deteriorate. When this ischemic phase has lasted a few hours, spiral arteries here and there start to open up and blood pours into the superficial capillaries, rupturing their weakened walls, so that there is extravasation of blood into the tissues beneath the epithelial lining. In a very brief time the now necrotic superficial tissue, the extravasated blood which remains unclotted, additional blood oozing from the freshly denuded surface, and the secretion from the opened mouths of the glands all start to come away together as the menstrual discharge. Once started, this process proceeds rapidly in a given area, but by no means is the entire uterine lining simultaneously affected. During the early part of the period, area after area is involved until by the third day the uterine surface has been essentially denuded. Repair, beginning initially in the areas which were the first to be affected, starts promptly, and a new proliferative phase is under way almost before menstruation has ceased.

Hormonal Regulation of the Female Sexual Cycle

From the earliest inquiries into the nature of the reproductive process, it was apparent that the functions and responses of the components of the female reproductive system are closely intertwined. It was soon recognized that hormones are the coordinating agents, but it has taken decades of painstaking work to understand the intricate interactions involved in the hormonal control of reproductive events. The initial flowering of the field of reproductive endocrinology occurred between 1920 and 1940, when a group of talented investigators delineated the fundamental hormones and pathways of control between the pituitary gland and the reproductive organs during various phases of the sexual cycle. This work is well summarized in a treatise edited by Young (1961). The next major developments were the recognition of the important relationship between the brain and the pituitary gland and the development of radioimmunoassays, which allowed extremely precise levels of hormones to be determined in body fluids. With the recent identification of cellular receptor molecules for hormones, it is now possible to ask significant questions about the mechanisms of hormones at the molecular level.

There are several levels in the hormonal control of reproduction (Table 2-1; Fig. 2-9). The first is in the brain itself. The brain receives and processes many

TABLE 2-1

MAJOR HORMONES INVOLVED IN MAMMALIAN REPRODUCTION

Hormone	Chemical nature	Function
Hypothalamus		
Gonadotropin-releasing hormone (GnRH, or LHRH)	Decapeptide	Stimulates release of LH and FSH by anterior pituitary
Prolactin-inhibiting factor	Dopamine	Inhibits release of prolactin by anterior pituitary
Anterior Pituitary		
Follicle-stimulating hormone (FSH)	Glycoprotein (alpha and beta subunits) MW ~35,000	Stimulates follicle cells to produce estrogen
Luteinizing hormone (LH)	Glycoprotein (alpha and beta subunits) MW ~28,000	Male: stimulates Leydig cells to secrete testosterone. Female: stimulates follicle cells and corpus luteum to produce progesterone. Promotion of lactation
Prolactin	Single-chain polypeptide (198 amino acids)	
Posterior Pituitary		
Oxytocin	Oligopeptide (MW ~1100)	Stimulates ejection of milk by mammary gland
Ovary		
Estrogens	Steroid	Multiple effects on reproductive tract, breasts, body fat, bone growth, etc.
Progesterone	Steroid	Multiple effects on reproductive tract, breast development
Testosterone	Steroid	Precursor for estrogen biosynthesis. Induces follicular atresia
Testis		
Testosterone	Steroid	Multiple effects on male reproductive tract, hair growth, and other secondary sexual characteristics
Placenta		
Estrogens	Steroid	Like ovarian estrogens
Progesterone	Steroid	Like ovarian progesterone
Human chorionic gonadotropin (HCG)	Glycoprotein (MW ~30,000)	Maintains activity of corpus luteum during pregnancy
Human placental lactogen (somatomammotropin)	Polypeptide (MW ~20,000)	Promotes development of breasts and corpus luteum during pregnancy

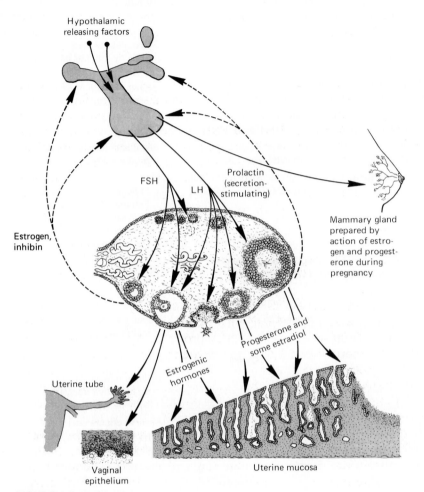

FIGURE 2-9
Diagram indicating the interplay between the hormones of the anterior lobe of the
hypophysis and the ovary in relation to major events of the human reproductive
cycle.

types of stimuli. Those of relevance to reproduction are channeled into an area
near the base of the brain known as the *hypothalamus*. For example, sheep are
hormonally stimulated by a gradually decreasing photoperiod, whereas in other
animals, such as the rabbit (Fig. 2-7), copulation is a prime stimulus that
ultimately leads to ovulation. Other sensory stimuli, such as olfactory sensa-
tions, can also affect endocrine function. It is now apparent that most of these
exogenous stimuli are translated into changes in endocrine activity via the
influence of the hypothalamus on the hypophysis. Cells of the hypothalamus
are also sensitive to levels of certain sex hormones in the blood. In response to
these various stimuli the hypothalamus produces a number of *releasing factors*,

as well as inhibiting factors, which are transported to the anterior lobe of the pituitary gland (hypophysis) via a specialized set of blood vessels known as the *hypothalamohypophyseal portal system* and stimulate it to secrete its hormones.

The hypophysis represents the second level of hormonal control (Fig. 2-9). Stimulated by the hypothalamic releasing factors and also responding directly to sex hormones in the blood, the anterior pituitary releases two *gonadotrophic hormones, luteinizing hormone* (LH) and FSH, glycoproteins with molecular weights of about 28,000 and 35,000, respectively. A single decapeptide-releasing factor from the hypothalamus promotes the release of both LH and FSH from the hypophysis. A third pituitary hormone is *prolactin*, a protein with a molecular weight of about 30,000. Formerly called *luteotrophic hormone* (LTH), prolactin is involved in a wide variety of regulatory functions that often differ greatly among the vertebrates.

The third level of hormonal control is embodied in the ovaries and, during pregnancy, in the placental tissues. The steroid hormones, 17β-estradiol[1] and progesterone, are produced by the follicles within the ovaries. These hormones are secreted into the blood.

The fourth level of hormonal control involves the effects of the ovarian steroid hormones on a wide variety of tissues throughout the body (Fig. 2-9). For example, circulating estrogen, which is produced by the preovulatory follicle, acts on the reproductive tissues to prepare them for gamete transport. As will be discussed in greater detail in Chap. 3, more cilia form on the epithelium of the uterine tube, and the smooth-muscle activity of the tube increases. The uterine mucosa begins the buildup that is required for implantation of the fertilized egg, and the cervical mucus becomes less viscous, thereby facilitating the passage of spermatozoa from the vagina into the uterus. After ovulation, progesterone, which is produced by the corpus luteum, further primes the endometrium for receiving the implanting embryo. Beyond the primary reproductive tissues, the breasts are highly responsive to both estrogen and progesterone (see Chap. 10), and other secondary sexual characteristics, such as the distribution of body fat, are due to the effects of the ovarian steroid hormones.

In the normal sexual cycle a set of ovarian follicles (Figs. 3-19 and 3-20) begin to mature, probably because of a slight rise in pituitary FSH, just before the menstrual period begins. As a result of both FSH and LH stimulation, the follicles begin to produce estradiol. All the follicles except one ultimately degenerate, and the remaining preovulatory follicle secretes increasing amounts of estradiol late in the follicular phase of the cycle (Fig. 2-10). The sharp increase in estradiol secreted by the ovarian follicle acts on the hypothalamohypophyseal

---

[1]Estradiol is one of a family of closely related steroids which have a similar physiological action. These substances are collectively known as *estrogens*. Estrone and estriol are other natural estrogens. The synthetic product, diethylstilbestrol, because it acts in a similar manner, is included in the same category.

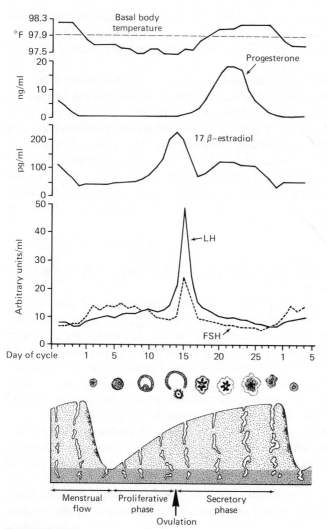

FIGURE 2-10
Diagram of representative curves of basal body temperature, and daily serum concentrations of gonadotropins and sex steroids in relation to the normal 28-day human menstrual cycle. (Redrawn from A. R. Midgley et al., 1973. in Hafez and Evans, eds., *Human Reproduction*, Harper & Row, New York.)

axis, and 1 day after the peak concentration of estradiol in the blood the pituitary, probably responding to increased hypothalamic releasing factor, puts out a sharp peak of both LH and FSH (Fig. 2-10). The LH peak is the final stimulus required for follicular maturation, and ovulation then occurs within 24 hours. Even before ovulation, estrogen production by the follicle falls, possibly because of decreased sensitivity of the follicular cells to gonadotropins.

After ovulation the remains of the follicle soon become transformed into the *corpus luteum* (Fig. 3-19), mainly through the actions of LH. The corpus luteum then secretes estradiol and especially progesterone in gradually increasing amounts until their levels in the blood reach a broad peak, which occurs about midway through the luteal phase (Fig. 2-10). With increased production of steroid hormones and an inhibitory substance called *inhibin*, which is released into the blood, a feedback inhibition acting on the hypothalamohypophyseal axis results in very low blood levels of the gonadotropins, FSH and LH. Later in the luteal phase the corpus luteum begins to regress. The cause of this regression has not been completely defined, but there is some evidence of a changing sensitivity of the luteal cells to LH. With the regression of the corpus luteum, production of estradiol and progesterone by the ovary falls, stimulating the production of the releasing factor by the hypothalamus and of the gonadotropins by the pituitary. Then the cycle begins again with the gonadotropic stimulation of a new crop of follicles shortly before the onset of the next menses.

Emphasis in the previous paragraphs was placed on interactions between pituitary and ovarian hormones, but these interactions would have no purpose without the effects of the ovarian steroids on other parts of the female reproductive tract. During the proliferative phase of the cycle, estrogen (estradiol) is the dominant hormone, and its actions on the reproductive tissues seem designed to facilitate the transport of gametes and fertilization. In the uterine tubes, estrogen causes an increase in the number of ciliated cells and changes in the oviductal fluid. The rapid fall in estradiol levels just before ovulation stimulates increased motility in the smooth muscle of the uterine tube, an adaptation that allows more rapid transport of the ovulated egg.

Estradiol stimulates mitosis of the endometrial cells in the uterus and also promotes early growth of the uterine glands. Progesterone, on the other hand, prepares the lining of the uterus for implantation of the embryo, should fertilization occur, by causing the uterine lining to become thicker and more richly vascularized. The uterine endometrium is in its most receptive state about 7 days after ovulation, which is the time when implantation would occur after fertilization of the ovum.

Both the cervix and the vagina are also responsive to hormones, and around the time of ovulation the pH of the upper vagina rises from its normally low levels to a level somewhat less inhospitable to spermatozoa. During much of the menstrual cycle the physical properties of cervical mucus act as a barrier to the passage of spermatozoa, but at the time of ovulation the viscosity of the mucus lessens and allows greater numbers of spermatozoa to pass through the cervix.

If the ovulated egg becomes fertilized, a new series of hormonal events is initiated. Of principal interest at this point is the continued maintenance of the corpus luteum by a gonadotropic hormone, *chorionic gonadotropin*, produced by the extraembryonic tissues associated with the embryo. The corpus luteum continues to grow and secrete large amounts of estrogens and progesterone. These not only help to maintain the uterine lining, they also begin to prepare the mammary gland for the eventual secretion of milk. Further details of hormonal relations during pregnancy are covered on page 284.

Hormonal Regulation of Reproduction in the Male

The principal hormone involved in sexual preparation in the male is *testosterone*, a steroid hormone produced and secreted by the *interstitial (Leydig) cells*, which are located in small clusters among the seminiferous tubules in the testes. It has been estimated (Fawcett, 1985, p. 832) that the average Leydig cell in the rat can produce about 10,000 molecules of testosterone per second. Testosterone has a local effect in maintaining spermatogenesis, but it is also secreted into the blood, through which it acts on a number of target organs, including the brain. In many target tissues testosterone is locally converted to a more potent form, *dihydrotestosterone*.

The interstitial cells are stimulated to produce testosterone by the pituitary gonadotropin LH, commonly called interstitial-cell-stimulating hormone (ICSH) in the male. LH is secreted from the human pituitary in pulses at roughly 90-minute intervals mainly during the night. FSH is specifically taken up by the Sertoli cells within the seminiferous tubules (Fig. 3-9), but the function of this hormone in spermatogenesis remains obscure. Under the influence of FSH, Sertoli cells synthesize an *androgen-binding protein* that helps to maintain a high concentration of testosterone in the seminiferous tubule. In the body, testosterone acts on the hypothalamohypophyseal axis so that there is a constant balancing between the blood level of testosterone and the production and release of the pituitary gonadotropins FSH and LH (ICSH). The regulation of the pituitary gonadatropins is mediated by the effects of inhibin, which is produced by the Sertoli cells of the testes and released into the bloodstream. As in the female, inhibin reduces the activity of the hypothalamohypophyseal axis. In general, high levels of testosterone are associated with low secretion of gonadotropins, and low testosterone levels with the stimulation of gonadotropin secretion.

GAMETOGENESIS AND FERTILIZATION

GAMETOGENESIS

The reproductive cells which unite to initiate the development of a new individual are known as *gametes*—the *ova* of the female and the *spermatozoa* of the male. The gametes themselves and the cells that give rise to them constitute the individual's *germ plasm*. The other cells of the body, which take no direct part in the production of gametes, are called *somatic cells* or, collectively, the *somatoplasm*. From a phylogenetic standpoint the germ plasm is of paramount importance because it constitutes the hereditary endowment that is passed on from one generation of the species to the next (Fig. 1-2). The somatoplasm can thus be regarded as the material that protects and nourishes the germ plasm.

In species such as frogs and a number of invertebrate forms, the germ plasm can be recognized very early in the life of an individual—sometimes as regions in the vegetal pole cytoplasm of the zygote or as specific cells during the cleavage stages. Irradiation of this region of frog embryos with ultraviolet light results in the development of an embryo lacking germ cells (Smith, 1966). The early segregation of a readily recognizable germ plasm has not been demonstrated in most vertebrate embryos. Only at later stages do cells of the germ line become apparent.

Gametogenesis (*oogenesis* in the female and *spermatogenesis* in the male) is a broad term that refers to the processes by which germ plasm is converted into highly specialized sex cells that are capable of uniting at fertilization and producing a new being. Commonly, gametogenesis is divided into four major phases: (1) the origin of the germ cells and their migration to the gonads, (2) the multiplication of the germ cells in the gonads through the process of mitosis, (3) reduction of the number of chromosomes by one-half by meiosis, and (4) the final stages of maturation and differentiation of the gametes into spermatozoa or ova that are capable of fertilizing or being fertilized.

The Origin of Primordial Germ Cells and Their Migration to the Gonads

Although much of the early history of the germ plasm is still unknown, the cells that are destined to give rise to the gametes are recognizable at a surprisingly early stage in development. In species (i.e., anuran amphibians) with recognizable germ plasm, it is possible to trace continuously the developmental lineage of this material from a circumscribed area in the unfertilized egg, through cleavage (in cells near the vegetal pole), and finally into certain endodermal cells during gastrulation (Fig. 3-1). It has not been possible to identify germ plasm in the early embryos of amniotes (birds, reptiles, and mammals), but in all these forms as well as in the anurans, future gametes can be identified among the endodermal cells of the yolk or yolk sac. These cells, called *primordial germ cells*, can be recognized by their large size and clear cytoplasm (Fig. 3-2) and by certain histochemical characteristics, such as

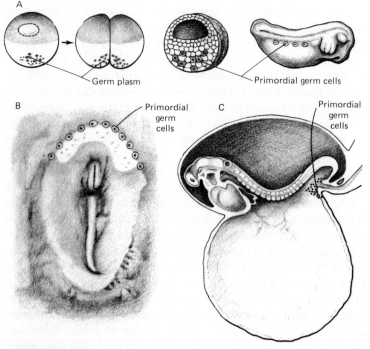

FIGURE 3-1
The origin of primordial germ cells in (A) anuran amphibians, (B) the chick embryo (after Swift), and (C) the 16-somite human embryo (after Witschi).

A   B

FIGURE 3-2
Primordial germ cells in human embryos. (After Witschi, 1948. *Carnegie Cont. to Emb.*, vol. 32). (A) Cross-section through a 16-somite human embryo (Carnegie Collection, 8005), showing primordial germ cells (large cells indicated by arrows) in the splanchnopleure of the yolk sac. (B) A primordial germ cell (large cell with clear cytoplasm) in the splanchnopleure at the coelomic angle in an embryo of 32 somites (Carnegie Collection, 7889; photomicrograph ×1100).

high alkaline phosphatase activity in mammals and a high glycogen content in birds.

Recent work has shown that although the primordial germ cells are usually associated with the endoderm, they need not originally arise within the endodermal germ layer. In the urodeles (salamanders), primordial germ cells arise from mesodermal cells, which form through the inductive influence of the ventral endodermal yolk mass on the overlying surface layer of cells. This mode of origin is so strikingly different from that of the anurans and higher vertebrates that Nieuwkoop and Sutasurya (1976) used it as the principal basis for a proposal that the urodeles have a phylogenetic origin separate from that of the anuran amphibians. More recently, Eyal-Giladi et al. (1981) presented evidence suggesting that before they enter the endodermal germ layer, avian primordial germ cells may originate in the ectodermal germ layer (epiblast). There is some evidence that mammals may follow a similar pattern, but the question is still open (Eddy et al., 1981).

Much remains to be learned about how germ cells come into being and about the relationships between germ cells and somatic cells. In animals with recognizable germ plasm (many invertebrates and anuran amphibians), only cells that contain germ plasm will become germ cells. However, transplantation of primordial germ cells of *Xenopus* into early blastulas has shown that cells containing germ plasm are not irrevocably determined to be germ cells but can differentiate into a number of cell types from all three germ layers (Wylie et al.,

1985). Although there is no evidence that in *Xenopus* somatic cells can become germ cells, ectodermal cells from the mammalian gastrula and even cells from a *teratocarcinoma*, a malignant germ cell-derived tumor, have been shown to be capable of forming functional germ cells (Gardner, 1978; Mintz and Illmensee, 1975).

When the primordial germ cells are first recognizable in embryos, the gonads either are very poorly developed or are not developed at all (Fig. 3-1). That these cells ultimately colonize the gonads has been determined by three means. One of the earliest methods was to extirpate the endodermal area in which the primordial germ cells are found. When Willier (1937) did this in the chick, gonads without gametes formed. A second way was to trace the course of the primordial germ cells during development. This could be done because of the large size and distinctive histochemical characteristics of these cells. Studies on a number of species demonstrated the migration of these cells to the gonadal primordia. The third way made use of a mutant strain of sterile mice. Mintz and Russell (1957), using histochemical staining for alkaline phosphatase, showed that only small numbers of primordial germ cells were present in the mutant embryos.

Why do the primordial germ cells originate so far from the gonads, and how do they get to the gonads? We cannot yet answer the first part of the question, but it should be noted that germ cells are not unique in taking their origin from the area of the yolk sac. Later in development the first blood cells also come from the yolk sac, but in contrast to the primordial germ cells, they arise from the mesoderm of the yolk sac rather than from the endoderm.

It is now well established that there are two principal routes by which primordial germ cells travel to the gonads. In mammals the primordial germ cells become capable of ameboid movements and migrate up through the dorsal mesentery and into the gonads. A different route is followed in birds (Fig. 3-3). At about the time when the earliest circulation to the yolk sac is established, the primordial germ cells make their way into the blood vessels and are passively carried to the body. Initially these cells are distributed at random throughout the body, but at later periods most are found in the gonads. There is some experimental evidence of chemical attractants that stimulate the migration of primordial germ cells to the gonads, but further work is necessary to confirm these results. There is also evidence that the orientation and physical characteristics of the substrate help guide the germ cells of amphibians and mammals to the gonads. Primordial germ cells that are deposited in extragonadal sites apparently die, but it is thought that misplaced primary germ cells occasionally develop into *teratomas*. Teratomas are bizarre tumors which often contain scrambled mixtures of highly differentiated tissues, sometimes including hair and teeth!

Careful quantitative studies have shown that the number of primordial germ cells increases during their migration to the gonads. For example, in the mouse the number of primordial germ cells rises from less than 100 to about 5000 during the period of migration (Mintz and Russell, 1957). However, the major increase in the number of germ cells occurs in the gonads.

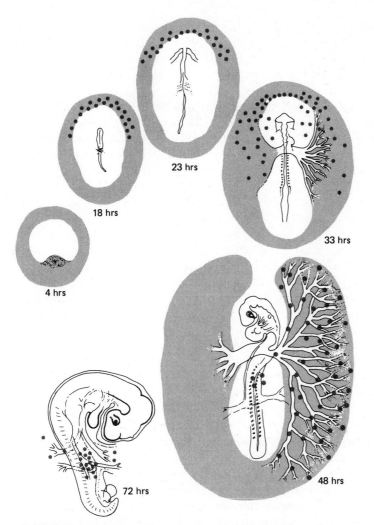

FIGURE 3-3
The migration of primordial germ cells (dark circles) in the avian embryo. 4 hours—no identifiable germ cells before the primitive streak is
formed; 18 and 23 hours—passive accumulation of primordial germ
cells in the anterior germinal crescent; 33 hours—active penetration
into blood islands and their entry into the circulation; 48 hours—circulation of germ cells and their early egress into the gonadal primordia; 72
hours—colonization of the gonads. (Redrawn from Nieuwkoop and Sutasurya, 1979.)

Proliferation of Germ Cells by Mitosis

The embryonic gonads are initially populated by a relatively small number of
migrating primordial germ cells. Once settled in the gonads, however, the germ
cells enter a proliferative phase in which their numbers increase greatly by

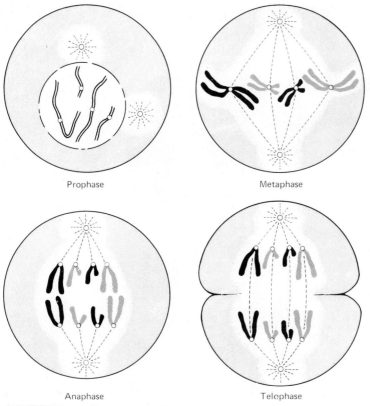

Prophase

Metaphase

Anaphase

Telophase

FIGURE 3-4
Schematic summary of the principal stages in mitotic cell division.

means of mitosis. (For a review of mitosis, see Fig. 3-4.) Mitotically active germ cells in the female are called *oogonia*; in the male they are known as *spermatogonia*.

The pattern of mitotic activity of the germ cells in the gonads differs widely between males and females. In the human female, intense mitotic activity between the second and fifth months of pregnancy brings the population of oogonia from a few thousand to about 7 million (Fig. 3-5). The number of oocytes then falls sharply, mainly because of *atresia* (natural degeneration), and by the seventh month most of the oocytes have entered the prophase of their first meiotic division. This brings to an end the proliferative phase of gametogenesis in the female. In contrast to the human, populations of oogonia in most of the lower vertebrates are capable of dividing throughout the reproductive life cycle. When one considers that some fishes release several hundred thousand eggs at one spawning, the need for the mitotic capability of oogonia throughout life is understandable.

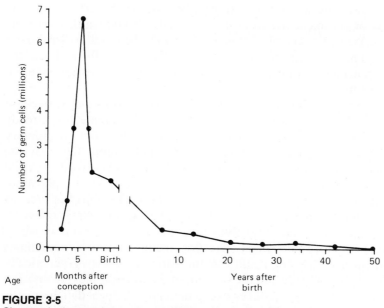

FIGURE 3-5
Changes in the population of germ cells in the human ovary with
increasing age. (Modified from T. G. Baker, 1970. In Austin and
Short, eds., *Reproduction in Mammals*, vol. 1, p. 20.)

The germ cells of the male follow a pattern of mitotic proliferation considerably different from that of the female. Mitosis begins in the gonad of the early
embryo, but it commonly persists throughout the life span of the male. The
testes always retain a germinative population of spermatogonia. Beginning at
puberty, periodic waves of mitosis produce subpopulations of spermatocytes
that enter meiosis as synchronous groups. This activity continues as long as the
male is capable of reproduction.

Meiosis

One of the fundamental requirements in the sexual reproduction of any species
is that the normal number of chromosomes must be maintained from one
generation to another. This is accomplished by the reduction of the chromosomal complement of the gametes from the *diploid* (2n) to the *haploid* (1n)
condition during gametogenesis. From the genetic standpoint, meiosis is
essentially the same in both male and female gametes. Because of this, the
general process of meiosis will be treated first. Later, the features of meiosis
specific to males and to females will be discussed.

A major requirement of meiosis is that each haploid gamete must acquire a
complete set of chromosomes. Yet meiosis is the phase during which new
combinations of genetic material, some arising from maternal genes and others

from paternal genes, are assembled. (Remember that each cell in the body contains one set of chromosomes from the maternal side and another set from the paternal side.) Genetic recombination occurs by (1) the random distribution of maternal or paternal chromosomes to the daughter cells and (2) the exchanging of portions of homologous chromosomes by crossing over at specific phases of meiosis.

Strictly speaking, meiosis does not involve the synthesis of DNA, because when meiotic prophase begins, DNA replication in the gamete has already taken place. The signal that tells a cell to make a change from an ordinary mitotic division to a meiotic division is poorly understood. Thus at the beginning of meiosis the cell can be described as 2n, 4c; in other words, the cell contains the normal number (2n) of chromosomes, but because of replication, its DNA content (4c) is double the normal amount (2c). The object of meiosis is to produce haploid gametes with a 1n, 1c complement of genetic material. The reduction of genetic material involves two *maturation divisions* in which new DNA synthesis does not occur. The first meiotic division, sometimes called the *reductional division*, results in the formation of two genetically dissimilar daughter cells (1n, 2c). In the second, or *equational*, meiotic division each of the previous two cells produces two genetically identical daughter cells (1n, 1c) that can now be properly called gametes. In contrast to mitosis, which is usually measured in terms of minutes or hours, meiosis may last from several days to as long as 45 to 50 years (in the human female).

First Meiotic Division   *Prophase I* is a complex stage that is usually divided into five substages: leptotene, zygotene, pachytene, diplotene, and diakinesis. Many events of both genetic and general developmental importance occur during the first meiotic prophase.

In the *leptotene stage* (Fig. 3-6) the chromosomes are threadlike and are just beginning to coil. Each chromosomal thread actually consists of two identical sister *chromatids*, which are joined somewhere along their length by a common *centromere*. One of the chromatids is an original DNA strand, whereas the other was newly synthesized just before the start of meiosis. It is usually not possible to resolve the individual chromatids by standard light microscopy.

The *zygotene stage* is the period during which the homologous paired chromosomes (one set of sister chromatids from the maternal side and one set from the paternal side) come together and become closely apposed along their entire length on a point-for-point basis. This precise lining up is called *synapsis*; it forms the basis for the crossing over of genetic material that occurs later in the first meiotic division. The area of contact between the paired chromosomes has a specialized ultrastructure and is called the *synaptinemal complex* (Moses, 1968). The synaptinemal complex is involved in the pairing up and possibly also in the crossing over of chromosomes.

During the early *pachytene stage* synapsis is completed; the two aligned chromosomal pairs are collectively referred to as a *bivalent*. A prominent feature of this stage is the thickening of the chromosomes owing to their coiling.

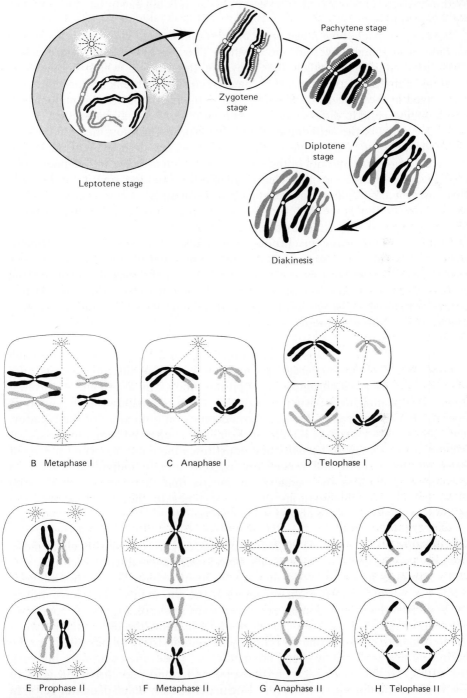

FIGURE 3-6
Schematic summary of the major stages of meiosis in a generalized germ cell.

Within a single chromosome, the sister chromatids appear to be held together by a centromere.

Late in the pachytene stage and during the *diplotene stage* portions of the paired chromosomes overlap one another. These areas of contact are called *chiasmata*, and it is thought that they are the sites at which homologous portions of maternal and paternal chromosomal strands break and are mutually exchanged in the process of crossing over. This is one of the two major ways of producing genetic differences between individuals. A major feature of the diplotene stage is the separation of the two paired chromosomes by splitting along the synaptinemal complex. The splitting occurs along most of the length of the chromosomes but not at the chiasmata, which are now well defined. The individual chromatids are also clearly visible by this time, and it can be seen that each bivalent consists of four distinct chromatids. It can now be called a *tetrad*. During this stage the chromatids uncoil slightly, and in eggs, at least, RNA synthesis occurs in areas of uncoiling.

In the *diakinesis stage* the chromosomes shorten even more. The splitting of the chromosome pairs continues, and one component of the splitting process that is characteristic of this stage is the moving of the chiasmata toward the ends of the chromosomes. This process is known as *terminalization*. At this point the nucleolus disappears, the nuclear membrane breaks, and the spindle apparatus becomes apparent.

Metaphase I The tetrads line up along the metaphase plate so that for each chromosome pair, the maternal chromosome is on one side of the equatorial plate and the paternal chromosome is on the other. The alignment of chromosomes is accomplished in such a way that there is a random distribution of maternally and paternally derived chromosomes on both sides of the equator. The random assortment of the chromosomes thus affected forms the basis for Mendel's second law of heredity and constitutes the most important means of ensuring that genetic differences are present among individuals. With 23 chromosome pairs in the human, this means that there are 2^{23} (8,337,408) different possible combinations of chromosomes in the haploid cells from random assortment of the chromosomes alone.

Anaphase I At this point the individual paired chromosomes begin to move toward opposite poles of the spindle. Both sister chromatids of each chromosome continue to be held together by the centromere. As the homologous chromosomes move away from one another, the chiasmata, which have moved to the ends of the chromosomes, are pulled apart and crossing over is complete. The events in anaphase I are crucial in understanding the difference between the first meiotic division and an ordinary mitotic division. In mitosis, after the chromosomes are lined up along the metaphase plate, the centomere between the sister chromatids of each chromosome splits and one chromatid goes to each pole of the mitotic spindle, resulting in genetically equal daughter cells. In contrast, the migration of the entire maternal chromosome to one pole and the paternal chromosome to the other pole during meiosis results in genetically unequal daughter cells.

Telophase I and Interphase In telophase I, the two daughter nuclei are separated from each other and nuclear membranes may re-form. Each nucleus now contains the haploid number of chromosomes (1n), but each chromosome still contains two sister chromatids (2c) connected by a centromere. Because the individual chromosomes of the haploid daughter cells are still in the replicated condition, there is no new replication of chromosomal DNA during the interphase between meiotic division I and meiotic division II.

Second Meiotic Division Except for the fact that the cell is haploid (1n, 2c), the second meiotic division behaves in most respects like an ordinary mitotic division. After an atypical prophase, a mitotic spindle apparatus is set up and the chromosomes line up along the equatorial plate at metaphase II. Then, in contrast to the first meiotic division but similar to a mitotic division, the centromere between the sister chromatid of each chromosome divides, thus allowing the sister chromatids to separate from each other during anaphase. With the completion of telophase II, meiosis is complete and the original diploid germ cell has produced four haploid daughter cells (1n, 1c). In the male, all four of the haploid cells will go on to form viable gametes, but because of asymmetric meiotic division in the female, only one viable ovum results. The remaining three daughter cells are much reduced in size and are represented only as the apparently functionless polar bodies.

Polyploidy

In rare instances animals may start their development with chromosomes in multiples of the normal number for their species. This condition is called *polyploidy*. When it occurs the individual may, for example, show three times the haploid number of chromosomes rather than the usual diploid number resulting from the union of two haploid gametes. The triploid condition does not always arise in the same manner, but it is known to be established when the reduction division fails to occur in a maturing oocyte. The most carefully studied instances of triploidy occur in Amphibia. Exposure of the eggs of newts or salamanders to pressure or extreme temperatures (from 0 to 3°C for 20 hours or 37°C for as little as 10 minutes) may inhibit the carrying through of the second meiotic division. When fertilization is completed in such an ovum, the unreduced diploid chromosomal number of the ovum, joined to the haploid number of a normal sperm, results in the triploid count.

Polyploidy is not always triploid, although that is by far the most common type. More rarely the count may be *tetraploid*, that is, four times the haploid number for the species. It seems probable that this condition may arise during early cleavage stages as a result of chromosomal division not followed by reorganization of daughter nuclei and cytoplasmic division. However polyploidy may be initiated, cells with a greater than normal chromosome number are proportionately larger than normal in size. Surprisingly, the body size of the embryo remains essentially normal, apparently because its large cells are correspondingly fewer in number.

Spermatogenesis and Oogenesis Compared

Despite similarities in the genetic aspects, differences are prominent when one compares spermatogenesis and oogenesis. Figures 3-7 and 3-8 summarize a number of parallels between the two processes. This section will point out some of the major differences between spermatogenesis and oogenesis in the human.

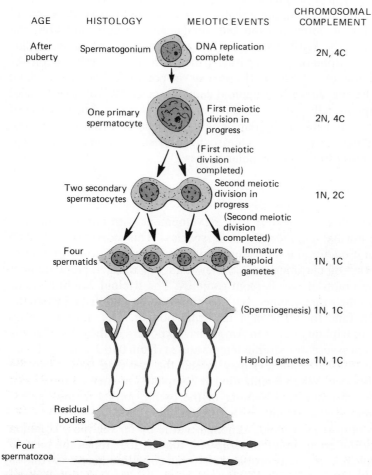

AGE	HISTOLOGY	MEIOTIC EVENTS	CHROMOSOMAL COMPLEMENT
After puberty	Spermatogonium	DNA replication complete	2N, 4C
	One primary spermatocyte	First meiotic division in progress	2N, 4C
		(First meiotic division completed)	
	Two secondary spermatocytes	Second meiotic division in progress	1N, 2C
		(Second meiotic division completed)	
Four spermatids		Immature haploid gametes	1N, 1C
		(Spermiogenesis) 1N, 1C	
		Haploid gametes 1N, 1C	
Residual bodies			
Four spermatozoa			

FIGURE 3-7
Summary of the major events in human spermatogenesis.

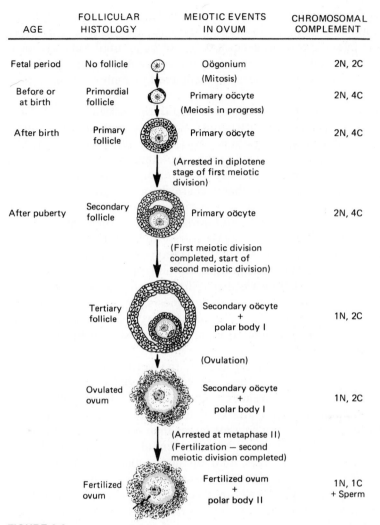

AGE	FOLLICULAR HISTOLOGY	MEIOTIC EVENTS IN OVUM	CHROMOSOMAL COMPLEMENT
Fetal period	No follicle	Oögonium (Mitosis)	2N, 2C
Before or at birth	Primordial follicle	Primary oöcyte (Meiosis in progress)	2N, 4C
After birth	Primary follicle	Primary oöcyte	2N, 4C
		(Arrested in diplotene stage of first meiotic division)	
After puberty	Secondary follicle	Primary oöcyte	2N, 4C
		(First meiotic division completed, start of second meiotic division)	
	Tertiary follicle	Secondary oöcyte + polar body I	1N, 2C
		(Ovulation)	
	Ovulated ovum	Secondary oöcyte + polar body I	1N, 2C
		(Arrested at metaphase II) (Fertilization — second meiotic division completed)	
	Fertilized ovum	Fertilized ovum + polar body II	1N, 1C + Sperm

FIGURE 3-8
Summary of the major events in human oogenesis.

In contrast to spermatogonia, each of which gives rise to four functional spermatozoa as the result of two meiotic divisions, an oogonium produces only one viable ovum. After the first meiotic division one cell is left with the bulk of the cytoplasmic material, whereas the other, called the *first polar body*, is left with little cytoplasm. The polar body often degenerates without taking part in the second meiotic division. During the second meiotic division there is also an unequal apportionment of cytoplasm between the daughter cells, and the ovum retains the bulk of the cytoplasm, leaving the other cell to fall by the wayside as the *second polar body*.

In the human female, the first meiotic division begins in the embryo and meiosis is not completed until the onset of puberty at the earliest or just before menopause at the latest. Spermatogenesis does not begin until puberty but is then continuous throughout life. There are no prolonged meiotic arrests during spermatogenesis, and the entire process is completed in somewhat more than 2 months.

Arrests in the process of meiosis are prominent during oogenesis. Although there is considerable interspecies variation, especially among the invertebrates, it is common for an egg to experience two periods of meiotic arrest. The first occurs during prophase I, especially in the diplotene phase. Such a pause seems necessary to allow the egg to build up its stores of yolk as well as to make preparations for the synthetic activity that necessarily occurs after fertilization. The diplotene arrest in meiosis can be very prolonged—over 40 years in some human ova. The first meiotic arrest is often broken by hormonal changes, and meiosis resumes, only to be arrested again (often at metaphase II in vertebrates). The second arrest is released with fertilization or artificial activation of the egg. Some eggs, e.g., those of the sea urchin, undergo a completion of the second meiotic division before they are shed. Although all phases of meiosis do not occur at the same rate during spermatogenesis, prolonged periods of meiotic arrest are not the rule. A general rule is that if the second meiotic division in eggs (or sperm) is completed before fertilization, no further DNA synthesis will occur until the egg and sperm have met.

Spermatogenesis is quite similar throughout the vertebrates. The mature spermatozoon is markedly smaller than the spermatogonium and is highly motile. Because of the wide differences both in the amount of yolk that is formed and in reproductive habits, the structural aspects of oogenesis vary considerably. Although mature ova become larger than oogonia, their relationship to the body of the mother varies with the amount of yolk that is produced. During the deposition of yolk, eggs take up large quantities of materials produced by the liver via the follicular cells. Aside from a possible contribution by Sertoli cells, developing sperm cells receive little formed material. During development the egg stores both energy sources and precursors of proteins and nucleic acids. Sperm cells, by contrast, shed most of their cytoplasm and must rely on the seminal fluid as an energy source. In anticipation of future requirements, the egg produces and stores up much RNA, whereas there is little or no RNA synthesis during the later stages of spermatogenesis.

SPERMATOGENESIS

The transition from mitotically active primordial germ cells to mature spermatozoa is called *spermatogenesis*, and it involves a sweeping series of structural transformations. Although there is a wide variety in the morphology of mature spermatozoa, the overall process of spermatogenesis is much the same throughout the vertebrate classes. This process can be broken down into three principal phases: (1) mitotic multiplication, (2) meiosis, and (3) spermiogenesis.

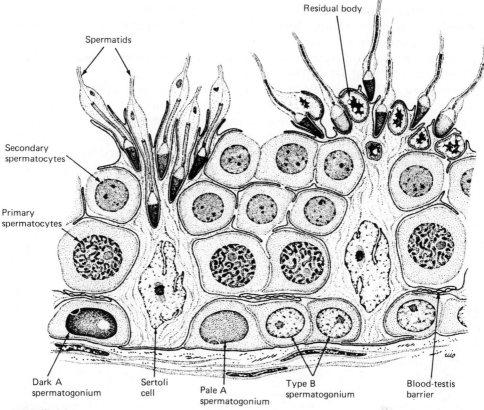

FIGURE 3-9
Drawing of a section of seminiferous epithelium, showing the relationships between Sertoli cells and developing sperm cells. (Courtesy of Y. Clermont, from Dym, 1977. In Weiss and Greep, eds., *Histology*, 4th ed., McGraw-Hill Book Co., New York, p. 984.)

Mitosis of sperm-forming cells occurs throughout life, and the mitotically active cells within the seminiferous tubules are known as *spermatogonia*. These cells are concentrated near the outer wall of the seminiferous tubules (Fig. 3-9). Spermatogonia have been subdivided into two main populations. *Type-A spermatogonia* represent the stem-cell population. Within this population is a group of dark, noncycling cells (Ad) that may be long-term reserve cells. Some of these cells become mitotically active pale cells (Ap), which ultimately give rise to *type-B spermatogonia*. These are cells which have become committed to leaving the mitotic cycle and which go on to finish the process of spermatogenesis.

After the final round of DNA duplication, the type-B cells are called *preleptotene spermatocytes* and are ready to pass through the meiotic phase of spermatogenesis. During the first meiotic division each *primary spermatocyte* divides into two equal daughter cells. With the onset of the second meiotic division these cells are known as *secondary spermatocytes*. In the human the first meiotic division lasts for several weeks, whereas the second one is completed in about 8 hours. Four haploid *spermatids* result from the meiotic phase of spermatogenesis.

Although they no longer divide, the spermatids undergo a profound transformation from relatively ordinary looking cells to the extremely specialized *spermatozoa*. The third phase in spermatogenesis is called *spermiogenesis*, or *spermatid metamorphosis*.

In the metamorphosis of a spermatid many radical changes occur. At the end of the second maturation division the nucleus is in typical interphase condition, with dispersed, finely granular chromatin and a reconstituted nuclear membrane (Fig. 3-10A). Almost immediately the nucleus begins to lose fluid, with a

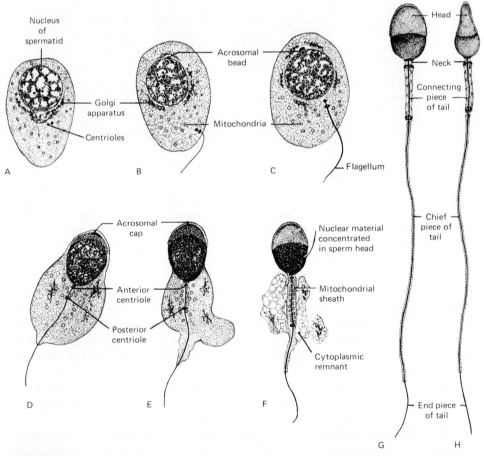

FIGURE 3-10
Stages in maturation of spermatids. (Modified from figures by Gatenby and Beams, 1935. *Quart. J. Micr. Sci.*, vol. 78.)

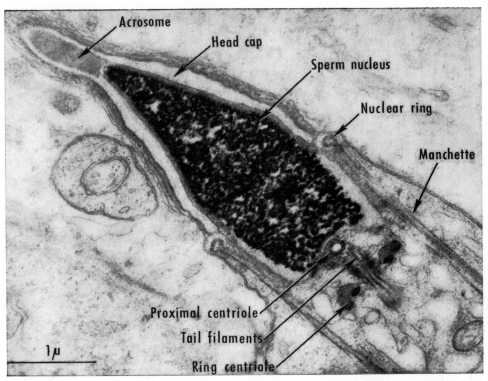

FIGURE 3-11
Electron micrograph of late spermatid of cat. Buffered osmic acid fixation. (From Bloom and Fawcett, *Textbook of Histology*, after M. H. Burgos and D. W. Fawcett, 1955, *J. Biophys. and Biochem. Cytol.*, vol. 1. Courtesy of the authors and the W. B. Saunders Company, Philadelphia.)

resultant decrease in its size and a concentration of its chromatin (Fig. 3-10C and D). This continues until the compacted chromatin comes to constitute the bulk of the head of the spermatozoon (Fig. 3-10D to 3-10F).

Concurrently there are changes in cell organization. The cytoplasm streams away from the nucleus, which will become the sperm head, leaving only a thin layer covering the nucleus (Fig. 3-10D and E). Part of the cytoplasm containing the Golgi apparatus becomes concentrated at the apical end of the developing sperm head, and the *acrosome* takes shape (Fig. 3-11). Within the cytoplasm the centrioles become more conspicuous and appear to be a point of anchorage for the developing *flagellum* (Fig. 3-10B and C). The posterior centriole moves away from the anterior one and takes on the shape of a ring encircling the flagellum (Fig. 3-10D to 3-10F). *Mitochondria* begin to form a spiral investment about the proximal part of the flagellum (Fig. 3-12). As spermiogenesis continues, the remaining cytoplasm disintegrates (Fig. 3-10E and F), leaving the mature spermatozoon stripped of all nonessential parts. It consists of (1) a head containing the nucleus and acrosome, (2) a middle piece containing the proximal part of the

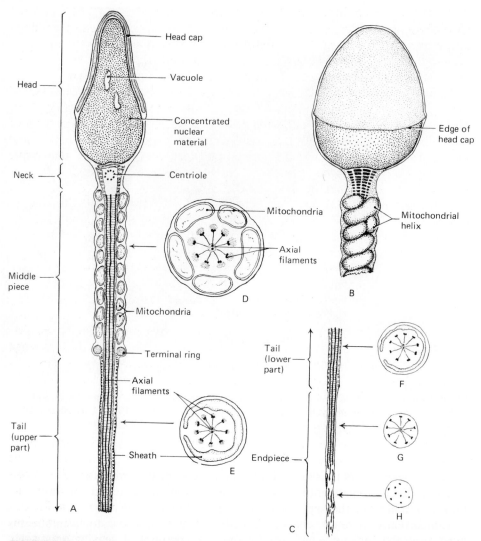

FIGURE 3-12
Diagrams showing the structure of human spermatozoa as revealed by electron micrographs. (A) Longitudinal section of head, neck, middle piece, and upper part of tail. (B) Head, viewed from its flattened surface, together with neck and part of middle piece. (C) Terminal part of tail proper and endpiece. Between the parts represented in B and C, a considerable portion of the tail has been omitted. (D)–(H), More highly magnified cross-sections of the middle piece and tail at the levels indicated by the arrows. (Schematized from Ånberg, 1957, and Fawcett, 1958.)

flagellum, the centrioles from which it arose, and the *mitochondrial helix*, which acts as an energy source, and (3) the tail (Fig. 3-12), a highly specialized flagellum.

As early as the mitotic divisions of the type-A spermatogonia, the daughter cells are connected to one another by fine *intercellular bridges* of cytoplasm that are the result of incomplete *cytokinesis (cell division)* after mitosis. Clusters of interconnected cells are found throughout spermatogenesis until the late spermatid stage (Fig. 3-7). These bridges probably facilitate the synchronous differentiation and division of the sperm-producing cells. At any given stage of spermatogenesis, dozens of cells may be so interconnected.

During spermatogenesis, the cells are also closely associated with *Sertoli cells,* which lie at regular intervals along the seminiferous tubule (Fig. 3-9). Until recently, the function of Sertoli cells, which constitute about 5 percent of the cells within the seminiferous tubules, was poorly understood. It is now known that they serve a wide variety of functions (Tindall et al., 1985), including (1) being the target cells for FSH, (2) synthesizing of an androgen-binding protein that maintains a high concentration of testosterone inside the seminiferous tubule, (3) maintaining the blood-testis barrier, (4) creating an environment that is important in the differentiation of sperm cells, (5) facilitating the release of mature spermatozoa, and (6) degrading the residual cytoplasm that is shed during spermiogenesis.

The *blood-testis barrier* is necessary to prevent the body's immune system from destroying the maturing sperm cells, which are antigenically different from the rest of the body. The blood-testis barrier consists of a continuously interlocking sheet of Sertoli-cell processes (Fig. 3-9), which are attached to one another by tight junctions. Outside the barrier are spermatogonia and spermatocytes that are just entering into meiosis. Spermatocytes in the zygotene stage of meiosis pass through the blood-testis barrier. Passage through the barrier is accomplished by the Sertoli cells' extending processes and forming an impermeable layer on the outer side of the leptotene spermatocytes. Concurrently, the original Sertoli-cell barrier on the inner side of the cells is disrupted as the Sertoli cells withdraw these processes.

Spermatogenesis is not a random process; rather, within the seminiferous tubules it is highly synchronized in regard to both time and location. The earliest stages of spermatogenesis occur at the periphery of the tubule, and progressively later stages are encountered closer to the lumen. However, all stages of spermatogenesis are not seen in the same section of the seminiferous tubule. Typically, several generations of developing sperm cells are present along any radial line drawn in a cross section of a tubule, and all cells within a given generation are in the same stage. In many mammals, such as the rat, well-defined waves of spermatogenesis can be delineated along the length of the seminiferous tubule. The distribution of cell associations in the human testis is more patchy than wavelike in nature (Clermont, 1972). In the human, the time required for a spermatogonium to develop into a spermatozoon is about 64 days.

Gene Expression During Spermatogenesis

To accomplish the structural reorganizations that take place during spermiogenesis, new proteins must be made. It is becoming clear that for some of these proteins, at least, the developing sperm cell uses a strategy similar to that employed by the egg, namely, to produce the appropriate messenger RNAs at an early stage, store them in an inactive form, and then, when needed, release the mRNA molecules that are used to form the needed proteins. This strategy involves what is often called *posttranscriptional control.*

A good example involves the synthesis of *protamines* in the trout. Protamines are small arginine-rich proteins which displace the histones and are involved in the high degree of compaction of the nuclear chromatin during the final stages of spermatogenesis. In situ hybridization studies with the cDNA to protamine mRNA have shown that protamine mRNA is first synthesized during the primary spermatocyte stage, but not until the spermatid stage almost a month later does translation of the protamine message into protein occur (Jatrou and Dixon, 1978). During the period between its synthesis and translation, the protamine mRNA is complexed with proteins in an inactive state.

There is also evidence that some genes in mice are transcribed as late as the spermatid stage (Palmiter et al., 1984). This phenomenon, called *haploid gene expression*, has been demonstrated by studying the appearance of cell-surface antigens during spermatogenesis of mice that are heterozygous with the t^{12} mutation in mice.

Sperm Maturation

Although in many invertebrates and in vertebrates that rely on external fertilization, sperm taken directly from the testes are capable of fertilizing eggs, those of mammals are not. Despite their appearance of morphological maturity, newly formed mammalian spermatozoa are minimally functional. During their leisurely transit from the seminiferous tubules to the tail of the epididymis, where they are retained until their ejaculation, the spermatozoa are exposed to a series of different humoral environments within the male genital duct system. Although the mechanisms are poorly understood, the metabolic apparatus of the spermatozoa becomes more capable of translating chemical energy into a certain degree of motility. In addition, the head of the sperm becomes covered with a glycoprotein coating which must be removed in the female reproductive tract before fertilization can occur. A final phase of sperm maturation in the male reproductive tract might better be called activation, after ejaculated spermatozoa have come into contact with the seminal fluid secreted by the seminal vesicle and prostate gland. The seminal fluid provides the functionally mature sperms with an external energy source which allows them to gain full motility. Further changes of sperm cells in the female reproductive tract will be described in the section on fertilization.

OOGENESIS

To a far greater extent than spermatogenesis, oogenesis varies in accordance with the reproductive habits of the animal. In species that rely on external fertilization in water, the number of mature eggs released at a single spawning ranges from hundreds to hundreds of thousands. Animals with internal fertilization are much more parsimonious in the production of eggs; commonly only 1 ovum, and seldom over 15 ova, matures at any one time. There is great variation not only in the number but also in the size of the eggs. Size depends principally upon whether the fertilized egg develops within or outside the body of the mother. The ova of mammals are very small, because it is not necessary to store in advance large amounts of yolk and other materials needed for development of the embryo. In contrast, the ova of animals developing outside the body are often quite large, since they contain within them the yolky materials needed for the embryo's development. Typically the eggs of aquatic animals are considerably smaller than the eggs of reptiles and birds. The differences in environment also necessitate different coverings which surround the egg. Another important aspect of oogenesis is preparation. There is an ever-increasing awareness of the extent of the maternal contribution to the control and maintenance of early embryonic development in many species. How the future needs of the early embryo are anticipated during oogenesis is another important theme in this section, where the development of three different types of eggs (amphibian, bird, and mammal) will be detailed.

Oogenesis in Amphibians

The reproductive patterns of amphibians, like those of most lower vertebrates, are geared towards the production of large numbers of eggs and their release at one time during the year. Typically, amphibians lay several hundred eggs at each spawning, whereas fishes as a rule release thousands or even hundreds of thousands of eggs. Because of both the massive number of eggs and the annual release of eggs, the pattern of oogenesis in lower vertebrates differs in some respects from that in many birds and mammals.

In amphibians the mitotic phase of oogenesis does not come to an early halt as it does in mammals. Rather, each year a new crop of eggs is generated by mitotic proliferation from a population of gametogenic stem cells. The maturation of frog (*Rana pipiens*) eggs requires 3 years (Fig. 3-13). The first batch of eggs begins to mature shortly after metamorphosis. During the first 2 years maturation is a relatively leisurely process, but during the third summer the eggs mature rapidly, ultimately attaining a diameter of about 1500 μm in the autumn of the third year. The female then hibernates, and the eggs, protected by a coating of jelly, are laid in the water early in the following spring. Because of the 3-year cycle of oogenesis, the ovary of a mature frog contains three batches of eggs at any one time.

Within the amphibian ovary, the eggs are arranged in individual follicles. They are surrounded first by a layer of follicular epithelium, next by a *theca* (a thin layer of ovarian connective tissue containing blood vessels), and then by a layer of ovarian epithelium.

Maturation of the Amphibian Egg

Maturation of the amphibian egg must anticipate the requirements of the embryo, because until the initiation of feeding the embryo develops as an essentially closed system; that is, everything that is required for early development must be contained within the fertilized egg. Therefore, a major requirement is a supply of yolk sufficient to provide substrates for synthetic activity of the embryo. In addition to accumulating yolk, the maturing egg synthesizes a large amount of RNA—enough to accommodate most of the needs of the embryo through the period of cleavage.

Maturation of the amphibian egg can be divided into two broad phases: *previtellogenic* (before the deposition of yolk) and *vitellogenic* (the major period of yolk deposition). The previtellogenic phase usually includes the period up to the early diplotene period of meiosis. The immature (prediplotene) egg is a relatively unspecialized cell, with few cytoplasmic organelles or inclusions (Fig. 3-14A). Among the earliest maturational changes are an increase in the number of mitochondria and the beginning of increased RNA synthesis, initially transfer RNA (tRNA) and the small (5S) ribosomal RNA (rRNA) molecules.

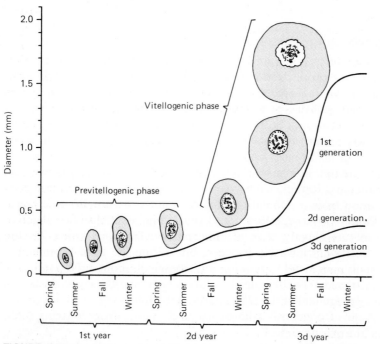

FIGURE 3-13
Growth of oocytes in the frog during the first three years of life. Ultimately, three generations of oocytes are found in the ovary. The drawings show the development of oocytes in the first generation. (Modified from P. Grant 1953. *J. Exp. Zool. 124*:513.)

As the oocyte enters the prolonged diplotene phase of meiosis, maturational changes begin in earnest. The nucleus becomes a site of intense synthetic activity, and with this its diameter increases by a factor of up to 7 or 8 until it reaches a diameter of about 350 μm at the end of the diplotene phase. For many years the enlarged nucleus of the amphibian egg has been referred to as the *germinal vesicle.*

Within the nucleus the chromosomes have begun to spread out from their tightly coiled configuration in the previous (pachytene) stage of meiosis, and they soon begin to form large numbers (up to 20,000 in the newt) of symmetrical loops; this has led to their being called *lampbrush chromosomes* (Fig. 3-15). The loops of lampbrush chromosomes are areas in which the DNA has been stretched out to a length equivalent to that of many individual genes. Along the loop the polarized synthesis of RNA occurs (Gall and Callan, 1962), and it has been estimated that during the lampbrush stage approximately 5 percent of the genome of the oocyte is exposed and acts as a template for RNA synthesis.

The early diplotene nucleus is also characterized by the formation of large numbers of nucleoli (up to 1500 in *Xenopus*), which soon become distributed beneath the nuclear membrane. These nucleoli are the morphological expression of a phenomenon known as *specific gene amplification* (Brown and Dawid, 1968). Gene amplification is an adaptive response for meeting the synthetic requirements of the egg, in this case the formation of a population of ribosomes sufficiently large to last throughout the period of cleavage in the embryo. The nucleolus is the principal structure involved in the synthesis of high-molecular-weight rRNAs and the assembly of ribosomes. It has been estimated that with only the number of nucleoli normally found in cells, several hundred years would be required to produce the number of ribosomes found in the mature amphibian egg. However, selective replication of those portions of the genome which must account for the formation of ribosomes reduces the time required to fulfill the production requirements of the oocyte to only a few months.

While synthetic activity is occurring in the nucleus, major changes are also occurring in the cytoplasm (Fig. 3-14D). These changes are principally concerned with the formation of yolk; hence one can say that the vitellogenic phase of oogenesis has now begun. Yolk is not a specific chemical but is a collective term for several classes of chemical substances which are stored in the cytoplasm to provide nutrition for the developing embryo. In the amphibian egg, proteinaceous material is stored in the form of membrane-bound *yolk platelets*, lipid is stored as inclusions called *lipochondria*, and carbohydrate is accumulated as aggregates of glycogen granules.

Yolk platelets are the most prominent inclusions in the cytoplasm of the amphibian egg. For many years, both their origin and their mode of formation remained a mystery, but now it is known that the bulk of yolk protein is produced by cells of the liver (Fig. 3-16) and carried to the ovary via the blood (rev. by Wallace, 1985). The yolk precursor in the blood is a phospholipoprotein called *vitellogenin*, which actually represents a family of proteins. This

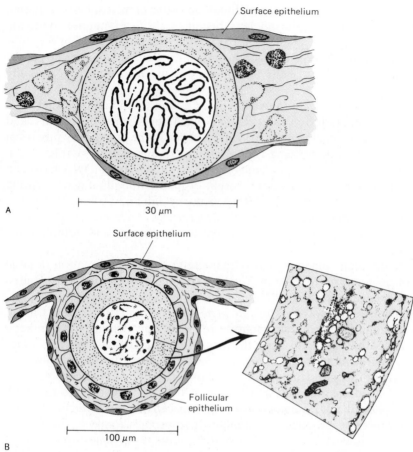

FIGURE 3-14
Structural changes in the amphibian oocyte during oogenesis. (A) Leptotene stage. (B)
Very early diplotene stage. (C) Diplotene stage. (D) Maturing oocyte in late prophase I.

material is cleared from the blood in the ovary and must pass through the
follicular epithelium to reach the egg. Because vitellogenin is too large for a
molecule to pass through the plasma membrane of the oocyte by diffusion, it is
incorporated into the oocyte by the process of *micropinocytosis*. In micropinocytosis, a membrane-bound vesicle filled with the yolk precursor forms at the
surface of the egg and then pinches off into the interior of the cell. Vitellogenin
cannot be detected within the mature oocyte. Instead, the protein yolk is
represented by two molecules: *phosphovitin*, a protein (MW 35,000) with a high
phosphorus content, and *lipovitellin*, a lipoprotein (MW ~400,000). These two
proteins are packed in crystalline form within a membrane to form the yolk
platelets. Although a small percentage of yolk platelets consists of the yolk
proteins crystallized within mitochondrial membranes, the majority of the yolk

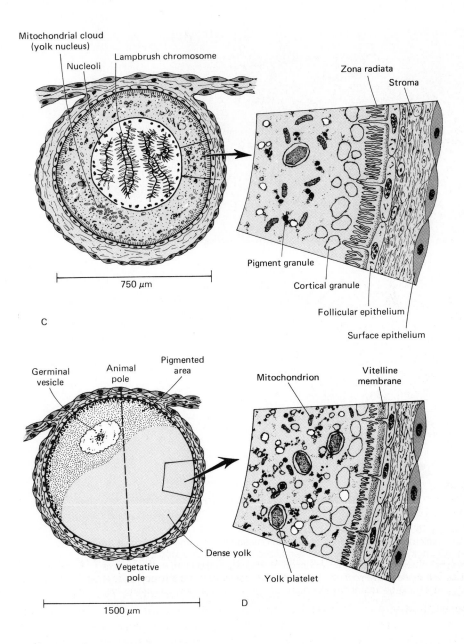

Mitochondrial cloud (yolk nucleus)
Lampbrush chromosome
Nucleoli
Zona radiata
Stroma
750 μm
C
Pigment granule
Cortical granule
Follicular epithelium
Surface epithelium

Germinal vesicle
Animal pole
Pigmented area
Mitochondrion
Vitelline membrane
Dense yolk
Vegetative pole
Yolk platelet
1500 μm
D

platelets apparently arise from the coalescence of smaller pinocytotic vesicles containing the yolk proteins.

For many years, before the origin of yolk proteins in the liver was understood, yolk formation was thought to be a function of a *yolk nucleus* (*Balbiani body*), which is quite prominent in the developing eggs of frogs. This structure, now recognized to be a dense cloud of mitochondria, is no longer

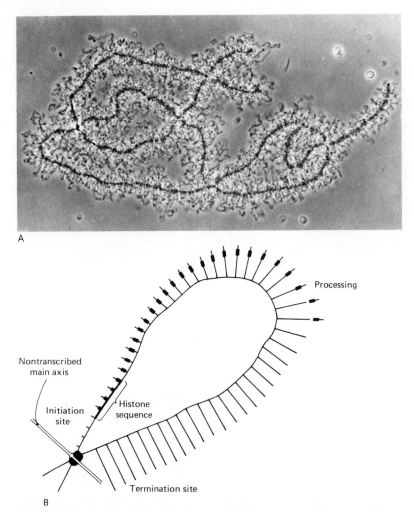

FIGURE 3-15
(A) Photomicrograph showing lampbrush chromosomes from the oocyte of a newt. (Courtesy of J. G. Gall.) (B) Schematized drawing of one loop of a lampbrush chromosome, showing transcription of RNA along the loop. In this loop, a histone sequence is transcribed. As synthesis proceeds around the loop (clockwise), processing occurs and the portions of the RNA molecules containing the histone transcript are cleaved off. (Modified from R. W. Olds, H. G. Callan, and K. W. Gross, 1977. *J. Cell Sci.*, *27*:57–80.)

thought to produce significant quantities of yolk material, but its real function remains obscure.

Cortical granules also begin to form in the cytoplasm at about the same time as the yolk platelets. Cortical granules (Fig. 3-14D) are membrane-bound inclusions composed principally of protein and mucopolysaccharide material. These structures have an irregular distribution throughout the animal kingdom.

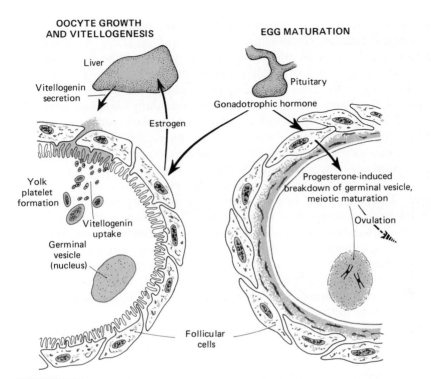

FIGURE 3-16
Scheme of growth (left) and final maturation of the amphibian oocyte. (Adapted from Browder, *Developmental Biology* (1984), Saunders, and Gilbert, *Developmental Biology* (1985), Sinauer.)

For instance, they are found in the eggs of sea urchins, frogs, and humans but not in those of salamanders. They appear to originate from the Golgi complex and ultimately become dispersed around the outer layer of cytoplasm (cortex) of the egg, just beneath the plasma membrane. As will be seen later in this chapter, the cortical granules play an important role in the reaction of the egg to penetration by a spermatozoon.

By the late diplotene or the early diakinesis stage, the oocyte is in most respects mature. The chromosomes lose their lampbrush configuration and begin to condense near the center of the nucleus. The nucleoli also move from the area adjacent to the inner nuclear membrane and become distributed around the chromosomal aggregate. These intranuclear changes presage the ultimate dissolution of the nuclear membrane (breakdown of germinal vesicle). This results in the mixing of nuclear contents with the cytoplasm, which in many species is an essential prelude to cleavage when the egg is fertilized.

The cytoplasm of the mature egg is packed with various organelles and inclusions. There are commonly well over a million mitochondria, in contrast to the several hundred found in most somatic cells. The large number of mitochondria is due to their autonomous division, which is based on the replication of the circular, single-stranded mitochondrial DNA. As a result, a high proportion of the DNA of the mature oocyte is of the mitochondrial variety. Other cytoplasmic organelles are centrioles, the Golgi apparatus (which is involved in the final packaging of carbohydrate and protein molecules), and a specialized form of endoplasmic reticulum (*annulate lamellae*) consisting of stacks of porous membranes that contain some RNA. The function of annulate lamellae is not known (rev. by Kessel, 1985).

In addition to yolk inclusions, the mature oocyte contains numerous pigment granules. Arising later during oogenesis than the other inclusions, the pigment granules are concentrated in the half of the egg known as the *animal hemisphere*. The other, less heavily pigmented but more yolk-laden half is known as the *vegetal hemisphere* (Fig. 3-14D). The transitional zone between these two hemispheres is commonly known as the *marginal zone*. With the appearance of an asymmetrical pigmentation of the egg, a tentative form of polarity is established. If an axis is drawn from roughly the apex of the animal hemisphere to the opposite point in the vegetal half, the *animal pole* in salamander eggs corresponds to the approximate position of the future mouth, and the *vegetal pole* to that of the anus.

The final stages of oocyte maturation consist of a hormonally induced release of the egg from its first meiotic block; completion of the first meiotic division, with formation of the first polar body; and breakdown of the germinal vesicle (the nucleus). Late maturation begins with the gonadotropin-induced secretion of progesterone by the follicular cells surrounding the oocyte (Fig. 3-16). In a manner unusual to steroid hormones, which commonly interact with cytoplasmic receptors, the progesterone acts upon the surface of the oocyte, causing an electrical depolarization of the plasma membrane. This results in an increase in Ca^{2+} and a decrease in cAMP within the egg. That progesterone acts upon the surface of the egg was demonstrated by injecting progesterone directly into the egg. Maturation did not occur, but the same oocytes could be stimulated to mature by immersing them in progesterone (Smith and Ecker, 1977).

The intermediate steps leading from Ca^{2+} release and the decrease in cAMP are complex and incompletely understood (Maller, 1985) and will not be covered here. These steps lead to the formation of a cytoplasmic *maturation-promoting factor* (Masui and Markert, 1971). An injection of a small amount of maturation-promoting factor into an immature egg will result in maturation of the egg and the synthesis of a new supply of this factor. Several steps later RNA synthesis in the maturing egg is shut down (Fig. 4-26), and within a day of exposure of the egg to progesterone *germinal vesicle breakdown* occurs. This allows the mixing of nucleoplasm and cytoplasm that is required for the egg to be able to undergo cleavage.

Another late development in the maturation process is the appearance of another cytoplasmic factor, a *cytostatic factor* (Meyerhof and Masui, 1979). This factor has been postulated to cause the second meiotic arrest of the egg in the metaphase II stage. Injection of a tiny amount of oocyte cytoplasm containing cytostatic factor into one blastomere of a two-cell embryo causes the nucleus of that cell to become arrested at the metaphase stage during the next mitosis.

Gene Expression and Other Preparatory Events during Oogenesis

In dealing with gene expression during oogenesis, the reader should be aware of a fundamental problem facing the embryos of many species. Things happen so rapidly in early development (cleavage) that the synthetic activities of the embryo alone are not sufficient to accommodate its needs. The common strategy adopted by most species within the animal kingdom is to anticipate future requirements by making molecules or organelles in advance and storing them in the egg until needed. The consequence of this strategy is that much of what happens in the early embryo is controlled by products created in the prefertilized egg. This means that the maternal genome actually controls many of the early postfertilization events in the embryo. Table 3-1 provides a graphic illustration of the amount of advance synthesis that occurs during oogenesis in *Xenopus*.

One of the major future requirements of the embryo will be protein synthesis. However, it would be highly impractical to deposit many structural and enzymatic proteins in the egg. Instead, the egg prepares for a major increase in protein synthesis after fertilization by building up the protein synthetic apparatus, namely, ribosomes, mRNA, tRNA, and RNA polymerase enzymes.

Ribosomes are complex structures consisting of tightly packed RNAs and proteins. Three classes of RNA are known by their sedimentation properties as 5S, 18S and 28S RNAs. The synthesis of large amounts of 18S RNA and 28S

TABLE 3-1

STORAGE OF COMPONENTS IN THE OOCYTE OF *Xenopus* FOR USE LATER IN DEVELOPMENT

Component	Approximate excess in amount over that in larval cells
Mitochondria	100,000
RNA polymerases	60,000–100,000
DNA polymerases	100,000
Ribosomes	200,000
tRNA	10,000
Histones	15,000
Deoxyribonucleoside triphosphates	2,500

Source: From Gilbert (1981), after Bull.

RNA is made possible by specific gene amplification and the formation of numerous nucleoli (described previously) during the diplotene stage of meiosis. Even earlier, the synthesis of large amounts of 5S RNA occurs, but the synthesis of large numbers of these molecules is made possible by the presence of large numbers of tandemly repeated genes already present in the RNA. Each of the four chromosome strands in the egg during meiotic prophase I contains 24,000 5S RNA genes, for a total of 96,000 (Brown and Sugimoto, 1973). The synthesis of ribosomal proteins is correlated with the synthesis of 18S and 28S RNAs, so that during midvitellogenesis 20 to 30 percent of the protein made in the oocyte is ribosomal protein.

Like 5S RNA, tRNA molecules are also synthesized early in oogenesis (Fig. 4-26). While they are not being used, these RNA molecules are stored, along with some 5S RNA molecules, in large particles. Later in oogenesis these molecules are released into the cytoplasm. Messenger RNAs are synthesized in differing amounts throughout oogenesis. These are commonly complexed with protein molecules and stored in the cytoplasm of the egg. The strategy of inactivating the mRNAs formed in the egg has received a great amount of attention, especially in the sea urchin [e.g., the "masked messenger" hypothesis (Spirin, 1966)].

The problem of meeting the requirements for DNA synthesis in the early embryo must be dealt with in a different fashion, because DNA cannot be synthesized in advance. Instead, the egg is prepared to provide the requisite nucleotides, along with synthesizing in advance a great excess of DNA polymerases over its current needs (Table 3-1). Another anticipation of the need for the rapid increase in the amount of functioning genetic material during cleavage is a major period of histone synthesis in the vitellogenic phase of oogenesis.

This summary presents some of the highlights of the biochemical strategies employed by the egg to accommodate the needs of the early embryo. Further details, as well as the experimental evidence behind many of the generalizations made above, can be found in many recent developmental biology textbooks.

Oogenesis in Birds

The part of the hen's egg commonly known as the yolk is a single cell, the female sex cell or ovum. Its enormous size compared with the other cells is due to the food material it contains. This stored food, which is destined to be used by the embryo in its growth, is gradually accumulated within the cytoplasm of the ovum before it is liberated from the ovary. The other components of the egg (the egg white, the shell membrane, and the shell itself) are all noncellular secretions which invest the ovum in protective layers as it passes down the reproductive tract of the hen.

The early ovum is about 50 μm in diameter and grows gradually until it is about 6 mm. After this the rate of growth accelerates greatly, increasing the diameter by about 2.5 mm per day. By the time ovulation occurs, the ovum may have reached a diameter of 35 mm. This great increase in size is due largely to the accumulation of yolk materials. As in the amphibian, these are not synthesized in the ovum but are produced in the liver and transported via the blood to the follicular cells surrounding the ovum. The follicular cells then transfer the yolk materials to the ovum, where final morphological structuring of the transferred yolk materials occurs (Bellairs, 1964, 1965).

Under the microscope, yolk has the appearance of a viscid fluid in which granules and globules of various sizes are suspended. In the eggs of birds (Romanoff and Romanoff, 1949) the yolk is about 50 percent water, 33 percent fatty material, 16 percent protein, and only about 1 percent carbohydrate. The water carries electrolytes such as sodium chloride and the calcium salts utilized in bone formation. The yolk contains several classes of proteins which are also found in the serum of the hen. The principal ones are *vitellin* (lipovitellin) and *lipovitellenin*, proteins which bind much of the lipid found in the yolk; *phosvitin*, which binds most of the phosphorus; and a water-soluble class called *livetin*, which includes proteins identical to many of the normal serum proteins. The fatty substances are primarily in the form of neutral fats, with the remainder being phosphatides, phospholipids, and cholesterol. In addition, there are vitamins A, B_1, B_2, D, and E. As the stored materials that constitute the yolk increase in amount, the nucleus and cytoplasm are forced toward the surface; eventually the yolk comes to occupy nearly the entire cell.

Figure 3-17 shows a section of a hen's ovary which includes several young ova and one ovum which is nearly ready for liberation. The very young ova lie deeply embedded in the substance of the ovary. As they accumulate more and more yolk, they crowd toward the surface and finally project from it, maintaining their connection by means of a constricted stalk of ovarian tissue. The protuberance containing the ovum is known as an *ovarian follicle*. The bulk of the mature ovum itself is made up of the yolk. Except in the neighborhood of the nucleus, the active cytoplasm is only a thin film enveloping the yolk. The region of the ovum containing the nucleus and the bulk of the active cytoplasm is known as the animal pole; fertilization occurs at the animal pole. The region opposite the animal pole is called the vegetative, or vegetal, pole.

Like the ova of many classes of vertebrates, the ova of birds have a highly irregular plasma membrane that is often called the *zona radiata*. The term was applied to this membrane because when viewed under high power with a light microscope, it shows delicate radial striations. From electron microscopic studies, we now know that the striations are due to closely packed microvilli. As in amphibian oocytes (Fig. 3-24), the functional significance of the irregular membrane is the great increase in membrane surface it affords, thereby enhancing the rate of the metabolic interchanges that can take place at the cell surface.

If one compares the avian follicle with the mammalian, one cannot fail to be impressed by the similarities in the investing structures and the basic relations. In both, the ovum is enclosed in a zone of follicle cells which in turn is covered by a two-layered vascular connective-tissue theca. In both, the nearly ripe follicles bulge out on the surface of the ovary. The striking difference is that whereas the large yolk-laden ovum of the bird occupies the entire follicle (Fig. 3-17), the small

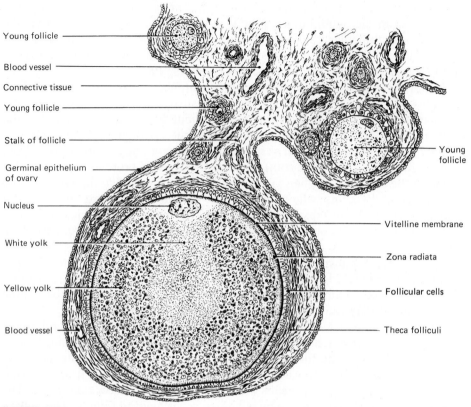

Young follicle

Blood vessel

Connective tissue

Young follicle

Stalk of follicle

Germinal epithelium
of ovary

Nucleus

White yolk

Yellow yolk

Blood vessel

Young
follicle

Vitelline membrane

Zona radiata

Follicular cells

Theca folliculi

FIGURE 3-17
Diagram showing the structure of a bird ovum still in the ovary. The section shows a follicle containing a nearly mature ovum, together with a small area of the adjacent ovarian tissue. (Modified from Lillie, after Patterson.)

mammalian ovum comes nowhere near filling it. The unoccupied territory within the mammalian follicle is the antrum filled with its liquor folliculi (Fig. 3-19).

When the full allotment of yolk has accumulated in the ovum of a bird, ovulation occurs by means of the rupture of an avascular band (the *stigma*) that surrounds the follicle (Fig. 3-18). Bands of smooth muscle, which run from the stalk into the follicle, contract, thereby releasing the ovum from the follicle. As is the case with the mammalian ovum, completion of the first maturation division is almost coincident with ovulation and the second maturation division does not occur unless the ovum is penetrated by a sperm cell.

Oogenesis in Mammals

In contrast to the lower vertebrates, mammals do not replenish the stores of oocytes present in the ovary at birth. At birth the human ovaries contain about 2 million oocytes (many of which are already degenerating) that have been

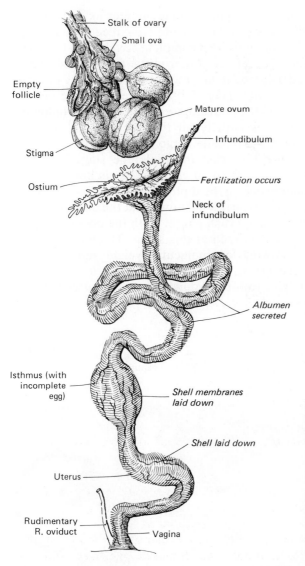

Stalk of ovary

Small ova

Empty follicle

Mature ovum

Infundibulum

Stigma

Fertilization occurs

Ostium

Neck of infundibulum

Albumen secreted

Isthmus (with incomplete egg)

Shell membranes laid down

Shell laid down

Uterus

Rudimentary R. oviduct

Vagina

FIGURE 3-18
Drawing of the female reproductive tract of a hen. After rupture of the follicle through the stigma, the ovum passes into the left oviduct. As it passes down the genital tract, the ovum becomes successively covered with albumen, the shell membranes and finally the shell. In the bird, the right oviduct is rudimentary. (Modified from Bellairs, after Romanoff and Romanoff.)

arrested in the diplotene stage of the first meiotic division. These oocytes are already surrounded by a layer of follicular cells, or *granulosa cells*, and the complex of the ovum and its surrounding cellular investments is known as a *follicle*. Of all the germ cells present in the ovary, only about 400 (one per menstrual cycle) will reach maturity and become ovulated. The remainder develop to varying degrees and then undergo atresia (degeneration). Why so many oocytes are produced when so few actually leave the ovary remains a mystery.

Oocytes first become associated with follicular cells in the late fetal period, when they are going through the early stages of prophase of the first meiotic division. In humans, meiosis begins in some oocytes as early as the fourth month of embryonic life. The primary oocyte (so called because it is undergoing the first meiotic division) plus its incomplete covering of flattened follicular cells is called a *primordial follicle*. Later in the fetal period, when the follicular cells have formed a complete layer around the primary oocyte, the complex is called a *primary follicle* (Fig. 3-19). By this time the oocyte has entered the first of its two periods of developmental arrest, the diplotene stage of meiosis. In the human, essentially all the oocytes, unless they degenerate, remain arrested in the diplotene stage at least until puberty. Some of these cells may not progress past the diplotene stage until the woman's last reproductive cycle (age 45 to 50 years). Less is known about what happens in the human oocyte during the diplotene stage compared with the oocytes of some lower vertebrates, but it is known that the chromosomes in the human oocyte go through a lampbrush phase. Both the oocyte and the follicular cells develop numerous microvilli at this stage. The microvilli are connected by gap junctions, allowing low-molecular-weight substances to pass from one cell type to the other. The *zona pellucida*, a translucent noncellular membrane, also begins to form around the oocyte after the first layer of follicular cells is complete.

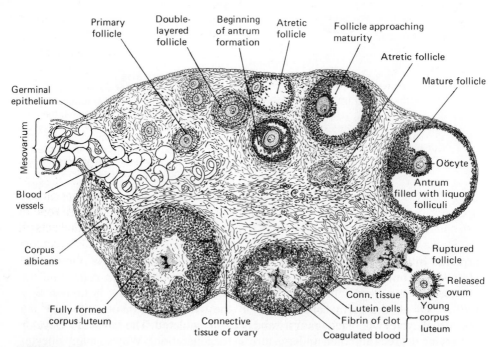

FIGURE 3-19
Schematic diagram of ovary showing sequence of events in origin, growth, and rupture of ovarian (Graafian) follicle and in formation and retrogression of corpus luteum. Follow clockwise around ovary, starting at mesovarium.

In the prepubertal years many of the follicles enlarge as a result of (1) an increase in the size of the oocyte, (2) formation of the zona pellucida, and (3) an increase in the size and number of follicular cells. A thin basement membrane called the *membrana granulosa* (Fig. 3-20F) forms around the granulosa cells of the follicle. No blood vessels are found inside the membrana granulosa, and both the oocyte and the granulosa cells must rely on diffusion for nourishment and oxygen.

The earliest stages of development of primary follicles (up to several layers of follicular cells) appear to occur without the mediation of sex hormones. Beyond this stage the differentiation of mammalian follicles involves the closely orchestrated interplay of several different hormones acting on the follicular cells. The next step in follicular development is the formation of a fluid-filled cavity called the *antrum* within the layers of granulosa cells (Figs. 3-20 and 3-21). This step depends on the presence of pituitary gonadotropic hormones, particularly FSH (follicle-stimulating hormone), and the probable mediation of estrogens produced within the follicle. When the antrum has formed, the follicle is known as a *secondary follicle*, but the oocyte within the follicle is still a primary oocyte and remains arrested in the diplotene stage. The fluid within the antrum is called the *liquor folliculi*. Initially it is formed by the secretions of granulosa cells, but later most of it arises as a *transudate* from the capillaries on the other side of the membrana granulosa.

The secondary follicle becomes further enveloped in a layer of modified ovarian connective tissue (*stromal*) cells. When first formed, this layer is known as the *theca folliculi* (Figs. 3-20E and 3-21), but it continues to differentiate into two layers. The inner layer, the *theca interna*, is glandular in nature and is highly vascularized, whereas the outer *theca externa* retains the characteristics of a connective-tissue covering.

Hormonal control of maturation of the secondary follicle preparatory to ovulation is complex (Richards, 1979). The thecal cells of the early secondary follicle contain LH (luteinizing hormone) receptors. When stimulated by LH, the thecal cells produce *testosterone*, which crosses the membrana granulosa into the granulosa cells (Erickson et al., 1985). The granulosa cells contain receptor proteins for FSH and an active cAMP-generating system. The latter in particular seems to be important in stimulating the enzymatic conversion of testosterone to *estradiol*, a potent estrogen. In addition to its systemic effects, the estradiol acts on the nuclei of the follicular cells and also stimulates the formation of LH receptors in the granulosa cells. In this way it coordinates the receptivity of the ovarian follicle with the LH surge in the blood just before ovulation.

The hormonally stimulated follicle now rapidly increases in size and is known as a *tertiary* (Graafian) *follicle*.[1] The enlarging follicle moves toward the surface of the ovary (Fig. 3-19), and the increasing liquor folliculi ultimately

---

[1]Formerly, tertiary follicles were designated *Graafian follicles*, after the Dutch anatomist Reijnier de Graaf (1641–1673), who first described them. This term was commonly used in the older literature and is still seen in clinical reports.

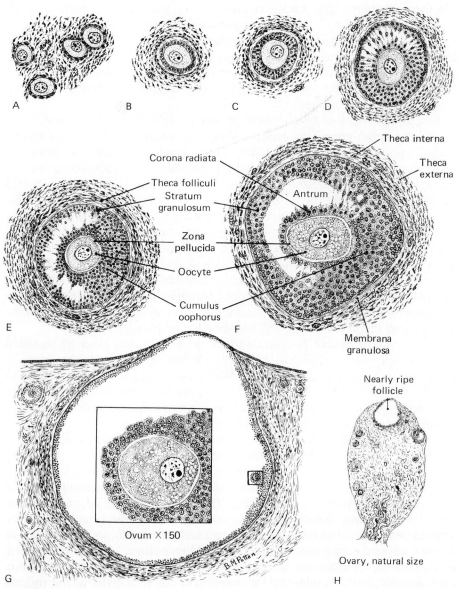

FIGURE 3-20
Drawing showing a series of stages in development of the human oocyte and ovarian follicle. (A–D) Primary follicles. (E–F) Secondary follicles, with the formation of an antrum. (G–H) Mature follicles. (Projection drawings in A–F, ×150; in G the follicle is magnified ×15, but the inset detail of the oocyte is magnified ×150.)

causes it to protrude above the general surface contour (Fig. 3-20G and H). When an ovary is exposed surgically, a nearly ripe follicle looks like a water blister. Such a follicle is nearly ready to rupture and release its contained ovum.

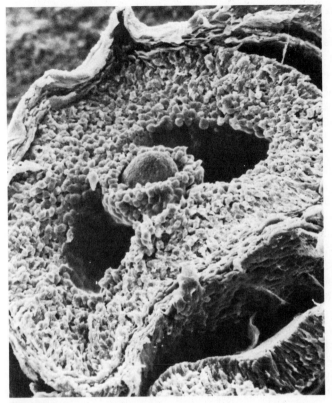

FIGURE 3-21
Scanning electron microscope of a mature follicle in the rat,
showing the spherical oocyte (center) surrounded by smaller
cells of the corona radiata, which projects into the antrum. ×840.
(Courtesy of P. Bagavandoss.)

Within the follicle the egg, surrounded by several layers of granulosa cells,
protrudes into the antrum as the *cumulus oophorus* (Fig. 3-20E and F). Just
before ovulation the follicle is producing large amounts of estradiol. LH
receptors are plentiful on both the thecal and granulosa cells, and the granulosa
cells also contain a high concentration of FSH receptors. The ovum has been
released from its first meiotic block in the diplotene stage and goes on to finish
its first meiotic division, releasing the first polar body as it does so. The follicle
is now ready to respond to the preovulatory LH and FSH surge and complete
the first stage of its cycle by releasing the ovum. In contrast to amphibian
oocytes, which are stimulated to maturation by progesterone, maturation of
mammalian oocytes is little affected by steroid hormones. Instead, luteinizing
hormone, along with other influences, is an important stimulus for maturation.

   Atresia of Follicles   Only a minute percentage of the ova and follicles in the
ovary reach maturity. The others undergo various degrees of maturational

changes and then begin to degenerate. This process is known as *follicular atresia*, and a follicle that is involved in degeneration is said to be *atretic* (Fig. 3-19). The regulatory factors underlying atresia of follicles have not been completely defined, but there is increasing evidence that atretic follicles are deficient in receptors for gonadotropins or estradiol.

Ovulation   The LH peak in the blood in conjunction with FSH, and perhaps with estrogen, sets in motion the final stages of follicular maturation that lead to ovulation. The follicle continues to swell as a result of both increased amounts of follicular fluid and growth of the follicle itself. The apex of the follicular protrusion is called the *stigma*, and within a day after the LH surge some characteristic changes take place in this area. The final preovulatory events begin with hemostasis of blood in the area around the stigma. Within an hour, the follicular wall in the stigma breaks down and the antral fluid, along with the ovum, surrounded by the cells of the cumulus oophorus, is expelled from the follicle (Fig. 3-22).

The precise mechanism that precipitates the rupture of the ovarian follicle is not completely understood. In all probability several factors are involved. An early hypothesis held that increased antral fluid pressure within the follicle causes bursting of the follicular wall, but measurements have shown no increase in fluid pressure. More recent hypotheses involve weakening of the follicular wall because of ischemia or possibly the action of local lytic enzyme activity, the latter being stimulated by pituitary hormones (LH). Demonstration of smooth-muscle-like properties of the ovarian stromal cells has raised the possibility that their contractile activity may play a role in ejecting the egg from the follicle (Schroeder and Talbot, 1985). No single hypothesis seems to account adequately for all the events known to occur at ovulation. This is particularly apparent when one considers ovulation in a broad range of animals, some of which (e.g., the insectivores) do not possess fluid-filled antra at the time of ovulation.

Corpus Luteum   The history of an ovarian follicle by no means ends when the follicle has liberated its contained ovum. Cells of both the stratum granulosum and the theca interna become involved in the formation of the *corpus luteum*. The corpus luteum, so called because of its yellow color in fresh material, grows rapidly and becomes an endocrine organ, secreting both estrogen and progesterone.

When the ovarian follicle ruptures, escape of most of the contained fluid and contraction of the stroma of the ovary reduce the size of its lumen (Fig. 3-19). Bleeding of the small vessels injured in the rupture of the follicle may partially fill the antrum with clotted blood. Several concomitant changes occur during the transformation of the ruptured follicle to the corpus luteum. The granulosa cells swell and develop the cytological characteristics of cells that secrete large quantities of steroid hormones. The central clot is reduced by phagocytic activity while there is a large invasion of the formerly avascular granulosa layer

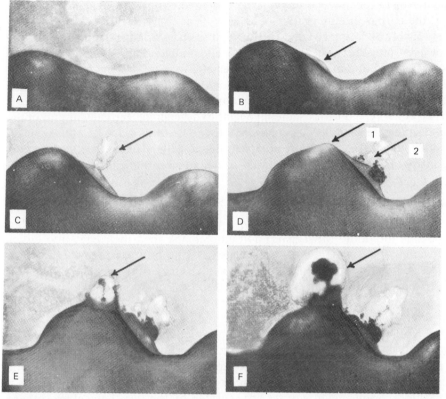

FIGURE 3-22
Enlargements of single frames of a time-lapse motion picture showing ovulation in the rabbit. (A) Profile view of two follicles about 1½ hours before rupture. (B) Same follicles about ½ hour before rupture. (C) Exudation of clear fluid in early phases of rupture. (D) At arrow 1, a new follicle becomes conical as the time of its rupture approaches. At arrow 2, the exudate from the follicle shown starting to rupture in (C) has become more abundant and contains some blood (dark). (E) The follicle indicated by arrow 1 in (D) is now beginning to rupture. The blood-tinged exudate from the follicle which started to rupture in (C), and showed more vigorous exudation in (D) (arrow 2) can be seen partly behind the more recently rupturing follicle. (F) The rupture of the follicle which is indicated by the arrow in E. Time elapsed between the photographs shown in (E) and (F), 8 seconds. The ovum is carried out with this final gush of fluid from the ruptured follicle. (From Hill, Allen, and Kramer, 1935 *Anat. Rec.*, vol. 63.)

by blood vessels. These vessels bring in with them small cells of thecal origin, which become packed in among the more conspicuous cells originating from the stratum granulosum.

Formation and maintenance of the corpus luteum in the human require the continuous presence of LH from the pituitary. Hormonal relations vary among the mammals; for instance, in rats and sheep both LH and prolactin are required. The corpus luteum produces large amounts of progesterone and some estrogens. One of the major functions of progesterone is to prepare the lining of the uterus to receive and implant the fertilized ovum. If pregnancy does not occur, the corpus luteum gradually loses its sensitivity to pituitary gonadotropins, probably by losing LH and FSH receptors on its cells, and it then regresses. If, however, pregnancy occurs, the corpus luteum undergoes a greatly prolonged period of growth and may attain a diameter of 2 to 3 cm in humans. The *corpus luteum of pregnancy* is maintained by *chorionic gonadotropin* secreted by the cells of the embryo and its surrounding membranes (Fig. 2-8).

When either type of corpus luteum begins to degenerate, the cellular part of the organ disintegrates and fibrous connective tissue takes its place. As this connective tissue grows older and more compact, it gradually takes on the characteristic whitish appearance of scar tissue and is called a *corpus albicans* (Fig. 3-19).

ACCESSORY COVERINGS OF EGGS

After ovulation, eggs are released into a variety of environments both within and outside the body. Most fishes and amphibians shed their eggs into either fresh water or salt water, where the eggs must be protected from predators, disease, and environmental factors such as extremes of osmotic pressure and pH. Animals that lay their eggs on land (reptiles, birds, primitive mammals) must prevent dessication as well as support and protect the ova. In all the species mentioned here the ova are protected by layers of secretions that are applied as they pass down the female reproductive tract. The eggs of higher mammals and viviparous forms within the other classes of vertebrates remain in a physiological environment, but they too are typically covered with one or more accessory coats, some of which have less to do with protection than with other factors. Some egg coverings are secreted by the ova themselves, others by the surrounding follicular cells, and still others by the female reproductive tract after the egg has left the ovary. In this section we shall deal with the accessory coverings of four general types of eggs: those of sea urchins, amphibians, birds, and mammals.

Coverings of the Sea Urchin Egg

The sea urchin egg is covered with two noncellular layers (Fig. 3-23). Adjacent to the plasma membrane is the *vitelline envelope*, a tough membrane composed of several varieties of glycoproteins, among which are inserted species-specific receptors for spermatozoa. Surrounding the vitelline envelope is a bulky *jelly coat*, which has a high concentration of fucose sulfate polysaccharides, along with glycoproteins and small peptides. When the eggs are shed, the jelly coat hydrates water and expands. It is virtually transparent. Fertilization of the sea urchin egg will be treated in detail later in this chapter.

The Membranes Surrounding the Amphibian Egg

Throughout its period of development in the ovary, the amphibian egg is surrounded by a follicular epithelium consisting of ovarian cells. In the very

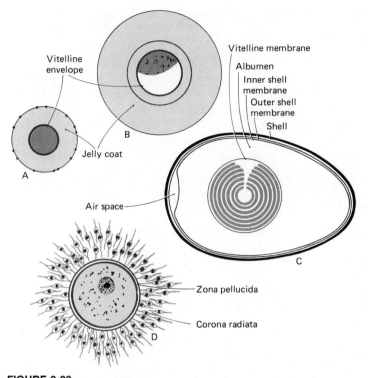
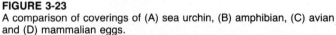

FIGURE 3-23
A comparison of coverings of (A) sea urchin, (B) amphibian, (C) avian and (D) mammalian eggs.

early oocyte, both the plasma membrane of the oocyte and the inner cell membranes of the follicular cells are quite smooth and are closely apposed. As the oocyte grows during its first year, the plasma membranes and follicular cells begin to form numerous minute projections, which are called *microvilli* and *macrovilli*, respectively. These structures increase in prominence, and the narrow space between the oocyte and its follicular epithelium becomes filled with a homogeneous, noncellular basement-membrane-like material that appears to be secreted by both the egg and the follicular cells. This noncellular membrane, the equivalent of the zona pellucida (Fig. 3-20F) in mammals, is commonly called the vitelline envelope in amphibians (Fig. 3-23). As the oocyte approaches the diplotene phase of meiosis, both the size and the number of microvilli and macrovilli increase, and the vitelline envelope becomes thicker (Fig. 3-24). These extreme specializations of the cell membranes, along with the gap junctions which join the villous processes, attest to an active exchange of materials between the follicular epithelium and the egg. After the bulk of the yolk has been laid down in the egg and as the egg approaches ovulation, the microvilli become smaller (Fig. 3-14D). At the time of ovulation a fluid-filled

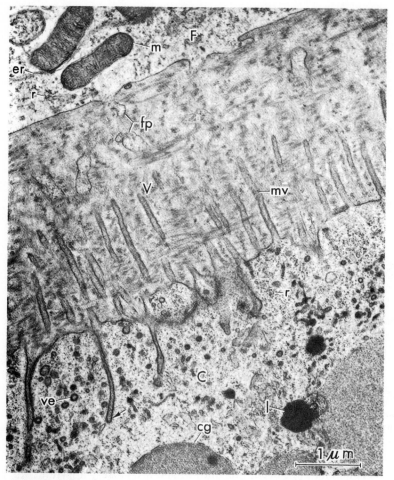

FIGURE 3-24
The outer cytoplasmic zone C of an oocyte of the frog *Rana pipiens* is in-folded at frequent intervals (arrow points to bottom of fold). It also extends microvilli *mv* outward into the substance of the vitelline membrane V. In the cytoplasm are large cortical granules *cg*, occasional lipid bodies *l*, many small vesicles *ve*, and abundant free ribosomes *r*. Follicular epithelial cells *f* extend processes *fp* downward into the vitelline membrane. The portion of the follicle cell shown here contains mitochondria *m*, sparse endoplasmic reticulum *er*, and clusters of ribosomes *r*. (Electron micrograph by N. E. Kemp.)

space known as the *perivitelline space* forms between the vitelline envelope and the plasma membrane of the eggs.

As the ovulated eggs enter the oviducts, the vitelline envelope undergoes some alterations, and material is added to it. These changes facilitate the later penetration of this layer by sperm. As the eggs continue through the oviduct, they become coated with three or more jelly layers consisting largely of polysaccharides. Like that of the sea urchin, the jelly coat swells upon contact with water. It allows the eggs to adhere to one another and to water plants and other submerged objects.

Formation of the Accessory Coverings of Bird Eggs

At ovulation, the ovum is surrounded by a vitelline membrane, which contains an interwoven meshwork of relatively coarse, noncollagenous protein fibrils. Fertilization normally takes place just as the ovum is entering the oviduct (Fig. 3-18). The remainder of the *accessory coverings*, as the other components of the egg are called, are secreted about the ovum during its subsequent passage toward the cloaca.

First an *outer vitelline membrane* composed of finer protein fibrils than those of the original *inner vitelline membrane* is laid down around the ovum when it is in the part of the oviduct adjacent to the ovary. Fibrils of this layer project along the oviduct from opposite ends of the ovum midway between the animal and vegetal poles and become enwrapped in *albumen* that is secreted in the upper oviduct. Because of the spirally arranged folds in the walls of the oviduct, the egg is rotated as it moves toward the cloaca. This rotation twists the adherent albumen into the form of spiral strands projecting at either end of the yolk; these are known as the *chalazae* (Fig. 3-25). Additional albumen, which has been secreted abundantly in advance of the ovum by the glandular lining of the oviduct, is caught in the chalazae and during the further descent of the ovum is wrapped around it in concentric layers. The albumen-secreting region of the oviduct constitutes about half its entire length.

One of the major albuminous proteins of egg white is *ovalbumin* (MW 43,000). The synthesis of ovalbumin and other secretions by the oviduct is a striking example of a specific response to the action of hormones. Normally the oviduct of a chicken does not become capable of secreting the components of egg white until the bird becomes sexually mature, but if a chick is treated with estrogen shortly after birth, the immature oviduct undergoes a series of rapid and profound maturational changes. The first major change is a five- to sixfold increase in mitosis that reaches a peak within 18 hours of estrogen injection. With continued injections of estrogen, tubular glands differentiate from the epithelium about 4 days after the start of hormone treatment. A couple of days later, these glands synthesize substantial amounts of ovalbumin and *lysozyme*, a bacteriostatic agent added to the egg white for protection of the embryo. At the same time ciliated cells become prominent in the oviductal epithelium. The next change is the differentiation of some of the epithelial cells into *goblet cells*. In response to a single injection of progesterone, the goblet cells rapidly begin to secrete large amounts of *avidin*, a major protein component of egg white (Fig. 3-26).

The *shell membranes*, which consist of two sheets of matted organic fibers, are added farther along in the oviduct. The *shell* is secreted as the egg is passing through the shell-gland portion of the oviduct (uterus). The entire passage of

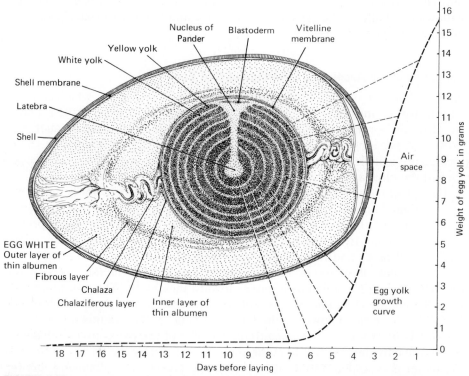

FIGURE 3-25
Diagram showing the structure of the hen's egg at the time of laying. The graph indicates the rate of growth of the egg during 18 days preceding its laying. The lines leading from the various layers of the yolk to the growth curve emphasize the time at which these layers were formed. (Redrawn, with slight modifications, after Witschi, 1956. *Development of Vertebrates.* Courtesy of the author and W. B. Saunders Company, Philadelphia.)

the ovum from the time of its discharge from the ovary to the time when it is ready for laying has been estimated to occupy about 25 to 26 hours. If the completely formed egg reaches the cloacal end of the oviduct during the middle of the day, it is usually laid at once; otherwise, it is likely to be retained until the following day. This overnight retention of the egg is one of the factors accounting for the variability in the stage of development reached at the time of laying.

Structure of the Hen's Egg at the Time of Laying The arrangement of structures in the egg at the time of laying is shown in Fig. 3-25. Most of the gross relationships are already familiar because they appear so clearly in eggs which have been boiled. If a newly laid egg is allowed to float freely in water until it comes to rest and is then opened by cutting away the part of the shell that lies uppermost, a circular whitish area will be seen to lie atop the yolk. In

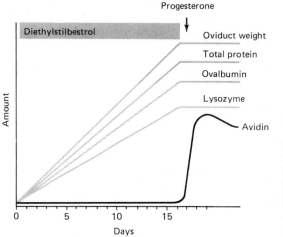

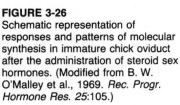

FIGURE 3-26
Schematic representation of responses and patterns of molecular synthesis in immature chick oviduct after the administration of steroid sex hormones. (Modified from B. W. O'Malley et al., 1969. *Rec. Progr. Hormone Res.* 25:105.)

eggs which have been fertilized, this area is somewhat different in appearance and noticeably larger than it is in unfertilized eggs. The differences are due to the development of an aggregation of cells known as the *blastoderm* in fertilized eggs. The blastoderm will be considered at greater length in Chap. 4.

Close examination of the yolk will show that it is not uniform throughout either in color or in texture. Two kinds of yolk can be differentiated, *white yolk* and *yellow yolk*. The principal accumulation of white yolk lies in a central flask-shaped area, the *latebra*, which extends toward the blastoderm and flares out under it into a mass known as the *nucleus of Pander*. In addition to the latebra and the nucleus of Pander, there are thin concentric layers of white yolk between which lie much thicker layers of yellow yolk. The concentric layers of white and yellow yolk indicate the daily accumulation during the final 7 or 8 days before ovulation. In this period the formation of yolk goes on day and night, but during the late hours of the night the yolk laid down contains only small amounts of fat but has a high protein content. The yolk deposited in the daytime has a high fat content. Its color is due to yellow carotenoids concentrated in it. Thus during the last week before ovulation, a thin layer of white yolk and a heavy layer of yellow yolk are added each day. The outermost yolk immediately under the vitelline membrane is always white.

The albumen, except for the chalazae, is nearly homogeneous in appearance, but near the yolk it is somewhat more dense than it is peripherally. The chalazae serve to suspend the yolk in the albumen.

The two layers of shell membrane lie in contact everywhere except at the large end of the egg, where the inner and outer membranes are separated to form an *air chamber*. This space appears only after the egg has been laid and cooled from the body temperature of the hen [about 41°C (106°F)] to ordinary temperatures. In eggs which have been kept for any length of time the size of the air space increases because of evaporation of part of the water content of

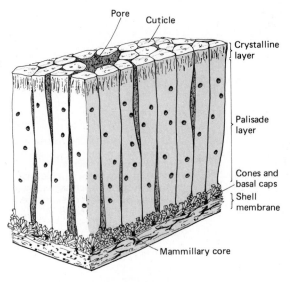

Pore
Cuticle
Crystalline layer
Palisade layer
Cones and basal caps
Shell membrane
Mammillary core

FIGURE 3-27
Diagram of a small portion of an avian egg shell and its underlying membranes, with emphasis upon the crystalline structures in the shell. (Redrawn from Dumont and Brummett (1985) in Browder, *Oogenesis*, Plenum Press, after Becking.)

the egg. The familiar method of testing the freshness of eggs by ''floating'' them is done in recognition of this fact.

The egg shell is composed largely of calcareous salts, mainly in the form of calcite, a crystalline form of calcium carbonate (Fig. 3-27). The calcium is ultimately derived from the food of the mother, but en route from the mother's intestinal tract it becomes involved in a curious structural adaptation found only in birds actively engaged in egg laying. The calcium is incorporated into specialized bony masses located within the marrow cavities of the long bones. These bony masses are called *medullary bone*. When the egg shell is being formed, the masses of medullary bone are rapidly broken down and serve as the major source of calcium for the formation of the egg shell. In the absence of medullary-bone deposits, the calcium of the egg shell would be derived from the other skeletal structures of the bird. If this were to occur, the skeleton would become seriously weakened. An interesting safeguard has been provided against this possibility. If lime-containing substances are not provided in the mother's diet, she stops producing eggs within a few days (Taylor, 1970). This is apparently due to an inhibition of the production of gonadotropic hormones by the pituitary gland. The normal egg shell is porous (about 7000 pores per egg), enabling the embryo to carry on an exchange of gases with the outside air by means of specialized vascular membranes arising in connection with the embryo but lying outside it, directly beneath the shell (Fig. A-42).

The Coverings of Mammalian Eggs

While in the ovary, mammalian eggs become surrounded by a noncellular membrane that forms between the plasmalemma of the ovum and the surrounding follicular cells. This membrane, commonly known as the zona pellucida, is composed of a number of components, including three distinct classes of glycoproteins (80 percent of its total mass), sulfated mucopolysaccharides, hyaluronic acid, and sialic acid. The glycoproteins, given names ZP-1, ZP-2, and ZP-3 (Bliel and Wassarman, 1980b), have different functional properties during fertilization. Protein ZP-3, which is found as a billion copies in a single mouse zona pellucida, acts as the sperm receptor and also plays a role in inducing the acrosome reaction (Wassarman et al., 1986). These functions will be discussed in connnection with fertilization. The zona pellucida begins to form when the ovum is surrounded by a single layer of cuboidal follicular cells. Current evidence (Bliel and Wassarman, 1980a) indicates that most of the zona pellucida is synthesized by the oocyte itself. At ovulation, the mammalian ovum remains surrounded by several layers of follicular cells known as the *corona radiata*. It is not known whether the corona radiata plays a protective role, but in some mammals (e.g., the rabbit) this layer is required for transport of the egg down the oviduct. The cells of the corona radiata continue to secrete steroids and prostaglandins in the ovulated egg, and it has been suggested (Schuetz and Dubin, 1981) that these hormones may support development of the egg or embryo until the corpus luteum becomes fully functional.

FERTILIZATION

There is perhaps no phenomenon in the field of biology that touches so many fundamental questions as the union of the germ cells in the act of fertilization; in this supreme event all the strands of the webs of two lives are gathered in one knot from which they diverge again and are rewoven in a new individual life-history. . . . The elements that unite are single cells, each on the point of death; but by their union a rejuvenated individual is formed, which constitutes a link in the eternal process of Life. (F. R. Lillie, *Problems of Fertilization*)

Many biological factors must work in concert before actual union of the male and female gametes can be effected. The growth cycle of the sex cells must be such that both the eggs and sperms mature and are released from the gonads within a closely coordinated time interval. We have already seen how the act of copulation itself stimulates ovulation in some animals. In other animals changes in the length of daylight serve to coordinate both the development of gametes and the period of sexual activity. The characteristic behavior of an animal in *estrus* ("heat") is another mechanism which increases the likelihood of a functional sperm cell meeting a mature ovum.

Once the spermatozoa are released from the male, other factors play a role in ensuring the approximation of sperms and eggs. The simplest mechanism is that of marine invertebrates and most fishes: The female merely extrudes the eggs into the water, and the male "floods" the area with millions of spermatozoa. The number of eggs that will be fertilized is a matter of chance and the water currents. Other aquatic vertebrates, such as salamanders, have adopted an elaborate breeding ritual in which the male deposits a package of sperm cells (spermatophore) on the bottom of a pond and the female, during the courtship ritual, takes up the sperm package with the lips of her cloaca. A more efficient mechanism is the internal fertilization characteristic of mammals. In this process the male deposits the spermatozoa directly into the female genital tract during coitus. But even in this case many critical factors intervene before the final merging of the male and female gametes. Some of these call for special consideration because they are so important in human reproduction.

Fertilization is a process, not a single event, that begins when a sperm cell first makes contact with the plasma membrane of the egg and ends with the intermingling of maternal and paternal chromosomes at the metaphase plate prior to the first cleavage division. Understanding the biology of fertilization, however, involves a knowledge of several preparatory events, such as the release of gametes from the gonads, their transport so that eggs and sperms are brought together, alterations of the spermatozoa so that they are capable of fertilizing the egg (capacitation and the acrosomal reaction), and the penetration of spermatozoa through the protective layers that surround the egg.

The fertilization process itself has a number of important components. These include (1) initial membrane contact between egg and sperm, (2) entry of the sperm cell into the egg, (3) prevention of *polyspermy* (entry of more than one sperm cell into the egg) by the egg, (4) metabolic activation of the egg, (5) completion of meiosis by the egg, and (6) formation and fusion of male and female pronuclei, leading to the final cleavage division.

In this section we shall deal in detail with the events surrounding fertilization in two forms, the sea urchin and mammals. These were selected not only because they are well understood but because they exemplify some important differences in the strategy of fertilization. Sea urchins are typical of invertebrates that play the numbers game. Because they use external fertilization, they have had to evolve effective means of getting eggs and sperms to meet in the open ocean. At the cellular level sea urchin eggs are covered by a noncellular jelly coat, and the eggs complete their second meiotic division before the entry of spermatozoon. In contrast, mammals employ internal fertilization, which places different constraints on eggs and sperms. The eggs are surrounded by a cellular layer, not a jelly coat, and it is necessary to complete the second meiotic division before the union of maternal and paternal genetic material can be consummated.

THE SEA URCHIN

Gamete Release and Transport

The sea urchin's mode of external fertilization is enormously expensive in terms of the metabolic cost of producing sufficient gametes to ensure the survival of the species. A single female *Arbacia* releases an estimated 4 million eggs and a male releases up to 100 billion spermatozoa at a single spawning.

Along with producing immense numbers of gametes, adult sea urchins improve the chances of a sperm meeting an egg by moving into dense aggregates before spawning. After this, a successful fertilization depends largely on coordinated timing in the release of gametes and the water conditions at that time.

Sperm Penetration of the Egg in Invertebrates and the Acrosomal Reaction

The egg of the sea urchin is covered by a layer of largely carbohydrate material which becomes hydrated and expands upon contact with seawater. The thick outer investment, called the *jelly coat,* is composed of a mixture of small peptides, glycoproteins, and a polysaccharide containing fucose sulfate units (Fig. 3-23). When spermatozoa encounter the jelly coat, they undergo a number of important changes. In the presence of a large number of eggs, spermatozoa tend to cluster together and their motility increases.

Direct contact with the jelly coat stimulates the *acrosomal reaction* of the sperm. This reaction is triggered by contact with a fucose sulfate polysaccharide of the jelly coat. A biochemically defined peptide, *speract*, found in the egg jelly is responsible, at least in part, for the increased motility and activated respiration that occur when the sperm contacts the jelly coat (Hansbrough and Garbers, 1981; Suzuki et al., 1981). The immediate reaction is an increased permeability of the plasma membrane of the sperm to Ca^{2+} and an increased concentration of intracellular Ca^{2+}. This leads to localized fusion of the outer acrosomal membrane with the plasma membrane and their ultimate breakdown. Concomitantly, an influx of Na^+, coupled with an efflux of H^+, leads to an increased intracellular pH of the sperm. The elevated pH stimulates the polymerization of *g-actin* (a globular form) to *f-actin* (the filamentous form). The f-actin forms the basis of the *acrosomal process*, which protrudes from the head of the sperm (Fig. 3-28). The acrosomal process is covered by a layer of *bindin*, a protein derived from the acrosomal contents.

As the spermatozoa pass through the jelly coat, they next encounter the *vitelline envelope*, a tough noncellular layer interposed between the jelly coat and the plasma membrane of the egg. The vitelline envelope is composed to a large extent of glycoprotein molecules, one of which serves as a species-specific binding site for spermatozoa. In this way the vitelline coat acts as the main discriminator that allows only sperm of the same species to fertilize the egg. A firm union between the sperm and the vitelline envelope is effected by bonds linking the sperm receptors on the vitelline coat to the bindin molecules that coat the acrosomal process (Fig. 3-29). There appear to be from 1500 to 6000 sperm-binding sites on the vitelline envelope of the sea urchin.

After having completed the acrosomal reaction and binding themselves to the vitelline envelope, the attached sperm digest their way through the vitelline envelope with the aid of acrosomal enzymes, which are generically called *lysins*. Although large numbers of spermatozoa attach to the vitelline envelope, few actually penetrate it.

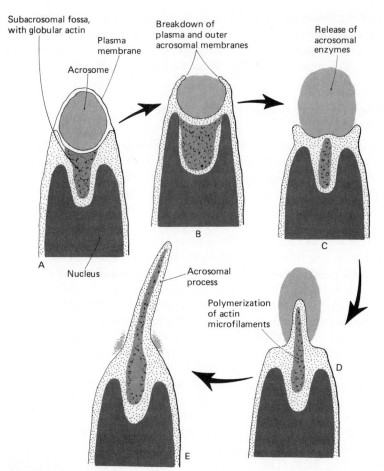

FIGURE 3-28
Acrosomal reaction of the sea urchin spermatozoon. Following fusion of the
acrosomal membrane with the plasma membrane and their subsequent
breakdown (*B*) acrosomal enzymes are released (*C*). Then globular actin
polymerizes to filamentous actin (*D*) and the acrosomal process forms (*E*).

Initial Contact between Sperm and Egg

After making its way through the vitelline coat, a sperm cell makes contact with
the plasma membrane of the egg. Initial contact is made between the acrosomal
process and microvilli that protrude from the surface of the egg. In the sea
urchin, there seems to be no particular site to which the spermatozoon
preferentially attaches. Aided by its swimming movements, the spermatozoon
presses harder against the egg until fusion begins between its plasma membrane
and that of the egg. Both the sperm and the egg are primed for fusion, and
fusion occurs readily. In the initial stages of fusion, a group of microvilli near
the sperm head seem to engulf the sperm. This action creates a small bulge in

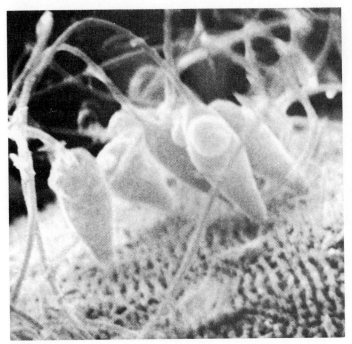

FIGURE 3-29
Scanning electron micrograph of sea urchin spermatozoa attaching perpendicular to the vitelline membrane surrounding the egg. (Courtesy of G. Schatten, from Schatten and Mazia, 1976. *J. Supramol. Struct. 5*:343).

the area of fusion known as the *fertilization cone* (Fig. 3-30). At the level of egg and sperm there is relatively little interspecies discrimination. Hybrid fusions occur readily in eggs from which the vitelline envelope has been removed. This type of experiment underscores the important role of the vitelline envelope in maintaining species specificity during the fertilization process. The plasma membrane of the sperm cell is antigenically different from that of the egg. Immunocytochemical studies have shown that sperm surface antigens can be detected on the plasma membrane of the egg some time after initial sperm contact. This had led to the interpretation that the plasma membrane of the zygote is a mosaic membrane with contributions from both egg and sperm. In addition, direct fluorescent labeling of the surface of sperm cells has shown that just after fertilization and into the cleavage period a discrete patch bearing the label is readily discernible on the surface of the zygote (Shapiro et al., 1980; Fig. 3-31).

Blocks to Polyspermy

Once the first spermatozoon has made contact with the egg, it is important for the egg to prevent any other sperm cells from fusing with it, for the normal consequence of *polyspermy* (the fertilization of the egg by more than one

FIGURE 3-30
(A) Sea urchin sperm with the head halfway embedded into the egg. Microvilli of the egg elongate near the sperm head whereas elsewhere over the surface of the egg, the microvilli are short and knob-like. (B) The head and midpiece of the sperm are completely embedded inside the egg, leaving only the tail protruding from the surface. (Courtesy of G. Schatten, from Schatten and Mazia, 1976. *J. Supramol. Struct.*, 5:343).

sperm) is an early disruption of development and the death of the embryo. Many species, including the sea urchin, have evolved two blocks to polyspermy (rev. by Schuel, 1984). The first is an extremely rapid but temporary depolarization of the plasma membrane. This can be viewed as an adaptation for quickly cutting off access to the egg by sperm that are not far behind the first one in penetrating the vitelline envelope. This *fast block to polyspermy*, as it is often called, buys a small amount of time for the egg to set up a more complex but permanent block, the *slow block to polyspermy*.

The fast block to polyspermy is a membrane event which is set in place within 2 to 3 seconds and lasts for about 60 seconds, by which time the permanent (slow) block to polyspermy is established. The plasma membrane of the egg, like virtually all cells, generates a difference in electrical potential called a *resting membrane potential*. The resting membrane potential of the unfertilized egg is about −70 mV, with the inside of the plasma membrane negative to the outside. As the acrosomal process of the sperm fuses with the plasma membrane of the egg, it sets into motion a rapid depolarization of the plasma membrane (caused by the rapid influx of Na^+ into the cell). This causes an almost immediate change of the local membrane potential from −70 to +10 mV, and within 2 to 3 seconds the membrane potential of the entire egg has changed to +10 mV. With a positive membrane potential, the egg no longer permits the fusion of other spermatozoa to its plasma membrane. This is the basis for the fast block to polyspermy.

Experimental manipulations (Jaffe et al., 1982; Lynn and Chambers, 1984) have supported the electrical basis for the fast block to polyspermy. If the

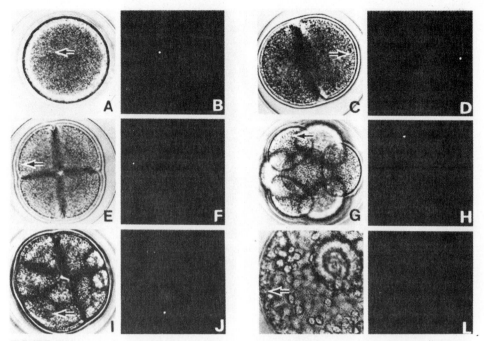

FIGURE 3-31
Preservation of components of fluorescently labelled surface components of spermatozoa during early embryogenesis in the sea urchin. The white dots in the dark squares show the fluorescent components of the sperm cells. The photomicrographs on the left are shown for reference. The points of the arrows show where, on the embryo, the fluorescent patch is located. (A and B) Single-cell zygote; (C and D) 2-cell embryo; (E and F) 4-cell embryo; (G and H) 8-cell embryo; (I and J) 16-cell embryo; (K and L) gastrula. (From Gabel et al., 1979. *Cell 18*:207, Courtesy of B. M. Shapiro.)

membrane potential of an unfertilized egg is raised to $+5$ to 10 mV, spermatozoa swarm around it but are unable to fuse with it. Conversely, if the membrane potential of an egg which has already been penetrated by a spermatozoon is reduced to a level approaching the resting membrane potential, additional spermatozoa enter it and polyspermy results.

Events associated with the fast block to polyspermy initiate the slow block. The first step is the mobilization of Ca^{2+} from stores bound within the egg. Ca^{2+} is first released at the site of sperm entry, and during the next minute a wave of free Ca^{2+} passes through the egg (Fig. 3-32). As it sweeps through the egg, the wave of released calcium ions initiates the *cortical reaction*—the rupture of cortical granules and the release of these contents into the space surrounding the egg (the *perivitelline space*). The sea urchin egg contains about 15,000 cortical granules, each having a diameter of 1 μm, in a layer just beneath the plasma membrane of the egg (Fig. 3-33). Each cortical granule contains a mixture of enzymes, structural proteins, and sulfated mucopolysaccharides (glycosaminoglycans).

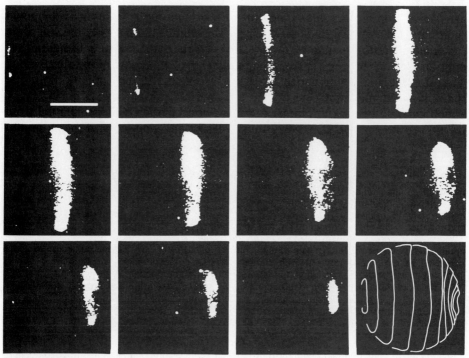

FIGURE 3-32
Demonstration by aequorin luminescence of a wave of free calcium propagated across a sperm-activated medaka (fish) egg. Starting at the upper left-hand box, a white band of luminescence passes across the egg from left to right. Each successive box was photographed at the 10 second interval. The drawing at the lower right illustrates a composite series of 11 wave fronts. The bar in the upper left-hand box represents 500 μm. (From Gilkey et al. 1978 *J. Cell Biol. 76*:451. Courtesy of L. F. Jaffe.)

Responding to the free calcium ions, the cortical granules move to the inner surface of the plasma membrane, fuse with it, and then open up. The fusion of the membranes of the cortical granules with the plasma membrane of the egg almost doubles the area of surface membrane of the egg. Some of this membrane goes into the production of microvilli, and other parts of the membrane are internalized by the egg.

A regular sequence of events follows the release of the contents of the cortical granules (Fig. 3-34). A proteolytic enzyme breaks the molecular bonds that bind the vitelline envelope of the unfertilized egg to the plasma membrane. At the same time, the sulfated mucopolysaccharides, which have a high affinity for water, begin to swell, forcing the vitelline envelope away from the plasma membrane. In the argot of classical embryology, this process has been called "raising the fertilization membrane." *Fertilization membrane* is merely a new name given to the vitelline envelope after it has undergone the changes set in motion by the cortical reaction. The hydrated mucopolysaccharides form a

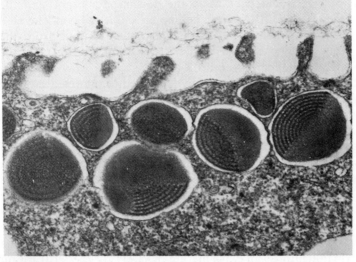

FIGURE 3-33
Transmission electron micrograph showing a group of cortical granules located just beneath the plasma membrane of the sea urchin egg. (Courtesy of H. Schuel; from Schuel, 1984. *Biol. Bull.*, *167*:271. Photograph provided by Drs. B. L. Hylander and R. G. Summers.)

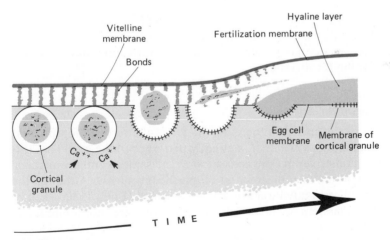

FIGURE 3-34
Diagrammatic representation of the sequence of events leading from the cortical reaction to the formation of the fertilization membrane in the sea urchin egg. Responding to Ca^{2+}, the cortical granules release their contents into the space between the plasma membrane and the vitelline membrane. This breaks the bonds holding down the vitelline membrane, which subsequently becomes raised as the fertilization membrane. Mucopolysaccharides released by the cortical granules form the hyaline layer. (Adapted from Gilbert (1985) *Developmental Biology*, Sinauer and Austin (1965) *Fertilization,* Prentice-Hall).

demonstrable *hyaline layer* located between the plasma membrane and the fertilization membrane (Fig. 3-34). As the vitelline envelope is being elevated from the plasma membrane, it is acted upon by another enzyme from the cortical granules. This enzyme alters the sperm receptors on the vitelline envelope, causing any attached sperm to drop off.

A final step in the slow block to polyspermy is related to the release of the enzyme *ovoperoxidase* from the cortical granules. Hydrogen peroxide (H_2O_2), a powerful oxidizing agent, is released by the egg at the time of the cortical reaction. At the fertilization membrane the chemical breakdown of H_2O_2 mediated by ovoperoxidase results in the cross-linking of the tyrosine groups of the proteins. This results in a hardening of the fertilization membrane into a tough envelope that surrounds the early embryo. Another likely effect of the H_2O_2 given off by the egg is to kill any spermatozoa that have penetrated the vitelline envelope. These sperm cells would have been kept out of the egg by the fast block to polyspermy, but after that subsided (after about a minute), the spermatozoa would be able to enter the egg and cause polyspermy. Thus the spermicidal H_2O_2 reaction affords an extra margin of safety for the fertilization process. Interestingly, the mechanism by which H_2O_2 inactivates sperm is very similar to the way in which phagocytic cells in the mature vertebrate body dispose of invading pathogens, such as bacteria.

The blocks to polyspermy are an effective way of maintaining the genetic integrity of the fertilized egg. Although they are by far the best understood in the sea urchin, there is good reason to believe that similar mechanisms operate in other species, as well. A major exception is found in certain vertebrate groups, such as the urodele amphibians and birds. In these species, polyspermy is normal and other means have been devised for inactivating and removing excess sperm from the fertilized egg.

Metabolic Activation of the Egg

The main function of the sperm in the early stages of the fertilization process is to activate a program of events that is already patterned in the egg. That this is true is readily demonstrated by the activation of the same events by artificial means, such as a pinprick (see the section on parthenogenesis). Although the major events of metabolic activation are well established (Epel, 1980; Shapiro et al., 1981), the connections between these events are not completely understood (Fig. 3-35).

Activation of the egg begins with the influx of Na^+ associated with the membrane depolarization of the fast block to polyspermy. This leads to the release of intracellular Ca^{2+}, which appears to be the main stimulus for the next major series of events. In addition to the cortical reaction (already discussed), these events include a three- to fivefold increase in oxygen consumption (probably related to the formation of H_2O_2); the activation of the enzyme NAD kinase, which may facilitate the biosynthesis of new membrane lipids; and a second influx of Na^+ coupled with an efflux of H^+ from the cell, leading to an

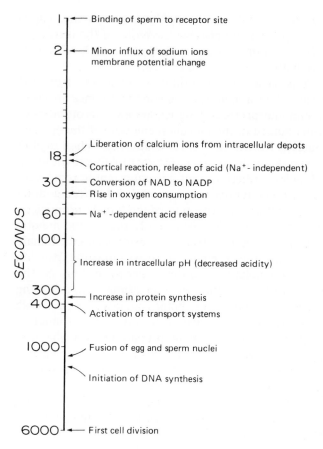

FIGURE 3-35
Time course of events in fertilization of the sea urchin egg. (After Epel, 1980. *Endeavour* 4:29.)

increase in intracellular pH, which occurs between 1 and 5 minutes after initial sperm contact with the egg. The increased pH in turn leads to an increase in protein synthesis, the activation of transport systems within the egg and ultimately the initiation of DNA synthesis in preparation for the first cleavage division. All these metabolic events prepare the egg for the main event in fertilization: the fusion of genetic material from the egg and sperm.

Penetration of the Spermatozoon into the Egg and Fusion of the Genetic Material

As the head of the sperm becomes incorporated into the fertilization cone of the egg, the nuclear membrane begins to disintegrate. The nuclear material interacts with the cytoplasm of the egg, and the chromatin begins to disperse from its formerly highly condensed state. As the phase of chromatin dispersion nears completion, a new membrane forms around what can now be properly called the *male pronucleus*. Of the other cytoplasmic components brought in with the sperm, the mitochondria and tail piece probably play no further role in development, but the centrioles persist and provide the basis for the formation of the *sperm aster*, which is very important in getting the male and female pronuclei together (rev. by Schatten, 1982; Longo, 1984).

The sperm aster, a radiating array of microtubules emanating from the original centrioles of the sperm, plays a major role in guiding the migrations of the pronuclei. According to one interpretation, the expanding microtubules of the aster push against the inner suface of the plasma membrane of the egg and help displace the male pronucleus toward the center of the egg. When rays of the sperm aster reach the female pronucleus, this structure rapidly moves along the rays toward the male pronucleus. (At this point it is important to remember that in the sea urchin the second meiotic division is completed before contact of the egg by the sperm. This simplifies the reaction of the female pronucleus in comparison to this stage of meiosis in mammals, where the second meiotic division is not completed until the sperm has entered the egg.) When the female pronucleus has reached the center of the sperm aster, the continued expansion of the sperm aster pushes both pronuclei toward the center of the egg. As the two pronuclei come into contact with each other, their membranes fuse, leading to both maternal and paternal chromosomes being enclosed in a single membrane. This process is called *pronuclear fusion*. Shortly after pronuclear fusion, the chromosomes replicate their DNA in preparation for the first cleavage division. As the chromosomes prepare to line up at the metaphase plate in preparation for the first cleavage division, the process of fertilization is completed and the period of cleavage is about to begin.

MAMMALIAN FERTILIZATION

Sperm Transport in the Female Reproductive Tract of Mammals

Much remains to be learned about the manner in which the spermatozoa make their way from the vagina through the uterus and the uterine tubes. In most common mammals, including humans, the spermatozoa are deposited in the upper vagina at insemination, but in many rodents the site of *insemination* is the uterus. From the standpoint of the individual spermatozoon, the journey from the site of insemination to the upper uterine tube, where fertilization occurs, is an arduous one, and only a minute fraction of the spermatozoa that are deposited in the female reproductive tract ever reach the vicinity of the ovulated egg. In comparison with their size, the distance the spermatozoa must travel is great. The route may be beset with chemical hazards in the form of strong acid secretions or with mechanical obstacles, such as a crooked and compressed cervical canal or uterine tubes narrowed or occluded by disease. Nevertheless, because of the enormous number of spermatozoa contained in an ejaculate of semen (200 to 300 million in humans), it is probable that under normal conditions some of them will reach the uterine tube while they are still capable of penetrating and fertilizing the ovum.

The first barrier the spermatozoa face is the natural acidity of the upper vagina. The apparent function of this acidity is to act as a bacteriostatic medium. The seminal fluid, however, acts as an effective buffer against the acidity, and within 8 seconds of insemination the vaginal pH can rise from 4.3 to 7.2 (Fox et al., 1973). In rodents, the semen coagulates shortly after insemination and forms a characteristic plug which prevents the backflow of spermatozoa. In embryological studies of rodents, pregnancy is usually timed from the appearance of the plug.

From the upper vagina, some spermatozoa are transported extremely rapidly up the female reproductive tract, and in many mammals, including humans, they reach the uterine tube in less than 30 minutes. This form of transport is far too rapid to be accounted for by the swimming movements of the spermatozoa themselves (estimated rates are from 2 to 4 mm per minute). In fact, experimental studies have shown that nonmotile spermatozoa reach the uterine tube as quickly as do motile ones during the early, rapid phase of sperm transport. For rapid sperm transport, there is evidence that some component(s) of seminal fluid stimulates contractions of the upper vagina, which may help to propel spermatozoa into the cervical canal.

There also appears to be a slower phase of sperm transport in which spermatozoa enter the cervix, possibly aided by their swimming movements, and lodge in the numerous irregular crypts that line the cervical canal. Normally, thick mucus fills the cervical canal, but hormonally induced changes at the time of ovulation reduce the viscosity of the mucus and allow better penetration by the spermatozoa. Once in the cervical crypts, the spermatozoa are slowly released into the uterine cavity.

The movement of spermatozoa through the uterus is less well understood, but at the height of sexual orgasm in many female mammals there are spasmodic contractions of the smooth muscle of the uterus, which may immediately draw some of the freshly deposited semen from the vagina into the uterus. Although uterine contractions may be an accelerating factor in sperm transport, they are certainly not an indispensable one, for there are innumerable well-authenticated cases, clinical and experimental, of pregnancy occurring in the absence of orgasm in the female. In such instances the traversing of the uterus must depend primarily on the activity of the spermatozoa themselves.

The next barrier in the path of the spermatozoa is the entrance into the uterine tubes. In species that ovulate only one egg at a time, one of the barriers is purely statistical. For spermatozoa that enter a uterine tube that does not contain an egg, mere chance prevents the realization of the goal of their journey. The uterotubal junction may also act as a valve which permits or prevents the passage of spermatozoa into the uterine tube. This function is more strongly expressed in some species (e.g., mice) than in others. Once within the uterine tube, the spermatozoa continue their upward path by some combination of muscular contractions and ciliary currents of the tube and swimming of the spermatozoa themselves.

The heightened muscular activity of the uterine tubes at the time of ovulation has already been mentioned. It seems probable that this increased activity is important in sperm transportation as well as in the journey of the ova toward the uterus. Careful observation of the activity of surgically exposed tubes in living experimental animals indicates that temporary rings of contractions tend to divide the tube into a series of compartments. At any given moment in any compartment, the downward-beating cilia along the outer walls also tend to create back eddies. In such currents and countercurrents the spermatozoa in the lumen of the tube would be scattered rapidly throughout the area between the two adjacent contraction rings. When the zones of contraction relaxed at one level and formed at another, some spermatozoa would be crowded back toward the uterus but others would find themselves in a new compartment nearer the ovary. The formation and re-formation of such compartments by temporary rings of contraction at shifting levels disperse the spermatozoa throughout the length of the tube.

Only at the upper end of the uterine tube does the swimming activity of the spermatozoa assume prominence. There is evidence that spermatozoa orient themselves so that they move against a gentle current, thus exhibiting what is called a *positive rheotactic response*. The downward ciliary currents in the uterine tube serve as an effective orienting stimulus.

As the spermatozoa are transported through the female reproductive tract, they are subjected to a poorly understood influence by the maternal tissues which enables them better to penetrate the membranes surrounding the egg. This phenomenon is called *capacitation* of the spermatozoa, and in its absence fertilization does not take place in many species of animals. The need for capacitation has been strikingly demonstrated in attempts to fertilize mammalian eggs in vitro. The ability of freshly obtained spermatozoa to fertilize eggs in vitro is often poor. If, however, the spermatozoa are first incubated for several hours close to female reproductive tissues, their success rate improves markedly. The time required for capacitation of spermatozoa varies from less than 1 hour in the mouse to 5 to 6 hours in primates and humans. The change brought about by capacitation remains obscure, but there is evidence that capacitation involves the removal of glycoproteins which cover the spermatozoa while they are stored in the male genital tract. There is also some evidence of changes in the plasma membrane of the spermatozoa.

The spermatozoa that are not directly involved in fertilization are ultimately removed from the female reproductive tract. Those in the uterine cavity are eventually swept through the cervix and into the vagina; those in the uterine tubes are ingested by phagocytic cells.

Egg Transport

The freshly ovulated egg, surrounded by the corona radiata, lies free in the peritoneal cavity. As a means of increasing the chances that the ovum will enter the uterine tube, hormonal changes preceding ovulation result in greater muscular activity of the fimbriated ostium of the uterine tube and an increased ciliary current leading down the uterine tube. This combination causes strong fluid currents around the ovary in the direction of the ostium of the uterine tube, and in the great majority of cases the ovum is efficiently swept into the tube. This phase of ovum transport has been vividly recorded in a film by Blandau (University of Washington Audiovisuals). Within the uterine tube ciliary currents appear to be the main driving force for the ovum, for if the muscular activity of the tube is blocked by pharmacological agents, downward transport of the egg continues at a normal rate (Halbert et al., 1976). The corona radiata (cells of the cumulus oophorus) surrounding the ovum is very important in egg transport; without this layer the ovum makes little progress. To a large extent, this is a function of mass rather than intrinsic motility, because inert objects of the same size are also effectively moved down the uterine tube. It takes approximately 3 days for an unfertilized egg to pass through the human uterine tube.

The Viability of Ova and Spermatozoa

Both ovum and spermatozoa have only limited viability once they are free in the female reproductive tract. When liberated from the ovary, the ovum at once begins to undergo certain changes which can be characterized as aging or deterioration. Among other things, its cytoplasm progressively becomes more coarsely granular. This is accompanied by a general depression of metabolic activity, the course of which is reversed only if fertilization occurs. In most mammals, including humans, the ovulated egg must be fertilized within 24 hours or it becomes "overripe" and nonviable.

There is still much misinformation about the feats of travel and length of life of spermatozoa. Persistence of motility used to be equated with fertilizing capacity. We now know that motility lasts much longer than the ability to fertilize. For example, spermatozoa of the rabbit lose their ability to fertilize after about 30 hours in the female genital tract, whereas their motility may last over 2 days. Estimates now place the fertilizing power of human spermatozoa in the female genital tract at 1 or 2 days, with motility persisting for perhaps double that length of time. Survival of spermatozoa in the female genital tract is unusually prolonged in some species. In some bats, insemination occurs in the autumn but the spermatozoa remain dormant throughout hibernation. Not until the following spring, several months later, do ovulation and fertilization occur. In the domestic chicken, spermatozoa are stored in crypts in the oviductal wall and are gradually released with the passing of eggs through the oviduct over a 3-week period.

It should be stressed that the previous statements apply to ejaculated sperm in the female genital tract. The viability of spermatozoa varies greatly in different environmental conditions. In the epididymis and ductus deferens, where they remain nonmotile, human spermatozoa retain their full capacities for many days. Their motility is aroused only when, at the moment of ejaculation, they are mixed with secretions of the seminal vesicles and the prostate and bulbourethral glands. Much of the increased metabolism of the activated spermatozoa is due to substrates supplied by the seminal fluid, such as fructose, which is produced by the seminal vesicles (Mann, 1964).

That their life after activation depends in large measure on the rate at which spermatozoa expend their limited store of potential energy is clearly indicated by experimental work in artificial insemination. The motility of the spermatozoa in freshly ejaculated semen can be checked by chilling. In these conditions the spermatozoa do not immediately dissipate their available store of energy, thus allowing the semen of pedigreed stock to be shipped thousands of miles by airplane and introduced into females by a syringe, with the successful production of offspring. The rapidly advancing techniques of cryobiology have made banks of stored human sperm a reality. Even early mammalian embryos can be frozen for extended periods of time and later resume normal development upon thawing.

Union of Gametes

By far, the majority of studies of mammalian fertilization have been carried out on the mouse egg fertilized in vitro (Wassarman, 1987). Results to date have shown remarkable parallels with the major events in sea urchin fertilization. It remains to be seen whether additional factors beyond those discovered through in vitro studies operate in fertilization within the reproductive tract.

In mammals, fertilization occurs in the upper part of the uterine tubes (Fig. 3-36). The spermatozoa must first penetrate the cells of the corona radiata and then the zona pellucida before they can make contact with the plasma membrane of the egg. To get through the zona pellucida the spermatozoa must undergo the *acrosome reaction*. This reaction, for which capacitation is a prerequisite, is the means by which *lytic enzymes* stored within the acrosome of the spermatozoon (Fig. 3-37) are released so that they can facilitate the passage of the sperm through the egg coverings. The first step in the acrosome reaction is the localized fusion of portions of the outer acrosomal membrane with the overlying plasma membrane of the spermatozoon. These areas soon break down, allowing the release of soluble enzymes contained within the acrosome.

Evidence from in vitro studies suggests that binding of the spermatozoon and initiation of the acrosomal reaction are both associated with contact between the plasma membrane of the spermatozoon and several tens of thousands of the ZP-3 glycoprotein at the outer margin of the zona pellucida. There is still some discussion concerning the extent to which the mammalian acrosome reaction may be initiated before a spermatozoon attaches to the zona pellucida.

In vivo, spermatozoa must pass through the cellular corona radiata before they reach the zona pellucida. Certainly, the lashing of the sperm tail plays an important role in this phase of approaching the egg. Traditionally it has been taught that the enzyme *hyaluronidase*, released from the acrosome, facilitates

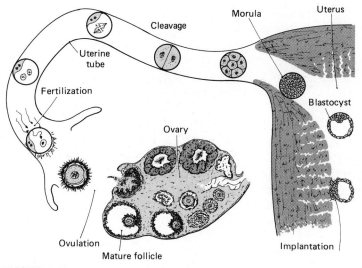

FIGURE 3-36
Summary diagram illustrating the sites of ovulation, fertilization, early
cleavage and implantation (at about 5–6 days after fertilization) in the
human.

sperm penetration through the corona radiata by dissolving extracellular matrix
material around cells of the corona radiata. However, if the acrosome reaction
does not occur until the spermatozoon has reached the zona pellucida, there
should be no apparent need at that time for the sperm to release hyaluronidase if
the main function of this enzyme is to enable the spermatozoon to penetrate the
corona radiata. Understanding of this apparently contradictory set of events
should be clearer by the time the next edition of this text is prepared.

After the acrosome reaction, the spermatozoon digests a narrow pathway
through the zona pellucida. This is accomplished through the actions of a
trypsinlike proteinase called *acrosin* that is bound to the inner acrosomal
membrane, which after the completion of the acrosome reaction is exposed to
the surface of the head of the sperm. Such chemical digestion apparently works
hand in hand with the actions of the sperm tail in propelling the spermatozoon
through the zona pellucida.

Once through the zona pellucida, the spermatozoon enters the fluid-filled
perivitelline space between the zona and the plasma membrane of the ovum.
The acrosome reaction appears to cause a change in the plasma membrane of
the sperm that allows it to fuse with other cell membranes. Contact is quickly
established between the egg and sperm membranes and may be faciliated by the
microvilli projecting from the ovum. Where the sperm has made contact with
the egg, the cytoplasm of the egg in many species bulges out in an elevated
process called the *fertilization cone*. The plasma membranes of the egg and
sperm fuse, and then the fertilization cone retracts, carrying the sperm head
into the ovum and completing the phase of penetration (Fig. 3-38).

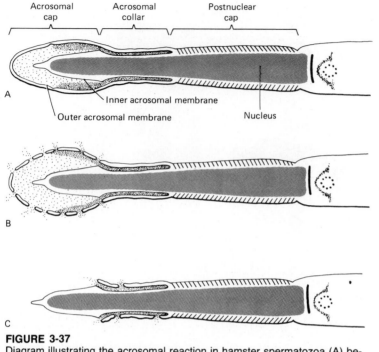

FIGURE 3-37
Diagram illustrating the acrosomal reaction in hamster spermatozoa (A) before, (B) during, and (C) after the acrosomal reaction. (Modified from Yanagimachi and Noda, 1970. *Am J. Anat. 128*:429.)

As in the sea urchin, it is important for the egg to prevent polyspermy once
it has been penetrated by the first sperm cell. This is accomplished by means of
blocks to polyspermy similar to those discussed earlier in this chapter. Because
it is more difficult to study many aspects of fertilization in mammals than it is
in sea urchins, most of the research on intracellular responses during fertilization is still carried out on sea urchins. However, recent studies of mammalian
eggs have shown a remarkable similarity in the two types of fertilization (Fig.
3-39). A slow block to polyspermy mediated by cortical granules definitely
occurs, but whether an effective fast block occurs is still an open question.

Development and Fusion of Pronuclei

Once the sperm has fused with the egg, a regular sequence of events leads to
the ultimate joining of the male and female pronuclei. With sperm entry, the
block to the second meiotic division of the egg is rapidly lifted and the second
polar body is released, leaving a haploid female nucleus. Shortly after entry
into the egg, the nuclear membrane of the sperm breaks down, allowing the
interaction between the nuclear contents of the sperm and the cytoplasm of the
egg. This results in decondensation of the tightly packed nuclear chromatin and

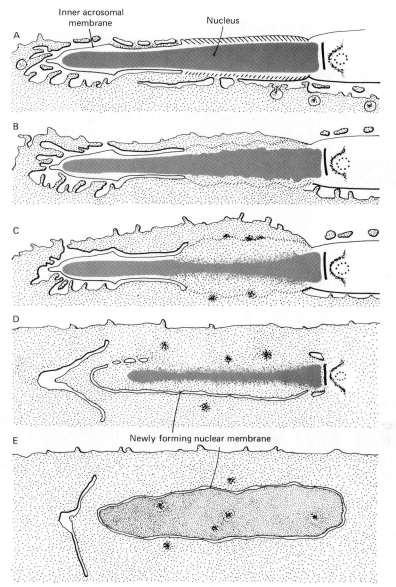

FIGURE 3-38
Diagram of stages of incorporation of a hamster spermatozoon into the egg.
(A, B) Fusion of the sperm head with the cytoplasm of the egg. (C) Swelling
of the sperm nucleus. (D, E) Formation of the nuclear envelope around the
swelling sperm nucleus. (Modified from Yanagimachi and Noda, 1970. *Am.
J. Anat. 128*:429.)

the replacement of the protamine-like proteins that were bound to the condensed sperm DNA with histones, probably derived from the oocyte. In
mammals as well as in trout (see p. 92), there is considerable evidence that

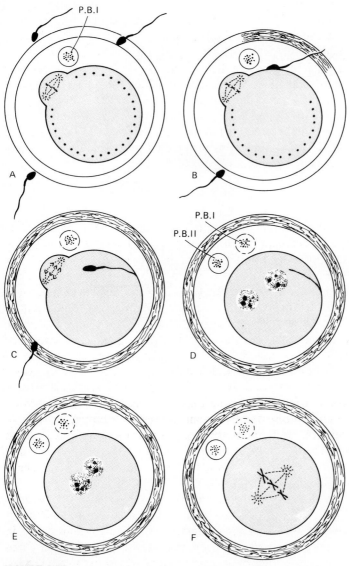

FIGURE 3-39
Diagrams illustrating the process of fertilization and the formation of
polar bodies. (A) Passage of spermatozoon through the zona pellucida.
(B) Initiation of the cortical reaction (*disappearance of black dots*) and
fertilization changes (*shading*) beginning in the zona pellucida. (C) Incorporation of spermatozoon into the egg. (D) Release of second polar
body and formation of male and female pronuclei. (E) Approximation of
pronuclei. (F) Metaphase of first cleavage division.

during spermiogenesis the histones, which normally bind to the DNA, are
replaced by protamine-like proteins. The protamines appear to facilitate the
extremely dense condensation of the nuclear chromatin that is required for

proper packing of the nucleus of the mature sperm cell. Decondensation of the sperm head is more complex than a mere exchange of protamines for histones. In the early stages of decondensation, the enzymatic breakdown of numerous disulfide bonds that keep the sperm chromatin condensed also occurs.

A new pronuclear membrane soon forms around the now decondensed nuclear material from the sperm. As the male and female pronuclei migrate toward each other, DNA synthesis occurs on the DNA of the haploid chromosomes. In contrast to the sea urchin, the chromosomes condense within the pronuclei so that when the membranes of the closely apposed male and female pronuclei break down, the chromosomes quickly become arrayed along the metaphase of the developing mitotic spindle.

In Vitro Fertilization

Many of the major advances in our understanding of the events of mammalian fertilization and early embryogenesis are due to the relative ease with which it is now possible to achieve in vitro fertilization. In vitro fertilization has three basic requirements: (1) an adequate supply of sperm, (2) at least one mature egg, and (3) appropriate conditions for penetration of the egg by the sperm.

Obtaining an adequate supply of spermatozoa is the easiest requirement to satisfy. In mammals, including humans, it is preferable to use ejaculated sperm rather than sperm taken directly from the testis because of the maturation that occurs in the epididymis. It is now possible to obtain far greater numbers of eggs than was formerly the case by employing techniques leading to *super-ovulation*. This can be accomplished by means of hormonal manipulations or the administration of certain fertility-promoting drugs. If the eggs are allowed to be ovulated, they can be collected by washing out the female reproductive tract shortly after ovulation. In humans, ripe eggs about to be ovulated can be collected directly from the ovary by the use of *laparoscopy*, a technique that allows direct visualization of the ovary. It is common to use ultrasonic images to determine when ovulation is ready to occur.

Early attempts to fertilize eggs in vitro were unsuccessful, largely because the phenomenon of capacitation was not understood at that time. When it became apparent that the sperm of most mammalian species have to be in contact with female reproductive tissues for a period of time before they can undergo the acrosome reaction, pieces of oviduct were added to the medium in the culture dish along with the eggs and spermatozoa. This resulted in a dramatic improvement in the success of "test-tube" fertilizations.

Parthenogenesis

Not all eggs require penetration by sperm for the initation of embryonic development. In a number of invertebrate groups as well as in scattered vertebrate species (some fishes, a few lizards, and even turkeys), unfertilized eggs may become activated and develop into viable individuals as part of the normal life cycle. This process is known as *parthenogenesis*. Artificial parthenogenesis of eggs can be stimulated in the laboratory by a variety of means. Pricking with a blood-dipped needle is a classic way of producing parthenogenesis in frog eggs. A large proportion of eggs stimulated to undergo artificial parthenogenesis fail to develop normally. More than likely this is due to the unmasking of deleterious recessive genes in the haploid embryos. Parthenogenetic embryos which do survive are commonly found to be diploid, probably owing to retention of the second polar body. In mammals, parthenogenetic individuals are all genetically female because of the XX chromosomal complement of the female. On the other hand, the female in both birds and reptiles is heterogametic, allowing the development of both males and females by parthenogenesis.

SEX DETERMINATION

Just before the turn of the century Henking (1891), while studying the chromosomal pattern of certain insects, was impressed by the fact that one pair of chromosomes lagged behind the others in moving toward the poles of the spindle in the maturation divisions. This was intriguing, but its significance was not at first apparent. In the manner of mathematicians, who give the unknowns in their problems the last letters of the alphabet as noncommittal designations, the members of this chromosomal pair were christened X and Y. Several years later McClung (1902) and Wilson (1905) came to the conclusion that this pair of chromosomes is involved in sex determination. Practically all the animals that have been critically studied since that time show consistent differences between the sexes in one of the pairs of chromosomes in both somatic and germ cells. In one sex, all the pairs of chromosomes are symmetrically mated (X-X). In the opposite sex, the members of the chromosomal pair are quite different from each other in size and shape (X-Y). It is now apparent that in mammals something carried on the Y chromosome but not on the X accounts for maleness. In some groups of vertebrates (e.g., birds), the chromosomal pairs of the male are alike whereas those of the female show sexual differences.

The union of gametes and the genetic determination of sex constitute just the beginning of a long process of sexual differentiation. Many aspects of this process are just beginning to be understood. The steps involved in the translation of genotypic sex into sexual phenotypes will be covered in Chap. 17. Recent experimental evidence has shown that phenotypic sex may not be as completely fixed at the time of fertilization as was first believed. There is little doubt that the chromosomal combination begins the trend of development toward one sex or another, but this trend may be inhibited or modified by massive doses of hormones or by the absence of appropriate hormone receptors. Our awareness of occasional discrepancies between genotypic and phenotypic sex stems largely from studies on the presence or absence of sex chromatin in cells.

In 1949 Barr and Bertram first presented evidence of sexual differences in fixed and stained nuclei of nondividing somatic cells. They found that in the

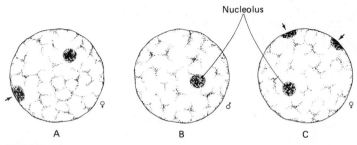

FIGURE 3-40
Drawings of nuclei of human epithelial cells to show the sex chromatin
(arrows). (A) Nucleus from a normal *XX* female, with one inactivated *X*-
chromosome. (B) Nucleus from a normal *XY* male, with no *X*-chromo-
somal inactivation. (C) Nucleus from a female with an *XXX* trisomy, with
two inactivated *X*-chromosomes.

nuclei of cells from females there was usually a conspicuous, characteristically
located mass of chromatin that was absent in the nuclei of cells taken from a
male (Fig. 3-40). This chromatin mass is called the *sex chromatin*. The sex
chromatin is now believed to represent one of the X chromosomes, which
remains highly condensed in the interphase (G_1) nucleus. Lyon (1961) postulated that the sex chromatin mass represents the inactivation of one of the X
chromosomes. The activity of only one X chromosome is required or permis-
sible for normal development in either males or females. Additional X
chromosomal activity is effectively eliminated by condensation of the extra X
chromosome. The sex chromatin, then, represents the morphological expression of a genetic control mechanism.

Sex chromatin bodies are typically not seen during early embryonic cleavage. The evidence to date suggests that both X chromosomes are actively
functioning during early cleavage, but as the trophectoderm and later the
primitive endoderm form during the early blastocyst stage (see Chap. 4), the
paternal X chromosome is selectively inactivated in each of these extraembryonic
tissues (Fig. 3-41). As the inner cell mass (which will form the embryo proper)
takes shape in the very late blastocyst stage, one X chromosome per cell becomes
inactivated, but in a random fashion, i.e., either maternal or paternal in a given
cell. The same chromosome is inactivated in all cells descended from the first cell
in which X chromosome inactivation occurs. It appears that differentiation of a cell
requires the inactivation of one X chromosome, but the mechanism and reason are
obscure. Equally important but unanswered questions are: (1) how the one X
chromosome remains inactivated through successive mitotic divisions and (2) how
the condensed X chromosomes in the oogonia are reactivated during oogenesis
(Gartler and Riggs, 1983).

ESTABLISHMENT OF POLARITY IN THE EMBRYO

The vertebrate body is bilaterally symmetrical and can be viewed in the context
of three polar axes: a *craniocaudal* axis, a *dorsoventral* axis, and a *mediolateral*

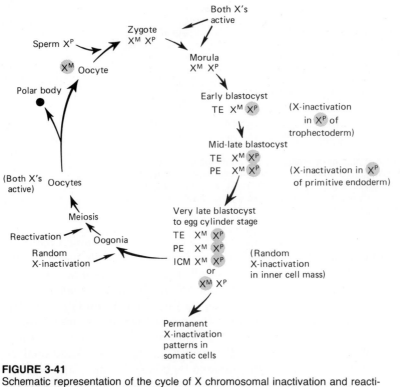

FIGURE 3-41
Schematic representation of the cycle of X chromosomal inactivation and reactivation in the human life cycle. X^m—maternal X chromosome; X^p—paternal X chromosome; ICM—inner cell mass; PE—primitive ectoderm: TE—trophectoderm. Chromosomes (X^m or X^p) in gray circles are inactivated. (Adapted from Gartler and Riggs [1983]).

axis. How these axes become imprinted upon the spherical egg remains one of the major mysteries of embryology.

The ovum developing within the amphibian ovary is already strongly polarized morphologically into animal and vegetal halves (the *primary polarity* of the egg). This is obvious to the naked eye because of the denser concentration of pigment granules in the animal half. There are gradients of other structures as well. The nucleus (germinal vesicle) is located near the animal pole, and there is a gradient of increasing density of both ribosomes and glycogen granules toward the animal pole. Conversely, both the size and the concentration of yolk platelets increase markedly toward the vegetal pole. In urodeles the craniocaudal axis of the future embryo nearly coincides with a line drawn through the animal and vegetal poles, but in anurans the two axes are not quite the same. Nevertheless, a useful generalization for Amphibia is that the region of the animal pole will ultimately form the head and the vegetal pole will form the tail. Although it is apparent that the craniocaudal axis is established before fertilization, the factors that lead to its being set in the ovary are obscure.

Fertilization is the next milestone in the establishment of polar axes within the amphibian egg. Shortly after fusion of the sperm to the egg some major reorganizations of cytoplasmic regions of the egg begin. One is a general convergence of cytoplasm beneath the thin (<10 μm) cortical region toward the sperm entry point. The other is a 30° shift between the subcortical cytoplasm and the overlying cortex. The most obvious consequence of these cytoplasmic rearrangements is a reduction in the density of dark pigment granules in the region of the animal hemisphere along the equatorial zone opposite to the sperm entry point. In many species of amphibians, such as *Rana*, this region of reduced pigmentation takes the shape of a crescent and is called the *gray crescent* (Fig. 3-42). The location of the gray crescent is determined by the site of sperm entry, for it appears on the opposite site of the egg.

Descriptive and experimental work (Gerhart et al., 1986) has shown that the middle point of the gray crescent is equivalent to the middorsal point of the

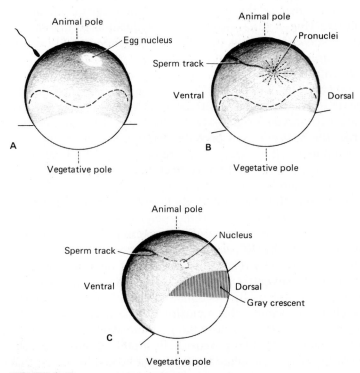

FIGURE 3-42
Diagram illustrating fertilization and gray crescent formation in amphibian egg. (A) Sperm contacting egg. (B) Approximation of egg and sperm pronuclei and early cortical shifting. (C) Formation of the gray crescent as the result of cortical shifting.

body. This determines the dorsoventral axis of the future embryo. With the superimposition of the dorsoventral axis upon the previously existing cephalocaudal one, the remaining mediolateral axis is also established simply by geometrical considerations. Thus, even before cleavage has begun, the three principal axes of the amphibian embryo are established and *secondary polarization* is completed.

The gray crescent has long captivated the interest of developmental biologists (Brachet, 1977) not only because of its importance as a structural landmark in the precleavage embryo but because it represents the future site of formation of the *dorsal lip of the blastopore* (Fig. 5-8C). The dorsal lip has been called the *embryonic organizer* because of its importance in controlling development. A variety of experiments (Fig. 3-43A to 3-43C) have suggested that in young embryos secondary polarization and subsequent normal development cannot occur in the absence of the gray crescent.

The technically elegant experiments of Curtis (1960, 1962) seemed to provide definitive proof of a central role of the gray crescent in establishing the dorsoventral axis and in generating the embryonic organizer. He grafted tiny pieces of gray crescent cortex into the prospective ventral region of a precleavage embryo and obtained the formation of a new dorsal lip and a secondary embryonic axis (Fig. 3-43E). However, subsequent experimentation (Kirschner et al., 1980) showed that it is possible to dissociate the site of the gray crescent from the future dorsal lip of the blastopore—in effect rearranging fixation of the dorsoventral axis so that the original gray crescent is on the ventral side of the embryo. This was accomplished simply by maintaining the precleavage embryo in such a position that the gray crescent was located downward with respect to gravity (Fig. 3-44). In the cortical transplantation experiments by Curtis, the host embryos probably would have been tipped in a manner similar to that illustrated in Fig. 3-44. Therefore, the positioning of the host alone and not the cortical graft could have been the factor leading to the formation of the secondary embryo illustrated in Fig. 3-43E.

How, then, can one account for the establishment of the dorsoventral axis? According to recent work in Gerhart's laboratory (Gerhart et al., 1986), the following is a likely sequence of events: The entering sperm provides a cue that serves to stimulate and orient the 30° displacement of cortex in relation to subcortical cytoplasm that ultimately results in formation of the gray crescent. The keys to fixation of the dorsoventral axis are (1) the activation of a specific region of vegetal cytoplasm and its displacement toward the cortex of the vegetal pole, (2) as cleavage occurs, the inclusion of the activated vegetal cytoplasm in cells (*blastomeres*) located near the gray crescent (future dorsal midline) area, and (3) an inductive effect of these cells on the neighboring cells in the equatorial region (between the animal and vegetal hemispheres) so that they prepare for gastrulation movements and formation of the dorsal lip of the blastopore.

The powerful influence of blastomeres containing the activated cytoplasm has been demonstrated in two ways by Gimlich and Gerhart (1984). Irradiation

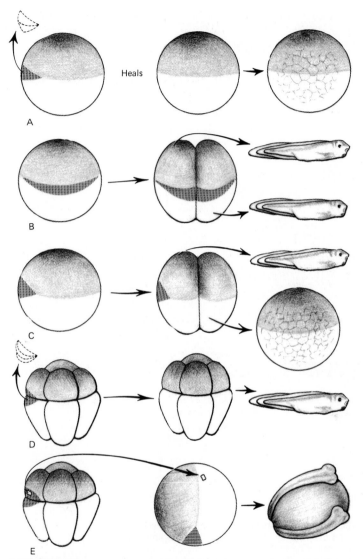

FIGURE 3-43
Experiments involving manipulations of the gray crescent. (A) Excision
at the one-cell stage blocks gastrulation. (B) Separation of blastomeres
at the two-cell stage of a normal embryo results in the development of
twins. (C) Separation of blastomeres after an abnormal first cleavage
leaves the gray crescent entirely in one blastomere. The blastomere
containing the gray crescent goes on to develop normally. Development
of the other blastomere is arrested. (D) Removal of the gray crescent in
an eight-cell embryo does not interfere with normal development. (E)
Grafting of a piece of gray crescent from an eight-cell stage to a one-
cell stage results in the production of a secondary embryonic axis.
(After Curtis.)

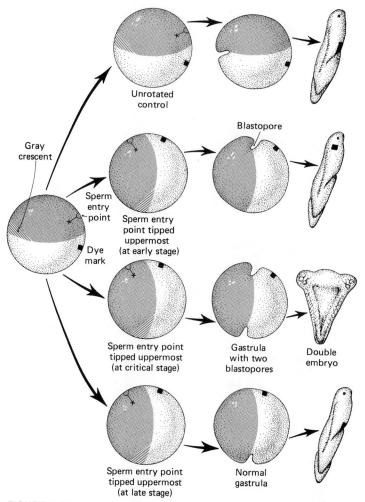

FIGURE 3-44
Experiments demonstrating that there is not necessarily a direct connection between the gray crescent and the location of the blastopore in the amphibian embryo. (First row) In normal control embryo the dorsal lip of the blastopore arises in the area of the gray crescent. (Second row) If a recently fertilized egg is tipped 90° so that the sperm entry point is uppermost, the blastopore forms there, 180° away from the gray crescent. (Third row) If a fertilized egg is tipped 90° at a critical stage, two blastopores form—one at the gray crescent and the other at the uppermost pole of the egg. A double embryo forms. (Bottom row). After the critical period, a 90° tipped embryo forms a blastopore in the normal location and a normal embryo forms. (After experiments of Kirschner and Gerhart, 1981. *Bioscience 31*:381.)

of the vegetal hemisphere of early postfertilization eggs with ultraviolet rays is known to block the formation of the body axes (presumably by preventing internal cytoplasmic rearrangements), leading to the formation of a featureless

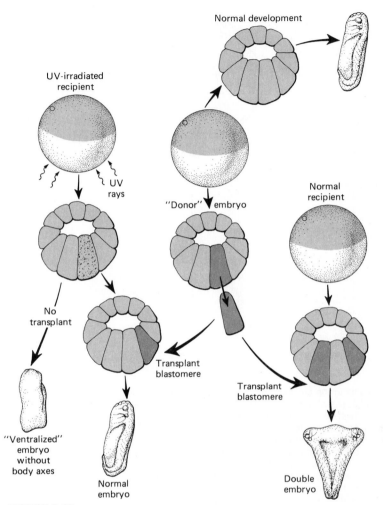

FIGURE 3-45
Experiments demonstrating the role of vegetal cells beneath the future dorsal lip of the blastopore in stimulating the initiation of gastrulation. Left—If the vegetal hemisphere of an amphibian embryo is irradiated with UV rays, gastrulation fails to occur and a "ventralized" embryo results. Middle—If a normal vegetal blastomere is transplanted to a UV-irradiated embryo, the embryo is "rescued" and normal development ensues. Right—If the vegetal blastomore is grafted into a normal recipient, a second dorsal lip of the blastopore forms and a double embryo results. (Adapted from the experiments of Gimlich and Gerhart, 1984).

"ventral" embryo (Fig. 3-45). When activated vegetal blastomeres from a normal donor are grafted into an irradiated embryo, axiation can be initiated and normal development ensues. Similarly, the grafting of an activated ventral blastomere into the prospective ventral region of a normal embryo leads to the formation of a secondary dorsal axis and a duplicated embryo (Fig. 3-45).

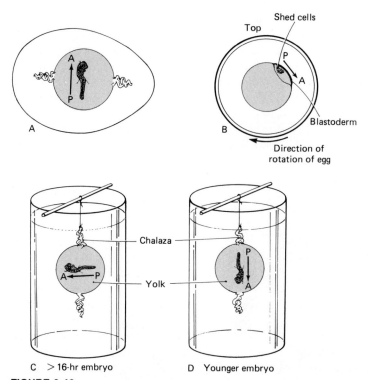

FIGURE 3-46
Effect of gravity of polarity in the chicken embryo. (A) In a normal egg,
the anteroposterior (A-P) axis of the embryo forms perpendicularly to the
axis between the two chalazae. (B) Cross-section of egg rotating in the
uterus (outside arrow). As the yolk tends to right itself, the blastoderm is
tipped at a slight angle. Cells shed from the under surface of the uppermost part of the tipped blastoderm mark the future posterior end of the
embryo. (C) If an older embryo is suspended by a chalaza, the A-P axis
of the embryo forms at the usual right angle to the chalazae. (D) If a
yolk is similarly suspended before the A-P axis is fixed, the A-P axis is
fixed parallel to the chalazae, with the posterior end uppermost.
(Adapted from Kochav and Eyal-Giladi, 1971).

Analysis of the mechanism of determination of polarity in avian embryos
begins with an old observation of von Baer (1828) that when most avian eggs
are observed with their pointed end at the right and the blunt end at the left, the
embryo is oriented perpendicular to the long axis, with its head directed away
from and its tail directed toward the observer (Fig. 3-46A). Although it was first
thought that axial determination takes place in the ovary, later research has
shown that it occurs during early cleavage, while the fertilized egg is in the
uterus (Fig. 3-18). The cleaving embryo is represented by a flat disk of cells on
the surface of the yolk (Fig. 4-10), and the cells on the outer surface become the
dorsal part of the embryo while those closest to the yolk become ventral. Thus,

the dorsoventral axis can be predicted by the position of cells or cell layers relative to the yolk.

Determination of the craniocaudal (anteroposterior) axis is the critical event in polarity determination in the bird. Vintemberger and Clavert (1960) showed that the craniocaudal axis becomes fixed after the fertilized egg has been in the uterus from 14 to 16 hours. What are the circumstances that surround the fixation of polarity in the uterus?

In the 25 hours between ovulation and laying in the chicken, the egg, which is fertilized in the region of the ostium of the oviduct (Fig. 3-18), spends 5 hours passing through the oviduct, where the albumen surrounding the yolk is secreted. The next 20 hours are spent in the uterus, where the egg, with its pointed end usually facing the cloaca, rotates at about 10 to 15 revolutions per hour as the shell is laid down. The albumen rotates within the shell membrane, but the yolk does not. This accounts for the spiral twisting of the chalazae located on either end of the yolk (Fig. 3-25). Although the yolk does not rotate, it is tipped slightly as a result of the pull of the albumen. This causes the embryo proper (blastoderm) to be tipped with respect to the flat gravitational surface (Fig. 3-46B). During the 14- to 16-hour period in the uterus when the craniocaudal axis of the embryo is determined, cells are shed from the part of the blastoderm that is uppermost with respect to the source of gravity. The area from which these cells fall becomes the caudal end of the embryo (Fig. 3-46B).

That the orientation of the blastoderm with respect to gravity is an important factor related to the determination of the craniocaudal axis was demonstrated by Kochav and Eyal-Giladi (1971). These investigators removed early uterine eggs and strung them up by the chalazae in abnormal orientations (Fig. 3-46C). Invariably, the caudal end of the embryonic axis appeared at the uppermost end of the blastoderm. The results of both this experiment and normal development point to the conclusion that the orientation of the blastoderm in relation to the earth's gravitational field is a critical factor in determining the direction of the craniocaudal axis of the embryo. At this juncture, it seems that the only way to definitively test this possibility—particularly the role of cell shedding from the uppermost region of the blastoderm—would be to allow birds' eggs to undergo the intrauterine phase of development in the microgravity of space.

Little is known about the establishment of polarity in mammalian embryos. Polarity in the mammalian embryo does not seem to be fixed until relatively late in development (around the time of implantation) and may be a consequence rather than a cause of differentiation.

CLEAVAGE AND FORMATION OF THE BLASTULA

Fertilization transforms the egg from a metabolically depressed state to one of extreme vigor, characterized by a sharp increase in respiratory and synthetic activity. One of the most immediate and spectacular consequences is the initiation of cleavage. During cleavage, waves of cell division follow one another almost without pause, subdividing the unmanageably large zygote into progressively smaller cellular units, resulting in a tightly knit mass of cells of more normal dimensions (Fig. 4-1). The cells resulting from the first few cleavage divisions are for the most part morphologically unspecialized. They remain metabolically unspecialized as well; their synthetic activity is geared toward the production of the DNA and proteins required for cell division rather than toward specialized molecular activity.

After a few synchronous cell divisions, during which the number of cells in the embryo doubles with each cycle, the embryo looks like a small mulberry (hence the commonly applied term *morula*), and the cell divisions begin to lose their synchronous character. As will be seen subsequently, many of the changes in the cleavage pattern within a given embryo, as well as differences in cleavage patterns among embryos of the various vertebrate classes, are related to the amount of yolk originally present in the egg. In time the cleaving embryo develops a central cavity (*blastocoel*) and enters the *blastula* stage. Beneath its rather unimposing and homogeneous exterior, major changes are beginning to take place. The synthesis of RNAs coding for specialized proteins gets under way in many species, and the cells of the embryo begin to communicate with one another in new and different ways. The cells themselves begin to assume different, nonequivalent properties, although their appearance does not show these developments. In many species the cytoplasm of the egg is not homoge-

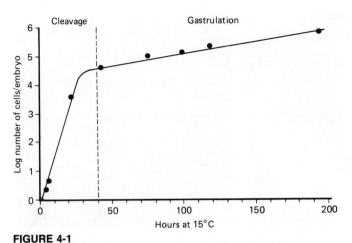

FIGURE 4-1
Changes in the number of cells in embryos of the frog, *Rana pipiens*. (After Sze, 1953. *J. Exp. Zool. 122*:594.)

neous, and according to one long-standing hypothesis, different cytoplasmic
areas of the early embryo act on the individual cleavage nuclei in different
ways, thus initiating their development along different lines. At the level of the
embryo as a whole, final polarity becomes firmly fixed and the unseen
developmental guidelines that keep the massive tissue displacements of gastrulation under tight control are set up. All these changes lead to a greater
integration of the embryo as a whole and a reduction in the ability of its
corresponding parts to compensate for damage or experimentally created
defects.

THE CELL DURING CLEAVAGE

In the normal embryo a cleavage division consists of nuclear division
(*karyokinesis*) followed by cell division (*cytokinesis*). Cleavage brings to the
forefront several issues of general importance in cell biology. Prominent
among these are the mechanisms of cytokinesis and the formation of new
membranous structures in dividing cells (Fig. 4-2). Much of our knowledge in
this area has arisen from investigations carried out on invertebrate forms, but
the cleaving amphibian egg has also been a favored research object. Although
many of the basic questions and answers in this field were set forth decades
ago, modern research methods in both cytology and biophysics have allowed
the identification of some concrete mechanisms by which these processes
take place.

Cytokinesis occurs after karyokinesis, and the plane of the cleavage furrow
is the same as that of the metaphase plate of the preceding mitotic figure. That
there is some sort of causal relationship between the mitotic apparatus and
furrow formation has been demonstrated by observations on both normal and

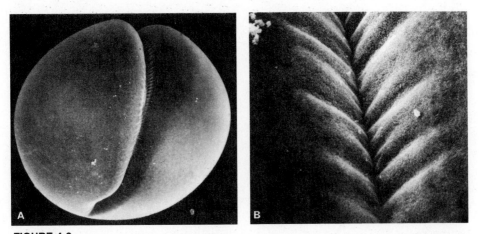

FIGURE 4-2
(A) Scanning electron micrograph (×92) of a frog embryo during the first cleavage division.
(B) Higher power view (×805) of the cleavage furrow, showing folds in the walls of the furrow.
(Courtesy of H. W. Beams and R. G. Kessel, 1976. from *Am. Sci. 64*:279.)

experimentally altered eggs in which the mitotic apparatus has been displaced.
The cleavage furrow forms first in the region of the cortex nearest to the
mitotic spindle and then moves around the cell (Fig. 4-3). If the mitotic
apparatus of a zygote is mechanically displaced toward the cell membrane,
the cleavage furrow begins to form in the region closest to the mitotic
apparatus. According to Rappaport (1974), the position of the cleavage
furrow is established by the time of anaphase. After anaphase the mitotic
apparatus can be removed or destroyed without altering the location or
course of cleavage furrow formation.

The *asters* are the structures that interact with the cell cortex to stimulate
the formation of the cleavage furrow. The exact nature of the stimulus is not
known, but it can be transmitted from the mitotic apparatus to the surface of
the egg within a few minutes. Ca^{2+} seems to be an important factor in
transmitting the cleavage stimulus from the asters to the cell cortex.

At the site of the cleavage furrow, the cortex of the cell contains a band of
microfilaments which becomes better defined as the furrow forms. The
microfilaments belong to the *actin* family of contractile proteins, and in
conjunction with myosin molecules in the same area their contractile behavior
appears to be the basis for the constriction of the cell during cleavage.

When a spherical cell divides into two cells, geometric necessity dictates
that the total surface area of the two daughter cells will be greater. In
confirmation of earlier hypotheses, recent investigations of early amphibian
embryos have shown that within 7 or 8 minutes of the establishment of the
cleavage furrow, areas of new membranous material form along both sides of
the contractile ring. The new membrane is smooth, pale in color, and highly
permeable to ions, and it seems to be deposited only in the cleavage furrow.

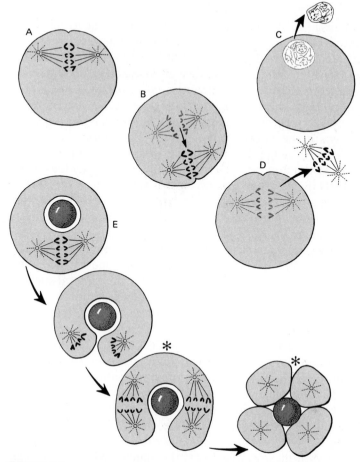

FIGURE 4-3
Experiments on the relationship between the mitotic apparatus and formation of the cleavage furrow. (A) In normal development the furrow
forms parallel to the metaphase plate in the region of the egg closest to
the mitotic apparatus. (B) If the early mitotic apparatus is pushed toward
the opposite side of the egg, the cleavage furrow begins there. (C) If the
prophase nucleus is removed from the egg, a cleavage furrow does not
form. (D) If a later stage mitotic apparatus is removed from the egg, a
cleavage furrow still forms in the normal location. (E) If the cleavage furrow is interrupted by a tiny glass ball, a horseshoe-shaped cell is created
with a nucleus at each end. When these nuclei divide again, cleavage
furrows form not only between the two pairs of daughter nuclei, but between the two dividing nuclei (asterisk), where there is no mitotic apparatus. This shows that the asters, and not the mitotic spindle, are the effective agent in initiating cleavage furrow formation. (E, based on the experiment of Rappaport, 1971).

DISTRIBUTION OF YOLK AND ITS EFFECT ON CLEAVAGE

The morphology of cleavage differs considerably among some of the major animal groups. An important factor accounting for these differences is the amount and distribution of yolk contained in the eggs. Some of the types of eggs (as classified by their yolk content) and the patterns of cleavage are outlined in Table 4-1.

The cells that arise from cleavage are known as *blastomeres.* In eggs with little yolk (*oligolecithal* eggs) the cleavage divisions produce blastomeres of roughly the same size. The moderate amounts of yolk in *mesolecithal* eggs tend to be concentrated toward the vegetal pole, thereby displacing the nucleus toward the animal pole. As noted in the previous section, the eccentric nucleus results in the initiation of the first cleavage furrow close to the nucleus at the animal pole (Fig. 4-9). As the cleavage furrow spreads toward the vegetal pole, its rate of formation decreases as a result of the retarding effect of the yolk. The net result of this process is the appearance of later and larger blastomeres at the vegetal pole than at the animal pole.

Telolecithal eggs possess a large amount of yolk, which commonly displaces the embryo-forming cytoplasm into a small disk on one edge of the ovum. When cleavage begins, membranous material appears along the lateral surfaces of the newly forming blastomeres, but initially it does not separate their inner borders from the underlying yolk. This is the basis for designating this type of cleavage as *meroblastic,* or incomplete.

The wide variation in amounts and types of yolk is closely correlated with the early life histories of the embryos and larvae. Relatively small animals that need to produce immense numbers of eggs to ensure propagation of the species (e.g., the sea urchin) must for reasons of size alone produce small eggs, which can contain little yolk because the number of eggs required simply would not fit inside the body of the mother. However, to survive, the embryos must rapidly develop to the feeding stage before the minimal stores of yolk are used up. Among the amphibians with mesolecithal eggs, the species with smaller eggs (e.g., *Xenopus*) develop into feeding larvae more rapidly than do those with larger eggs (e.g., most salamanders).

Among the terrestrial vertebrates (e.g., birds and reptiles) that lay eggs there is a closer correlation between the amount of yolk and the size of the individual that hatches from the egg than there is between the length of gestation and the amount of yolk. Both the leisurely pace of development and the possibility of obtaining some nutrition by diffusion from maternal fluids permit early development to occur in eggs containing very little yolk, but it is not until a functional circulation and the rudiments of a placenta have developed that the phase of extremely rapid growth of the embryo can begin.

CLEAVAGE AND FORMATION OF THE BLASTULA IN
Amphioxus

The oligolecithal eggs of *Amphioxus* (*Branchiostoma*) undergo a very regular form of equal holoblastic cleavage in which there is relatively little difference in

TABLE 4-1
CLEAVAGE PATTERNS IN RELATION TO YOLK CONTENT OF EGGS

Type of egg (based on yolk content)	Type of cleavage	Pattern of cleavage	Representative animal groups	Blastula	Blastula cavity
Oligolecithal-isolecithal (little yolk, evenly distributed)	Holoblastic (blastomeres completely separated)	Radial Bilateral Spiral Rotational	Echinoderms, *Amphioxus* Ascidians Mollusks, annelids Mammals	Sphere; wall a single layer	Large central sphere
Mesolecithal (moderate amount of yolk)	Holoblastic	Radial	Amphibians, lampreys, lungfish	Sphere; wall layered and of nonuniform thickness	Small eccentric sphere
Telolecithal (large amount of yolk)	Meroblastic (blastomeres incompletely separated)	Discoidal	Most fishes, reptiles, birds	Cell disk on surface of yolk	Flat space between epiblast and hypoblast
Centrolecithal (yolk concentrated in center of egg)	Meroblastic	Superficial (blastomeres on outside)	Insects and other arthropods	Yolk-filled cylinder	None

Source: Modified from Gilbert.

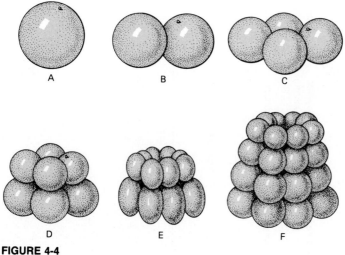

FIGURE 4-4
Cleavage in *Amphioxus*.

size among the blastomeres. The basic pattern of cleavage followed by *Amphioxus* occurs with some variation in many animal groups. The first cleavage division cuts the egg in half from the animal pole to the vegetal pole in the plane of the metaphase plate of the dividing nucleus (Fig. 4-4B). The nuclei of the resulting two blastomeres form mitotic spindles at right angles to the plane of the first division, and the second cleavage division proceeds again from animal pole to vegetal pole, resulting in the formation of a four-cell embryo (Fig. 4-4C). The third cleavage division, sometimes called an equatorial division, cuts each of the four blastomeres in half, roughly midway between the animal and vegetal poles (Fig. 4-4D). The next cleavage division produces simultaneous vertical planes of cleavage that cause the embryo to double its cell numbers from 8 to 16 (Fig. 4-4E). This stage and the 32-cell stage resulting from the next cleavage represent the morula stage. Subsequently, a cavity forms, and the multicelled embryo is now called a blastula (Fig. 4-4F).

CLEAVAGE AND FORMATION OF THE BLASTULA IN SEA URCHINS

Cleavage in the sea urchin is so rapid a process that before one cleavage cycle is completed the next has begun. Specifically, the DNA-synthetic (S) phase of a cleavage cycle begins before the M (mitosis) phase of the previous cell cycle is finished (Fig. 4-5). This eliminates the G_1 (interphase) component of the cell cycle, which can be very long in many cell populations.

The eggs of sea urchins are oligolecithal, but their pattern of cleavage is more complex than that of *Amphioxus* (Czihak, 1975). The first two divisions are meridional (from animal pole to vegetal pole), and the third is equatorial,

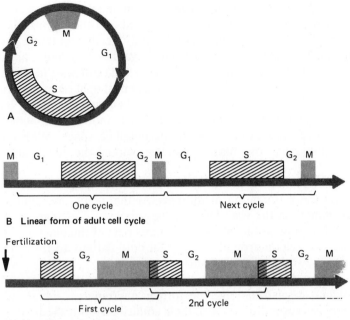

FIGURE 4-5
Cell cycle in mature and cleaving cells. (A) Circular representation of the cell cycle in a mature cell. M—mitosis; G_1—Gap 1; S—DNA synthesis; G_2—Gap 2. (B) Linear representation of the cell cycle in a typical adult cell. (C) The earliest cycles after fertilization in the sea urchin. The S phase of the first cell cycle is very short, and the S phase of succeeding cycles begin before mitosis (M phase) has ended, thereby eliminating the G_1 phase in these cycles. (After Wessels, 1977, *Tissue Interactions and Development.* Benjamin/Cummings.)

producing an eight-cell embryo with blastomeres of almost equal size. The fourth cleavage, however, results in the production of three distinct types of blastomeres (Fig. 4-6). The four blastomeres of the animal tier undergo a meridional cleavage, resulting in a single animal tier of eight cells called *mesomeres*. The four vegetal blastomeres divide asymmetrically, forming a tier of four large *macromeres* and, below that, a cluster of four small *micromeres*. Formation of the micromeres appears to be dependent on the presence of cytoplasm that was located at the vegetal pole of the egg. If the vegetal cytoplasm is removed, micromeres do not form. In normal development the three types of blastomeres have different fates. These will be dealt with later in this chapter and in Chap. 5.

During the entire period of cleavage, the embryo is enclosed in the fertilization membrane and the outer surfaces of the blastomeres are closely associated with the hyaline layer, which formed as a result of the emptying of the contents of the cortical granules into the perivitelline space during fertilization. As the embryo goes into the seventh and eighth cleavage cycles, the central cavity (*blastocoel*) becomes well established and the embryo has entered the blastula phase. The wall of the blastula is only a cell layer thick, and all cells have an apical surface exposed to the outer hyaline layer and a basal surface exposed to the blastocoel, although a basal lamina covers the entire inner surface of the blastula wall. The early blastula is almost spherical, and for a time it can be very difficult to locate the original animal and vegetal poles. By about the tenth cleavage cycle, the blastomeres form motile cilia which penetrate the hyaline layer and extend into the perivitelline space. With the coordinated beat of the cilia, the blastula rotates within the fertilization membrane.

Soon the cells of the blastula secrete into the perivitelline space a hatching enzyme which digests the fertilization membrane. At this point, the ciliated blastula is freely swimming in the sea. The next morphological change is the formation of a tuft of long, nonmotile cilia at the animal pole of the blastula. At about the same time, the region around the vegetal pole flattens to form the micromere-containing *vegetal plate* (Fig. 4-6I).

The final major event during the blastula phase presages the sweeping changes that will occur at gastrulation. The former micromeres, which are tightly integrated into the vegetal plate, lose their affinity for both the hyaline layer and their neighboring cells; passing through the basal lamina, they make their way into the blastocoel (Solursh, 1986; Fink and McClay, 1985). These cells, which will form the *primary mesenchyme*, first become elongated by the elaboration of parallel bundles of microtubules. As the cells elongate further, their apical surfaces detach from the hyaline layer; with the disappearance of the desmosomes, their lateral surfaces separate from the adjacent cells of the vegetal plate. They next pass through the basal lamina and into the blastocoel, where they rest along the inner surface of the basal lamina and become readily recognizable as the primary mesenchyme (Figs. 4-7 and 4-8). With this pregastrulation movement of cells, the embryo is often called a *mesenchyme blastula*.

CLEAVAGE AND FORMATION OF THE BLASTULA IN AMPHIBIANS

The period of cleavage and blastula formation in amphibians is rapid, usually being completed within 24 hours. The first cleavage division begins at the animal pole and bisects the gray crescent (Figs. 4-2 and 4-9). In the axolotl, the cleavage furrow elongates at a rate of about 1 mm/minute in the animal hemisphere but slows to 0.02 to 0.03 mm/minute as it nears the vegetal pole (Hara, 1977). The second division also begins at the animal pole, with its plane perpendicular to that of the first cleavage plane. The third cleavage plane is horizontal and passes nearer to the animal pole, dividing the embryo into four smaller blastomeres at the animal hemisphere and four larger blastomeres at the vegetal pole. Successive cleavage divisions follow one another rapidly and

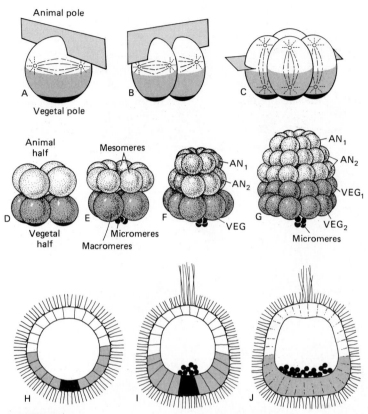

FIGURE 4-6
Cleavage in the sea urchin embryo. The fourth cleavage division leads to
an asymmetrical organization of the embryo, with an animal hemisphere of
8 cells and two vegetal tiers, with a row of 4 macromeres and 4
micromeres. With further divisions (F, G), both animal and vegetal tiers of
blastomeres subdivide. (H–J) Sections through blastulae. The black circles
in I and J are primary mesenchymal cells.

synchronously. In amphibians, an embryo between the 16- and 64-cell stages is
commonly called a morula. In subsequent cleavage cycles the waves of
cleavage begin to lose their synchrony because of a lengthening of the cell cycle
in the cells of the vegetal hemisphere. By the fifteenth cleavage cycle in the
axolotl, cleavage at the vegetal pole compared with cleavage at the animal pole
is delayed about two cycles (Hara, 1977).

After the morula stage, a prominent cavity (blastocoel) appears in the
animal hemisphere above the mass of yolk (Fig. 4-10). Formation of the
blastocoel in vertebrate embryos has long been of interest to embryologists, but
only recently has there been any real understanding of the mechanisms
involved. Kalt (1971) has traced the formation of the blastocoel in *Xenopus*
back to the first cleavage division. A specialization of the cleavage furrow at the

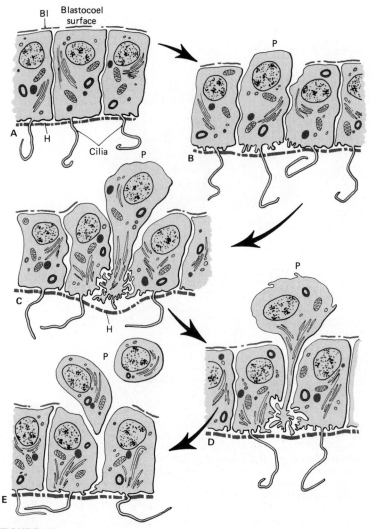

FIGURE 4-7
Ingression of primary mesenchyme cells in the sea urchin embryo. (A) Wall of
blastula before pressing has begun. (B) Primary mesenchyme cells (P) begin
to elongate into the blastocoel through an incomplete basal lamina (BL). (C)
The apical surface of the primary mesenchymal cell detaches from the hyaline
layer (H). (D) Separation of a primary mesenchymal cell from the wall of the
blastocoel. (E) Rounding-up of separated primary mesenchymal cell (P).
(Adapted from Katow and Solursh, 1980.)

animal hemisphere leaves a small intercellular cavity which is sealed off from
the exterior by specialized close junctions between the two blastomeres. This
small cavity is preserved and enlarged during subsequent cleavage cycles. The
maintenance and accumulation of fluid in the blastocoel seems to be due to the

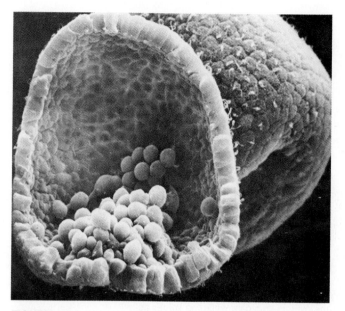

FIGURE 4-8
Scanning electron micrograph of an early gastrula of the sea urchin, *Lytechnius pictus*. The embryo in the forefront has been fractured so that the area of invagination is seen. Note the mass of invaginated primary mesenchyme cells lying in the ventral part of the blastocoel of the fractured embryo. The invagination is seen on the surface of the embryo in the background. (Courtesy of M. Solursh, from H. Katow and M. Solursh, unpublished data.)

pumping of Na^+ from the blastomeres into the emerging blastocoel. With the wall of the blastula acting as a semipermeable membrane, water enters the blastocoel to maintain an ionic balance, thus causing the blastocoel to expand (Slack et al., 1973).

When the embryo has completed the cleavage cycle that takes it from the 64-cell morula to the 128-cell stage, it is commonly called a *blastula*. It remains in the blastula stage until successive cleavage cycles have increased the cell number to between 10,000 and 15,000 blastomeres. At this time the massive morphogenetic movements characterizing gastrulation begin. The amphibian blastula can be conveniently subdivided into three main regions:

1 A region around the animal pole, roughly including the cells forming the roof of the blastocoel. These cells correspond roughly to the future ectodermal germ layer.

2 A region around the vegetal pole, including the large cells in the interior which constitute the yolk mass. These are the future endodermal cells.

3 A marginal ring of cells in the subequatorial region of the embryo, including the region of the gray crescent. Cells of this zone normally form the embryonic mesoderm.

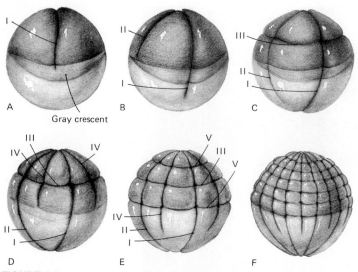

FIGURE 4-9
Cleavage in the frog's egg. The darkness of the upper hemisphere is due to the presence of pigment in the cytoplasmic cap, while the lighter appearance of the lower hemisphere is caused by the massing of yolk in that part of the egg. The cleavage furrows are designated by Roman numerals indicating the order of their appearance. In (A) and (B) note the retarding effect of the yolk on the extension of the cleavage furrows toward the vegetative pole. In (C) observe the displacement of the third cleavage furrow away from the yolk-laden pole, toward the center of the mass of active cytoplasm. The displacement of the center of activity from the geometrical center of the egg and the mechanical retardation of cleavage at the vegetative pole—both due to the yolk mass—result in the formation of a morula with many small blastomeres in the animal hemisphere and fewer large blastomeres in the vegetative hemisphere.

From the phylogenetic standpoint, the blastula (and later stages up through the tail bud larva) of the lungfish shows a remarkable morphological similarity to that of the amphibian. In contrast, early embryos of the coelocanth, a popular candidate for the transitional form between fishes and amphibians, resemble typical fish embryos. Whether these embryological characteristics are sufficient to restore lungfish to preeminence as links between fishes and amphibians remains to be determined.

For many years pregastrulation in the amphibian embryo was considered to be a relatively unexciting period, with the main activity being the increase in number of cells. It is now known, however, that major organizational changes take place during the late cleavage period. Nieuwkoop (1973) performed recombination experiments in which cells from the animal half of the embryo above the blastocoel were directly apposed to the yolk mass from the vegetative hemisphere. Under an inductive influence by the yolk mass, the cells from the animal hemisphere formed mesodermal structures. Normally the mesoderm comes from a zone of surface cells located closer to the yolk mass

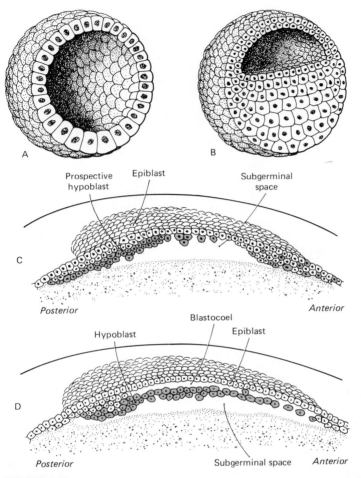

FIGURE 4-10
Schematic sections comparing the blastulae of (A) *Amphioxus*,
(B) amphibian, and (C, D) the chick.

(Fig. 4-11). Nieuwkoop has postulated that one of the functions of the
blastocoel may be to restrict the interaction between future endodermal and
ectodermal cells to the marginal ring surrounding the edges of the blastocoel.

One factor that facilitates communication among the cells of cleaving
embryos is the widespread distribution of low-resistance intercellular junctions which favor the passage of electrical currents and small molecules from
one cell to another (Warner, 1985). Although antibodies to gap junction
proteins injected into cleaving embryos can result in morphological defects,
the specific relationship between cell coupling and the phenomena demonstrated by classical experimental techniques of embryology has not been
established.

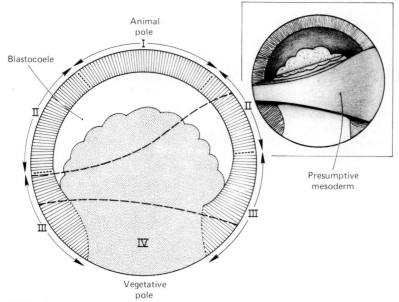

FIGURE 4-11
Subdivision of the blastula of the urodelean amphibian into formative zones.
Zones I and II from the animal hemisphere form ectoderm upon isolation. Zone
IV, from the vegetative pole, contains yolk-laden endodermal cells. Cells from
the ring-shaped subequatorial zone (III) form mesodermal cells upon isolation.
(Modified from P. Nieuwkoop, 1973. *Adv. Morphogen. 10*:1.)

CLEAVAGE AND FORMATION OF THE BLASTULA IN BIRDS

Cleavage in avian eggs has received less attention than cleavage in amphibian
eggs because the entire period of cleavage occurs as the egg is passing down the
oviduct (Eyal-Giladi, 1984). By the time the egg is laid, early gastrulation has
already begun.

The newly fertilized egg, which is about to undergo cleavage, contains a
whitish disk (*blastodisk* or *germinal disk*) of active protoplasm about 3 mm in
diameter at the animal pole. Around this disk is a darker-appearing marginal
area known as the periblast, but there is no distinct line of demarcation between
the two.

The first cleavage furrow begins to appear near the center of the blastodisk
during late anaphase of the first mitotic division after fertilization. As in
amphibian embryos the cleavage furrow lies in the plane of the chromosomal
plate during metaphase, and microfilaments are found at the base of the
cleavage furrow. The sequence of avian cleavage is not always regular, and
after about the third cleavage division it is not synchronous. Nevertheless, the
mitotic spindles align themselves so that the subsequent cleavage furrow forms
at right angles to the preceding one (Fig. 4-12). The fourth cleavage furrow is
a circumferential one which cuts a central row from a peripheral row of
blastomeres.

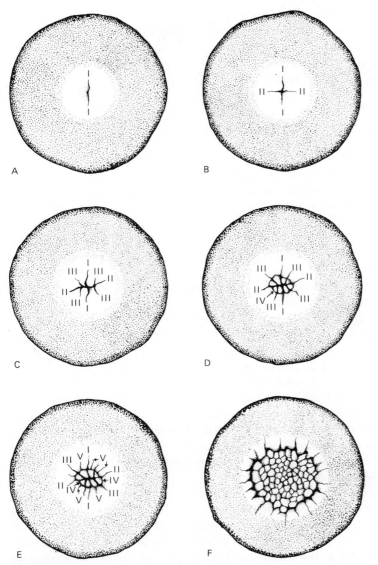

FIGURE 4-12
Surface aspect of blastoderm of bird's egg at various stages of cleavage.
The blastoderm and the immediately surrounding yolk are viewed directly
from the animal pole, the shell and albumen having been removed. The
order in which the cleavage furrows have appeared is indicated on the dia-
grams by Roman numerals. (A) First cleavage. (B) Second cleavage. (C)
Third cleavage. (D) Fourth cleavage. (E) Fifth cleavage. (F) Early morula.
(Based on Blount's photomicrographs of the pigeon's egg.)

The blastomeres formed by the first few cleavage divisions are unusual in
having their top and sides bounded by plasma membranes but their basal
surfaces open to the underlying yolk. Further cleavage in the early disk of

embryonic cells, now called a blastoderm, results in the radial extension of the embryo toward the periblast region. Nuclei are rarely seen in the open cells at the expanding periphery of the blastoderm, and it has been suggested that the presence of "accessory sperm nuclei"[1] during early cleavage may trigger cytoplasmic division in this region (Bellairs et al., 1978).

In addition to surface cleavages, the 32-cell embryo shows cleavage planes of an entirely different character. These cleavages appear below the surface and parallel to it. They establish a superficial layer of nucleated cells which are completely delimited by plasma membranes. These superficial cells rest on a layer of cells which on their deep faces are continuous with the yolk. Continuous divisions of the same type eventually establish several strata of superficial cells. The division progresses centrifugally as the blastoderm increases in size but does not extend to its extreme margin. The peripheral margin, where the blastoderm abuts the periblast, remains a single cell in thickness, and the cells there lie unseparated from the yolk. By the time the embryo contains about 100 cells, the blastoderm is underlain by a *subgerminal cavity* (Fig. 4-10C).

After a number of cleavage cycles, the shedding of individual cells begins from the undersurface of the area of the blastoderm that is farthest away from the source of gravity (Fig. 3-46B). As was described in Chap. 3, the area from which the cells are first shed becomes fixed as the posterior (caudal) end of the embryo. The cell shedding spreads toward the future anterior end of the embryo. The central portion of the blastoderm, thinned out by the shedding of cells and underlain by the subgerminal cavity, is called the *area pellucida*. Surrounding the area pellucida is the *area opaca*, a region where the cells of the blastoderm still abut directly onto the yolk.

At about the time of laying, individual cells or aggregates of cells shed from the lower surface of the blastoderm by a process of *delamination* coalesce to form a thin disklike layer called the *primary hypoblast*[2] (Fig. 4-10D). This process occurs first, and to a greater extent, at the posterior end of the embryo. In addition, cells moving in from the posterior margin of the blastoderm may play a role in filling out the primary hypoblast. The primary hypoblast is separated from the outer layer of the blastoderm, called the *epiblast*, by a thin cavity, the blastocoel. The primary hypoblast, which ultimately forms extraembryonic endoderm, possesses an inherent polarity, which it confers on the embryo proper, which is represented at this stage by the early epiblast. The polarity and location of the primary hypoblast determine the location and direction of the future primitive streak (Fig. 5-12) by a form of inductive interaction.

---

[1]Even though just a single spermatozoon takes part in fertilization in birds, other spermatozoa sometimes become lodged in the cytoplasm of the blastodisk. The nuclei of these spermatozoa migrate to the periphery, where they are recognizable for some time before they degenerate.

[2]In contemporary descriptive and experimental literature, the terms *epiblast* and *hypoblast* are almost universally used to designate the inner and outer layers of the blastoderm of birds. This convention will be followed here in the designation of the layers in early bird embryos.

The two-layered blastoderm of the bird has been compared with a flattened amphibian blastula. The epiblast is considered the equivalent of the animal hemisphere, and the primary hypoblast shares many common properties with the vegetal hemisphere of the amphibian embryo, particularly its intrinsic polarity and the ability to induce the formation of mesoderm from the early epiblast. The epiblast, on the other hand, remains competent to react to the influence of the hypoblast by forming the mesoderm (*mesoblast*).

CLEAVAGE AND FORMATION OF THE BLASTULA IN MAMMALS

Despite their presumed evolutionary origin from species that laid eggs with large amounts of yolk, mammals produce extremely small eggs with almost no yolk. Freed from the encumbrance of yolk, the mammalian egg has reverted to the simple type of cleavage seen in many primitive forms. Early cleavage divisions are practically unmodified mitoses, or in more technical terms, *equal holoblastic cleavage of an isolecithal egg*. Only later in development does the pattern of morphogenetic movements of the mammalian embryo reveal a persistence of traits characteristic of large-yolked embyros.

Historically, early descriptive studies on cleavage were largely conducted in pig embryos, which were easily obtained from slaughterhouses. In recent years, however, attention has been directed toward primate and mouse embryos, the former because of the importance of in vitro fertilization techniques in humans and the latter because of the development of methods allowing studies of cell lineages and cell determination. Unfortunately for the student, several aspects of cleavage and gastrulation in mice are topographically different from those in other mammals, but because of the great importance of mice in studies of early mammalian development, the early development of both mice and more typical mammals will be presented here.

Cleavage in mammals is much slower than it is in most other vertebrates. Commonly, the first cleavage division is not completed for 24 hours, and subsequent early divisions take about 10 to 12 hours each. The point at which the polar bodies are given off at least establishes a point of reference, and the plane of the first cleavage division includes the polar bodies (Fig. 4-13A). The second cleavage division may not occur simultaneously in both blastomeres. This results in the temporary appearance of a three-cell stage. In many mammals, the mitotic spindle of one of the blastomeres rotates 90° during the second cleavage division (Gulyas, 1975). This results in a crosswise arrangement of the blastomere at the four-cell stage (Fig. 4-14). In vitro studies of both monkey and human embryos have shown that early cleavage in primates follows the general mammalian pattern.

A critical stage called *compaction* takes place at the eight-cell stage in the mouse. During compaction the blastomeres flatten and become tightly joined so that they cannot become distinguished from one another with the light microscope. The intercellular connections, which are the dominant feature of compaction (Fig.

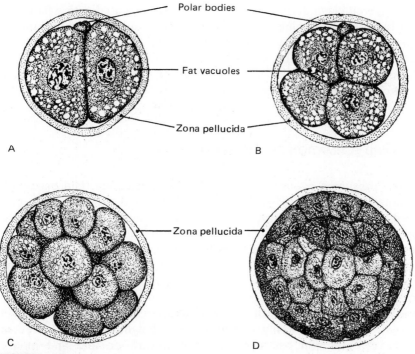

FIGURE 4-13
Cleavage of the pig ovum. (A) 2-cell stage. Specimens secured from the oviduct of a
sow killed 2 days, 3½ hours after copulation. (B) 4-cell stage. Probable age about 2½
days. (C) A morula of about 16 cells. Drawn from unsectioned specimen, probable age
about 3½ days. (D) Blastula stage, drawn from a specimen secured from the uterus of
a sow killed 4¾ days after copulation. Note the lighter center area indicating the beginning of the formation of the segmentation cavity (blastocoel) by cell rearrangement
(cf. Fig. 4-15). (A, B, and D, drawn from preparations loaned by Streeter and Heuser;
C, after Assheton. All figures ×400.)

4-16), serve two purposes. Tight junctions prevent the free exchange of fluid
between the inside and the outside of the embryo, allowing the accumulation of a
fluid with special properties inside the embryo (Ducibella and Anderson, 1975).
Gap junctions couple all the blastomeres of the compacted embryo and permit the
exchange of ions and small molecules from one cell to the next.

By the 16-cell stage the embryo, still enclosed in the zona pellucida, is said
to be in the morula stage (Fig. 4-13C). Although the term *cleavage* is not
ordinarily applied to cell divisions which occur after the morula stage, cell
division occurs with unabated rapidity. In the morula the internal secretion of
fluid by the blastomeres leads to the formation of a well-defined central cavity
called the blastocoel, or more properly, the *blastocyst cavity* (Denker, 1983)
(Fig. 4-15). The transition from morula to blastula is marked by two obvious
changes. The first is a rapid enlargement of the blastocyst cavity; the second is
the emergence of distinctly different types of cells within the embryo.

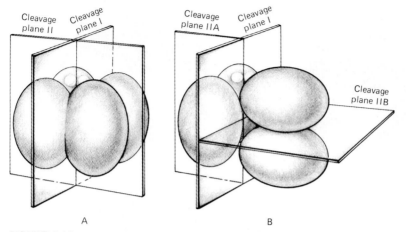

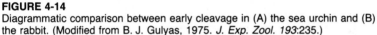

FIGURE 4-14
Diagrammatic comparison between early cleavage in (A) the sea urchin and (B) the rabbit. (Modified from B. J. Gulyas, 1975. *J. Exp. Zool. 193*:235.)

The fluid that is secreted into the cavity of the *blastocyst*, as the mammalian blastula is called, is derived from cytoplasmic vesicles which accumulate in the blastomeres. Studies of the composition of this fluid reveal differences in the concentrations of many inorganic ions and other substances between the fluid inside and the fluid outside the embryo, an indication of active transport processes (Borland et al., 1977). In addition, *uteroglobin* and other uterine proteins have been found within the blastocyst fluid of rabbit embryos (Beier and Maurer, 1975). The early mammalian blastocyst remains enclosed within the zona pellucida, but the overall size of the embryo increases to some extent because of the accumulation of fluid.

As the blastocyst becomes filled with fluid, it becomes readily apparent that the embryo consists of two distinct populations of cells. The cells that constitute the outer wall of the blastocyst, collectively called the *trophoblast*, have assumed the configuration and many of the properties of epithelial cells. Functionally, the cells of the trophoblast can both pump fluid and induce special changes in the uterine lining upon implantation. One unusual feature of trophoblastic cells is that maternal genes are preferentially expressed and paternal genes are inactivated (Fig. 3-41). On the inner surface of the tropho- blast is a small group of cells called the inner cell mass (Fig. 4-15). These cells are joined to one another by communicating gap junctions (Fig. 4-16), and they retain the ability to reaggregate if separated or mixed with the cells of other embryos. Cells of the inner cell mass can neither pump fluid nor evoke the decidual reaction (see p. 277), as can the cells of the trophoblastic layer. They are destined to form the embryo plus some of the membranes associated with it, whereas the cells of the trophoblast form a large part of the placenta (Fig. 5-24).

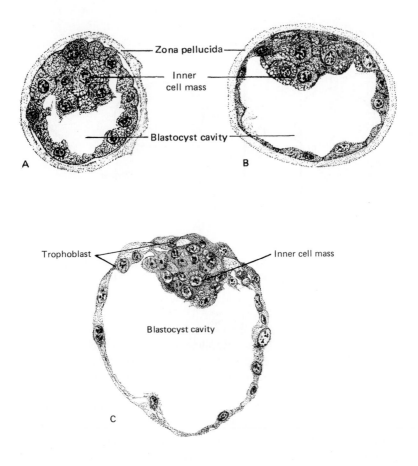

FIGURE 4-15
Three stages of the blastodermic vesicle (blastocyst) of the pig, drawn from sections to show the formation of the inner cell mass. (A) Removed from uterus of sow 4¾ days after copulation (cf. Fig. 4-13D). (B) Copulation age, 6 days, 1¾ hours. (C) Copulation age, 6 days, 20 hours. (A, B, from embryos in the Carnegie Collection; C, after Corner, all ×375.)

The results of experimental investigations have led many embryologists to conclude that the position of a blastomere in the morula determines whether it will become part of the trophoblast or the inner cell mass. According to the "inside-outside" hypothesis (Tarkowski and Wróblewska, 1967), cells of the morula which have no contact with the exterior develop in a unique microenvironment created by the external cells (Fig. 4-17). These cells develop into the inner cell mass, whereas the cells located at the surface of the morula, presumably because of physiological functions required by their superficial

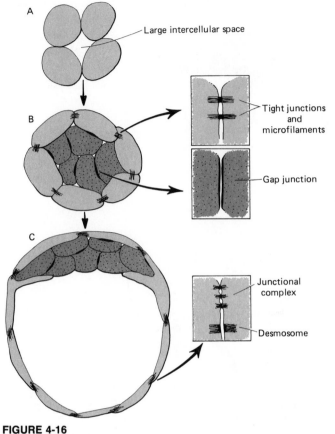

FIGURE 4-16
Cell junctions during cleavage of the mouse embryo. (Adapted from
Denker, 1983.)

position, are channeled into becoming trophoblastic epithelium. Tight junctions
between outer cells are thought to preserve the environmental differences
between outer and inner cells (Fig. 4-16). An alternate hypothesis, the
polarization or cytoplasmic segregation hypothesis, relates the differentiation
of the two cell types to gradients of cytoplasmic determinants which become
segregated into internal or external blastomeres as cleavage divisions proceed
(Gardner, 1983).

In keeping with the relationship of mammals to birds, the embryos of
mammals form a layer of cells beneath the inner cell mass that seems to be
equivalent to the primary hypoblast of avian embryos. This layer, which is
called the *primitive endoderm*,(Fig. 4-18), does not contribute to the embryo
proper, but cells from the primitive endoderm in the mouse contribute to the
formation of the yolk sac (Fig. 5-31).

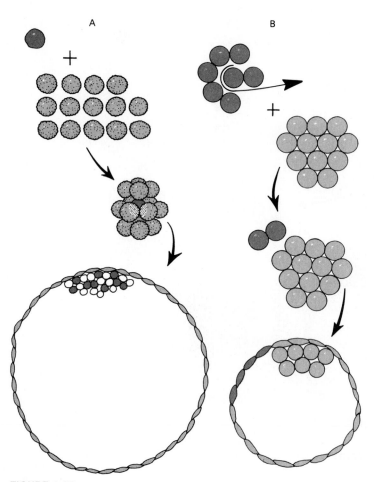

FIGURE 4-17
Experiments illustrating the importance of position in determining cell fate in the early mouse embryo. (A) If an entire 8- to 16-cell embryo (dark gray) is surrounded by 14 other embryos (light gray), the labeled single embryo becomes part of the inner cell mass of the giant blastocyst that results. (B) If two labeled blastomeres (dark gray) are placed on the periphery of another embryo, the donor cells normally become part of the trophoblast, but not inner cell mass. (Adapted from Denker, 1983.)

ORGANIZATION AND PROPERTIES OF THE EMBRYO DURING CLEAVAGE AND BLASTULATION

During cleavage and blastulation the embryo is simple in structure, and alterations in its cell arrangement are straightforward. However, many properties of the embryo and its constituent cells change during this period. These changes are not for the most part detectable by observation alone. To learn

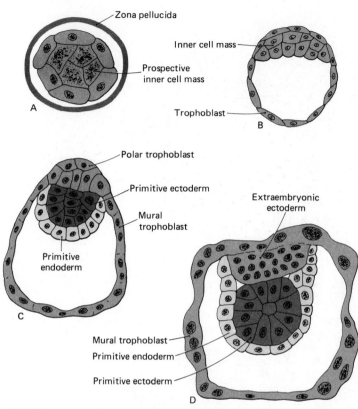

FIGURE 4-18
Early development of the mouse. (A) Morula. (B) Blastocyst stage. (C) Stage of first appearance of primitive endoderm (hypoblast). (D) Early appearance of extraembryonic ectoderm.

about the organization of early embryos, the investigator must be able to ask incisive questions and carry out delicate manipulations.

Following the epoch-making experiments of Roux (see Chap. 1), other researchers experimentally investigated the developmental capacities of single blastomeres of embryos. One of the most informative early analyses was carried out by the German biologist Han Driesch in the early years of this century. Driesch isolated blastomeres from two- and four-cell sea urchin embryos and found that a perfect pluteus larva formed from each blastomere (Fig. 4-19). From these observations Driesch formulated the important developmental principle that the *prospective potency* of the early blastomeres is greater than their *prospective fate*; in other words, a given blastomere possesses the developmental potential to form a wider array of structures than it would if it were left in its normal place in the embryo. Even in contemporary studies on cell lineages in mammalian embryos, investigators must take this principle into account in interpreting their results. The property of a part being

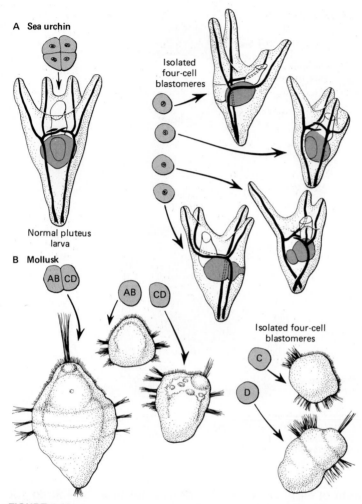

A Sea urchin

Isolated
four-cell
blastomeres

Normal pluteus
larva

B Mollusk

AB CD

AB   CD

Isolated four-cell
blastomeres

C

D

FIGURE 4-19
(A) Regulative properties of the 4-cell sea urchin embryo. Left—normal
development. Right—Isolated blastomeres of the 4-cell embryo each give
rise to a normal pluteus larva. (After Hörstadius, 1939). (B) Mosaic properties of the mollusk, *Dentalium*. Left—normal development. Right—Isolated blastomeres of 2- or 4-cell embryos give rise to the incomplete embryo. (After Wilson, 1904).

able to form a whole is a good example of embryonic *regulation*, and Driesch
called systems possessing this property *harmonious equipotential systems*.
Trying to explain the basis for regulation on a mechanistic basis so frustrated
Driesch that he turned toward *vitalism* (the control of life processes by
intangible forces) to explain the phenomenon. He ultimately left experimental
biology to become a professor of philosophy, but he remains one of the giants
of early experimental embryology.

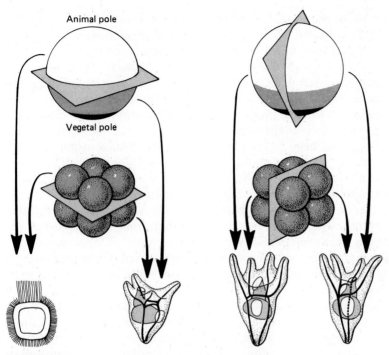

FIGURE 4-20
Mosaic and regulative properties demonstrated in the sea urchin embryo.
Left—Equatorial division of the egg or 8-cell embryo into animal and vegetal
halves results in the formation of unequal progeny (mosaic properties). The
animal half (far left) forms a ciliated *Dauerblastula* (permanent blastula),
whereas the vegetal half forms a mildly abnormal embryo. Right—Meridional
division of the egg or 8-cell embryo into equal halves results in complete regulation and two normal progeny. (Modified from Gilbert, *Development Biology,*
Sinauer, 1985, after Hörstadius).

Other investigators, working on the embryos of a variety of invertebrates,
found that not all types of embryos are capable of compensating for defects by
regulation. Another quite different response to defects is exhibited in groups
such as mollusks, in which an isolated blastomere forms little more than what
it would have formed if left in situ (Fig. 4-19). This type of development is
known as *mosaic* development. Although initially it was tempting to categorize
embryos as being either regulative or mosaic, it soon became apparent that the
same embryo can exhibit both properties. An excellent example is the sea
urchin. Simply slicing either the egg or the embryo into two parts through a
plane including the animal and vegetal poles results in the parts undergoing
good regulation and giving rise to normal larvae (Fig. 4-20). If, however, either
eggs or embryos are bisected through the equatorial plane, the progeny of the
two halves are different and exhibit mosaic properties (Fig. 4-20). This suggests
that there are major differences in the intracellular environment between the
animal and vegetal hemispheres.

Further analysis of the organization of the 64-cell stage of the sea urchin (Hörstadius, 1939) revealed some fascinating interrelationships within the embryo. If the cells of the animal hemisphere are separated from the vegetal cells and allowed to develop, they form a highly ciliated *"Dauerblastula"* (permanent blastula), which does not develop much further. If, however, the animal hemisphere is combined with single rings of mesomeres or the micromeres, development becomes progressively more complete the closer the origin of the vegetal cells is to the vegetal pole (Fig. 4-21). The most extreme combination, the animal hemisphere plus micromeres, results in the development of a remarkably normal looking pluteus larva (Fig. 4-21). These and similar experiments were interpreted as indicating the presence of a double gradient system within the cleaving embryo. An animalizing gradient with the highest concentration at the animal pole pushes the system toward forming "animal" structures, whereas a vegetalizing gradient that is strongest at the micromeres has a converse effect. When various tiers of blastomeres are recombined, the morphology of the resulting larva is based on the proportional strengths of the animalizing and vegetalizing gradients brought together by the recombined blastomeres.

The blastomeres of young vertebrate embryos, like those of the early sea urchin, are characterized by an astonishing amount of plasticity in their developmental fate. Some of the most incisive experimental work demonstrating this point has dealt with the properties of nuclei rather than with those of entire cells. In an early attempt to determine whether the nuclei of blastomeres retain the potential to guide the entire course of development or whether their capacities become restricted, Spemann (1928) constricted a fertilized amphibian egg with a hair loop so that one lobe of the egg contained the nucleus and some cytoplasm and the other lobe contained only cytoplasm. When the nucleated portion had developed to the 16-cell stage, he loosened the ligature and allowed a nucleus to move over into the lobe of nonnucleated cytoplasm (Fig. 4-22). The secondarily nucleated lobe of cytoplasm then went on to develop, thus demonstrating that the nucleus of a 16-cell embryo had still not lost its capacity to direct embryonic development from the stage of the single cell. More recent work (Briggs and King, 1952) involving the transplantation of nuclei taken from blastomeres during cleavage into enucleated eggs has confirmed this point. The furthest extension of this approach has been the work of Gurdon (1974), who obtained viable frogs from nuclei of adult frogs transplanted into enucleated eggs (Fig. 1-27).

The capacity of isolated entire blastomeres to continue development and to regulate into entire individuals has been demonstrated in several ways. Identical twins can result from the regulation and normal development of blastomeres that have become separated during early cleavage. Blastomeres isolated from mammalian embryos show a steadily decreasing ability to form complete individuals from the two-cell stage to the eight-cell stage (Pederson, 1986). In the mouse, blastomeres of four-cell embryos have been shown to be totipotent in experiments in which individual blastomeres or pairs of cells

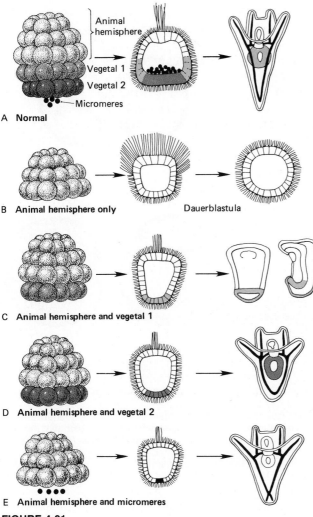

FIGURE 4-21
Experiments demonstrating the importance of a vegetal influence
in the completeness of regulation of components of sea urchin
embryos. (B) The animal hemisphere alone forms a ciliated ball
called a *Dauerblastula*. (C–E) When progressively more vegetal
rings of cells are added to the animal hemisphere, development
becomes progressively more complete. (Adapted from Hörstadius,
1939).

derived from a single blastomere have been combined with blastomeres of
genetically different hosts (Kelly, 1977). Totipotency in this case means that
these cells have been shown to be capable of entering either the trophoblastic
or the inner cell mass lineage. At least some cells of the early inner cell mass
remain totipotent, but for technical reasons it has not been possible to

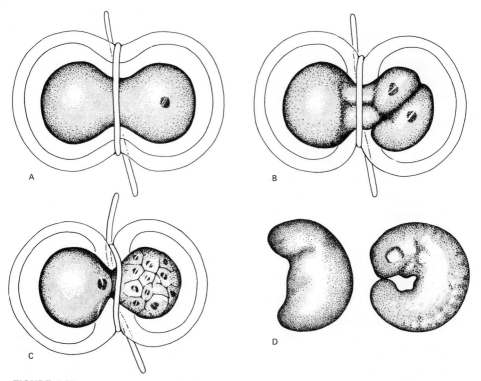

FIGURE 4-22
Summary of Spemann's constriction experiment on newt eggs. Shortly after fertilization the egg is constricted with a hair loop so that one segment of cytoplasm contains a nucleus and the other does not (A). During early cleavage, the nucleated segment divides (B). If, after several divisions a nucleus is allowed to migrate to the nonnucleated segment of cytoplasm (C), this half of the constricted egg also begins to form an embryo which is essentially normal (D, left) but less mature than the embryo arising from the originally nucleated segment of the constricted egg (D, right). (Parts B–D after Spemann, 1938. *Embryonic Development and Induction*, Yale University Press, New Haven.)

determine whether early trophoblastic (or *trophectoderm* cells, as they are called in the mouse) are capable of forming inner cell mass derivatives.

At a different level of organization, pieces of many vertebrate embryos are able to reconstitute entire individuals by regulation. Spratt and Haas (1964) cut chick blastoderms containing as many as 30,000 cells almost in half. Each half underwent regulation and formed a complete chick embryo (Fig. 4-23). Once the mammalian blastocyst has become subdivided into the trophoblast and inner cell mass, these two regions cease to be equivalent, but partial or complete separation of portions of the inner cell mass results in twinning (Fig. 1-21). The nine-banded armadillo regularly produces quadruplets from a single early embryo. At the blastocyst stage, the inner cell mass forms four buds, each of which becomes organized into a separate embryo (Patterson, 1913).

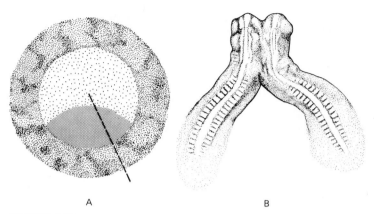

A   B

FIGURE 4-23
Regulation in the early chick blastoderm. If the posterior part of the blasto-
derm is cut (A) and does not heal, partially conjoined twin embryos form
(B). (Drawn from Spratt and Haas, 1967. *J. Exp. Zool. 164*:31.)

A valuable experimental model for both embryological and genetic studies makes use of another regulative property of mammalian embryos. Both Tarkowski and Mintz have produced *tetraparental* (*allophenic*) mice by removing the zona pellucida from cleaving embryos and then combining the two embryos. The cells become reorganized into a single sphere, which is then transferred to the uterus of a foster mother. The rest of pregnancy proceeds normally, and mosaic embryos containing the genetic input of four parents are produced (Fig. 4-24). In this case the regulative properties of the embryos allowed the harmonious integration of two potentially separate individuals into one embryo. More recently, adult sheep-goat chimeras (Fig. 4-25) have been produced by combining blastomeres from early sheep and goat embryos and injecting them into the trophoblastic shell (from which the inner cell mass had been removed) of another individual—even a pig (Fehilly et al., 1984; Mei-necke-Tillmann, 1984).

At the 64-cell stage, the inner cell mass of a mouse blastocyst contains about 15 cells. Statistical studies based on the frequency of appearance of mosaic characteristics in tetraparental mice have provided strong indications that the body of the mouse arises from as few as three cells of the inner cell mass. These inferences are supported by other experiments in which chimeric mice have been produced by injecting single cells from one embryo into the blastocyst of another. The injected cells often fuse with the inner cell mass and become integrated into the body of the host embryo. The use of techniques such as those mentioned here is providing an immense amount of new information on the organization of early mammalian embryos (Rossant and Pederson, 1986).

Another unseen but important aspect of the organization of embryos during cleavage and the blastula stage is their polarity. Aspects of the initial development of polarity in amphibians and avian embryos were introduced in Chap. 3.

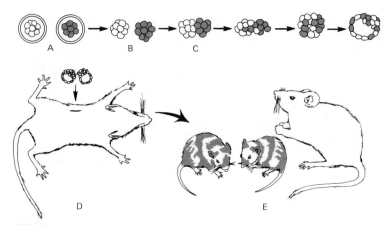

FIGURE 4-24
Diagram of the experimental procedure for producing tetraparental (allophenic) mice. Two cleavage stage embryos are obtained (A). After removal of the zone pellucida (B), the embryos are allowed to fuse (C). The fused embryos are implanted into a foster mother (D) who later gives birth to the chimeric offspring (E). (Modified from B. Mintz, 1962. *Proc. Natl. Acad. Sci. 58*:334.)

In both amphibians and birds initial control of polarity appears to reside in the vegetative regions of the embryos (Nieuwkoop, 1977). In birds, early control of polarity resides in the primary hypoblast. Experimental reorientation of the hypoblast in relation to the overlying epiblast results in an orientation of the embryo that corresponds to the position of the hypoblast. In contrast to amphibians and birds, the mammalian embryo shows no signs of polarity (except for the position of the polar body) until well into the stage of the blastocyst. The circumstances and mechanisms of fixation of polarity in mammalian embryos have received relatively little attention.

MOLECULAR EVENTS DURING CLEAVAGE AND BLASTULATION

Well before the era of molecular biology it became apparent that cleavage can take place in the absence of expression of the embryonic genome. This was most definitively demonstrated by experiments in which enucleated fragments of sea urchin eggs were produced (Harvey, 1936). When these egg fragments, called *merogones*, were parthenogenetically activated, cleavage divisions showing a remarkable degree of regularity ensued, and structures showing the external features of blastulae, but without a blastocoel, developed. Similar experiments have been conducted on frog eggs with similar results.

The next major step in understanding the overall molecular dynamics occurred in the early 1960s when a number of specific molecular inhibitors derived from soil-dwelling funguslike organisms became available. When

FIGURE 4-25
A sheep-goat chimaera produced by combining blastomeres of early sheep and goat embryos. By surrounding goat blastomeres with sheep trophocetoderms before introducing the mosaic embryo into the uterus of a sheep host, the incidence of spontaneous abortion (host rejection of the mosaic fetus) is reduced. (Courtesy of C. Fehilly-Willadsen, from Fehilly et al., 1984.)

actinomycin D, which in certain doses acts as a general inhibitor of RNA synthesis, was administered to sea urchin eggs, cleavage continued unabated until the embryo had approached the early blastula stage (Gross and Cousineau, 1964). In contrast, when sea urchin or amphibian embryos were exposed to *puromycin* or other inhibitors of protein synthesis, cleavage quickly ceased. The general conclusion that could be derived from these experiments was that RNA synthesis is not required for early cleavage divisions to occur but that in the absence of protein synthesis further development comes to a rapid halt. The logical inference from these early results was that in the fertilized eggs there must be preformed stores of RNAs which contain sufficient information to guide development through the early stages of cleavage.

This general conclusion was supported by studies on an anucleolate mutant in *Xenopus*. As the name implies, homozygous embryos lack nucleoli and are consequently unable to synthesize ribosomal RNA. Despite their inability to make ribosomes, these anucleolate embryos develop normally until the hatching stage and are able to begin swimming movements, at which time they begin to die. These embryos survive as long as they do because of the immense supply of maternally derived ribosomes that was generated in the egg before fertilization.

Evidence that maternally derived rather than embryonic RNAs preside over cleavage has been derived from genetic studies on a number of invertebrate groups. A classic example is the maternal influence on the orientation of the mitotic spindle during cleavage of the snail (*Limnea*), which in mutants ultimately results in the formation of left-handed (*sinistral*) coiling shells rather than the usual right-handed (*dextral*) variety (Boycott et al., 1930). Well over 30 mutants in *Drosophila* show effects of the maternal genome upon early cleavage.

A number of investigations of early development in hybrid embryos, particularly sea urchin embryos, showed that many features, such as the rate of cleavage, follow the instructions of the maternal rather than the paternal genome (rev. by Davidson, 1976). In echinoderm hybrids, there is little evidence of expression of the paternal genome until early gastrulation, when the primary mesenchyme has formed.

Direct studies of RNA synthesis in amphibians seemed to confirm the conclusions derived from the indirect studies summarized here (rev. Gurdon, 1974). Very little RNA of any type is produced until just before the blastula stage, when there is a burst of synthesis of both mRNA and tRNA (Fig. 4-27). As gastrulation begins, members of the ribosomal RNA family become

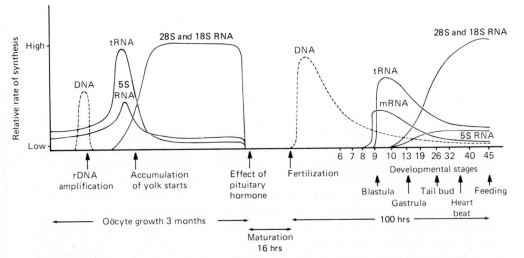

FIGURE 4-26
Diagram of relative rates of nucleic acid synthesis during amphibian development. 5s, 18s, and 28s RNAs are forms of rRNA. The different patterns of RNA synthesis during development constitute evidence in favor of controls operating at the level of transcription. (Modified from J. B. Gurdon, 1974. *The Control of Gene Expression in Animal Development*, Harvard University Press, Cambridge.)

synthesized in increasing amounts. These findings correlate closely with the results of studies with inhibitors and the anucleolate mutant in amphibians.

Direct studies of RNA synthesis in echinoderm embryos, however, revealed a quite different set of events. As early as the third or fourth cleavage division, the synthesis of mRNA and heterogeneous nuclear RNA begins. New rRNA and tRNA appear around the time of the blastula. Overall, there is a steady increase in embryonic RNA synthesis until the blastula stage, at which point it begins to level off.

Despite the relatively early synthesis of embryonic RNAs in the sea urchin, it is the maternal mRNAs that are expressed during cleavage. Maternal mRNAs become unmasked within 30 minutes after fertilization. They soon join with stored-up ribosomes to form polysomes upon which proteins needed for cleavage can be synthesized. One class of such proteins is the *cyclins* (Evans et al., 1983). These proteins, which are coded for by maternal RNAs, are synthesized after each cleavage division and are accumulated at high levels by the middle of the cell cycle, only to be destroyed with the succeeding round of cleavage divisions. As the time of blastulation approaches, the prominence of the cyclins decreases. Another prominent class of proteins that is synthesized on maternal messages during early cleavage is the histones.

At the blastula stage in both sea urchins and amphibians, the activity of maternal RNAs declines and translation from embryonic mRNAs begins. The factors that regulate the translation of mRNAs in these embryos are incompletely understood. Several possibilities are currently under investigation. One, called the masked messenger hypothesis, posits that the mRNAs are combined with proteins in such a manner that the message is physically blocked. Another possibility is that either the mRNAs or the ribosomes are somehow sequestered so that they cannot combine to form polysomes. A third possibility is that changes in physical conditions within the cell, such as pH or ionic composition, affect the efficiency of the translational mechanism.

One interesting mutant in amphibians, the *o* mutant of the axolotl, provides a tool for investigating the transition between the maternal and embryonic control of early development. In this maternal effect mutant, females with the *o/o* genotype produce eggs that undergo normal cleavage and blastulation, but their cleavage slows and development stops shortly after the first traces of the dorsal lip of the blastopore are seen in gastrulation (Fig. 4-27). However, if a small amount of cytoplasm from a normal blastomere is injected into a mutant egg, the developmental block is overcome and normal development ensues. This suggests that some maternal gene product is involved in stimulating or permitting the switch from maternal to embryonic control of development.

The pattern of molecular activities of early development in mammals differs from that in the sea urchin and other vertebrates that have been studied (Schultz, 1986). As a generalization, one can say that there is less advance storage of macromolecules during oogenesis and a correspondingly earlier activation of the embryonic genome in mammals than there is in other forms. This pattern can be accommodated because of the leisurely pace of early

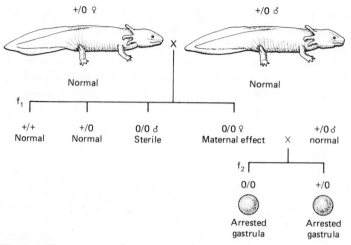

Figure 4-27
Genetics of the recessive maternal effect *o* mutant in the axolotl. Regardless of their own genotype (*o/o* or *+/o*), eggs from *o/o* females (f₂) become arrested during gastrulation. In contrast, all
eggs derived from *+/o* females develop into adults (f₁) even though the embryos themselves
may have an *o/o* genotype. *o/o* males are sterile. (Adapted from Brothers, 1979).

mammalian cleavage. Whereas in the first 24 hours after fertilization *Xenopus*
embryos have developed to a gastrula containing approximately 10,000 cells
and the sea urchin embryo is a 500-cell blastula ready to hatch, the mouse
embryo has only passed through the first cleavage division to the 2-cell stage.
With such a rate of cleavage, the mammalian blastomeres have sufficient time
to synthesize components needed for the next cleavage division from the
embryonic genome; this is impossible in rapidly cleaving embryos.

In the mouse, enough informational macromolecules are accumulated in the
developing oocyte to guide the zygote through the first cleavage division, but a
rapid switching over from control by the maternal to the embryonic genome
begins at the two-cell stage. This conclusion was reached by demonstrations of
(1) a high rate of degradation of maternal RNAs, (2) a burst of synthesis of new
RNAs, (3) blockage of development past the two-cell stage by the transcriptional inhibitors actinomycin D and *alpha-amanitin*, (4) changing patterns of
polypeptide synthesis, and (5) results of crosses of parents with differing
variants of key enzymes. Correlated with this pattern of early embryonic
synthesis is the presence of nucleoli during the early cleavage period.

Patterns of protein synthesis in the early mammalian embryo are complex.
During the first cleavage division, there is evidence for protein synthesis that
would occur whether or not fertilization had occurred and the synthesis of
other proteins that occurs only as a result of fertilization. Most of these early
proteins are made from maternal mRNAs. As might be expected, the synthesis
of *histones* is prominent during cleavage. There is little evidence that any
maternal mRNAs guide protein synthesis after the four-cell stage. This means

that the process of cavitation, compaction, and formation of the blastocyst is under the control of the embryonic genome. Overall, the rates of protein synthesis are low until about the eight-cell stage. Thereafter there is a sharp increase in the production of ribosomes and in protein synthesis.

Mammalian embryos differ from amphibian embryos in their energy requirements. Of necessity, the amphibian embryo must be a self-contained unit with respect to energy sources, but mammalian embryos depend on a continuing supply of energy from their surroundings in both the uterine tubes and the uterine cavity.

WHAT IS ACCOMPLISHED BY CLEAVAGE AND BLASTULATION?

It is easy to consider the period of cleavage as a relatively simple phase of development during which the embryo is simply increasing its number of cells. Certainly the increase in cell number is the dominant manifestation of cleavage. Without a critical number of cells, the morphogenetic movements that form the basis of gastrulation, the next major period in embryonic development, could not be accomplished in a normal fashion. Implicit in the increase in cell number is the fact that there is a corresponding increase in the amount of genetic material (DNA) that the embryo has at its disposal. The formation of many separate cells also allows the segregation of different types of cytoplasm which may have been present in the egg (e.g., the germ plasm in frogs; see Fig. 3-1). Cytoplasmic influences can be important in evoking or allowing differences in expression of the embryo's genetic content. This, however, cannot be accomplished until the embryo has undergone the transition from relying on maternally derived mRNAs to producing transcripts from its own genome. The cytodifferentiation that follows the formation of embryonic mRNAs is facilitated by the decrease in cell size toward a more typical nucleocytoplasmic ratio. This allows greater control over intracellular events.

At a higher level of complexity, the major embryonic axes of most embryos are definitively fixed by the late blastula stage, although the axes of mammalian embryos remain labile longer than those of most other vertebrate embryos. Finally, during this period the cells in some embryos make major decisions regarding their future fate. The most striking example is the mammalian embryo, in which the cells become segregated very early into those which will form the embryo proper and those which will produce the extraembryonic structures. As the blastula becomes more mature, its activities are increasingly directed toward making preparations for gastrulation. These preparations are reflected not only in changing patterns of synthesis but also in surface properties and interactions of cells. In some cases they are even reflected in the movements of cells in relation to others. Thus, as with most phases of development, no sharp demarcation exists between the last blastula stages and the start of gastrulation.

GASTRULATION AND THE ESTABLISHING OF THE GERM LAYERS

GASTRULATION AS A PROCESS

Following the period of cleavage and the formation of a blastula, the embryo embarks on one of the most critical periods in its development—the stage of *gastrulation* (*gastrula*, diminutive from the Greek word *gaster*, meaning *stomach*). Up to this time the course of development in most animals has been directed by maternally derived instructions and processes that took place in the egg before fertilization. The cells of the cleaving embryo, although increasing greatly in number, have not for the most part begun to express specific properties which are unique or irreplaceable. During the period of blastulation the genetic reservoirs of the cells are activated by poorly understood mechanisms that direct the synthesis of substantial amounts of new RNA and proteins, and the transition from control by the maternal genome to the embryonic genome begins. At about the same time the rate of cleavage divisions slows considerably (Fig. 4-1). In most embryos there is little growth or change in mass as gastrulation begins.

Gastrulation is characterized by profound but well-ordered rearrangements of the cells in the embryo. One of the major changes that occurs during early gastrulation is the acquisition by the cells of the capacity for undergoing directed *morphogenetic movements*, which often result in a reorganization of the entire embryo or smaller regions within it. The term *morphogenetic movements* encompasses a variety of specific behavioral patterns of cells. Some of the more common types of morphogenetic movements are outlined in Table 5-1. The major morphogenetic movements during gastrulation result in the rearrangement of the embryo from the blastula to a stage characterized by

TABLE 5-1
MAJOR TYPES OF MORPHOGENETIC MOVEMENTS SEEN IN EARLY EMBRYOS

Type of movement	Description	Example
Invagination	Inpocketing of a sheet of cells	Archenteron formation in Amphioxus (Fig. 5-1A-D)
Evagination	Outpocketing of a sheet of cells	Exogastrulation (Fig. 6-3)
Involution	Rolling around a corner of an expanding outer layer of cells and spreading over an internal surface	Cell movements through the amphibian blastopore (Fig. 5-9)
Epiboly	Spreading of a cell sheet	Spreading of outer cells toward amphibian blastopore (Fig. 5-8B)
Ingression	Sinking of individual cells from a surface into an area	Primary mesenchyme formation in sea urchin blastula (Fig. 4-7)
Delamination	Separation of a second sheet from an original single sheet	Formation of primary hypoblast in avian embryos (Fig. 4-10C and D)
Ameboid motion	Migration of cells as single individuals through their own motility	Migration of neural crest cells (Fig. 6-16)

the presence of three germ layers, but other morphogenetic movements continue into later stages. One of the major consequences of this reorganization is that groups of cells which may have been far removed from one another are brought close enough together to undergo the inductive interactions that are involved in the establishment of the major organ systems.

Embryos have adopted two main strategies for dealing with gastrulation. The first is to carry out the gastrulation movements within the context of a sphere. This mode of gastrulation is seen in the embryos of very primitive vertebrates such as *Amphioxus*, which contain little yolk, and the amphibians, which contain moderate amounts of yolk. In *Amphioxus*, gastrulation consists of an inpocketing (invagination) of a blastula, forming a double-layered cup from a single-layered hollow sphere in much the same way that a hollow rubber ball can be pushed in with one's thumb (Fig. 5-1A to 5-1D). The new cavity in the double-walled cup is termed the gastrocoel or archenteron.

The opening from the outside into the gastrocoel has traditionally been called the *blastopore*. Thus in gastrulation the original single layer of the blastula has been rearranged to form two layers. The outer cell layer is known as the *ectoderm*. The inner layer of the early gastrula is composed primarily of cells belonging to the *endoderm*, but its upper surface also includes cells destined to form the middle germ layer (*mesoderm*).

The second principal way in which embryos handle gastrulation is dictated by the great amount of yolk found in large eggs, such as those of reptiles and birds. In embryos of this type the sheer bulk of the inert yolk mass precludes the simple inpocketing mechanism used by *Amphioxus*. Instead, gastrulation occurs by means of the elaboration of the three germ layers as two-dimensional

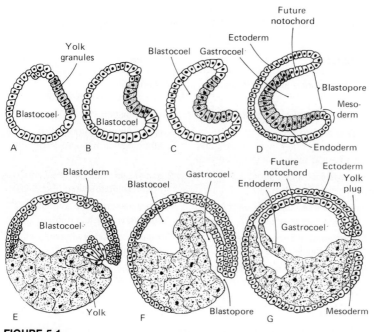

FIGURE 5-1
Schematic diagrams to show the effect of yolk on gastrulation. (A–D) *Amphioxus*. (E–G) Frog.

sheets upon one sector of an enormous sphere of completely passive yolk. Interestingly, mammalian embryos betray their origin from phylogenetic ancestors which laid highly yolky eggs by retaining the type of gastrulation movements common to birds and reptiles.

GASTRULATION IN SEA URCHIN EMBRYOS

The morphogenetic movements that characterize gastrulation begin in the late blastula stage with the separation of the primary mesenchyme from the wall of the blastula (Fig. 4-7). The 50 or so ingressed primary mesenchymal cells develop prominent projections called *filopodia*; by extending and retracting the filopodia, they move along the basal lamina that lines the blastocoel until they stop and form a loose ringlike structure near the base of the invaginating archenteron (Fig. 5-2). The behavior of these migrating cells strongly suggests that they are testing with their filopodia specific properties of the substrate over which they migrate (Gustafson and Wolpert, 1967). Transplantation studies have shown that primary mesenchymal cells injected into blastocoels of hosts of different species or different ages home toward the appropriate location unless the host is too old or of a distant species (Fig. 5-3). Studies with monoclonal antibodies have shown that unique antigens appear on the surface

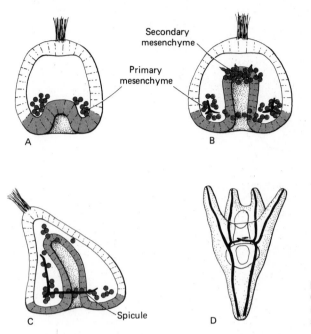

Secondary
mesenchyme

Primary
mesenchyme

Spicule

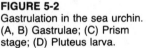

FIGURE 5-2
Gastrulation in the sea urchin.
(A, B) Gastrulae; (C) Prism
stage; (D) Pluteus larva.

of the primary mesenchymal cells as they are developing their specific patterns
of behavior (McClay and Wessel, 1985). During their migratory phase, the
primary mesenchymal cells show little affinity for one another, but after they
have moved into the subequatorial ring, their long filopodia join into cablelike
structures. Calcified spicules form on these cables, and they serve as the basis
for the internal skeleton of the sea urchin (Fig. 5-2).

The main feature of gastrulation in the sea urchin is the formation of the
archenteron, or primitive gut (Fig. 5-2). This occurs in two readily recognizable
phases. The first stage is dominated by the inpocketing, or *invagination*, of cells
at the vegetal pole to form the early archenteron (Fig. 5-2). The indentation
made by this inpocketing is called the *blastopore*. The early stages of
invagination have commonly been attributed to intrinsic changes in cell shape
along with a certain degree of motility of the cells involved. After a pause, the
second phase of archenteron formation is characterized by the presence of a
population of *secondary mesenchymal cells* which become distinguishable at
the innermost tip of the archenteron (Fig. 5-2). These cells extend long filopodia
toward the opposite wall of the blastula. It has long been recognized that the
filopodia probe the inner surface of the wall of the blastocoel and ultimately
make stable contacts at a given region near the animal pole (Fig. 5-2). The
filopodia then contract, presumably pulling the archenteron along behind them.
A recent study by Hardin and Cheng (1986) has cast doubts on the assumption
that the shortening of the filopodia provides the main motive force to complete
the second phase of elongation of the archenteron. Using biomechanical

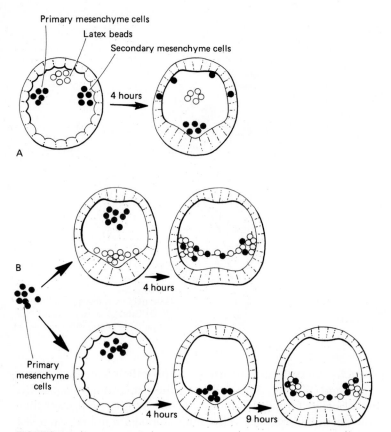

FIGURE 5-3
Transplantation experiments in sea urchin blastulas. (A) Primary and secondary mesenchymal cells and latex beads were injected into the blastocoel. After 4 hours, the primary mesenchymal cells had honed in to their
normal location on the vegetal pole; the latex beads had not moved; and
the secondary mesenchyme cells had undertaken nonspecific movements.
(B) Primary mesenchyme cells ingested into a late blastula quickly migrate
to join the primary mesenchyme of the host (upper row). When injected
into an early blastula, primary mesenchyme cells migrate to the vegetal
pole and remain there until the primary mesenchyme of the host forms;
then the two primary mesenchymes become integrated into a normal structure (lower row). (Adapted from D. McClay and C. Ettensohn. 1987. Cell
recognition during sea urchin gastrulation. In W. Loomis, ed. *Genetic Regulation of Development*. A. R. Liss.)

modeling and producing *exogastrulation* (causing the archenteron to form
outside instead of inside the blastocoel), they concluded that other forces in
addition to or instead of filopodial shortening must play an important role in the
inward extension of the archenteron.

Ultimately the tip of the archenteron reaches the opposite wall of the
blastula, forming a bilaminar layer which soon ruptures to form the oral
opening. At the other end, the blastopore becomes the anus. Meanwhile, the

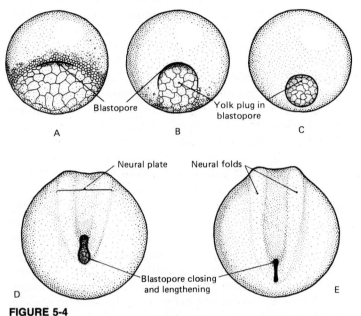

FIGURE 5-4
Caudal views of amphibian embryos, showing the changing configuration of the blastopore. (Five stages selected and modified from Huettner, *Fundamentals of Comparative Embryology of the Vertebrates.* (By permission of The Macmillan Company, New York.)

secondary mesenchymal cells move away from the archenteron proper to form two lateral pouches which will form the coelomic mesoderm. With this separation, the archenteron is composed entirely of endodermal cells. The embryo begins to assume a triangular appearance, and as the skeleton arises from the primary mesenchyme, armlike structures form and the embryo is well on the way to becoming a *pluteus* larva.

GASTRULATION IN AMPHIBIAN EMBRYOS

As was just pointed out, gastrulation in the amphibian embryo is modified to a certain extent by the presence of the large yolk-laden cells in the vegetal hemisphere of early embryos. Nevertheless, some essential features of early gastrulation in amphibians resemble quite closely those of forms having isolecithal eggs. The first external evidence of gastrulation is the appearance of a slightly curved groove (Fig. 5-4A). This groove, the blastopore, represents the site at which cells move into the interior of the embryo. Because movements of vegetal cells are impeded by the high content of yolk, the inpocketing of cells during gastrulation is originally restricted to a region that is dorsal to the yolk (Figs. 5-4A and 5-8A), and the upper margin of the groove is known as the *dorsal lip of the blastopore*. As the process gains momentum, the ingrowing margins or lips of the blastopore are gradually extended so that they assume a circular shape as they turn inward around the yolk (Fig. 5-4B and C). The mass of yolk left presenting at the blastopore is known as the *yolk plug*.

As the blastoporal groove is taking shape, a group of cells at the area of involution exhibit a pronounced change of shape. While retaining tight connections with the external surface, the cells elongate inwardly until their shape resembles an attenuated flask (Fig. 5-5). It has been commonly believed that the change of shape of these *bottle cells*, as they are called, is associated with an inward pulling movement which results in the formation of the blastoporal groove. On the other hand, Keller (1981) has suggested that the bottle cells play a more passive role in involution than was formerly thought.

Much of the process of gastrulation consists of surface cells moving into the interior of the embryo at the blastopore. As cells turn inward around the lips of the blastopore, they are followed by other cells which move over the surface of the embryo toward the blastopore. Even before the morphogenetic movements of gastrulation begin, certain groups of cells in the surface layer can be shown to be precursors of adult tissues or regions. With appropriate tracing methods, these cells can be assigned not only to definite germ layers but also to specific organ primordia within these layers.

Historically, the problem of tracing cell fates in early amphibian embryos was one of the first in which tracing methods were employed. Most of these methods involved the use of vital dyes (Vogt, 1929). Although a much wider array of sophisticated intracellular labels is now available (Slack, 1984), the findings of the early investigators have withstood the test of time remarkably well. The general principle behind most labeling experiments is to insert a stable label inside or on the surface of a cell. The embryo is then allowed to

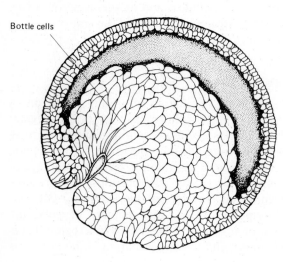

Bottle cells

FIGURE 5-5
Slightly schematized section through an advanced amphibian gastrula, showing the flask-shaped "bottle cells" moving into the interior of the embryo at the blastopore. (After Holtfreter, 1943. *J. Exp. Zool. 94*:261–318.)

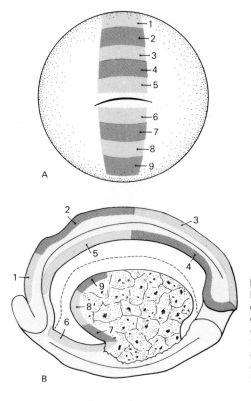

FIGURE 5-6
Diagram illustrating the principle of using vital
dyes as tracers in studying the displacement
of cells during gastrulation in the amphibian
embryo. (A) Small areas (1–9) along the midline of a blastula are marked with two different
dyes. (B) Morphogenetic movements during
gastrulation displace the stained areas to different regions of the embryo.

develop for a determined length of time, after which it is sectioned and treated
in a manner that allows the localization and identification of originally labeled
cells or their descendants. Figure 5-6 shows the extensive displacement of cells
during gastrulation in the amphibian embryo. The results of many such tracing
experiments can be summarized by *fate maps*, which show on an early embryo
the regions that are destined to give rise to specific structures or regions later
in development (Fig. 5-7).

Gastrulation movements are not uniform among the major amphibian
groups. Unless otherwise indicated, we shall in the following paragraphs
describe the morphogenetic movements which take place in the tailed amphibians (urodeles). The process is quite similar in the tailless amphibians (anurans), but some details, such as endodermal participation in the early formation
of the primitive gut (archenteron), differ between the two groups.

Around most of the ventral margins of the blastopore and extending down
onto the ventral part of the embryo the *prospective endoderm* is rolled into the
interior of the embryo and eventually comes to line its gastrocoel, or primitive
gut (Fig. 5-8A to C). Most of the cells passing over the dorsal lip of the
blastopore are often called *chordamesoderm* because they eventually give rise
to the notochord and cephalic mesoderm (Fig. 5-8C).

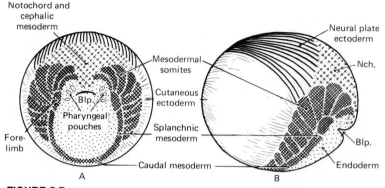

FIGURE 5-7
Prospective areas of the embryos of tailed amphibians at the stage when gastrulation is just beginning. (A) Caudal aspect; (B) lateral aspect. (Modified from Vogt, 1929. *Arch. f. Entwickl.-mech. d. Organ., vol. 120.*)

When involution has just begun, the early gastrocoel is lined by chordamesoderm on its dorsal surface and elsewhere by endoderm (Fig. 5-8A and B). The overall dorsal displacement of the entire early gastrocoel by the yolk-laden endoderm is quite evident. As involution continues, the gastrocoel increases in size and extends beneath the outer layers of cells toward the cephalic end of the embryo. The inertness of the yolk-laden endoderm finally seems to prove to be too much for the cells actively moving in around the blastopore, and some of these cells undercut ventrally the external portion of the yolk-filled endoderm (Fig. 5-8B and C). The result of this process is the persistence of the externally visible yolk plug, which is surrounded on all sides by invaginating cells. In urodele embryos, part of the dorsal surface of the gastrocoel remains lined by chordamesodermal material until late in gastrulation. Finally the invaginated endodermal cells spread beneath it to form a complete endodermal lining to the primitive gut.

Studies on gastrulation in *Xenopus* (Keller, 1981) have provided much new information on individual cell movements around the dorsal lip of the blastopore. In the early dorsal lip (Fig. 5-9A) the surface cells are underlain by a deeper marginal zone which consists of several layers of cells. According to Keller, the main impetus for movement of cells around the dorsal lip occurs in the marginal zone. The cells of the marginal zone interdigitate and form a single layer. This forces the marginal zone to expand toward the newly forming dorsal lip. As the cells of the marginal zone reach the dorsal lip, they undergo a pronounced change of shape and behavior, and they turn around and migrate away from the dorsal lip on the deep surface of the marginal zone (Fig. 5-9A and B). This active process is an example of involution. The involuted cells that were derived from the marginal zone will form the mesoderm of the embryo.

Recent evidence (Boucaut and Darribere, 1983) suggests that an extracellular matrix, with fibronectin as an important component, forms the substrate

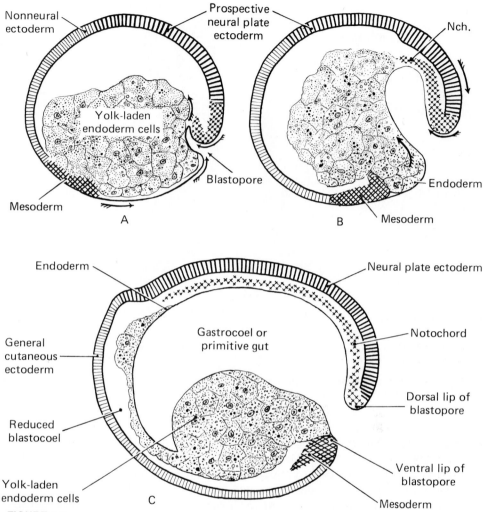

FIGURE 5-8
Diagrams indicating the cell rearrangements that occur in amphibian gastrulation. (Schematic sagittal sections based primarily on the work of Vogt, 1929, *Arch. f. Entwickl.-mech. d. Organ.*, vol. 120.)

upon which these mesodermal cells migrate. Johnson (1985) has shown that migrating cells of the gastrula acquire an increasing ability to adhere to fibronectin-coated beads. In contrast, cells from the blastula or cells from hybrid embryos that experience a developmental arrest during gastrulation do not adhere to the beads.

While these active changes are taking place in the deeper layers, the surface layer expands (epiboly) by means of a combination of cell division, flattening, and spreading. As the surface cells pass around the rim of the dorsal lip, they

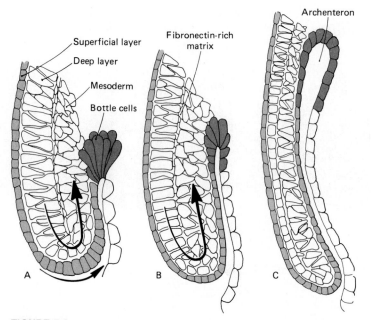

FIGURE 5-9
Cell behavior in the dorsal lip of the blastopore in *Xenopus* and formation of the mesoderm. Cells of the deep layer of the dorsal lip interdigitate into a single layer (A, B) and spread around the lip of the blastopore (curved arrow). Once inside, they migrate over the inner surface
of the deep layer toward the animal pole as the mesoderm. (Adapted
from Keller, 1981).

become the endodermal lining of the archenteron. The bottle cells, which are so
prominent in early gastrulation, decrease in height and ultimately become
indistinguishable from other endodermal cells of the endodermal lining.

The origin of the mesoderm has not received as much recent attention in
urodeles as it has in *Xenopus*. Nevertheless, it appears that premesodermal
cells move toward the blastopore, turn inward at its lips, and then migrate away
from the blastopore as the definitive middle mesodermal germ layer (Fig. 5-10).

As is the case with the other germ layers, the prospective ectoderm does not
remain static during gastrulation. Fate maps of early stages indicate that it can
be divided into areas with different futures. One region is destined to take part
in the formation of the central nervous system and is therefore designated as
neural ectoderm (Figs. 5-7B and 5-8A). The remainder of the ectoderm will
form the epidermis of the skin and is therefore called *general cutaneous
ectoderm*. By the end of gastrulation the entire outer surface of the embryo is
covered by neural and general cutaneous ectoderm, whereas the future
endoderm and mesoderm have come to lie entirely within (Fig. 5-8C). The
limited extent of some of these prospective areas is striking; for example, note
the prospective general cutaneous ectoderm shown in Fig. 5-7A. This area

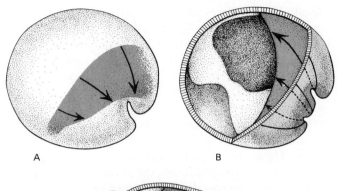

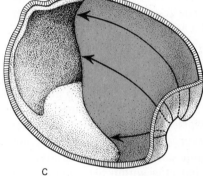

FIGURE 5-10
Diagrams showing the spread of mesoderm in embryos of tailed
amphibia. The arrows indicate (A) the migrations of future mesodermal cells toward the blastopore, (B) the involution of the cells
around the lips of the blastopore, and (C) away from the blastopore
as a discrete layer of mesodermal cells. Note that the mesoderm is
extended from the entire circumference of the lips of the blastopore,
but that its growth from the dorsal lip is most vigorous. (Based on
Vogt, 1919. *Arch. f. Entwickl.-mech. d. Organ., vol. 120.*)

must be expanded to cover the surface areas vacated by the involution of the
cells that form the inner germ layers (cf. Fig. 5-8A to 5-8C). Therefore, the
ectoderm must increase its area with the growing size of the embryo it clothes.

Experimental studies have shown that during gastrulation the cells of the
three germ layers develop properties characteristic of each layer and that much
of the process of gastrulation and the stability of the arrangements of the
resulting germ layers are due to these properties. Both ectodermal and
endodermal cells have acquired the propensity spread out into sheets adjacent
to the mesoderm, which is now interposed between these two germ layers. The
ectoderm, which has spread out over the entire outer surface of the embryo
after the involution of the endoderm and mesoderm, seems by its spreading
tendency literally to hold in the endodermal lining of the primitive gut. If the
ectoderm is removed, the endoderm takes advantage of its absence and tends

to spread outward over the outside of the mesoderm in what is almost a reversal of its normal tendency to remain in the interior, lining the gut (Holtfreter, 1944).

The massive tissue displacements resulting from the morphogenetic movements that occur during gastrulation represent just a part of the total developmental activity during this stage. Unseen by the eye, other processes are occurring which are vital for the future development of the embryo. In one of the most illuminating experiments in the history of embryology, Spemann and Mangold (1924) demonstrated that the cells that constitute the dorsal lip of the blastopore (Fig. 5-8C) possess the ability to "organize" much of the future development of the embryo. These investigators removed a section of the dorsal lip from an early gastrula and implanted it into the ventral portion of the blastocoel of another amphibian embryo. As development of the host embryo progressed, a secondary embryo, quite complete in most respects, was formed in Siamese-twin fashion along its ventral surface (Fig. 5-11). The secondary embryo was composed partly of cells derived from the graft and partly from cells of the host. As a result of its ability to direct normal development in an orderly fashion, Spemann called the dorsal lip of the blastopore the *organizer*. Since one of the earliest evident reactions of dorsal lip grafts is the formation of a secondary neural plate, it has often been assumed that primary neural induction (see Chap. 6) is an intrinsic component of the organizing function of the dorsal lip. Slack (1984), however, concluded that the "organizing" function of the dorsal lip is actually an influence, perhaps inductive, that causes the dorsalization of any mesoderm that falls within the sphere of influence of the dorsal lip. Once "dorsalized," the mesoderm then develops the properties of the primary neural inductor.

Recent molecular investigations (Dawid and Sargent, 1986) have related gene expression in amphibian gastrulae to the induction of mesoderm in the cleaving embryo (See Chap. 4). The genes in question have been called *Dg* (differentially expressed in gastrula) *genes*. The proteins coded from these genes represent the earliest expression of the embryonic (as opposed to the maternal) genome, and none of these gene products persists into the adult. The functions of most of these proteins are poorly understood.

Although detailed descriptions of specific *Dg* proteins is beyond the scope of this book, it is relevant that the gene products that are markers of the ectodermal or endodermal germ layers are autonomously expressed by animal or vegetal cells even in the absence of interactions with other cells of the embryo. Whether ectodermal or endodermal gene products are expressed by a given cell is related to the cell's location along the animal-vegetal pole axis and is thus a function of the polarity of the egg. In contrast, the expression of mesodermal proteins does not occur if the early inductive interaction between the vegetal (endodermal) and animal (ectodermal) parts of the embryo, which leads to formation of the mesoderm, is prevented. Within hours after the transient activation of the *Dg* genes, a new wave of gene expression begins in the early neurula stage. The products of these genes, in contrast to those of the earlier *Dg* genes, are permanent and persist into the adult.

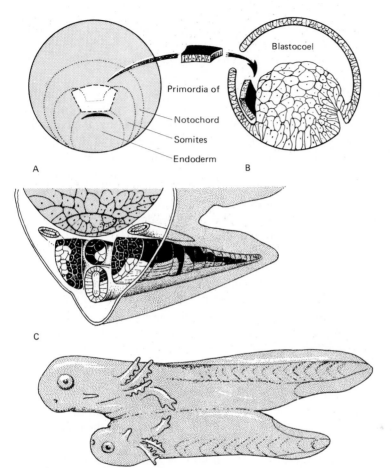

FIGURE 5-11
Transplantation of the upper blastoporal lip from one amphibian gastrula into another (A, B). This procedure produces a secondary embryo (D) composed (C) of self-differentiated tissues of the graft (*black*) and induced tissues of the host (*white*). (From Holtfreter and Hamburger in Willier, Weiss and Hamburger, 1955, *Analysis of Development*, W. B. Saunders Company, Philadelphia.)

GASTRULATION IN BIRDS

Gastrulation in birds is morphologically complex, and over the years it has been subject to a number of quite divergent interpretations. Compounding the inherent complexity of the process is the relative inaccessibility of early developmental stages in the chick, for almost all the pregastrulation stages occur before the egg is laid.

We have already traced the establishment of the blastula in the chick embryo as a two-layered structure consisting of an upper layer (the epiblast) and a lower layer (the primary hypoblast), with a thin blastocoel in between (Fig. 4-10). The embryo proper occupies the transparent area pellucida and is surrounded by the area opaca, where the cells of the blastoderm lie unseparated from the yolk (Fig. A-4C).

Next, a thin sickle-shaped mass of cells (*Koller's sickle*) takes shape at the posterior end of the embryo. From Koller's sickle, a second generation of hypoblastic cells (*secondary hypoblast*) pushes anteriorly, compressing and folding the primary hypoblast ahead of it (Fig. 5-15). The exact mode of formation of the secondary hypoblast is still not known. Neither the primary nor the secondary hypoblast seems to form any of the embryonic germ layers. Primordial germ cells are found in the primary hypoblast. Their crescentic distribution along the anterior border of the blastoderm (Fig. 3-1) may be due to compression of the primary hypoblast by the expanding secondary hypoblast. The secondary hypoblast forms extraembryonic endoderm, principally the yolk stalk.

Formation of the primary and secondary hypoblast can be considered a pregastrulation phenomenon. Gastrulation and formation of the definitive embryonic germ layers begin with the appearance of a condensation of cells in the posterior part of the epiblast. This condensation, seen in an embryo which has been incubated for 3 to 4 hours (Fig. 5-12A), gradually assumes a cephalocaudal elongation (Fig. 5-12B). By the seventh or eighth hour of incubation the elongation is still more definite (Fig. 5-12C), and by the end of the first half day the thickened area has assumed a shape which has led to its being called the *primitive streak* (Bellairs, 1986).

The appearance of the primitive streak is the result of an inductive interaction of the epiblast with the hypoblastic layer (Eyal-Giladi and Wolk, 1970), and its orientation is a reflection of the intrinsic polarity of the underlying hypoblastic layer. The latter was dramatically demonstrated by Waddington (1933), who altered the orientation of the primitive streak by changing the position of the hypoblast with respect to the epiblast. More recent experiments have shown that the organizational center of the early chick embryo is located in the posterior margin of the hypoblast (Azar and Eyal-Giladi, 1981).

The early primitive streak initially elongates in both a cephalic and a caudal direction. The carbon-marking experiments of Spratt (1946) have been especially instructive in demonstrating the processes involved in primitive-streak formation. He placed spots of carbon particles on chick blastoderms just as they were about to form the primitive streak. These marking experiments showed that throughout much of the posterior part of the blastoderm cell movements converge from the lateral areas toward the forming primitive streak (Fig. 5-13). As more cells enter the streak region, the primitive streak elongates in a cephalic direction. The cephalic extension of the primitive streak keeps pace with the expansion of the secondary hypoblast beneath it. Caudal extension moves the primitive streak into the area opaca.

After 16 hours of incubation the primitive streak becomes so prominent that embryos are characterized as being in the primitive-streak stage (Fig. 5-14). A central furrow called the *primitive groove* now runs down the center of the

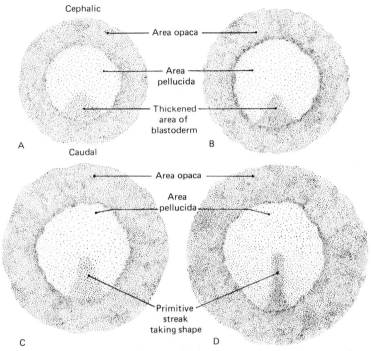

FIGURE 5-12
Chick embryos showing four stages in the formation of the primitive streak.
(A) 3 to 4 hours' incubation. (B) 5 to 6 hours' incubation. (C) 7 to 8 hours'
incubation. (D) 10 to 12 hours' incubation. (Based in part on the photomicrographs of Spratt, 1946. *J. Exp. Zool.*, vol. 103.)

primitive streak. Along both sides it is flanked by thickened margins, called the
primitive ridges (Fig. 5-14). At the cephalic end of the primitive streak closely
packed cells form a local thickening known as *Hensen's node* (Fig. 5-14).[1] After
the primitive streak has reached its full length at about the eighteenth hour of
incubation, the cephalic end begins to regress, leaving in its wake a structure
commonly referred to as the *head process*. This is a gross morphological term
referring to the area where the notochord has been recently laid down (Fig. A-3).

The part of the area pellucida adjacent to the primitive streak begins to
thicken and is said to constitute the embryonal area (Fig. A-3). Because of its
shape, the embryonal area is frequently spoken of as the *embryonic shield*.
Accompanying the formation and elongation of the primitive streak, the area
pellucida undergoes a change in shape from an essentially circular disk to a

[1]In the experimental embryological literature the term Hensen's node generally implies a
somewhat larger area than that defined as the node in most descriptive texts. Hensen's original
description of the node in rabbit embryos described it as the slightly enlarged anterior end of the
primitive streak. Recognition of varying definitions of the node is important in interpreting the
experimental literature.

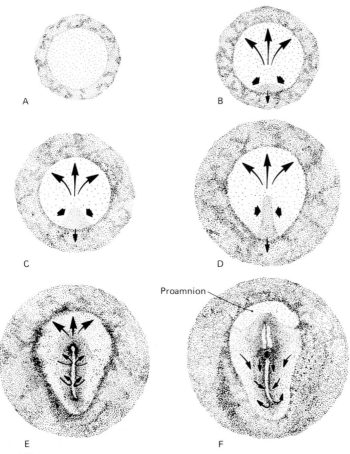

FIGURE 5-13
Diagrams illustrating the patterns of cell movements involved in the formation and elongation of the primitive streak (A–D), invagination of cells
through the streak (E), and regression of the streak (F) in the chick.
(Based on the data of Spratt and Haas.)

pear-shaped configuration. The long axis of the future embryonic body is
clearly established by the primitive streak.

With the establishment of the primitive streak and Hensen's node, the main
period of gastrulation begins. The embryonic germ layers are formed by the
migration of cells in the epiblast toward Hensen's node and the primitive
streak, and their ingression to form the middle and lower germ layers (the
embryonic mesoderm and endoderm[2]). The anterior portion of the primitive

---

[2]For consistency with the designations of the germ layers in amphibians and mammals, the
definitive germ layers resulting from gastrulation will be designated as ectoderm, mesoderm and
endoderm. Some authors continue to call these layers epiblast, mesoblast, and hypoblast).

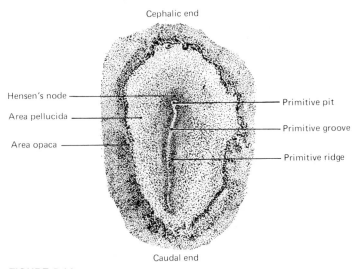

Cephalic end

Hensen's node

Area pellucida

Area opaca

Primitive pit

Primitive groove

Primitive ridge

Caudal end

FIGURE 5-14
Dorsal view (×14) of entire chick embryo in the primitive-streak stage
(about 16 hours of incubation).

streak and the node serve as a passageway for cells even while the streak is elongating anteriorly. In birds, gastrulation is accomplished by the coordinated passage of individual cells from the exterior into the interior of the embryo rather than the immigration of integral sheets of cells (Sanders, 1986).

The first cells to pass through the area of the anterior part of the primitive streak are future embryonic endodermal cells. After about 8 to 10 hours of incubation more than 80 percent of these cells are found in the endoderm; the remainder migrate into the middle mesodermal layer. As time goes on, a progressively greater percentage of the cells which pass through the node are destined to be incorporated into the mesoderm and a correspondingly smaller number lodge in the endoderm. The endodermal cells that are formed in this manner enter the original hypoblastic layer and steadily displace the cells of the hypoblast outward and cephalad toward the edge of the area opaca (Fig. 5-15). Although the bulk of the endoderm has passed through the nodal region during the early, formative stages of development of the primitive streak, increasing numbers of future endodermal cells migrate through the anterior part of the primitive streak as well. By about 22 hours of incubation, when regression of the primitive streak has commenced, essentially all the future endodermal cells have left the epiblast.

Little formative activity of the middle germ layer (embryonic mesoderm) occurs until around the fifteenth hour of incubation, when the primitive groove becomes well established within the primitive steak (Fig. 5-14). There are two principal areas of invagination and mesoderm formation in the early chick embryo. The most extensive invagination of mesodermal cells occurs along the

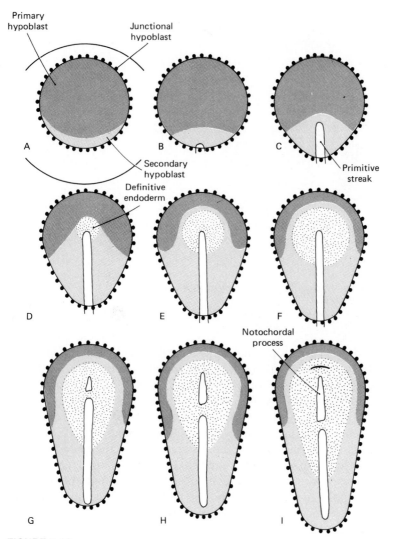

FIGURE 5-15
Successive stages in the formation of the lower layer in early chick embryos.
(After L. Vakaet, 1970. *Arch. Biol. 81*:387.)

length of the primitive streak, where the coherent layer of mesodermal cells that is formed expands parallel to the underlying layer of the embryonic endoderm (Figs. 5-17 and 5-18). The spread of the mesoderm is shown in Fig. 5-16. The other major site of mesoderm formation is through Hensen's node, where a rod of mesodermal cells directed cephalad lies in the midline of the embryo in the track of the regressing primitive streak. This mesodermal rod becomes the notochord (Fig. 5-18), which is essential for the next series of major changes that sweep over the embryo.

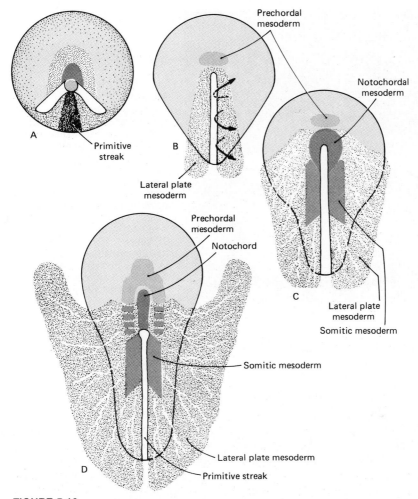

FIGURE 5-16
Successive stages in the formation of the mesodermal layer in early chick embryos. (Adapted from Pasteels and Vakaet.)

As the cells of the epiblast migrate toward and through the primitive streak and ultimately take their place among the other cells of the mesodermal layer, they undergo certain characteristic changes in form. The epiblast is composed of cuboidal to columnar epithelial cells. As in any typical epithelium, the apical surfaces are tightly bound to one another by specialized tight intercellular junctions which encircle the entire cell apex and act as sealing devices to preserve differences in the environment between the inside and the outside of the epithelial layer. In addition, the deeper parts of the cells are bound together by spotlike junctions known as *gap junctions*, which are involved in cell-to-cell communication.

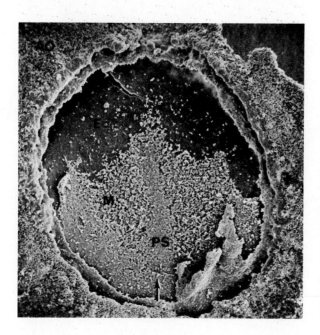

FIGURE 5-17
Scanning electron micrograph of the ventral surface of an early chick embryo (Hamburger-Hamilton stage 4). The endoderm has been removed, providing a good view of the mesoderm as it extends outward from the primitive streak. ×52. The arrow points from the posterior toward the anterior end of the embryo. (From M. A. England and J. Wakely, 1977. *Anat. Embryol.* *150*:291. Courtesy of the authors and publisher.) *Abbreviations*: AO, area opaca: E, ectoderm; M, mesoderm; PS, primitive streak.

When the cells of the epiblast enter the primitive groove, they undergo a pronounced change of shape, becoming to some extent bottle-shaped in a manner reminiscent of the bottle cells seen in amphibian gastrulation (Fig. 5-19). The change of shape of these cells is associated with the appearance of orderly arrays of intracellular microtubules, which are associated with changes of shape in many varieties of cells. During this change of shape, the tight junctions begin to break up and lose their circumferential arrangement at the apex of each cell. After they have passed through the primitive streak, the cells of the mesoderm assume the stellate appearance characteristic of mesenchyme (Sanders, 1986). These cells are connected to one another by small gap junctions.

Shortly after the first prospective notochordal cells are laid down, the primitive streak and Hensen's node undergo a regression toward the caudal end of the embryo. Accompanying this regression of the primitive streak is a corresponding elongation of the notochord (Fig. A-5). The notochord at this stage is often called the *head process*. Further morphological details of the development of chick embryos are given in the Appendix.

Prospective Areas in Chicks at the Primitive-Streak Stage

It has long been known that specific areas of early embryos typically contribute to the formation of characteristic tissues and organs in the adult. As a result of this knowledge embryologists have been able to construct *fate maps* for embryos of some of the more intensely studied species. A fate map of the early amphibian gastrula has already been presented (Fig. 5-7).

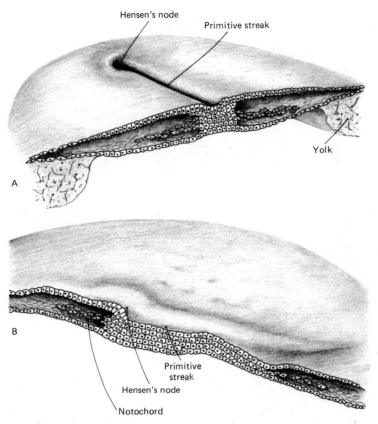

Hensen's node

Primitive streak

Yolk

A

B

Primitive
streak

Hensen's node

Notochord

FIGURE 5-18
Schematic three-dimensional models of early chick embryos, illustrating
the migration of mesodermal and endodermal cells through the primitive
streak. (A) Transverse section, showing anterior half of the embryo. (B)
Sagittal section, showing right half of the embryo. (Part A after Balinsky.)

Two properties of embryonic cells or groups of cells are important to
embryologists who attempt to study the organization of early embryos in
relation to later stages of development. One property is the *prospective
significance* (*prospective fate*), which can be defined as the fate of a cell or a
group of cells during the course of normal development. The other property is
the *prospective potency*, which is defined as the types of differentiation of
which a cell or group of cells is capable at a given stage of development.
Typically the prospective potency of a group of cells is greater than the
prospective significance, particularly at early developmental stages. As developmental restriction sets in, the prospective potency decreases until at the time
of final determination the prospective potency and prospective significance are
the same. Some cells always retain the potency to undergo alternate pathways
of differentiation (i.e., *metaplasia*).

FIGURE 5-19
Scanning electron micrograph through the primitive streak of a cross-fractured chick embryo (Hamburger-Hamilton stage 8). Cells from the epiblast entering the region of the primitive groove become flask-shaped (*arrow*) as they prepare to move into the interior of the embryo. ×2635. (From Solursh and Revel, 1978. *Differentiation 11*:185. Courtesy of the authors.)

Several techniques have been successfully used in the construction of fate maps for embryos. One is to mark certain cells with vital dyes and to follow the stained cells as long as possible during development. This method has been very successful in the mapping of early amphibian embryos. In avian embryos the marking of cells with carbon particles was very useful in early studies (Spratt, 1946). Later, experiments involving the transplantation of radioactively labeled pieces of early embryos into equivalent locations in unlabeled hosts (Rosenquist, 1966) allowed further refining of early fate maps of the chick embryo. The technique of interspecific marking, using cells of the quail embryo as markers, has added still more information about the future fates of cells in early chick embryos.

Another mapping technique consists of explanting small pieces of embryos as grafts onto the chorioallantoic membrane or into the coelomic cavity. Care must be taken in the interpretation of explantation experiments, for in these conditions the cells may differentiate according to their prospective potency rather than their normal prospective fate. If an area (e.g., the liver) where a particular potency has been located is explored in more detail, it is found that there is a central part from which practically all the explants exhibit the potency in question. Explants taken farther peripherally show the potency in decreasing

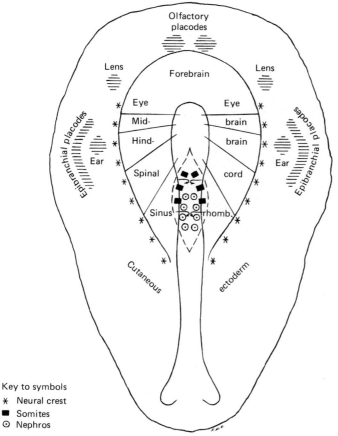

Key to symbols
* Neural crest
■ Somites
⊙ Nephros

FIGURE 5-20
Map of the prospective areas of the outer layer of the chick embryo
in the primitive-streak. (Modified from Rudnick, 1944. *Quart. Rev.
Biol.*, vol. 19.) *Abbreviation*: rhomb., rhomboid.

percentages of the grafts made. The territory where a specific potency is
regularly manifested is said to be the *prospective center* for the organ in
question.

Two classical maps of prospective areas for chicks of the primitive-streak
stage are reproduced as Figs. 5-20 and 5-21. It should be emphasized that the
sharpness of the boundaries between different prospective areas as shown in
such figures is an entirely artificial device for vivid graphical presentation. In
reality there are vague transition zones rather than anything like the sharp
delimitations of the schematic design.

Recent studies with cell adhesion molecules (CAMs; see p. 15) have shown
striking relationships to recently updated fate maps (Fig. 5-22). In the pregastrulation chick embryo, the epiblast and hypoblast contain both N-CAM

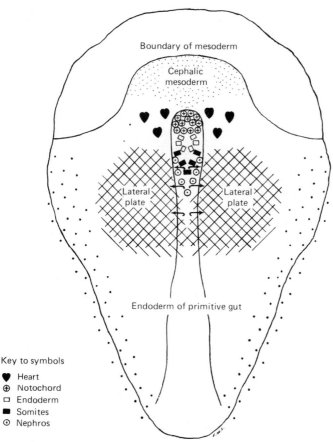

FIGURE 5-21
Map of the prospective areas of the invaginated layers of the chick in the primitive-streak stage. (Modified from Rudnick, 1944. *Quart. Rev. Biol.*, vol. 19.)

and L-CAM (Edelman et al., 1983). As mesodermal cells migrate through the primitive streak, neither CAM can be detected on their surfaces. Soon, however, a major change takes place. Cells of the future nervous system lose their L-CAM but retain N-CAM, whereas non-neural ectoderm retains L-CAM but loses N-CAM. The distribution of CAMs on a slightly later fate map of surface cells (Fig. 5-22) shows a central core of N-CAM–positive cells (nervous system, notochord, somites, and a number of mesodermal organs) surrounded by a ring of L-CAM–positive nonneural ectoderm and endoderm. As development progresses, cells that will form a number of different organs show dramatic shifts in their CAMs as the inductive interactions leading to their formation proceed. It is a general rule that whenever epithelial cells are transformed into mesenchymal cells, their surface CAMs are lost.

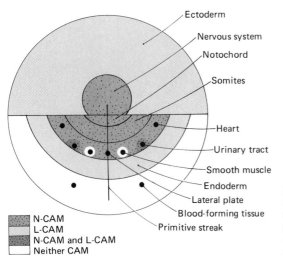

Ectoderm
Nervous system
Notochord
Somites
Heart
Urinary tract
Smooth muscle
Endoderm
Lateral plate
Blood-forming tissue
Primitive streak

N-CAM
L-CAM
N-CAM and L-CAM
Neither CAM

FIGURE 5-22
Fate map of the avian embryo with
the distribution of cell adhesion molecules (CAMs) superimposed.
(Adapted from the diagrams of Vakaet
and of Edelman.)

COMPARISON OF BIRD AND AMPHIBIAN DEVELOPMENT

There are a number of parallels between the early development of birds and
that of amphibians. At the blastula stage, which in the bird can be represented
by a two-layered structure containing the epiblast and hypoblast, the cells that
will later form the endodermal and mesodermal germ layers are located in the
surface layer of the embryo. Likewise, the surface layer of the amphibian
embryo at a similar stage contains cells destined to become part of the
endoderm and mesoderm. There are interesting parallels as well between the
induction of the mesoderm and control of polarity by the vegetative yolk mass
in amphibians and the effects of the hypoblast on the epiblast in early chick
embryos.

It also seems reasonable to regard the pre-primitive-streak thickened area of
the chick blastoderm as the symbolic homologue of a blastopore that could not
open because of the impeding effect of the enormous yolk mass. The movement
of surface cells toward the primitive streak in the chick is suggestive of the way
cells move by means of epiboly toward the amphibian blastopore. In both
forms, cells which will constitute the future endodermal layer migrate from the
exterior to the interior of the embryo first, and the inward movement of the
mesodermal cells follows later. Although the early gastrula in birds is compressed because of its disposition on top of the yolk, there is a similarity in both
the size and the fate maps of amphibian and avian embryos during this period
(Fig. 5-23).

If we follow the process of gastrulation in avian embryos into its later
phases, when the notochord and mesoderm are being established, the homology of the primitive streak with the fused lips of the blastopore becomes even
more apparent. Shortly after the primitive streak has formed and the endoderm
has been well established, cells of the chick embryo begin to push in from the

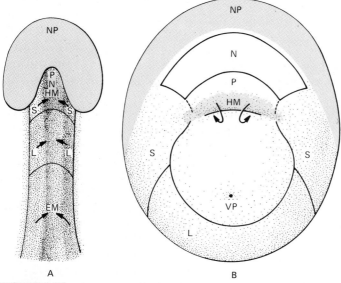

A   B

FIGURE 5-23
Comparison of presumptive areas in the vicinity of the primitive streak
in the chick (A) and the dorsal lip of the blastopore (*arrows*) in *Ambystoma* (B). *Abbreviations*: EM, extraembryonic mesoderm; HM, head
mesoderm; L, lateral plate mesoderm; N, notochord; NP, neural plate;
P, prechordal plate; S, somites; VP, vegetal pole. (Adapted from Nicolet, 1971. *Adv. Morphogen. 9*:231.)

region of Hensen's node to form the rodlike notochord in the midline beneath
the ectoderm (Fig. 5-18). The area where the chick notochord is formed clearly
corresponds with the dorsal lip of the blastopore where the amphibian
notochord arises (Fig. 5-7). Sections taken across the primitive streak caudal to
Hensen's node show the chick mesoderm extending out on either side between
the ectoderm and the endoderm (Fig. 5-18). These relationships are again very
similar to those seen in amphibian embryos, with the mesoderm arising from
cells turning in at the lips of the blastopore and extending between ectoderm
and endoderm (Figs. 5-8C and 5-10C). Interestingly, most reptiles form a
blastoporelike structure rather than a primitive streak (Bellairs, 1986).

ORIGIN OF THE GERM LAYERS IN MAMMALS

Although the mammalian embryo contains almost no yolk, the morphogenetic
movements and tissue displacements are nevertheless remarkably similar to
those which occur in birds. The inner cell mass can be compared with the cap
of blastomeres situated on the animal pole of the yolk sphere in large-yolked
forms such as the chick. In view of their phylogenetic relationships, it is not
surprising that the origin of the germ layers in mammals resembles the process
occurring in our more immediate large-yolked ancestors rather than the simple

infolding type of gastrulation seen in the small-yolked eggs of amphibians. Although the eggs of higher mammals have lost their endowment of yolk, the proliferating cells of the inner cell mass still behave as if they were crowded in on top of a larger yolk sphere and had restricted space in which to maneuver. As a result, mammals show more of a subtle emergence of their germ layers by cell migration and regrouping than the segregation of an endodermal layer by the primitive, infolding type of gastrulation.

Recent studies of early mammalian development have shown that the origin and mode of formation of a number of the intra- and extraembryonic germ layers differ considerably from those proposed by early descriptive mammalian embryologists. Schemes summarizing the origins of intraembryonic tissues of both primates and rodent embryos (Figs. 5-24 and 5-31) will be useful in understanding the material presented here and in Chap. 8. Because of the importance of the mouse as an experimental object and the major differences in the morphology of early development, a separate section will be devoted to gastrulation in rodents.

Chapter 4 described how cells of the mammalian blastocyst become segregated into an embryo-forming inner cell mass surrounded by a layer of trophoblastic cells (Fig. 5-25). The first cells to segregate out from this inner cell mass form a thin layer called the *hypoblast* (Fig. 5-26A and B). This layer forms

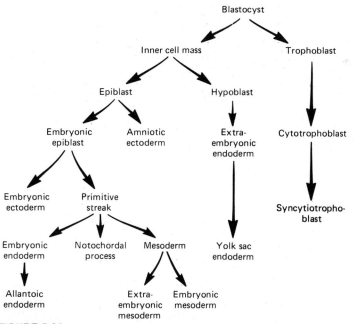

FIGURE 5-24
Scheme illustrating the origin and derivatives of tissues in presomitic human and rhesus monkey embryos. (Modified from W. P. Luckett, 1978. *Am. J. Anat. 152*:59.)

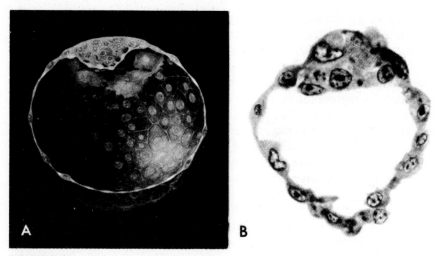

FIGURE 5-25
(A) Graphic reconstruction of the blastocyst of a monkey on ninth day after fertilization. The specimen has been drawn as if opened in the midline to show the early differentiation of hypoblast cells, here represented as being somewhat lighter in color than other cells of the inner cell mass. (After Streeter, 1938, *Carnegie Inst. of Washington Publication* 501.) (B) Section of the blastocyst of a human embryo estimated to be at about the fifth day after fertilization. (After Hertig, Rock, Adams, and Mulligan, 1954, *Carnegie Cont. to Emb.*, vol. 35.)

only extraembryonic endoderm, and it is regarded as equivalent to the hypoblast of chick embryos. The hypoblast contributes the cells that will line the yolk sac. As the hypoblast forms, the remainder of the inner cell mass can be called the *epiblast*. In addition to future ectodermal cells, the epiblast contains the cells that will ultimately migrate through the primitive streak and become the definitive endodermal and mesodermal germ layers of the embryo.

Formation of the primitive streak in mammalian embryos follows a pattern quite similar to that in bird embryos. Not long after the hypoblast is established, the remaining cells of the inner cell mass become more regularly arranged and are collectively called the *embryonic disk* (Fig. 5-26C). Soon one margin of the disk becomes thickened. The thickening occurs at the part of the disk that is destined to become the caudal end of the embryo. From this caudal thickening, a cephalad expansion of cells results in the formation of a longitudinal band of cells called the *primitive streak*.

The cell movements in the region of the primitive streak of mammalian embryos have not been studied in as much detail as they have been in amphibians and birds. Little is known about the details of formation of the definitive endodermal layer, but indirect evidence suggests that the cells forming this layer migrate through the primitive streak and become situated as the roof of the primitive gut (archenteron). It now appears that cells that

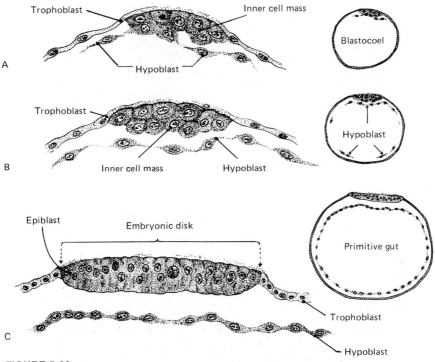

FIGURE 5-26
Sections of pig blastocysts showing the first appearance and subsequent rapid extension of the hypoblast. *Left*, Detailed drawings of inner cell mass (×375). *Right*, sketches of same sections, entire. The approximate ages of the embryos range from 7 to 8 days. (From embryos in the Carnegie Collection.)

constitute both the embryonic and extraembryonic mesoderm pass through the posterior part of the primitive streak (Fig. 5-27B and C). Much of the early mesoderm that is formed passes beyond the confines of the embryonic disk as extraembryonic mesoderm. Luckett (1978) presented evidence suggesting that probably all extraembryonic mesoderm in the early mammalian embryo migrates out from the primitive streak instead of differentiating from trophoblastic cells as was once believed. Later, the embryonic mesoderm arises by means of the migration of cells in the epiblast toward the primitive streak. These cells then pass through the streak and spread out laterally beneath the epiblast (Figs. 5-28 and 5-29).

Except for topographical differences resulting from the absence of yolk in mammalian embryos and the very early formation of the amnion, the basic processes occurring during the primitive streak stage of mammalian and chick embryos are strikingly similar (cf. Figs. 5-16 and 5-28). It is clear that in the young mammalian embryo as well as in the bird embryo, the primitive streak is the homologue of the fused lips of the blastopore of lower vertebrates.

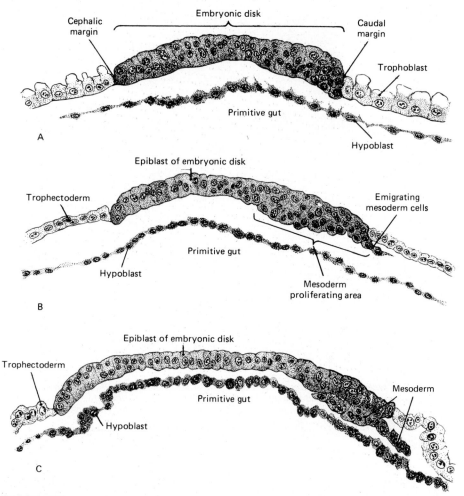

FIGURE 5-27
Longitudinal sections of the embryonic disk of the pig during the ninth day of development, showing three stages in the origin of the mesoderm. (Projection drawings, ×180, from sections of embryos in the Carnegie Collection.)

ORIGIN OF THE GERM LAYERS IN RODENTS

Early cleavage and cell segregation leave the mouse embryo with a blastocyst that is organized in a manner similar to that of other mammals (Fig. 4-16). The early hypoblast, which is called the *primitive endoderm* in the mouse, initially forms a single layer beneath the inner cell mass (Fig. 5-30). Cells from this layer soon spread out beneath the trophoblast (called *trophectoderm*) to form the endodermal layer of the *parietal yolk* sac (Fig. 5-30A). Once associated with the trophectoderm, the parietal endodermal cells secrete a thick basement membrane known as *Reichert's membrane*. Because of its thickness and accessi-

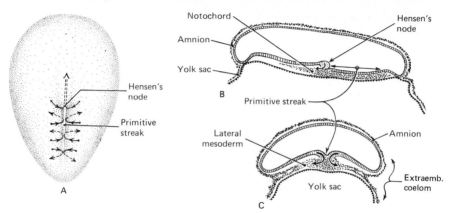

FIGURE 5-28
Diagrams indicating by arrows the probable paths of cell migration in the region of the mammalian primitive streak. (A) Surface plan. (B) Longitudinal section. (C) Cross-section through the primitive streak.

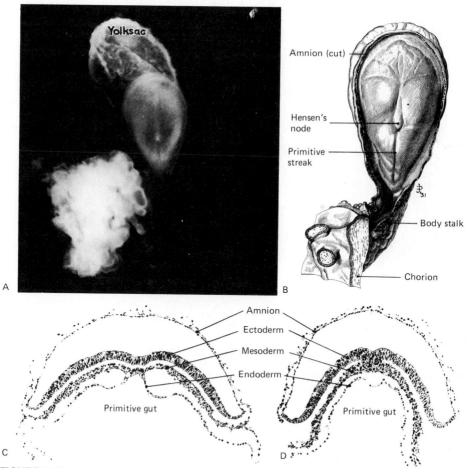

FIGURE 5-29
Human embryo in primitive-streak stage, probable fertilization age of 14 to 15 days. (A) Photographed (×18) before sectioning. (B) Reconstructed after serial sections (×25). (C) Section through neural plate. (D) Section through primitive streak. (After Heuser, 1932. *Carnegie Cont. to Emb.*, vol. 23.)

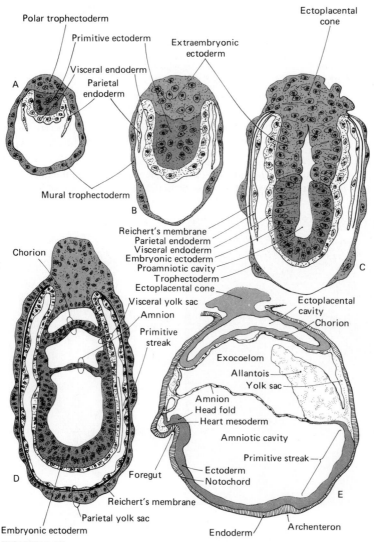

FIGURE 5-30
Gastrulation in the mouse. (A) Stage of early visceral endoderm (4½ days); (B) Early egg cylinder stage (5 days); (C) Advanced egg cylinder stage (6 days); (D) Head process stage (7 days); (E) Three cavity stage (8 days). (Adapted from several sources, mainly Snell and Stevens, 1966.)

bility, Reichert's membrane has often served as a source of material for analytical studies of basal laminae (Minot et al., 1976).

From an early stage, the trophectoderm can be subdivided into two sections. The part overlying the inner cell mass is called the *polar trophectoderm*, whereas the remainder, surrounding the blastocyst cavity, is the *mural trophectoderm*. According to Pederson (1986), who injected individual cells with markers, some cells of the inner cell mass become incorporated into the polar trophectoderm.

Cells of the polar trophectoderm are still able to undergo mitotic divisions, and their descendants contribute to the mural trophectoderm. In contrast, cells of the mural trophectoderm are unable to undergo normal mitotic divisions and instead are transformed into giant cells by becoming polyploid.

In rodents, the inner cell mass undergoes a transformation that is strikingly different from its course in other mammals. It protrudes deeply into the blastocyst cavity in the form of a tonguelike lobe. A cavity (the *proamnion*) forms within the lobe, and the cells surrounding it represent the primitive *ectoderm* (or epiblast, since its developmental properties are very similar to the early epiblast of avian embryos) (Fig. 5-30C). This unusual configuration has resulted in the embryo's being called an inverted *egg cylinder* at this stage. Immediately above the primitive ectoderm, the cells of the polar trophoblast form a relatively massive *ectoplacental cone*.

Soon three cavities are found in place of the original proamniotic cavity (Fig. 5-30D). Details of the developmental anatomy from this point are beyond the scope of this chapter. The interested reader is referred to the descriptions by Snell and Stevens (1966) and Theiler (1972). After giving rise to additional extraembryonic mesoderm, the primitive ectoderm becomes organized into a more typical gastrula and ultimately gives rise to the three embryonic germ layers through a primitive streak (Fig. 5-30). A detailed flow chart of the intra- and extraembryonic cell lineages in the mouse as they are currently understood is given in Fig. 5-31.

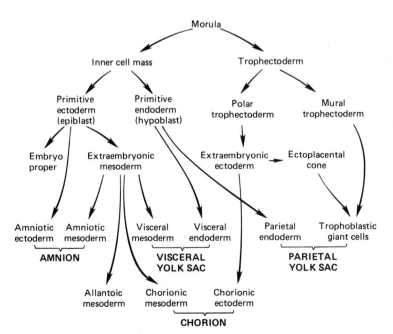

FIGURE 5-31
Scheme illustrating cell lineages in the early mouse embryo. (Based on Gardner, 1983).

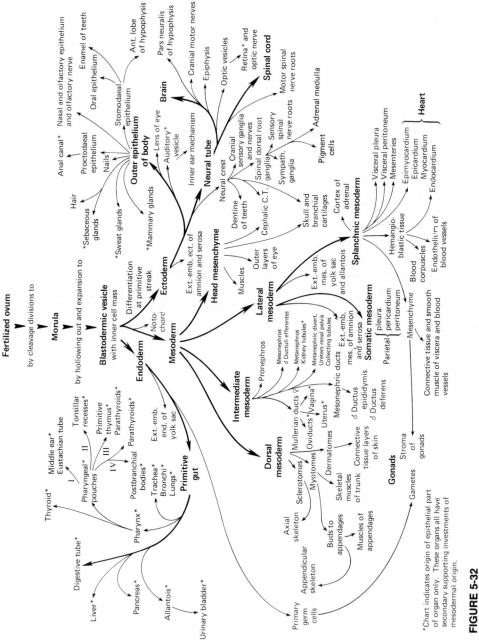

FIGURE 5-32

Chart showing derivation of various parts of the body by progressive differentiation and divergent specialization. Note especially how the origin of all the organs can be traced back to the three primary germ layers.

*Chart indicates origin of epithelial part of organ only. These organs all have secondary supporting investments of mesodermal origin.

220

EMBRYOLOGICAL IMPORTANCE OF THE GERM LAYERS

The formation of three primary germ layers is a common denominator in the early development of all vertebrates. Ideas concerning the significance and importance of the germ layers have undergone gradual modification over the years. Many early embryologists viewed the formation of germ layers as an irreversible segregation of the embryo into rigid compartments, with little interconvertibility among the germ layers. Evidence accumulated over the years, however, has shown that differentiation into a given phenotype is not always limited to cells from a single germ layer. A good example is cartilage. Although most of the cartilage in the body is derived from cells of the mesodermal germ layer, some cartilaginous elements of the head and neck differentiate from cells of the neural crest, which is an ectodermal derivative.

Nevertheless, the concept of germ layers is a very useful one for categorizing and interpreting many developmental phenomena, and it provides a good framework for students to use in organizing their knowledge about the formation of specific tissues and organs. A flow chart relating the differentiation of the major tissues and organs of the body to the primary germ layers is presented in Fig. 5-32. For students beginning the study of embryology, this chart will serve as a means of pointing out in a general way the direction in which the early processes with which we have been dealing are destined to lead. As the phenomena of development are followed further, it will be seen that each natural division of the subject centers more or less sharply on some particular branch of this genealogical tree of the germ layers.

NEURULATION, THE NEURAL CREST, AND SOMITE FORMATION

The morphogenetic movements that dominate the period of gastrulation not only result in the formation of the three primary germ layers but also cause groups of cells that were far apart in the blastula to become located close to one another. The future developmental fate of the embryo depends on inductive interactions between some of these newly associated groups of cells. The primary inductive event is the action of the chordamesoderm, or notochord, on the overlying ectoderm, resulting in the transformation of a band of unspecialized ectodermal cells into the primordium of the central nervous system. The initial response of the induced ectoderm is to form a plate of thickened cells, but soon this plate becomes transformed first into a longitudinal groove and ultimately into a tube. While this is occurring, other ectodermal cells from the junction between the neural and general cutaneous ectodermal tissues form segmentally arranged aggregations that are known collectively as the *neural crest*. Later, cells of the neural crest follow extensive and varied migration and differentiation pathways throughout the body of the embryo.

Following the changes leading to the formation of the neural tube, the mesodermal layer on either side of the notochord splits into longitudinal divisions. The blocks of mesoderm on either side of the notochord soon begin to form symmetrical pairs of bricklike masses called *somites* (Figs. 6-1 and 6-23), which are both major landmarks in the early embryo and the source of a number of important segmentally arranged mesodermal derivatives later in life. The somite pairs first take shape near the cranial part of the embryo. In successive stages additional pairs of somites are formed caudal to those already laid down. From the earliest stages of formation of the nervous system, differentiation of axial structures follows pronounced *cephalocaudal gradients*.

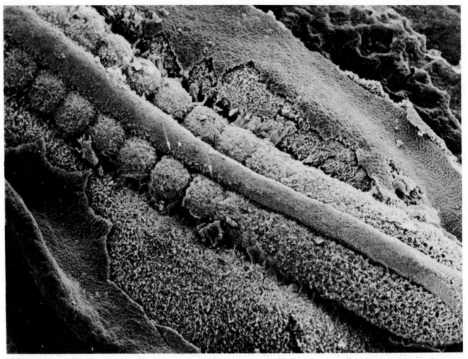

FIGURE 6-1
Scanning electron micrograph of the dorsal surface of a chick embryo after removal of the ectoderm. The upper left is cranial and the lower right is caudal. The long unsegmented structure is
the neural tube and the segmented structures on either side of the neural tube to the left are
somites. In the same line of tissue to the right of the somites is the unsegmented paraxial mesoderm. On the side of the neural tube just above the somites an irregular fringe of neural crest
cells is beginning to move out from the neural tube (arrows). (Courtesy of K. Tosney.)

Because of these gradients, processes which have already been completed in
the cranial part of the embryo may be just beginning in the caudal part.

Commonly, neurulation is considered to be the period of development
starting with the first traces of formation of the neural plate and ending with
closure of the neural tube. This chapter will outline the major events involved
in the development of the neural tube, the migration and differentiation of the
neural crest and the development of somites.

PRIMARY (NEURAL) INDUCTION

During late gastrulation, the chordamesoderm, involuting around the dorsal lip
of the blastopore in amphibians, and the notochordal process, which forms
from cells passing through Hensen's node in birds and mammals, pushes
cranially, just beneath the ectoderm. While the forward movement of the
notochordal tissue is taking place, the chordal cells act on the overlying
ectodermal cells, causing them to thicken and form the neural plate. This

reaction, which both initiates the formation of the central nervous system and causes the central longitudinal axis of the body to be established, is commonly called *primary induction*. The inductor is the *chordamesoderm* (future notochordal tissue), and the responding tissue is the ectoderm.

As in other inductive systems, it is essential that the inductor and the responding tissue be at the right place at the right time. Without the presence of the underlying notochord, the cells of the dorsal ectoderm do not form neural tissue but rather continue to differentiate as general cutaneous ectoderm. This has been demonstrated experimentally by transplanting to the ventral side of the embryo small pieces of prospective neural ectoderm before it has been acted upon by the notochord. The explants do not form neural tissue (Fig. 6-2). However, if the same operation is performed in the late gastrula, the grafted ectoderm forms a neural plate, as it would have done if it had remained in its original location. The necessity of the notochordal inductor has also been strikingly shown by the analysis of amphibian *exogastrulae*. Exogastrulae are embryos in which the normal inpocketing of the *archenteron* is exteriorized, commonly by the concentration or types of certain ions (e.g., Li^+) in the medium surrounding the embryos. The endodermal and chordamesodermal tissues lining the archenteron form an outpocketing from the posterior end of the embryo (Fig. 6-3). Although the tissues lining the everted archenteron

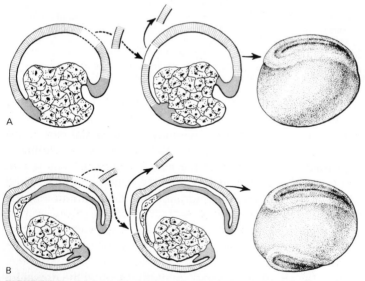

FIGURE 6-2
(A) Transplantation of a piece of presumptive neural plate, before it has been acted upon by chordamesodermal induction, to the ventral part of another embryo results in integration of the graft with its surroundings. (B) After primary induction has occurred, a similar graft produces a secondary neural plate on the ventral side of the embryo. (Modified from Saxén and Toivonen, 1962, *Primary Induction*, Logos Press, London.)

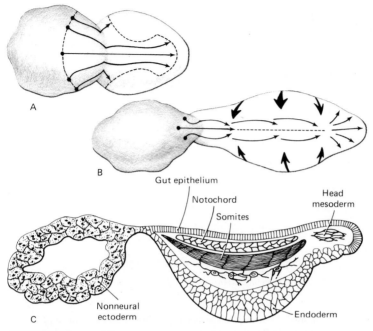

FIGURE 6-3
Exogastrulation in an amphibian embryo. (A, B) Drawings showing the
movement of cells (arrows) along the outer surface of the embryo toward
the region of the blastopore (constriction), but instead of involuting, these
future endodermal and mesodermal cells form a new vesicular structure (to
right of constriction). (C) Sagittal section through exogastrulated embryo,
showing empty ectodermal hull (left) and endodermal and mesodermal differentiation (right). (After J. Holtfreter, 1943, *J. Exp. Zool. 94*:261.)

undergo a certain degree of self-differentiation, the empty ectodermal hull
remains nonneural.

In order for neural induction to occur, the ectoderm overlying the notochordal process must be able (*competent*) to respond to the inductive stimulus.
During much of the period of gastrulation in amphibian embryos both the dorsal
and ventral ectoderm have the competence to form neural tissues when
subjected to the influence of inductors. Figures 5-11 and 6-4 illustrate experiments in both the amphibian and the bird in which neural inductors grafted
beneath the ectoderm induced secondary neural tubes in competent ectoderm
normally not destined to form neural tissue. Later in the gastrula period the
ectoderm farthest from the normal location of the nervous tissue begins to lose
its capacity to respond to neural inductors, and by late in the neurula stage most
nonneural ectoderm has lost its neural competence.

At least in amphibian larvae different regions of the chordamesodermal
inductor impose regional specificity upon the structures differentiating as a
result of this induction. This has been demonstrated by grafting different areas
of chordamesoderm beneath competent ectoderm. Grafts of anterior chorda-

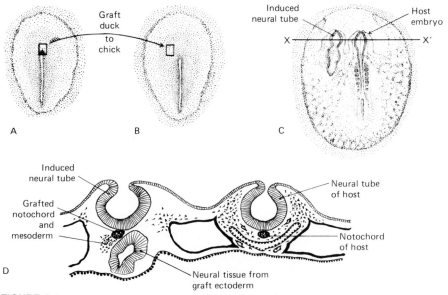

FIGURE 6-4
Semischematic drawings showing the induction of an accessory neural tube as a result of
grafting notochordal tissue from a duck donor into a chick host. (A) Duck embryo showing
the location from which the graft was taken. (B) Chick host showing the location where the
graft was implanted. (C) Embryo cultivated for 31½ hours after implanting of the graft,
showing the location of the induced accessory neural tube. (D) Section at level of the line
X-X′ in (C), diagrammed with the same conventional representation of the component layers as that employed in the other sectional diagrams in this text. (Based on the work of
Waddington and Schmitt, 1933. *Arch. f. Entwickl.-mech. d. Organ.*, vol. 128.)

mesoderm cause accessory heads to form (Fig. 6-5A), and grafts of posterior
chordamesoderm induce accessory tails (Fig. 6-5B). Numerous grafting experiments have led a number of embryologists to conclude that there are two or
three main inducing regions that evoke the formation of the nervous system and
other axial structures of the embryo (Saxén and Toivonen, 1962; Saxén, 1978).
The two major regional components of primary induction have been called
cephalic induction and *spinocaudal induction*. Cephalic induction, a neuralizing influence, results in the formation of head structures, and spinocaudal
induction, a mesodermalizing influence, results in trunk and tail structures.
Saxén and Toivonen have postulated that the character of the induced axial
tissues is a function of the interaction of two inducing gradients associated with
the roof of the archenteron, a dorsoventrally directed neuralizing gradient that
extends along much of the length of the embryo and a caudocephalic gradient
of decreasing mesodermalizing intensity (Fig. 6-6). Whether forebrain, hindbrain, or trunk and tail structures form in a given region depends on the relative
strengths of the two interacting gradients in that area.

Another influential viewpoint on the nature of neural induction is that of
Nieuwkoop (1966) and Leussink (1970). According to this viewpoint, neural

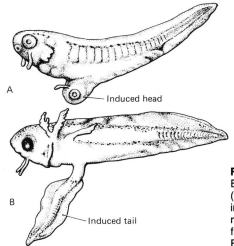

FIGURE 6-5
Examples of archencephalic (A) and spinocaudal (B) induction of newt embryos. Archencephalic induction is produced by grafting anterior chorda-mesoderm, and spinocaudal induction results from grafts of posterior chordamesoderm. (After Balinsky, from Mangold and Tiedemann.)

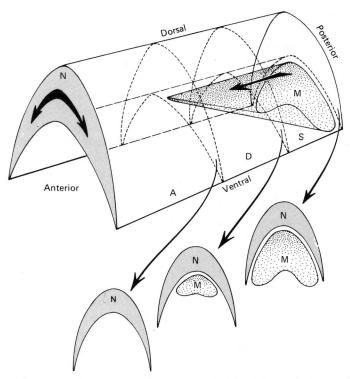

FIGURE 6-6
A schematic representation of the two-gradient hypothesis of primary induction. *Abbreviations*: M, gradient of mesodermalizing action; N, gradient of neuralizing action; A, archencephalic induction; D, deuterencephalic induction; S, spinocaudal induction. (Redrawn from Saxén and Toivonen, 1962. *Primary Induction*, Logos Press, London.)

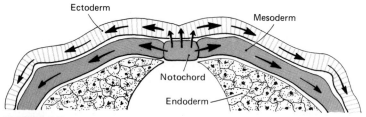

FIGURE 6-7
Representation of the hypothetical course of action events in the activation-transformation hypothesis of neural induction. The number and thickness of the arrows indicate the course and concentration of the activating factor. (Modified from J. Leussink, 1970. *Neth. J. Zool. 20*:1.)

induction involves two separate phases, an activation phase and a later transformation phase. Activation, which begins during gastrulation with the invagination of cells around the dorsal lip of the blastopore, is said to evoke neural differentation tendencies in the overlying ectoderm. Transformation, occurring somewhat later, results in the regional organization of the central nervous system. The lateral extent of the neural plate is considered to be a function of a decreasing concentration of activating factor as a result of inactivation with time and distance (Fig. 6-7).

Despite decades of intensive research, the mechanism of the inductive interaction and the nature of the inductive agent(s) remain unclear. Normally, close contact between notochordal tissues and ectoderm is the rule for neural induction, but some experiments conducted in vitro have shown that ectoderm placed in culture medium previously conditioned by the presence of mesodermal inducing agents will sometimes form neural structures in the absence of the inducing tissue itself. There is good evidence for the transfer of macromolecules from the inducing to the responding tissue (Toivonen et al., 1976), but one cannot rule out the influence of ions or products of cell damage (Barth and Barth, 1974).

Gallera (1971) has summarized well the major similarities and differences in primary induction between amphibians and birds. Too little is known about primary induction in mammals for meaningful comparisons to be made. In both classes of vertebrates the inducing substances are diffusible, and in both classes the time of the response to the inductive stimulus is determined by factors residing in the ectoderm. Both forms also show a similar pattern in the decline of competence of the ectoderm to the neural inductive stimulus. Initially neural competence declines slowly, but later the rate of decline changes rapidly. Cephalic parts of the nervous system are induced first, followed later by the spinal cord. In the chick, induction of the brain is a very early event, possibly carried out by the hypoblast. The notochord of the chick loses its neural inducing capacity by the time the first pair of somites appears, whereas in amphibians its inducing capacity is preserved later into development. For neural differentiation to occur in the chick, 6 to 8 hours of close contact between inducing and responding tissues is required. Amphibians show great variability in this regard. Finally, neural induction in amphibia can be obtained by a wide variety of tissues other than chordamesoderm. These tissues, such as liver and bone marrow from guinea pigs, are called *heteroinductors*. Bird embryos are much less responsive than are amphibian embryos to the effects of heteroinductors.

In response to the inductive stimulus, the ectodermal cells overlying the notochordal process proliferate, synthesize new mRNAs, and cross a restriction threshold, so that their future developmental course is channeled into the production of nervous tissues. Morphologically, the cells responding to neural induction change their shape from a cuboidal or low columnar configuration to a high columnar form. This causes the neural tissues to rise above the surrounding ectoderm as the flattened *neural plate* (Fig. 6-9). One of the molecular responses to specific inductions (or other changes in cell state) is the types of CAMs (see p. 210) expressed by the cells. At different stages in their life history different types of CAMs are expressed (Table 6-1). Much remains to be learned about the way in which the CAMs are involved in these changes of state.

NEURULATION IN AMPHIBIANS

Later stages of gastrulation in amphibians are dominated by the completion of cell movements leading to the formation of the primitive gut (Fig. 5-8) and, in urodeles, the spreading of the mesodermal mantle between the ectodermal and endodermal layers of the embryo (Fig. 5-10). Concomitant with these changes the blastopore decreases in prominence. During this period, the process of primary induction is coming to a close and the ectoderm overlying the notochordal process has begun to respond by thickening to form the neural plate.

Although anuran and urodelan embryos differ to some extent in regard to the manner in which the endodermal and mesodermal layers become established, their basic organization is essentially similar by early neurulation. In the interior of the embryo the primitive gut is becoming completely surrounded by endodermal cells (Fig. 6-8). Surrounding the endoderm is a complete layer of mesoderm. In the dorsal midline the notochord is now a discrete rod. On either side of the notochord the dorsal part of the mesodermal sheet (*epimere*) begins to thicken and simultaneously undergo segmentation. The paired mesodermal segments will become the *somites*. The broad mass of early mesoderm (*hypomere*) filling the lateral and ventral parts of the embryo with a thin layer of cells is called the *lateral plate mesoderm*. Ultimately the lateral plate mesoderm will split into two layers: an outer layer apposed to the ectoderm (*parietal* or *somatic mesoderm*) and an inner layer surrounding the endoderm (*visceral* or *splanchnic mesoderm*). The somatic mesoderm and its overlying ectoderm are collectively called the *somatopleure*, whereas the splanchnic mesoderm and the underlying layer of endoderm constitute the *splanchno-*

TABLE 6-1
CHANGES IN CELL ADHESION MOLECULES (CAMs) AS CELLS CHANGE THEIR STATE DURING EARLY EMBRYOGENESIS

Original cell type	Later stage	Final differentiation product
Mode I: Cells passing through a mesenchymal morphology as they develop		
N-CAM ⟶	Mesenchyme (no CAMs) ⟶ N-CAM	⟶ N-CAM disappears
Ectoderm		
Neural plate ⟶	Neural crest mesenchyme ⟶ Peripheral nerve ganglia	
Mesoderm		
Somite ⟶	Mesenchyme ⟶ Skeletal muscle (motor end plate) ⟶ Feather (dermal papilla) ⟶ Somite	⟶ Chondrocytes
Nephrotome ⟶	Mesenchyme ⟶ Gonadal epithelium and connective tissue	
Splanchnopleure ⟶	Mesenchyme ⟶ Parts of spleen Gut Mesenteries	
Somatopleure ⟶	Mesenchyme ⟶ Smooth muscle	

Original cell type	Later cell type	Final differentiation product
Mode II: Epithelial cell conversions		
N-CAM and L-CAM Ectoderm ⟶	N-CAM Neural tube Placode-derived ganglia Lens primordium	⟶ N-CAM lost ⟶ Lens
	L-CAM Nonneural ectoderm Basal layer of skin Apical ectodermal ridge of limb bud	⟶ Stratum corneum of epidermis
N-CAM Mesoderm ⟶	N-CAM and L-CAM Urogenital mesoderm	⟶ L-CAM Wolffian duct Mesonephric duct Müllerian duct
N-CAM and L-CAM Endoderm ⟶	L-CAM Epithelium of digestive tract Trachea Pharyngeal glands	

Source: Adapted from Crossin et al. (1985).

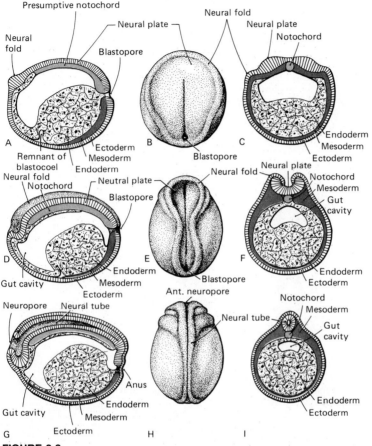

FIGURE 6-8
Representative stages of neurulation in the frog embryo. (A–C) Early neurula, (D–F) middle neurula, (G–I) late neurula. *Left column*, midsagittal sections; *right column*, transverse sections cut through the embryos illustrated in
the middle column. (After Balinsky.)

pleure. The cavity that forms between these two layers of mesodermal cells is
known as the *coelom*. Joining the somites and the lateral plate mesoderm is a
thin connecting link of mesodermal cells (*mesomere*) called the *intermediate
mesoderm*. From these cells the urogenital structures ultimately arise.

In the ectodermal layer, the cells of the future nervous system have begun
to follow a morphological course different from that of the rest of the ectoderm.
Initially all the ectodermal cells are arranged as a single sheet of low columnar
cells (Fig. 6-9). Subsequent to neural induction, the cells in an oval area
overlying the notochord and future somites undergo a pronounced elongation
to form the neural plate, whereas at the same time the presumptive epidermal
cells covering the rest of the embryo become flattened.

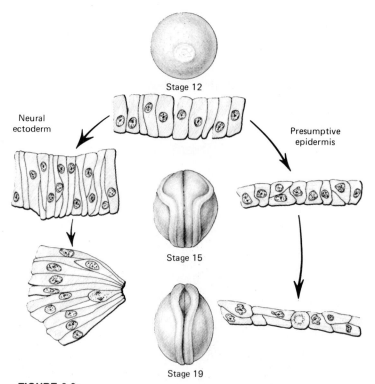

Stage 12

Neural
ectoderm

Presumptive
epidermis

Stage 15

Stage 19

FIGURE 6-9
Changes in shape of ectodermal cells during neurulation in the newt embryo. (Modified from Burnside, 1971. *Devel. Biol.*, *26*:419.)

Soon the edges of the neural plate become raised, forming an elevated *neural fold* on either side of a shallow, longitudinal *neural groove*. The shape of the neural plate then undergoes a pronounced transformation, becoming attenuated posteriorly, where the spinal cord will eventually form, and remaining broader anteriorly, where the brain will form (Fig. 6-8E). The deformation of the neural plate is the result of a set of morphogenetic movements caused principally by changes in the shape of cells composing the neural plate (Burnside and Jacobson, 1968). During its deformation, the surface area of the neural plate decreases, but its volume remains roughly constant as the shrinkage of the apical surfaces of the neuroectodermal cells is matched by a corresponding increase in their height. The shifts in the positions of cells are much more pronounced in the region of the future spinal cord than in that of the brain (Fig. 6-10), and the magnitude of cellular displacement is correlated with the degree of change in shape of individual cells. Jacobson and Gordon (1976) simulated the cell movements and changes in shape of the neural plate on a computer by taking into account both shrinkage of the apical surfaces of the cells and the anterior expansion resulting from an intimate association between the cells of the neural plate and the notochord, which

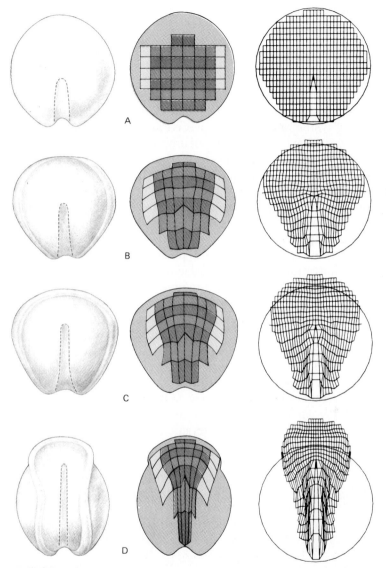

FIGURE 6-10
Cell shifts during the formation of the neural plate, in the newt embryo. Left
column, gross views of the embryo; middle column, changes in shape of the
neural plate region, as indicated by deformation of individual components of
the grid; right column, computer simulation of the deformation processes illustrated in the middle column. (Redrawn from papers by Jacobson and Gordon.)

continues to extend in a cephalic direction (Fig. 6-10C). Malacinski and Youn
(1982), however, have presented evidence suggesting that elongation of the
notochord may not be essential for cephalic extension of the neural plate.

FORMATION OF THE NEURAL TUBE

The neural plate does not remain flat very long. Soon after it has taken shape, its lateral borders become elevated, forming the neural folds, which flank the neural groove (Figs. 6-8 and 6-11). The two lateral edges of the neural folds eventually come together in the dorsal midline to form a complete *neural tube* (Fig. 6-11). Closure of the neural tube first occurs in the upper spinal cord levels and from there proceeds both cephalad and caudad. The last remaining open portions of the nervous system are small uncovered areas called the *anterior* and *posterior neuropores* (Fig. 6-8H). Both neuropores ultimately become obliterated without leaving behind any structures of note.

The mechanism of neural-tube formation has been the subject of much speculation over the years, and even now not all aspects of the process are well understood (Gordon, 1985). Modern investigations have confirmed earlier speculations that at least part of the process of neural folding can be attributed to intrinsic changes in shape of the neuroepithelial cells. Holtfreter (1974) observed that single cells isolated from the neural plate of salamander embryos complete their normal elongation in vitro. As these cells elongate, their apices constrict. Elongation of a neuroepithelial cell requires the presence of a series of intact microtubules running from the base to the apex of the cell (Fig. 6-12). The microtubules act like an internal skeleton of the cell, supporting its greatly increased height. Meanwhile, just beneath the apical surfaces of these cells are organized bundles of thin microfilaments, which can contract, resulting in the constriction of the apical end of the cell (Fig. 6-12). Integrity of the microtubules can be disrupted by exposing them to the drug *colchicine*, and the organized bundles of microfilaments can be replaced by dense granular masses through the action of *cytochalasin B*. After exposure to these inhibitors, neuroepithelial cells do not undergo their chacteristic changes in shape and the neural plate remains unfolded.

It is unlikely that changes of cell shape are sufficient in themselves to account for formation of the neural tube. One must also take into account tensions caused by the growth of structures (e.g., the notochord) underlying the

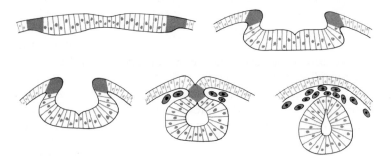

FIGURE 6-11
Cross-section illustrating the formation of the neural tube and neural crest in the amphibian embryo. Gray areas, neural crest. (After Balinsky.)

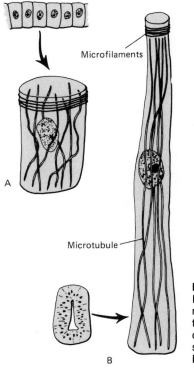

FIGURE 6-12
Intracellular correlates of changes in cell shape during neural tube formation in the salamander embryo. Elongation accompanies lengthening and alignments of the microtubules; whereas narrowing of the apex is due to constriction of bands of microfilaments. (Adapted from B. Burnside, 1973. *Am. Zool. 13*:989.)

neural plate as well as changes in the aggregative and mechanical properties of the neuroepithelial cells as they relate to adjacent structures in the embryo.

THE NEURAL CREST

As the raised lateral walls of the neural folds come together to form the neural tube, a new group of ectodermal cells, emanating from the junctions between neural and nonneural ectoderm, becomes evident (Fig. 6-13). These cells, originally arranged as a longitudinal pair of loosely aggregated cells that extend along either side of the dorsal midline between the neural tube and the superficial ectoderm, constitute the *neural crest*, one of the most remarkable primordia in the embryonic body (Hörstadius, 1950; Le Douarin, 1982).

From the time of their first appearance, the cells of the neural crest (Fig. 6-14) are endowed with the ability to undertake extensive but tightly controlled migrations throughout the body. In the head the cells of the neural crest begin their migration (actually, a significant component of this "migration" is tissue displacement) as a relatively cohesive unit (Noden, 1984; Fig. 6-15), whereas in the trunk the migrations of the cells are more individualized from the start.

In the trunk, cells leave the neural crest in two major streams (Fig. 6-16). One stream is superficial and dorsally directed; the other is more ventrally

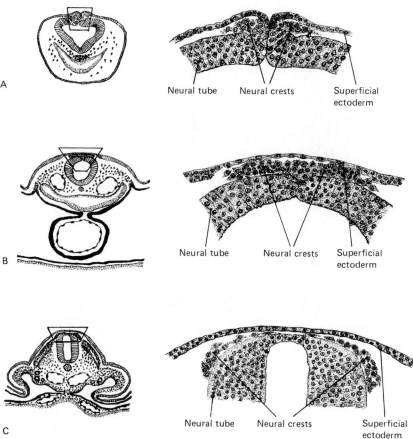

FIGURE 6-13
Drawings from transverse sections to show the origin of neural crest cells. The location of the area drawn is indicated on the small sketch to the left of each drawing. (A) Anterior rhombencephalic region of 30-hour chick. (B) Posterior rhombencephalic region of 36-hour chick. (C) Middorsal region of cord of 55-hour chick.

directed and leads through and around the somitic mesenchyme. Cells following the dorsal migratory pathway move into the ectoderm, where they differentiate into pigment cells, which ultimately settle down in either the epidermis or the dermis according to patterns characteristic of the species. The cells that traverse the ventral pathway develop into components of the autonomic nervous system as well as other structures. Still other cells remain close to the region of the original neural crest but soon become aggregated into segmental pairs: the dorsal root ganglia of the sensory nerves.

Cells from the cranial division of the neural crest have a wide repertoire of differentiative capacities, as can be seen in Table 6-2. In contrast to neural crest cells of the trunk, cranial crest cells form much of the skeleton and connective tissue of face (Fig. 6-17).

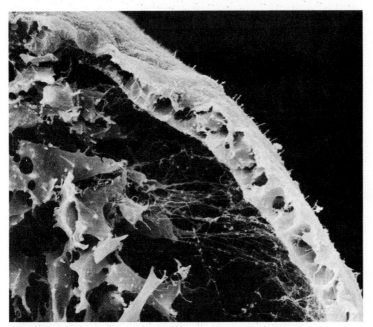

FIGURE 6-14
Scanning electron micrograph of neural crest cells (left) migrating beneath the ectoderm of a chick embryo. (Courtesy of K. Tosney.)

With the use of improved cell-tracing techniques, especially grafts of quail cells with their unique nuclear marker (see Chap. 1), embryologists have made

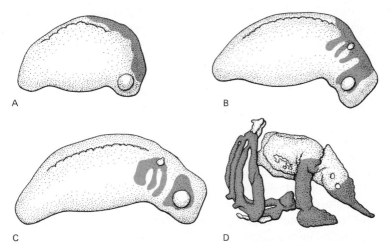

A

B

C

D

FIGURE 6-15
(A–C) Migration of neural-crest cells (gray areas) in branchial and head region of the salamander. (D) Skull of salamander, showing portions derived from neural crest (gray) and from mesoderm (light). (Modified from L. S. Stone, 1926. *J. Exp. Zool. 44*:95.)

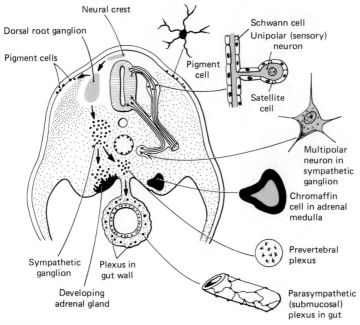

FIGURE 6-16
Schematic cross section through the trunk of an embryo, showing the superficial and deep pathways of migration (left side) and adult derivatives
(right side) of neural crest cells.

rapid strides in understanding the development of the neural crest. The basic
technique is to graft early neural crest cells or their precursors from quail
embryos into chick embryos. The grafted cells can then be followed throughout
their development.

In addition to determining the adult derivatives of neural crest cells,
investigators have been able to ask a number of penetrating questions about
their behavior and their differentiation. One set of questions concerns the
migration of neural crest cells: What causes neural crest cells to leave the
neural tube in the first place? What supports their migration and the path or
direction of migration? What causes migration of these cells to cease? Another
set of questions relates to the differentiation of neural crest cells: What
accounts for the wide array of cell types derived from the neural crest (Table
6-2)? Is there a common precursor cell or are the earliest neural crest cells
partially determined into several major groups? What is the role of intrinsic
determination versus environmental influences in the differentiation of specific
neural crest derivatives? A third question involves the determination of the
morphology, or pattern, of structures derived from the neural crest: Is the
information required for pattern formation inherent in the early neural crest or
do the neural crest cells respond to extrinsic morphogenetic cues? Many of

TABLE 6-2
MAJOR NEURAL CREST DERIVATIVES

	Cranial crest	**Trunk crest**
Nervous system		
Sensory nervous system	Ganglia of (V) Trigeminal nerve (VII) Facial nerve (root) (IX) Glossopharyngeal (superior ganglion) (X) Vagus nerve (jugular ganglion)	Spinal ganglia Rohon-Béard cells (amphibian larvae)
Autonomic nervous system	Parasympathetic ganglia Ciliary Ethmoidal Sphenopalatine Submandibular Visceral	Parasympathetic ganglia Pelvic plexus Remak's Visceral Sympathetic ganglia Superior cervical Paravertebral Prevertebral
Nonneural cells	Oligodendroglia Satellite cells of ganglia Schwann cells of peripheral nerves (relatively minor contribution)	Satellite cells of ganglia Schwann cells of peripheral nerves
Pigment cells	Melanophores (black) Xanthophores (yellow) Erythrophores (red) Iridophores (iridescent)	Melanophores Xanthophores Erythrophores Iridophores
Endocrine and paraendocrine cells	Calcitonin-producing cells Carotid body (type I cells) Parafollicular cells (thyroid)	Adrenal medulla Neurosecretory cells of heart and lungs
Mesectodermal cells		
Skeleton	Cranial vault (squamosal and part of frontal) Nasal and orbital Otic capsule (part) Palate and maxillary Sphenoid (small contribution) Trabeculae (part) Visceral cartilages	None
Connective tissue	Dermis, fat, and smooth muscle of skin Ciliary muscles of eyes Cornea of eye (fibroblasts of stroma and corneal endothelium) Connective tissue stroma of glands in head and neck Dental papilla (odontoblasts) Meninges of prosencephalon and part of mesencephalon Connective tissue and muscle in walls of aortic and arch-derived arteries	Dorsal fin mesenchyme (amphibians)

Note: This table is intended to be a reference that illustrates the diversity of cell types and structures generated by the neural crest. It was not included as a hurdle that must be memorized in full at this early stage in the study of embryonic development.

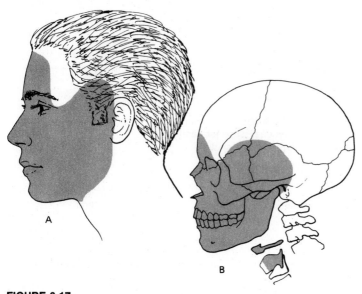

FIGURE 6-17
Presumed distribution of neural crest cells in the dermis (A) and skeleton
(B) of the human head. (Adapted from Johnston et al., 1973.)

these questions remain unanswered, but partial answers to some of them have
been discovered.

Mystery still surrounds the origin of neural crest cells and their emigration
from the neural tube. It is known that as they emerge from the neural tube,
neural crest cells become transformed from an epithelial to a mesenchymal
type of cell (Fig. 6-14). How they penetrate the basal lamina of the neural tube
is not known, but as they become mesenchymal, neural crest cells lose the cell
adhesion molecule N-CAM, which is characteristic of cells in the neural tube,
and develop migratory properties (Fig. 6-18). Although they originally migrate
into a cell-free space, neural crest cells show distinct preferences in their
pathways of migration (Newgreen and Erickson, 1986). Basal laminae, such as
that beneath the ectoderm, are favored substrates, and migrating neural crest
cells, which initially move at about 70 μm/hour, show a strong affinity for the
extracellular matrix molecule, fibronectin, and for loose spaces created by
hyaluronic acid (Fig. 1-14). Conversely, another matrix molecule, *chondroitin
sulfate*, does not support neural crest migration and serves as a barrier to the
advance of neural crest cells. This may explain why neural crest cells avoid the
area of the notochord and vertebral bodies, which are rich in chondroitin
sulfate (Fig. 6-19).

To a considerable extent, the pathway of migration of neural crest cells is
determined by the local environment rather than by intrinsic factors. For
example, Noden (1975) grafted neural crest cells from one part of the body to
another. In general, the grafted cells in the trunk migrate according to a pattern

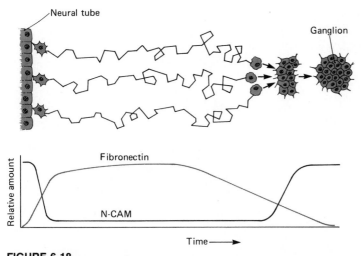

FIGURE 6-18
Changes in N-CAM expression and its relation to fibronectin in migra-
tion of avian neural crest cells from the neural tube. (After Edelman,
Sci. Am., April 1984, based on the work of Thiery.)

appropriate for the area of the host into which they were placed. Despite the
existence of favored channels available for cell migration, very few cell types
other than neural crest cells have the ability to migrate along them. Interestingly, in addition to neural crest cells, certain types of malignant cells can both
penetrate basal laminae (e.g., of the neural tube) and migrate along the normal
neural crest pathways.

Surprisingly little is known about what causes neural crest cells to differentiate into specific mature types. At one extreme, all neural crest cells could
have identical developmental potential, and their fate would be determined
entirely by environmental factors. At the other extreme, the seemingly
identical neural crest cells could all be determined for different fates before
leaving the neural tube. The truth is probably somewhere in between. Experimental evidence suggests that with time the developmental potential of neural
crest cells becomes progressively restricted (rev. by Weston, 1986). For
example, even when transplanted to the cephalic region, trunk crest cells are
unable to form skeletal structures or connective tissue. On the other hand,
there is considerable evidence from both in vivo and in vitro studies that the
transmitter substance produced by certain autonomic neurons can be influenced by the environment into which the neuron is placed (Black, 1982). Bunge
et al. (1982) propose that there are two major steps in the differentiation of
autonomic neurons from the neural crest. The first step is an early decision to
become an autonomic neuron. This commits the cell to becoming a defined type
of neuron, which already expresses a *neurotransmitter substance* (*acetylcholine* or *norepinephrine*). The choice of a transmitter, however, is not fixed. As

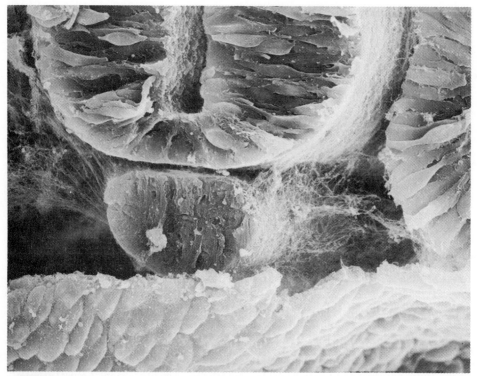

FIGURE 6-19
Scanning electron micrograph showing the neural tube (top), notochord (middle) and part of a somite (right margin) of a chick embryo. Alongside the neural tube and notochord a tangled web-bing of fibrillar components of the extracellular matrix plays an important role in cell migrations and cellular interactions in the area. (Courtesy of K. Tosney.)

the neuron matures and its processes sample the environment around the periphery, a final choice of transmitter is made. Under experimental conditions, a neuron that produced norepinephrine (*adrenergic*) early in development can be converted to an acetylcholine-producing (*cholinergic*) neuron later in its life history.

Studies on the morphogenesis of neural crest derivatives are still in their infancy, but a study by Noden (1983) provided evidence that much of the information determining the specific form of certain neural crest derivatives in the head is inherent in the neural crest cells before they leave the neural tube. In avian embryos Noden grafted the neural crest region that gives rise to the skeleton of the first branchial arch (see Chap. 15) into the regions formerly occupied by cells that migrate to the second or third arch. Although the grafted first arch crest cells followed the normal second or third arch pathways, they gave rise to bones recognizable as derivatives of the first branchial arch, including a beak arising from the neck.

THE MESODERM OF THE EARLY EMBRYO

During the periods of germ-layer formation and early neurulation the mesoderm of the embryo consists of mesenchymatous tissue sandwiched between the epithelial layers of the ectoderm and endoderm (Fig. 6-20). *Mesenchyme* is a morphological term that refers to tissues, regardless of their germ layer of origin, which consist of aggregates of spindle-shaped or stellate cells embedded in an intercellular matrix containing varying amounts of mucopolysaccharide (glycosaminoglycan) ground substance. The mesenchyme of early embryos contains very little matrix, whereas some forms of mesenchyme in later embryos (e.g., that of the umbilical cord) are dominated by matrix. Epithelia, on the other hand, have distinct apical and basal surfaces (Fig. 6-21). The lateral walls of adjoining epithelial cells closely adhere to one another by continuous tight junctions, which seem to be very important in regulating permeability and electrical properties of the epithelia. Except for very early embryos, the basal surfaces of epithelia rest upon a basal lamina, the nature of which varies with the epithelium that secretes it. Like the term *mesenchyme*,

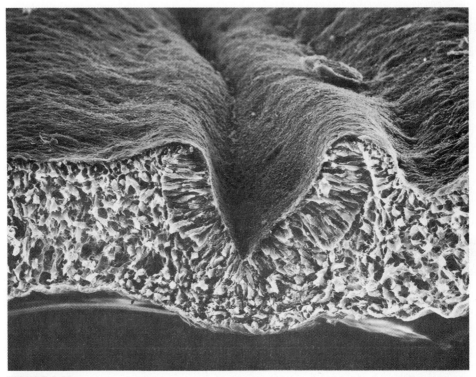

FIGURE 6-20
Scanning electron micrograph of a cross-sectioned chick embryo, showing the prominent neural folds (center). The regularity of the ectodermal cells of the thickened neuroepithelium stands out in sharp contrast to the meshwork of mesodermal cells beneath the surface epithelium. (Courtesy of K. Tosney.)

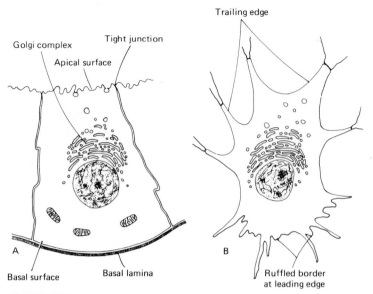

FIGURE 6-21
Diagram illustrating the major structural features of epithelial (A) and mesenchymal (B) cells in the embryo. (Modified from E. Hay, 1968, in Fleisch-
mayer and Billingham, eds., *Epithelial-Mesenchymal Interactions*, Williams
& Wilkins, Baltimore.)

epithelium is a structural classification. Cells of all three germ layers can form
epithelia.

Most modern studies of mesodermal structure have been carried out on
chick embryos; hence, the major emphasis in this section will be on avian
mesoderm. Mesoderm originates from cells derived from the epithelial epiblast
layer. These cells pass through the primitive streak and spread outward as the
definitive mesoderm. Hay (1968) described several phases in the early development of the mesoderm. When it first emerges from the primitive streak, the
mesoderm consists of migrating cells arranged as a mesenchymal layer. This
has been called the *primary mesenchyme*. The cells emigrating from the
primitive streak appear to be polarized, with a leading edge that possesses an
irregular border, from which filopodia project almost like microantennae to test
the local environment through which the cells move. The trailing edge of these
cells appears to be the original apical surface when the cells were part of the
epiblast layer. These mesenchymal cells are connected by numerous small gap
junctions (Revel et al., 1973), which may serve as the structural basis for
electrical coupling that has been demonstrated between cells in early chick
embryos (Sheridan, 1966).

Once it has completed its spread outward from the primitive streak, much of
the mesoderm of the chick becomes organized as an epithelium. The epithelial
organization of the somite is particularly pronounced (Fig. 6-23). When first

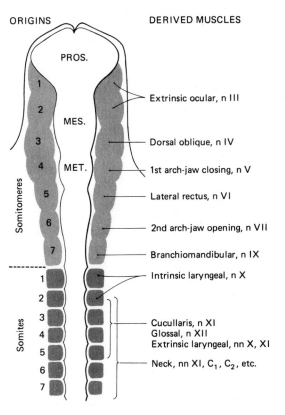

ORIGINS   DERIVED MUSCLES

Extrinsic ocular, n III

Dorsal oblique, n IV

1st arch-jaw closing, n V

Lateral rectus, n VI

2nd arch-jaw opening, n VII

Branchiomandibular, n IX

Intrinsic laryngeal, n X

Cucullaris, n XI
Glossal, n XII
Extrinsic laryngeal, nn X, XI

Neck, nn XI, C_1, C_2, etc.

FIGURE 6-22
Somitomeres and the origins of cranial muscles in the avian embryo. (Adapted from D. Noden, 1983. *Am. J. Anat. 168*:260.)

formed, each somite contains a central cavity, although in some species, such as the chick, it may be partially occluded with cells. The surface of the somitic cells facing the cavity of the somite is the apical surface of the somitic epithelium. Cilia are present on this surface, and the Golgi apparatus is located on this side of the cells. A basal lamina forms around the basal surfaces of these cells.

Later in development, following inductive activity by the neural tube and notochord, the somite loses its epithelial configuration, starting at its ventro-medial border. Cells in this area reacquire mesenchymal properties and migrate away from the main body of the somite (Fig. 6-23C and D). These cells are called *secondary mesenchyme*, and they are unlikely to ever again take on epithelial characteristics during their subsequent developmental course. A major distinguishing feature of secondary mesenchyme in contrast to primary mesenchyme is a greater prominence of matrix among the cells.

SECRETION OF EXTRACELLULAR MATERIALS IN THE EARLY EMBRYO

Virtually all cells in early embryos are in contact with some form of extracellular matrix, and properties of the matrix provide some of the microenviron-

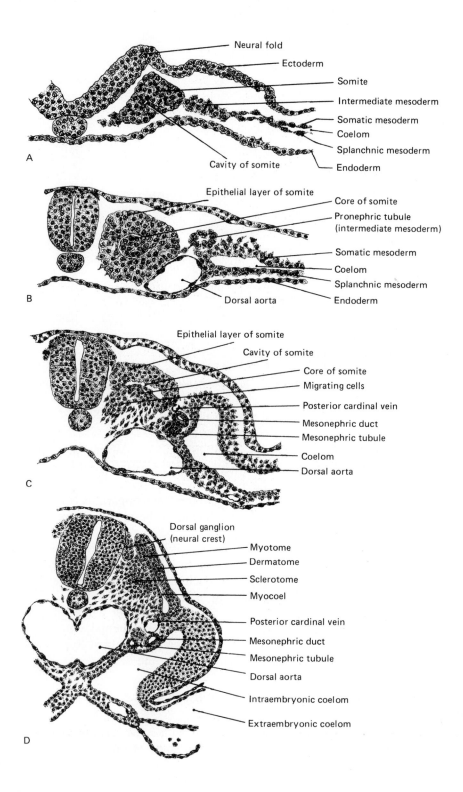

mental cues that initiate, stabilize, or change morphogenetic and differentiative processes in the embryo. Two major varieties of extracellular matrix have recently received considerable attention, particularly in the chick embryo. One is the *basal lamina*, a relatively dense sheetlike structure that is closely apposed to the basal surface of epithelia. There is increasing evidence that most epithelia secrete their own basal laminae, often as a function of a series of communications between the epithelium and the underlying extracellular material. The other major variety of matrix is an amorphous form which surrounds mesenchymal cells. The composition of this matrix varies in different regions of the embryo, particularly as the mesenchymal cells undergo specialization into various types of tissues.

The role of the extracellular matrix in facilating cell movements of the mesoderm of the amphibian embryo and of cells of the neural crest has already been discussed. Local variations in the amount and composition of the extracellular matrix can be important factors in determining the distribution of migrating cells. Other instances of the role of matrix molecules in promoting or inhibiting cell migrations (e.g., corneal development) will be covered in later chapters. The role of the extracellular matrix in supporting cell migrations persists into postnatal life. When human skin is cut, the epidermal cells that seal off the wound migrate over a substrate of extracellular matrix.

At the end of neurulation, both the neural tube and the notochord are arranged as epithelia, with their apical surfaces in the interior and their basal surfaces, surrounded by basal laminae and matrix material, at the periphery. There is now considerable evidence that both the neural tube and the notochord secrete collagen and glycosaminoglycans. At this stage the newly formed somites are also arranged as epithelial structures, each one surrounded by its own basal lamina. It is becoming increasingly evident that the later inductive actions of the neural tube and notochord upon the somites are mediated at least in part by the extracellular matrix lying between them (Fig. 6-19).

THE FORMATION AND DIFFERENTIATION OF SOMITES AND SOMITOMERES

After the mesodermal cells of the chick have passed through the primitive streak, they migrate laterad, using the inner surfaces of the epiblast and hypoblast as substrates to assist or guide their movements. The newly formed mesoderm is a continuous sheet several cell layers thick. The cells that will eventually form the somites are found in two thickened bands (*paraxial mesoderm* or *segmental plate*) running longitudinally along each side of the

---

FIGURE 6-23
Drawings from transverse sections to show the differentiation of the somites. (A) Second somite of a 4-somite chick (about 24 hours), (B) ninth somite of a 12-somite chick (about 33 hours), (C) twentieth somite of a 30-somite chick (about 55 hours), (D) seventeenth somite of a 33-somite chick (about 2½ days).

neural tube and notochord. For many years the segmental plate mesoderm was considered to be a relatively featureless structure in which unseen processes leading to the later formation of somites were beginning to take place. However, by viewing three-dimensional images achieved with stereo scanning electron microscopy, Meier (1984) was able to discern a barely visible but regular segmentation in the segmental plate of the chick embryo. These segments, which begin in the rostral part of the head, are called *somitomeres* (Fig. 6-22). As the primitive streak regresses caudad, additional pairs of somitomeres are laid down, with the most recently formed pair of somitomeres located just caudal to Hensen's node. Some investigators (Wachtler and Jacob, 1986) feel that there is still insufficient morphological evidence to support the conclusion that the paraxial mesoderm of the head is segmentally arranged into somitomeres.

The next stage in the development of the paraxial mesoderm is the transformation of somitomeres into blocklike *somites*. Interestingly, the first pair of somites forms from the eighth pair of somitomeres. The first seven somitomeres in the head retain their primitive organization.

Somitomeres become transformed into somites by taking on the characteristics of epithelial cells. As the cells become more compactly arranged around a central cavity, their apical surfaces become cemented together by continuous tight junctions, and a basal lamina is laid down around the outer surface of the somites. The early formation of somites is a very regular process, with one pair of somites taking shape about every hour in the chick embryo until 50 pairs are formed.

Over the years there have been numerous experimental and theoretical attempts to understand the mechanisms of somite formation. The presomitic segmental plate already contains the necessary information or a pattern for somite formation, for if a piece of segmental plate is removed and transplanted to another site, somites still form within it. There is some evidence that the somite-patterning mechanism is set in place during or even before gastrulation, but virtually nothing is known about the basis for this pattern. The recent explosion of knowledge about the genetic basis for segmentation in *Drosophilia* gives hope that a better understanding of the basis for the segmental pattern in vertebrates will be achieved in the near future.

More progress has been made toward understanding the expression of the somite pattern (Lash and Ostrovsky, 1986). Neither the neural tube nor the notochord is essential, for somites can form in their absence. Likewise, there is little evidence of inductive interactions being involved in either the genesis or the expression of the basic somite pattern. Some experimental evidence (Lipton and Jacobson, 1974) suggests that the mechanical shearing effect of Hensen's node and the regressing primitive streak may be sufficient to cause the formation of somites from the previous pattern. Bellairs (1985), however, casts doubt on this interpretation and cites evidence that in the absence of Hensen's node a median unpaired line of somites still forms.

Early in their formation, somites are committed to produce structures characteristic of their level along the body axis. If somites are removed and transplanted elsewhere, they still give rise to structures characteristic of their original position in the body.

Once the somites are established, the next major morphological change is the reflection of an inductive influence emanating from the notochord and neural tube. As a result of this induction, cells of the ventromedial wall of the somites undergo a burst of mitosis, lose their epithelial characteristics, and become transformed into a secondary mesenchyme. These cells, which are derived from the region of the somite known as the *sclerotome* (Fig. 6-23D), migrate away from the somite and in time surround the notochord and the ventral portion of the neural tube. They soon begin to secrete large amounts of chondroitin sulfate and other molecules characteristic of cartilage matrix. Ultimately, the cells of the sclerotome form the vertebrae, the ribs, and the scapulae.

The inductive role of the notochord and neural tube was demonstrated by extirpation experiments (Holtzer and Detwiler, 1953). If these structures are removed from an early embryo, cartilage cells do not differentiate from the somites. However, there is evidence that before the inductive stimulus, the cells of the sclerotome produce small amounts of molecules characteristic of cartilage (Lash, 1968). In this system, the role of the inductor is to increase greatly the amount of synthesis of these molecules. It now appears that various forms of collagen and proteoglycans act as the effective agents in this specific inductive process.

After the cells of the sclerotome have emigrated from the somite, the remaining dorsolateral wall gives rise to a new layer of cells which takes shape along its inner surface (Langman and Nelson, 1968). The new inner layer is called the *myotome*, and the lateral layer from which it arose is called the *dermatome*. Marking experiments have shown that cells of the dermatome give rise to the dermis, whereas those of the myotomes form the musculature of the body wall and the limbs (Chevallier et al., 1977).

With the persistence of somitomeres in the head and the presence of somites in the trunk, there are some striking comparisons between what can and cannot be formed by both mesoderm and neural crest in the head and trunk regions of the embryo. The cranial somitomeres do not contribute cells to the skeleton or connective tissues of the head, a function characteristic of the sclerotomal portion of somites. In contrast, the cellular precursors for these tissues in the head arise from the cranial neural crest, whereas the trunk neural crest is incapable of forming skeletal or connective tissue. A common function of both somitomeres and somites is the generation of the precursor cells of skeletal muscle (see Chap. 9).

EXTRAEMBRYONIC MEMBRANES AND PLACENTA

One of the most important evolutionary adaptations required for the independent terrestrial existence of vertebrates was the development of a means of preserving a moist, protective environment around the embryo. Some sharks and live-bearing fishes have evolved reproductive patterns involving the internal fertilization and development of their eggs, but in the vast majority of fishes the female lays a large number of eggs, which are fertilized externally by sperm emitted from the males. The eggs which are fertilized develop within the confines of simple spherical membranes which serve to keep the embryos in a germ-free, physiologically hospitable environment. Although the amphibians solved the problem of terrestrial locomotion by evolving limbs, two major factors have forced them to remain near the water. One, of less concern to embryologists, is the incomplete adaptation of their skin and kidneys to prevent excessive water loss in a dry environment. The other is the absence of any substantial change in their mode of reproduction and embryogenesis. Each spring amphibians must return to the ponds and streams to lay eggs which, upon fertilization, develop within simple noncellular membranes in a manner not unlike that of fish eggs.

A significant evolutionary step occurred when the first reptiles laid eggs capable of developing on land. This was made possible by the elaboration of a protective shell and a series of cellular membranes surrounding the embryonic body. These membranes, initially stemming from the body of the embryo itself, assist the embryo in vital functions, such as nutrition, gas exchange, and removal or storage of waste materials. In addition, they keep the embryo surrounded by an aquatic environment much like that of its cold-blooded forebears. Some reptiles and most mammals have dispensed with a shell in favor of intrauterine development, but the basic form and function of the extraembryonic membranes remain the same.

Four sets of extraembryonic membranes are common to the embryos of the higher vertebrates. The *amnion* is a thin ectodermally derived membrane which eventually encloses the entire embryo in a fluid-filled sac. The amniotic membrane is functionally specialized for the secretion and absorption of the amniotic fluid that bathes the embryo. So characteristic is this structure that the reptiles, birds, and mammals as a group are often called *amniotes*. The fishes and amphibians, lacking an amnion, are collectively called *anamniotes*.

The endodermal *yolk sac* is intimately involved with the nutrition of the embryo in large-yolked forms such as reptiles and birds. Despite the lack of stored yolk in mammalian eggs, the yolk sac has been preserved, probably because of its vital secondary properties. Two of the most important of these properties are (1) the source of primordial germ cells from the yolk-sac endoderm and (2) the source of the original circulating blood cells (both red and white) from the mesodermal cells lining the yolk-sac endoderm. In addition, the endoderm of the early embryo is a participant in a number of important inductive events that lead to the establishment of some of the major organs and organ systems in the body.

The *allantois* is an endodermally lined evagination originating from the ventral surface of the early hindgut. Its principal functions are to act as a reservoir for storing or removing urinary wastes and to mediate gas exchange between the embryo and its surroundings. In reptiles and birds the allantois is a large sac, and because the egg is a closed system with respect to urinary wastes, the allantois must sequester nitrogenous by-products so that they do not subject the embryo to osmotic stress or toxic effects. In mammals the role and prominence of the allantois vary with the efficiency of the interchange that can take place at the fetal-maternal interface. The allantois of the pig embryo rivals that of the bird in both size and functional importance, whereas the human allantois has been reduced to a mere vestige which contributes only a well-developed vascular network to the highly efficient placenta.

The outermost extraembryonic membrane, which abuts onto the shell or the maternal tissues and thus represents the site of exchange between the embryo and the environment around it, is the *chorion (serosa)*.[1] In species which lay eggs, the principal function of the chorion is the respiratory exchange of gases. The chorion in mammals serves a much more all-embracing function which includes not only respiration but also nutrition, excretion, filtration, and

---

[1]Throughout the years there has been considerable confusion regarding the most appropriate term for the outermost membrane surrounding the embryo. In early editions of this book it was designated the *serosa*, consistent with usage in comparative embryology. According to this system of terminology, the composite membrane resulting from the fusion of the allantois and the serosa is called the *chorion*. In contemporary usage, however, the outermost extraembryonic membrane is almost universally called the chorion (instead of the serosa) and the composite membrane formed where the allantois is fused to it is called the *chorioallantoic membrane*, particularly in the literature on the experimental embryology of birds.

synthesis—with hormone production being an important example of the last function.

The presence of the extraembryonic membranes introduces an added degree of complexity to the morphological study of amniote embryos. Particularly in early embryonic development it is difficult to dissociate the processes that result in membrane formation from those which are involved in the shaping of the gross form of the embryo itself. Extraembryonic membranes have different patterns of formation in different species. This chapter will describe the formation of extraembryonic membranes in the chick and in mammals.

EXTRAEMBRYONIC MEMBRANES OF THE CHICK

The Folding-off of the Body of the Embryo

In early chick embryos the somatopleure[2] and the splanchnopleure extend peripherally over the yolk, beyond the region where the body of the embryo is being formed. Distal to the body of the embryo the layers are termed *extraembryonic*. At first the body of the chick has no definite boundaries; consequently, embryonic and extraembryonic layers are directly continuous, having no definite point at which one ends and the other begins. As the body of the embryo takes form, a series of folds develop about it, undercut it, and finally nearly separate it from the yolk. The folds which thus definitely establish the boundaries between intraembryonic and extraembryonic regions are known as the *body folds*.

The first of the body folds to appear marks the boundary of the head. By the end of the first day of incubation, the head has grown forward and the fold originally bounding it appears to have undercut it and separated it rostrally from the blastoderm (Fig. 7-2). The *subcephalic fold* at this stage is crescentic, concave caudally. As this fold continues to progress caudad, its posterior extremities become continuous with folds which develop along either side of the embryo. Because these folds bound the body of the embryo laterally, they are known as the *lateral body folds*. The lateral body folds, which are at first shallow (Fig. 7-1A), become deeper, undercutting the body of the embryo from either side and further separating it from the yolk (Fig. 7-1).

A *caudal fold*, bounding the posterior region of the embryo, appears during the third day. It undercuts the tail of the embryo, forming a *subcaudal pocket* (Fig. 7-2C), just as the cephalic fold undercuts the head to form the subcephalic pocket. The combined effect of the development of the subcephalic, lateral body and the subcaudal folds is to constrict the embryo more and more from the yolk (Figs. 7-1, 7-2, and 7-4).

---

[2]*Somatopleure* is the name given to a layer of ectoderm underlain by mesoderm, and *splanchnopleure* refers to a bilayer of endoderm and mesoderm (Fig. 7-1A). In keeping with this terminology, the mesoderm associated with ectoderm is called *somatic mesoderm* and that associated with endoderm is called *splanchnic mesoderm*.

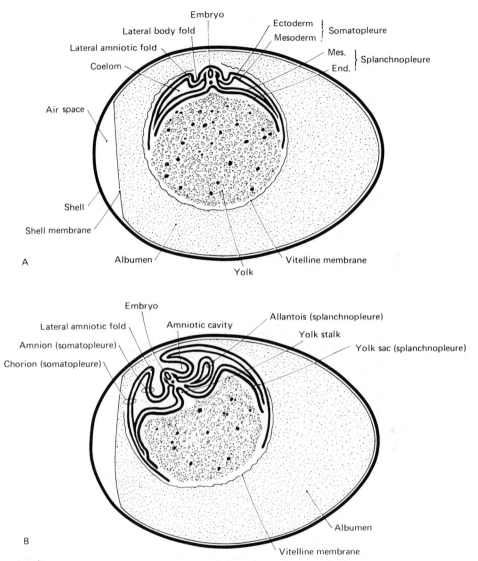

FIGURE 7-1

Schematic diagrams to show the extraembryonic membranes of the chick. (After Duval.) The diagrams represent longitudinal sections through the entire egg. The body of the embryo, being oriented approximately at right angles to the long axis of the egg, is cut transversely. (A) Embryo of about 2 days' incubation. (B) Embryo of about 3 days' incubation. (C) Embryo of about 5 days' incubation. (D) Embryo of about 14 days' incubation. (See colored insert.)

The Establishment of the Yolk Sac and the Delimitation of the Embryonic Gut

The yolk sac is the first extraembryonic membrane to make its appearance. The splanchnopleure of the chick, instead of forming a closed gut, as happens in

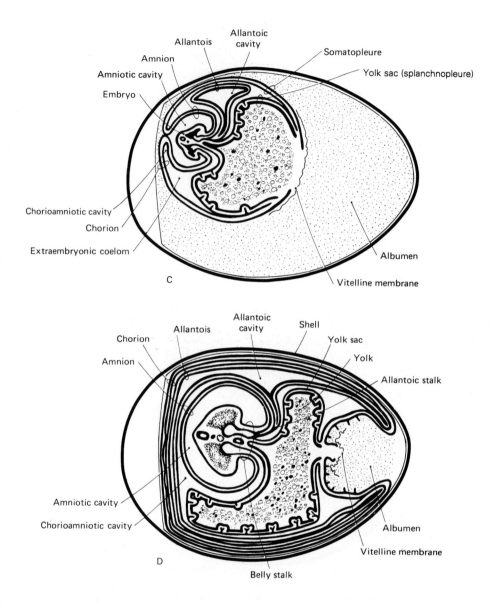

forms with little yolk, grows over the yolk surface. The primitive gut has only a dorsal cellular wall, and the yolk acts as a temporary floor (Fig. 7-2A). The extraembryonic extension of the splanchnopleure, derived originally from the primary and secondary hypoblast, eventually forms a saclike investment for the yolk (Figs. 7-1 and 7-4). Concomitantly with the spreading of the extraembryonic splanchnopleure about the yolk, the intraembryonic splanchnopleure undergoes a series of changes which result in the establishment of a completely walled gut in the body of the embryo.

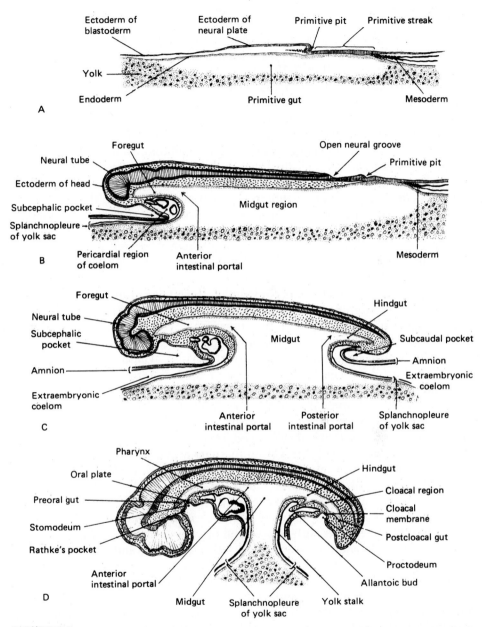

FIGURE 7-2
Schematic longitudinal section diagrams of the chick, showing four stages in the formation of the gut tract. The embryos are represented as unaffected by torsion. (A) Chick toward the end of the first day of incubation; no regional differentiation of primitive gut is as yet apparent. (B) Toward the end of the second day; foregut established. (C) Chick of about 2½ days; foregut, midgut, and hindgut established. (D) Chick of about 3½ days; foregut and hindgut increased in length at expense of midgut; yolk stalk formed.

The first part of the primitive gut to acquire a cellular floor is its cephalic region. Through a series of lateral folds (Fig. A-21A to A-21C) and anterior cephalic growth, the head and neck are separated from the underlying yolk by the *subcephalic pocket* (Fig. 7-2B and C). The same folding process that forms the head and neck causes the cephalic endoderm to form a tube. The part of the primitive gut that becomes tubular as the subcephalic fold progresses caudad is termed the *foregut* (Fig. 7-2B and C). During the third day of incubation the caudal fold undercuts the posterior end of the embryo. The splanchnopleure of the gut is involved in the progress of the subcaudal fold, so that a hindgut with an endodermal floor is established in a manner analogous to the formation of the foregut (Fig. 7-2C). The part of the gut that still remains open to the yolk is known as the *midgut*. As the embryo is constricted away from the yolk by the progress of the subcephalic and subcaudal folds, the foregut and hindgut increase in extent at the expense of the midgut. The midgut is finally diminished until it opens ventrally by a small aperture which flares out like an inverted funnel into the yolk sac (Fig. 7-2D). This opening is the *yolk duct*, and its wall constitutes the *yolk stalk*.

As the neck of the yolk sac is constricted, the *vitelline (omphalomesenteric) arteries* and the *vitelline (omphalomesenteric) veins*, caught in the same series of foldings, are brought together and traverse the yolk stalk side by side (Fig. A-42). The vascular network in the splanchnopleure of the yolk sac, which in young chicks can be seen spreading over the yolk, eventually nearly encompasses it (Fig. 7-5). The embryo's store of food material thus comes to be suspended from the gut of the midbody region in a sac provided with a circulatory arc of its own, the *vitelline arc*. Yolk does not pass directly through the yolk duct into the intestine; instead, absorption of the yolk is effected through enzymatic activity of the endodermal cells lining the yolk sac. These digestive enzymes change the yolk into soluble material which can then be absorbed through the lining of the yolk sac and passed on through the endothelial lining of the vitelline blood vessels to the circulating blood, by which it is carried to all parts of the growing embryo. In older embryos the splanchnopleure of the yolk sac undergoes a series of foldings (Fig. 7-1C and D) which greatly increase its surface area and thereby the amount of absorption it can accomplish. In addition to its absorptive function, the endoderm of the yolk sac is the sole site of synthesis of the serum proteins (transferrin, alpha globulins, and prealbumin) in the early embryo (Young et al., 1980).

During development the albumen loses water, becomes more viscous, and rapidly decreases in bulk. The growth of the allantois, an extraembryonic structure which we have yet to consider, forces the albumen toward the distal end of the yolk sac (Fig. 7-1D). The albumen, like the yolk, is surrounded by an extension of the yolk-sac splanchnopleure through which it is absorbed and transferred, by way of the extraembryonic circulation, to the embryo.

Toward the end of the period of incubation, usually on the nineteenth day, the remains of the yolk sac are enclosed within the body walls of the embryo. After the yolk sac's inclusion in the embryo, both the wall and the remaining contents of the yolk sac rapidly disappear; their resorption is practically completed in the first 6 days after hatching. The remaining yolk reserves are vital to the newly hatched chick while it is adapting to a free-living existence and is developing its feeding behavior.

The Amnion and Chorion

The amnion and chorion are so closely associated in their origin that they must be considered together. Both are derived from the extraembryonic somatopleure. The amnion encloses the embryo as a saccular investment, and the cavity thus formed between the amnion and the embryo becomes filled with a watery fluid. Muscle fibers of a primitive type develop in the amnion and by their contraction agitate the amniotic fluid. The slow rocking movement thus imparted to the embryo apparently aids in keeping its growing parts free from one another.

The first indication of amnion formation appears in chicks of about 30 hours' incubation. The head of the embryo sinks somewhat into the yolk, and at the same time the extraembryonic somatopleure anterior to the head is thrown into a fold, the *head fold* of the amnion (Figs. 7-3 and 7-4A). From a dorsal aspect the margin of this fold is crescentic in shape, with its concavity directed toward the head of the embryo. As the embryo increases in length, its head grows forward into the amniotic fold. At the same time, growth in the somatopleure itself tends to extend the amniotic fold caudad over the head of the embryo (Fig. 7-4B). By continuation of these two growth processes the head soon comes to lie in a double-walled pocket of extraembryonic somatopleure which covers it like a cap (Fig. 7-3). The free edge of the amniotic pocket retains its original crescentic shape as it covers more and more of the embryo in its progress caudad (Figs. A-26 and A-28).

The caudally directed limbs of the head fold of the amnion are continued posteriorly along either side of the embryo as the lateral amniotic folds. The lateral folds of the amnion grow toward and eventually meet at the midline dorsal to the embryo (Fig. 7-1A to 7-1C). During the third day, the tail fold of the amnion develops about the caudal region of the embryo. Its manner of development is similar to that of the head fold of the amnion, but it grows in the opposite direction (Fig. 7-4B and C).

Continued growth of the head, lateral, and tail folds of the amnion results in their meeting above the embryo. At the point where the folds meet, they become fused in a scarlike thickening termed the *amniotic raphe* (Fig. 7-4C). The way in which the somatopleure has been folded about the embryo leaves the amniotic cavity completely lined by ectoderm, which is continuous with the superficial ectoderm of the embryo at the region where the yolk stalk enters the body (Fig. 7-1D).

All the amniotic folds involve doubling the somatopleure on itself. Only the inner layer of the somatopleuric fold, however, is directly involved in the formation of the amniotic cavity. The outer layer of somatopleure becomes the chorion (Fig. 7-1B). The cavity between the chorion and amnion (chorioam-

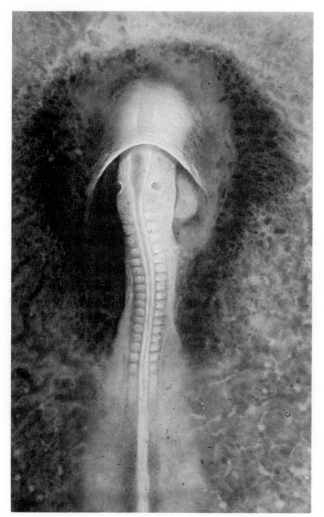

FIGURE 7-3
Unstained chick of about 40 hours' incubation, photographed by reflected light to show the cephalic fold of the amnion enveloping the head of the embryo.

niotic cavity) is part of the extraembryonic coelom. The continuity of the extraembryonic coelom with the intraembryonic coelom is most apparent in early stages (Fig. 7-1A and B). They remain, however, in open communication in the yolk-stalk region until relatively late in development.

The rapid peripheral growth of the somatopleure carries the chorion about the yolk sac, which it eventually envelops. The albumen sac also is enveloped by folds of chorion. The allantois, after its establishment, develops within the chorion, between it and the amnion (Fig. 7-6). Thus the chorion eventually encompasses the embryo itself and all the other extraembryonic membranes.

An important function of the mature chorion is the transport of Ca^{2+} from the egg shell into the embryonic circulation, where it is distributed to the

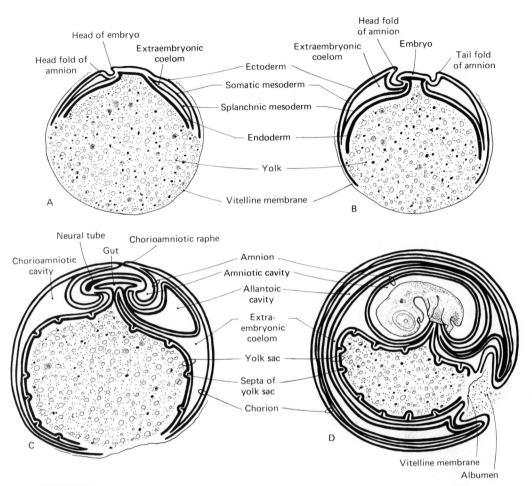

FIGURE 7-4
Schematic diagrams to show the extraembryonic membranes of the chick. The embryo is cut
longitudinally. The albumen, shell membranes, and shell are not shown; for their relations see
Figure 7-1. (A) Embryo early in second day of incubation. (B) Embryo early in third day of incubation. (C) Embryo of 5 days. (D) Embryo of 9 days. (See colored insert.)

developing beak and skeleton. A direct interaction with the overlying shell
membrane is required for maximal levels of Ca^{2+} transport by the chorion
(Dunn et al., 1981).

Despite its simple histological organization in the normal embryo, the
extraembryonic ectoderm retains considerable unexpressed developmental
potential. When placed in contact with appropriate dermis from chicks, ducks,
or even mice, the extraembryonic ectoderm of chick embryos not only can
become cornified, like mature epidermis, but also form feathers in response to
the influence of the foreign dermis (Dhouailly, 1978).

The Allantois

The allantois differs from the amnion and chorion in that it arises within the body of the embryo (Fig. 7-6A). Its proximal portion remains intraembryonic throughout development. Its distal portion, however, is carried outside the confines of the intraembryonic coelom and becomes associated with the outer extraembryonic membranes (Fig. 7-6B and C). As with the other extraembryonic membranes, the distal portion of the allantois functions only during the incubation period and is not incorporated into the structure of the adult body.

The allantois first appears late in the third day of incubation as a diverticulum from the ventral wall of the hindgut. Its walls are therefore splanchnopleure (Figs. 7-4 and 7-6).

During the fourth day of development the allantois pushes out of the body of the embryo into the extraembryonic coelom. Its proximal portion lies parallel to the yolk stalk and just caudal to it. When the distal portion of the allantois has grown clear of the embryo, it becomes enlarged (Figs. 7-4C and 7-6C). Its narrow proximal portion is known as the *allantoic stalk*, and its enlarged distal portion is referred to as the *allantoic vesicle*. Fluid accumulating in the allantois distends it, so that the appearance of its terminal portion in entire embryos is somewhat balloonlike (Fig. 7-5).

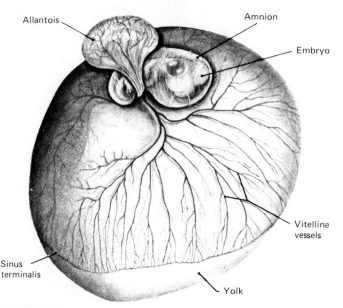

FIGURE 7-5
Chick of about 5½ days' incubation taken out of the shell with yolk intact. The albumen and the serosa have been removed to expose the embryo lying within the amnion. The allantois has been displaced upward in order to show the relations of the allantoic stalk. Compare this figure with Figure 7-1, C, which shows schematically the relations of the membranes in a section through an embryo of similar age. (Modified from Kerr, 1919, *Textbook of Embryology*, vol. II, The Macmillan Company.)

The allantoic vesicle enlarges very rapidly from the fourth day to the tenth day of incubation. Extending into the chorioamniotic cavity, it becomes flattened and finally encompasses the embryo and the yolk sac (Figs. 7-1C and D and 7-4C and D). In this process the mesodermal layer of the allantois becomes fused with the adjacent mesodermal layer of the chorion. Thus a double layer of mesoderm is formed, the chorionic component of which is somatic mesoderm and the allantoic component of which is splanchnic mesoderm (Fig. 7-6C). In this double layer of mesoderm an extremely rich vascular network develops which is connected with the embryonic circulation by the allantoic arteries and veins. It is through this circulation that the allantois carries on its primary function of oxygenating the blood of the embryo and relieving it of carbon dioxide. This is made possible by the position occupied by the allantois, close beneath the porous shell (Figs. 7-1C and A-42). This highly vascular fusion membrane, commonly called the *chorioallantoic membrane*, has been used effectively as a site for grafting small explants from younger embryos for the purpose of testing their developmental potencies. Through the chorioallantoic membrane and the shell, the chick embryo takes up about 5 liters of oxygen and gives off about 4 liters of carbon dioxide during the 21-day period before hatching (Wangensteen, 1972).

In addition to the respiratory interchange of oxygen and carbon dioxide, the growth of the embryo of course involves the metabolism of proteins with the formation of urea and uric acid. Averting toxic effects from the accumulation of these waste products in an embryo growing within the confines of a shell presents some interesting problems. The allantois is again involved, for it serves as a reservoir for the secretions coming from the developing excretory organs (Fig. A-42). In the early stages of development the chick excretes mostly urea. Later the excreted material becomes chiefly uric acid. This is a significant change, for urea is a relatively soluble substance and requires large amounts of water to hold it at nontoxic levels. Urea can be taken care of while the embryo is young and its excretory output is small, but it would present grave problems if it were produced in large quantities. In contrast, uric acid is relatively insoluble, and the large amounts of it produced by older embryos can be stored without ill effects. At the time of hatching, the slender allantoic stalk is broken and the distal portion of the allantois with its contained excretory products remains as a shriveled membrane adherent to the broken shell.

THE FORMATION OF EXTRAEMBRYONIC MEMBRANES AND PLACENTA IN MAMMALS

Mammals form and utilize the same extraembryonic structures as does the chick, but certain modifications are necessitated by their intrauterine mode of development. Despite the fact that virtually no yolk accumulates in the ovum, a yolk sac is formed just as if yolk were present (Figs. 8-10 and 8-11). Such persistence of a structure in spite of the loss of its original function is not an uncommon phenomenon in evolution and has given rise to the biological

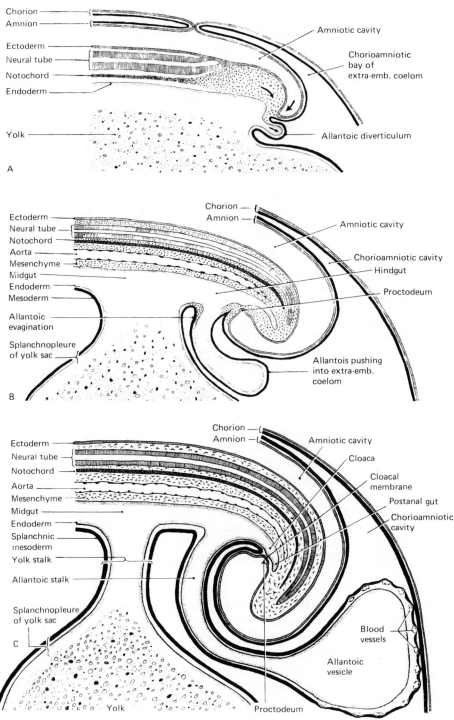

aphorism "Morphology is more conservative than physiology." Some highly important secondary functions of the yolk sac remain, however. The primordial germ cells arise from yolk-sac endoderm (Fig. 3-1), and the first blood cells originate in the splanchnic mesoderm that surrounds the yolk sac (Fig. 18-3). In addition, parts of the blood vessels that supply the yolk sac are incorporated into the circulatory system of the body as vessels that supply part of the intestinal tract.

In many mammals (e.g., the dog and the pig), the amnion arises as a layer of somatopleure which enfolds the developing embryo in much the same manner as already described for the chick (Fig. 8-10). Like that of the chick embryo, the mammalian amnion encloses a fluid-filled cavity in which the embryo is suspended. As the amnion takes shape, the body of the embryo both grows and curves, displacing the attachment of the amnion to the body as a ring surrounding the yolk sac and allantois on the ventral body wall.

Almost as soon as the hindgut of the embryo is established, a diverticulum known as the allantois arises from it (Fig. 8-10B and C). In many mammals the distal position of the allantois becomes dilated (Fig. 8-10D) and establishes a close relationship with the chorion in a manner similar to that in the chick embryo. The extraembryonic mesoderm that lines the allantois is richly vascular, and the allantoic capillary plexuses assume the function of mediating metabolic exchange between the fetus and the mother.

The outer layer of the mammalian blastocyst is given various names at different stages of development. At the stage of the inner cell mass it is most commonly called the *trophoblast* (Fig. 4-15C). After the formation of the hypoblast and mesoderm, it is given the more specific designation of *trophectoderm* because of the part this layer plays in the acquisition of food materials from the uterus. Still later, after the mesoderm has split and its somatic layer becomes associated with the ectoderm, this layer, now an extraembryonic somatopleure, is called the chorion (serosa). After completion of the amnion, the chorion completely surrounds the complex of the embryo and its other extraembryonic membranes and serves as the interface between the embryonic tissues and the uterus.

The *placenta* is the region where the interchange of food materials, oxygen, and wastes take place between the fetus and its mother. Although the chorion provides the interface between fetal and maternal tissues, the embryo itself is linked to the region by the body stalk and the allantoic blood vessels which course through the body stalk and then branch out into an extensive capillary network throughout the placenta.

In some mammals, such as the pig, the chorion and the uterine lining lie close together, but they can be peeled apart. This arrangement constitutes a *contact* (nondeciduous) *placenta* (Fig. 7-7). In most mammals, including

---

FIGURE 7-6
Schematic longitudinal-section diagrams of the caudal regions of a series of chick embryos to show the formation of the allantois. (A) At about 2½ days of incubation; (B) at about 3 days; (C) at 4 days. (See colored insert.)

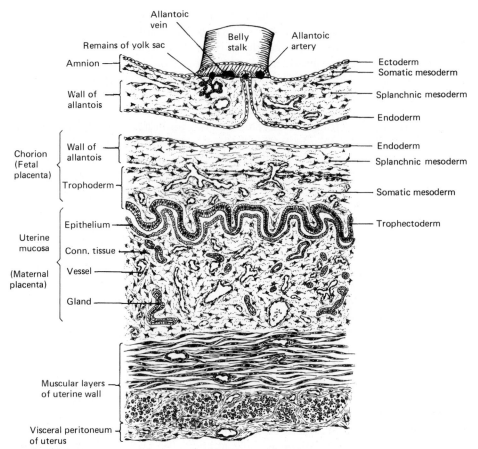

FIGURE 7-7
Semischematic diagram showing the structure of the chorion and its relation to the uterine wall.

humans, the fetal (chorionic) and maternal (uterine mucosal) portions of the placenta actually grow together so that they cannot be pulled apart without hemorrhage. So intimately do they become fused that a large part of the uterine mucosa is pulled away with the chorion when, shortly after the birth of the fetus, the extraembryonic membranes are delivered as the *afterbirth*. This type of placenta is called a *burrowing* (deciduous) *placenta*.

TYPES OF PLACENTAE

Several structurally different types of placentae have appeared among the mammals. These differ in the numbers and types of cell layers intervening between the blood of the mother and that of the embryo, and they are named accordingly.

In the pig the chorion rests upon an intact uterine epithelium. Such a placenta is called *epitheliochorial* (Fig. 7-8A). In some hooved mammals, for

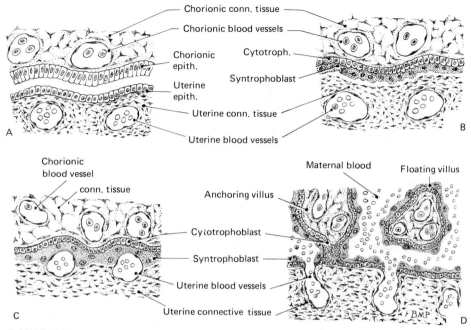

FIGURE 7-8
Schematic diagram to show four basic types of choriouterine relationships. (In part after Flexner et al., 1948, *Amer. J. Obstet. Gynec.*, vol. 55.) (A) Epitheliochorial. (B) Syndesmochorial. (C) Endotheliochorial. (D) Hemochorial. For explanation, see text.

example, deer, giraffes, and cattle, varying amounts of the uterine epithelium may be absent, thus bringing the chorion into contact with the connective tissue of the uterus. This type of placenta is called *syndesmochorial* (Fig. 7-8B) and is sometimes considered to be a variant of the epitheliochorial placenta.

When no maternal connective tissue intervenes between the endothelium of the maternal vessels and the chorionic epithelium, we say the placenta is *endotheliochorial* (Fig. 7-8C). The placenta of dogs and cats is endotheliochorial.

In many mammalian groups, including rodents, rabbits, bats, and primates, the endothelial continuity of the maternal blood vessels is disrupted and the chorion is actually bathed in maternal blood. This arrangement is designated as a *hemochorial placenta* (Fig. 7-8D).

Not only the histological arrangement of tissue layers but also the gross shape of the placenta vary widely among the various mammalian groups. Examples of the variety of gross placental form are shown in Fig. 7-9).

THE SPACING OF EMBRYOS IN THE UTERUS

In animals that produce multiple offspring in one reproductive cycle, ova are normally derived from each ovary. Despite the fact that sometimes one ovary

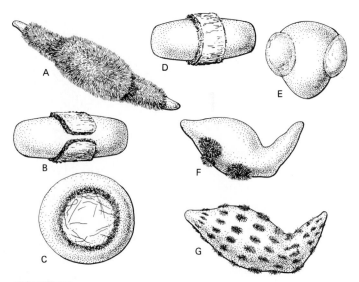

FIGURE 7-9
Different gross forms of the placenta in mammals. (A) Pig—diffuse;
(B) Raccoon—incomplete zonary; (C) Brown bear—subtype of zo-
nary; (D) Dog, cat, seal—zonary or annular; (E) Monkey—bidiscoidal;
(F) Mexican deer—cotyledonary; (G) Cow—cotyledonary. (From Ham-
ilton, Boyd and Mossman, 1972. *Human Embryology*, Williams and
Wilkins.)

is much more prolific than the other, the embryos are usually evenly spread
throughout the horns of the uterus (Fig. 7-10). The blastocyst of the pig is a
highly attentuated and elongated structure (Fig. 7-11A), and even at an early
stage the blastocysts are evenly spaced within the uterus.

The mechanism by which pig embryos are spaced within the uterus is not
known, but spacing apparently takes place before elongation of the blastocyst
occurs. The threadlike early blastocysts, as much as a meter long, follow the
extensive uterine folds so that they will occupy only about 10 to 15 cm of a
uterine horn. The attentuated condition of the blastocyst does not persist long.
With the increase in the extent of the allantois and growth of the embryo, the
blastocyst, now called a *chorionic vesicle*, is greatly dilated and somewhat
shortened. The allantois never grows to the entire length of the chorionic
vesicle, and there remains an abrupt narrowing at either end, where the
allantois ends (Fig. 7-12). Where the terminal portions of neighboring chorionic
vesicles lie close to each other, the uterus remains undilated, sharply marking
off the region (*loculus*) where each embryo is located (Fig. 7-11B).

Work by Boving (1971) has clarified considerably the mechanism of
blastocyst spacing in the rabbit. Early rabbit blastocysts are randomly arranged
within the uterine horn. Shortly before implantation the blastocysts undergo a
striking increase in diameter, and at this time the blastocysts, whatever their
number, become arranged equidistant from one another. Boving has observed

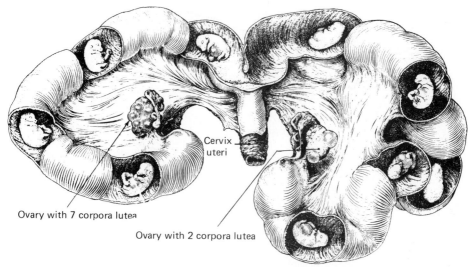

FIGURE 7-10
Uterus of pregnant sow opened to show distribution of embryos. Note that approximately uniform spacing of the embryos in the two horns of the uterus in spite of the fact that the corpora lutea indicate the origin of two ova from one ovary and seven from the other. (After Corner, 1921. *Johns Hopkins Hosp. Bull.*, vol. 32.)

that blastocyst spacing is effected by peristaltic waves of uterine contractions which are propagated not only from either side of the uterine horn but also in both directions from each area where the uterine horn is distended by a blastocyst. The peristaltic waves adjust the position of the blastocysts (as well as glass beads of similar diameter) until they become evenly spaced. As the blastocysts continue to increase in diameter, uterine contractions are no longer able to shift their positions, and implantation soon occurs.

HUMAN FETAL-MATERNAL RELATIONS

In both birds and mammals with a saccular allantois the term *chorioallantois* has been applied to the fetal membrane secondarily formed by the coalescence of the allantois with the chorion (Fig. 7-13 A and B). In primates, the lumen of the allantois is rudimentary and the allantoic endoderm does not reach the chorion (Fig. 7-13C). However, a fundamental component of the chorioallantoic association is preserved in the allantoic (umbilical) mesoderm and blood vessels, which spread out along the chorion to form the disklike placenta. The combination of allantoic vessels and the chorionic specializations (villi) in the placenta provides the anatomical basis for the exchange of materials between mother and fetus.

Implantation

Following fertilization, the embryo, still surrounded by the zona pellucida and the corona radiata, begins a weeklong free-floating trip down the uterine tube

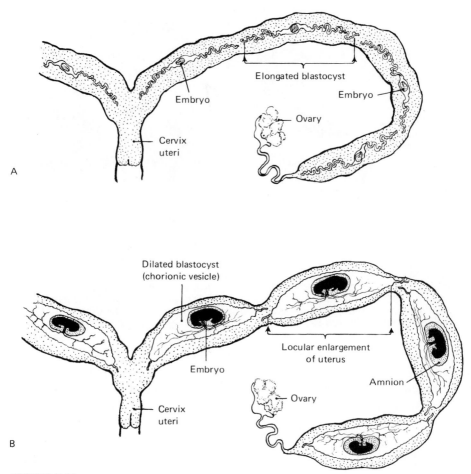

FIGURE 7-11
Schematic diagrams indicating the intrauterine relations of pig embryos and their membranes; (A) in the elongated blastocyst stage; (B) when the blastocysts have been dilated to form the chorionic vesicles characteristic of the later stages of pregnancy.

and into the uterine cavity (Fig. 3-36). While the embryo is still in the uterine tube, the cells of the corona radiata continue to secrete progesterone and prostaglandins (Schultz and Dubin, 1981). These secretions may help maintain the embryo before the corpus luteum becomes functional. By about the sixth or seventh day the human embryo, which has traversed the length of the uterine tube and has been freely floating in the uterine cavity for a couple of days, is ready to become attached to the uterine lining. The embryo itself is in the blastocyst stage, consisting of an inner cell mass and a trophoblast (Fig. 7-14A). The zona pellucida, which surrounds the embryo during the free-floating early phase of development, begins to break down just before implantation. In the monkey, embryos are noticeably sticky just prior to implantation. This

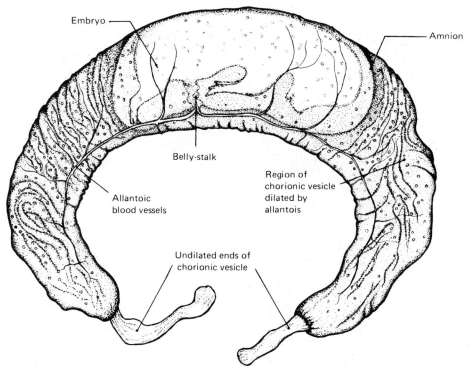

FIGURE 7-12
Drawing of pig embryo in unruptured chorionic vesicle. Compare with figure 7-11, B. (After Grosser.)

characteristic is most marked in the cells directly overlying the inner cell mass, and it is this part of the embryonic surface that initially adheres to the uterine lining (Fig. 7-17).

Implantation typically takes place in the upper part of the uterine cavity, on either the anterior or the posterior wall of the uterus. Upon making contact with maternal tissues, the cells of the trophoblast proliferate rapidly and erode the underlying uterine mucosa (Fig. 7-18). As it expands, the original trophoblast is subdivided into two distinct layers. The inner layer retains a distinctly cellular nature and is therefore called the *cytotrophoblast*. The outer layer is an irregular syncytium, formed by the fusion of large numbers of cells derived from the cytotrophoblast, and is appropriately called the *syntrophoblast* (*syncytiotrophoblast*). It appears likely that contact with the maternal tissues stimulates expansion of the trophoblast, because in the newly implanted embryo the portion of the trophoblast that is still uncovered in the uterine cavity remains a thin single-cell layer (Figs. 7-17A and D and 7-18C).

While in the process of implantation, the blastocyst essentially creates a wound in the uterine mucosa. By 11 or 12 days, the blastocyst has become almost completely embedded within the endometrium and the uterine epithe-

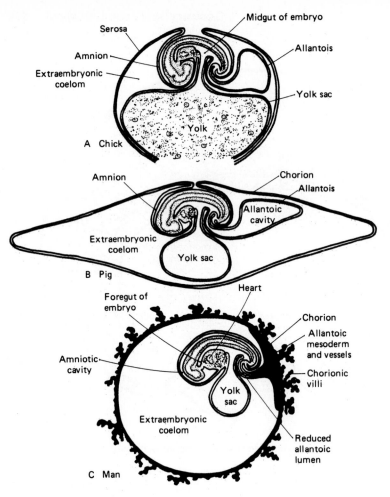

FIGURE 7-13
Diagrams showing interrelations of embryo and extraembryonic membranes characteristic of higher vertebrates. Neither the absence of yolk from its yolk sac nor the reduction of its allantoic lumen radically changes the human embryo's basic architectural scheme from that of more primitive types. (See colored insert.)

lium grows back over the embryo, thus healing the wound caused by the invading blastocyst (cf. Figs. 7-15 and 7-16). During the early phase of implantation, the embryo is apparently nourished by diffusion of materials from maternal fluids and debris resulting from the destruction of endometrial cells. Soon, however, irregular lacunae appear within the expanding syntrophoblast. These lacunae become filled with maternal blood emanating from eroded uterine blood vessels (Figs. 7-15 and 7-16). This marks the beginning of the intimate relationship between maternal blood and trophoblastic tissues which forms the basis for the hemochorial placenta.

Whatever the importance of diffusion of maternal breakdown products into the embryo for temporary nutrition, the invasion of the endometrium rapidly paves the way for the establishment of the type of vascular interchange on

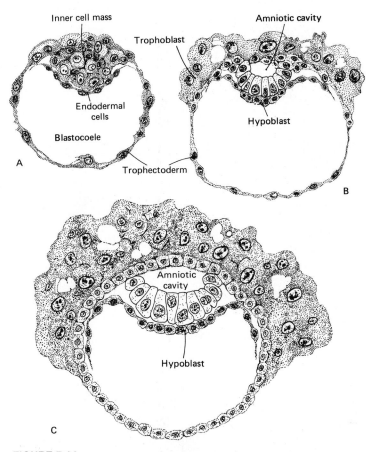

FIGURE 7-14
Schematic diagrams illustrating the early steps by which young human embryos attain their basic structural plan. (A) Diagram of late primate
blastocyst. (B) Diagram of primate blastocyst shortly after implantation has
begun (based on the Hertig-Rock 8-day human embryo). Formation of the
amniotic cavity has begun, and cells of the hypoblast have begun to
spread over the inner surface of the trophoblast. The expanded trophoblast
represents the area that is in contact with maternal endometrium. (C) Partially implanted human blastocyst at a stage slightly later than that illustrated in B.

which the embryo is to be dependent for the rest of its intrauterine life. As the
trophoblast spreads out into the uterine mucosa, it inevitably comes into
contact with small blood vessels and breaks down their walls. Although cells
from the cytotrophoblast tend to dam the opened vessels somewhat and check
excessive extravasation of blood, there must nevertheless be continued oozing
from the invaded vessels, for the trophoblast is known to produce a substance
which inhibits the coagulation of blood. In addition, there is transudation of
blood serum and lymph, so that the invading trophoblast comes to lie in eroded
areas of endometrium saturated in maternal blood and lymph. By this time,

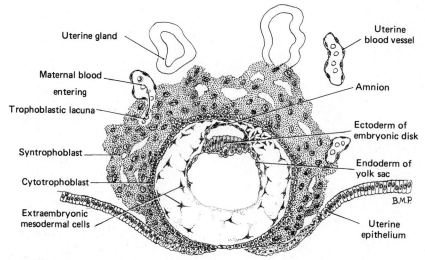

FIGURE 7-15
Human embryo of about 9 to 10 days fertilization age. (Schematized from Hertig and Rock, 1941. *Carnegie Cont. to Emb.*, vol. 29.)

highly branched projections of the trophoblast, called villi, have become vascularized by terminal vessels of the allantoic circulatory arc (Fig. 8-17B). It

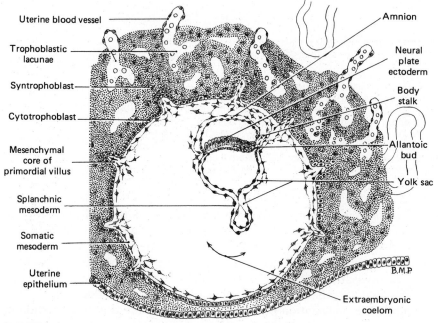

FIGURE 7-16
Human embryo of about 13 days fertilization age. (Schematized from several sources.)

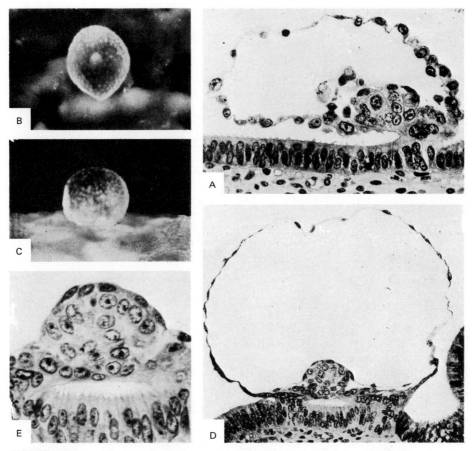

FIGURE 7-17
Photomicrographs showing stages leading up to implantation of monkey embryo. (After Heuser and Streeter, 1941. *Carnegie Cont. to Emb.*, vol. 29.) (A) Section (×350) of embryo after ninth day, showing its initial adhesion to uterine epithelium. (B) A 9-day embryo attached to uterine mucosa, viewed from above (×50). (C) Same embryo photographed from the side (×50). (D) Same embryo shown in B and C after sectioning (photomicrograph, ×200). (E) Inner cell mass of same embryo (photomicrograph, ×500).

remains only for the embryonic heart to start the blood circulating and the entire elaborately interlocking mechanism for the nutrition of the embryo is ready to go into operation early in the fourth week after fertilization, or approximately 2 weeks after implantation.

Formation of Extraembryonic Tissues in the Human Embryo

Because of the lack of well-preserved, early human material, it was difficult in the past to reconstruct the early events of human embryogenesis without recourse to inference from the findings of studies on series of primate embryos. Luckett (1974,

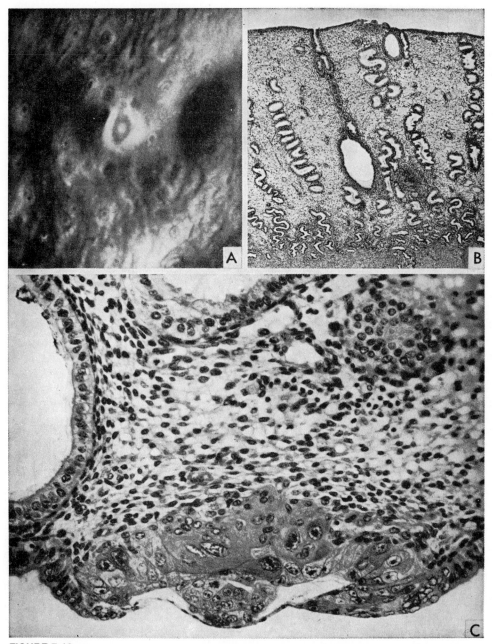

FIGURE 7-18
Human embryo of about 7½ days' fertilization age. (Carnegie Collection, embryo no. 8020, after Hertig and Rock, 1945. *Carnegie Cont. to Emb.*, vol. 31.) (A) Surface view of implantation site (×27). (B) Photomicrograph (×27) of section through the uterine mucosa to show the exceedingly superficial position occupied by the recently implanted embryo. (C) Section (×300) through center of embryo and immediately surrounding endometrium. Note that the position of the embryo, as compared with that in (B), has been reversed to bring it into conventional orientation with its future dorsal side through the top of the page.

1975, 1978) has reexamined many of the classical human preparations, as well as new embryos, and provided an interpretation of the early steps in the formation of extraembryonic tissues that is more consonant with the findings of comparative embryologists. Luckett's papers will serve as the basis for this section.

Formerly it was widely believed that all the extraembryonic tissues in the human arise from the trophoblast. Recent experimental studies on mammalian embryos, particularly rodents, have shown that the developmental potentialities of the trophoblast are far more restricted than the early descriptive studies suggested (Fig. 5-31). According to Luckett, the original trophoblast gives rise only to the cytotrophoblast and syncytiotrophoblast (Fig. 5-24). The remainder of the extraembryonic tissues come from the inner cell mass.

There are several stages in the early formation of the amniotic cavity. They are similar in the embryos of both humans and rhesus monkeys (Luckett, 1975). Shortly after implantation has begun, a primordial amniotic cavity forms by means of cavitation within epiblastic components of the inner cell mass (Fig. 7-14B). Then the roof of this cavity opens because of lateral spreading of the cells constituting the roof (Fig. 7-14C). During this transitory phase the primitive amniotic cavity is temporarily bounded in part by a region of cytotrophoblast. By means of unfolding and subsequent fusion of the lateral walls of the epiblast, the amniotic cavity again becomes completely bounded by epiblastic (ectodermal) cells (Fig. 7-15). The amniotic membrane is completed by the spreading of extraembryonic mesodermal cells, derived from the early primitive streak, around the amniotic ectoderm. Luckett (1975) suggested that the presence of the early amniotic cavity may be necessary in order for the primitive streak to form.

Even before the primordial amniotic cavity first appears, the inner cell mass gives rise to the *hypoblast*, a thin layer of endodermal cells which soon spread to line the entire inner surface of the trophoblast (Fig. 7-14). These extraembryonic endodermal cells constitute the primary yolk sac. By 12 to 13 days, the primary yolk sac collapses, leaving a smaller secondary yolk sac attached to the embryo (Fig. 7-16).

In human embryos, the caudal margin of the primitive streak develops precociously, giving rise to most or all of the extraembryonic mesoderm. These cells form the body stalk and continue to spread out between the extraembryonic endoderm and the overlying trophoblast. They also form the mesodermal lining layers of the amnion and yolk sac. Meanwhile, the definitive primitive streak is associated with gastrulation movements that result in the formation of the embryonic endoderm and mesoderm (see Chap. 5). The human allantois arises as an outpocketing from the hindgut much like that of the pig and chick, but the allantoic diverticulum remains a rudimentary structure embedded within the mesoderm of the body stalk (Fig. 7-13C).

Development of Chorionic Villi

Once implantation has occurred, expansion of the trophoblast continues. After the early establishment of lacunae, there is little in the arrangement of the trophoblast at these early stages to suggest the characteristically shaped branching villi that will be seen in later stages. Embryos that have a trophoblast in this sprawling, unorganized condition are commonly characterized as *previllous* (Figs. 7-14B and 7-15).

As embryos approach the end of the second week, the trophoblast begins to be molded into masses more suggestive of villi. These very young villi at first consist entirely of epithelium, with no connective-tissue core. In this stage they are referred to as *primary villi*. Their differentiation is very rapid, for even the previllous cell masses are already beginning to show two layers of cells. The inner layer of *cytotrophoblast* (Langhans' layer) is composed of a single, relatively regular layer of cells, each of which possesses distinct boundaries (Fig. 7-20B). Surrounding the cytotrophoblast is an outer syncytium of variable thickness containing irregularly disposed nuclei. This layer is known as the syntrophoblast (syncytiotrophoblast) (Fig. 7-19). Tracing studies using tritiated thymidine as a marker have demonstrated that nuclei of the syntrophoblast arise from the cytotrophoblastic layer. It is a general rule in development that nuclei in syncytial structures, such as the syntrophoblast or a skeletal muscle fiber, do not undergo mitotic divisions. In the case of the trophoblast, the cytotrophoblast serves as a germinative center providing both nuclei and cytoplasmic material to the syntrophoblast, which, because of its syncytial

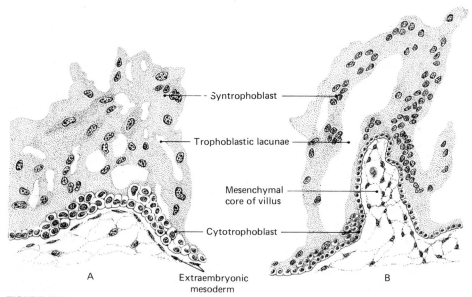

A   Extraembryonic mesoderm   B

Syntrophoblast

Trophoblastic lacunae

Mesenchymal core of villus

Cytotrophoblast

FIGURE 7-19
Early stages in development of chorionic villi. (A) Primitive trophoblastic projection without mesenchymal core. (Redrawn, ×225, from Streeter's photomicrographs of the Miller embryo.)
(B) Young villus just developing a mesenchymal core. (Redrawn to about the same scale as A, from Fischel's figure of an embryo in the primitive-streak stage.)

nature, is incapable of augmenting its own supply of nuclei as its volume expands (Tao and Hertig, 1965).

The phase in which the developing villi lack a mesenchymal core is very short-lived. While the primitive villi is forming, the inner face of the blastodermic vesicle receives an ingrowth of allantoic vessels and mesoderm. Early in the third week after fertilization, the mesoderm pushes into the primitive villi so that the trophoblastic cells, instead of constituting the whole structure, become a covering epithelial layer over a framework of delicate connective tissue derived from the mesodermal ingrowth (Fig. 7-19B). These are known as *secondary villi*. Blood vessels soon appear in the connective-tissue core of the villus and push out into its newly formed branches. Such villi, with a vascular connective-tissue core, are called *tertiary villi*. This condition in which the villi are prepared for their absorptive function is reached by about the end of the third week. The villi retain this same general structural plan throughout pregnancy, although as gestation advances their connective-tissue core and blood vessels become more highly developed and there are marked regressive changes in their epithelial covering (Fig. 7-20).

Formation of Placenta

Under the influence of the presence of an embryo, striking changes take place in the endometrium. These are most marked, naturally, at the site of implantation. Uterine stromal cells around the blastocyst undergo a pronounced transformation in which they enlarge and their cytoplasm becomes filled with glycogen and lipid droplets. This transformation is known as the *decidual reaction*, and the transformed stromal cells are called *decidual cells*. There is recent evidence (Kearns and Lala, 1982) that the decidual cells, which have a function in maintaining immunological compatibility between the immunologically different mother and fetus, are derived from the bone marrow as are other cells of the lymphoid system. Ultimately the decidual reaction spreads to the stromal cells throughout the endometrium. At the termination of pregnancy the endometrium containing these cells is extensively sloughed off and then rebuilt. This postpartum phenomenon of shedding and replacement has given rise to the term *decidua* (root meaning, *to shed*) for the endometrium of pregnancy.

The fact that the human embryo promptly burrows into the endometrium instead of becoming merely adherent, as is the case in certain other mammals, establishes at the outset positional relationships which shape the later course of events. As the chorionic vesicle grows, the overlying portion of the endometrium is stretched out over it, forming a layer known as the *decidua capsularis* (Fig. 7-21). The portion of the endometrium lining the walls of the uterus elsewhere than at the site of attachment of the chorionic vesicle is called the *decidua parietalis*. The area of the endometrium directly underlying the chorionic vesicle is termed the *decidua basalis*.

The ultimate absence of chorionic villi in the decidua parietalis leaves this part of the endometrium with no direct role to play in the nutrition of the

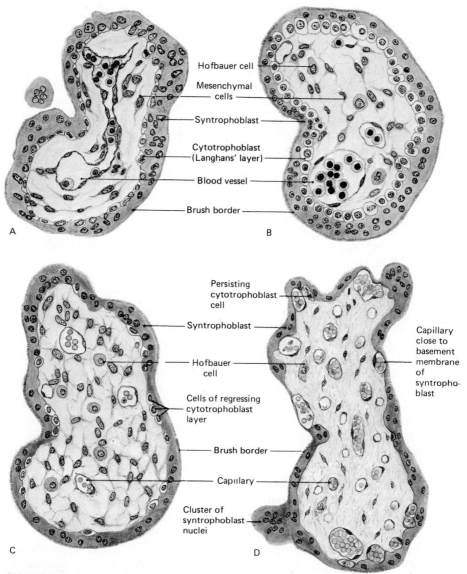

Hofbauer cell

Mesenchymal cells

Syntrophoblast

Cytotrophoblast (Langhans' layer)

Blood vessel

Brush border

A

B

Persisting cytotrophoblast cell

Syntrophoblast

Hofbauer cell

Cells of regressing cytotrophoblast layer

Brush border

Capillary

Cluster of syntrophoblast nuclei

Capillary close to basement membrane of syntrophoblast

C

D

FIGURE 7-20
Chorionic villi at various ages. (Camera lucida drawings, ×325.) (A) From chorion of 4-week embryo (crown-rump length, commonly abbreviated C-R, 4.5 mm). (B) Chorion from an embryo of about 6½ weeks (C-R 15.1 mm). (C) Placenta from a fetus of the fourteenth week. (D) Placenta at term. (From preparation loaned by Dr. Burton L. Baker.)

embryo. The maternal blood supply is most direct and abundant in the decidua basalis. Conditions in the decidua capsularis vary considerably at different ages. At first the chorion underlying this part of the decidua is as well supplied

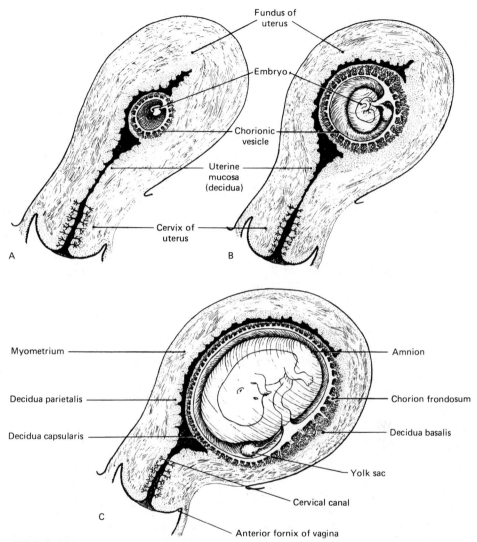

FIGURE 7-21
Diagrams showing uterus in early weeks of pregnancy. Embryos and their membranes are
drawn slightly smaller than actual size. (A) At fertilization age of 3 weeks; (B) 5 weeks; (C) 8
weeks.

with villi as any other region (Fig. 7-21A), but before long the growth of the
chorionic vesicle causes the decidua capsularis to be pushed away from the
maternal vascular supply. Moreover, the tissue of the decidua capsularis itself
becomes more and more attenuated as the chorionic vesicle increases in size.
These unfavorable conditions in the decidua capsularis are promptly reflected
in the less exuberant growth of the chorionic villi embedded in it (Fig. 7-21C).

At the end of the first trimester of pregnancy the decidua capsularis begins to undergo severe atrophy, and by midpregnancy most of the decidua capsularis has disappeared, leaving fetal chorionic tissue (chorion laeve; see next paragraph) in direct contact with the decidua parietalis of the opposite wall of the uterus (Fig. 7-28).

In contrast to the chorionic villi in the region of the decidua capsularis, the villi adjacent to the decidua basalis grow with increasing vigor. This is unmistakably the part of the mechanism most effectively situated for carrying on metabolic interchange between the fetus and the mother. By the third month, when the growth of the embryo and the expansion of the amnion begin to compress the decidua capsularis and the decidua parietalis against each other, the villi begin gradually to disappear altogether from this area. Thus the chorionic vesicle, at first uniformly villated over its entire surface, by the end of the fourth month becomes denuded of its villi everywhere except where they lie in the decidua basalis (Figs. 7-21 and 7-28). The part of the vesicle under the decidua capsularis which has thus lost its villi becomes known as the *chorion laeve* (smooth), and that part of the chorion next to the decidua basalis where the villi are highly developed is termed the *chorion frondosum* (tufted or bushy). The interlocked chorion frondosum of the fetus and decidua basalis of the mother constitute the *placenta*.

Later Changes in the Structure of the Uterus and Placenta

From their first invasion of the uterine lining, the chorionic villi lie in excavated spaces in the endometrium, bathed in maternal blood and lymph. Essentially, this relationship is retained throughout pregnancy, but the extent of the blood spaces, the relations of the villi to the endometrium, and the structure of the villi themselves all vary as development progresses. During the first few weeks after implantation, the invasion of the endometrium is exceedingly rapid and the area which is becoming the decidua basalis is progressively extended (Fig. 7-21). In this period, the syntrophoblast is very conspicuous, forming sprawling processes extending into the endometrium far beyond the main mass of the chorionic vesicle.

Once the chorion has become well established in the uterus, the invasive process becomes relatively slow, merely keeping pace with the growth of the embryo. The slower rate of invasion is reflected in a reduction of the syntrophoblast to form a more regularly arranged covering outside the cytotrophoblastic layer of the villus. Meanwhile, the mesenchymal core of the villus has become organized into a delicate connective tissue supporting the endothelial walls of the blood vessels, so that the entire villus takes on a much more definitely organized appearance (Fig. 7-20A). Scattered in the connective tissue there appear, in variable numbers, cells which are conspicuously larger than the ordinary connective-tissue cells. These have been given the name *Hofbauer cells* (Fig. 7-20), after the man who first described them. Their significance is not entirely understood, but they appear to be phagocytic and are commonly believed to act as a primitive type of macrophage.

In the established parts of the placenta, the invasive function of the epithelial covering of the villi ceases to be important and the epithelial layers become relatively thinner. The cytotrophoblastic layer reaches the height of its development during the second month (Fig. 7-20B). Thereafter, it gradually loses its completeness (Fig. 7-20C). It is as if this layer spent itself in the production of the syncytial layer. During the fourth and fifth months, the cytotrophoblast undergoes still further regression. Most of the villi come to be clothed in a reduced syntrophoblastic layer, with only occasional cytotrophoblastic cells persisting. During the last third of the period of gestation, attenuation of the syntrophoblast becomes more marked (Fig. 7-20D).

As pregnancy advances, the villi grow greatly in size and the complexity of their branching increases (Fig. 7-22). If we likened them to trees, we should find them growing over the discoidal area of the chorion frondosum, not quite

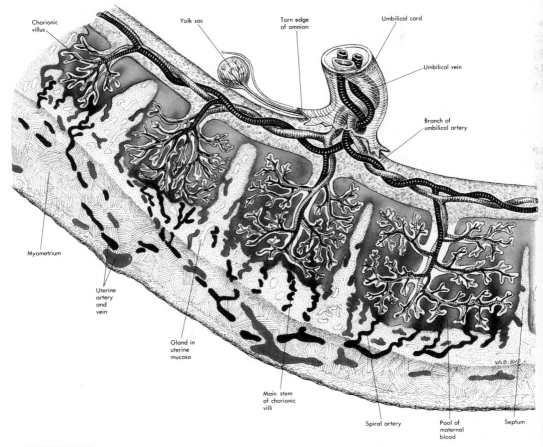

FIGURE 7-22
Schematic diagram to show interrelations of fetal and material tissues in formation of placenta. Chorionic villi are represented as becoming progressively further developed from left to right across the illustration. (See colored insert.)

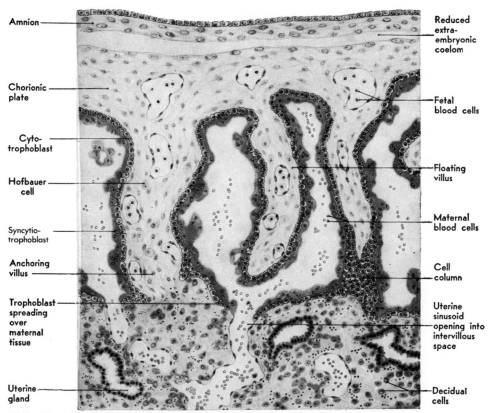

FIGURE 7-23

Semischematic drawing to show the relations of the chorionic villi and the trophoblast to the maternal tissues of the placenta. (Redrawn, with modifications, from J. P. Hill, 1931. Phil. Trans. Roy. Soc. London, ser. B., vol. 22.)

uniformly but in about 15 to 16 dense clumps. These main concentrations of villi are known as *cotyledons*. Between the cotyledons, the maternal tissue has been less deeply eroded and constitutes the "placental septa" (Fig. 7-22). Between the septa the tips of most of the villi lie free in the space which has been excavated in the uterine mucosa. The tips of other villi make contact with the uterine tissue at the bottom of the excavation. At this phase of their development the rapidly growing ends of these villi are richly cellular, consisting of a core of cytotrophoblast, with syntrophoblast forming an irregular covering. This modified portion of a villus is commonly spoken of as a *cell column* (Fig. 7-23). Where such cell columns are in contact with the uterine mucosa, the trophoblastic elements spread out to clothe the eroded surfaces of the maternal tissues with trophoblast. Thus the maternal blood entering the spongy areas of the placenta from blood sinuses opened by the invading chorion comes into a maze of irregular spaces clothed on both their fetal and maternal faces by trophoblast. Some of the villi are especially intimately related to the maternal tissue and are called *anchoring villi* (Fig. 7-23). The majority of the villi, however, continue to lie more or less free in spaces in the decidua basalis. Maternal blood enters the spaces about the villi from the small vessels which were opened in the excavating process. As this blood drains back into the uterine veins, it is replaced by blood supplied by the uterine arteries, so that the villi are continuously steeped in fresh maternal blood.

It should be emphasized that at no time during pregnancy is there any mingling of fetal and maternal bloodstreams. The fetal circulation is from its first establishment a closed circuit, isolated from the maternal blood which bathes the villi by the placental barrier. This barrier consists of (1) the trophoblast, (2) its underlying basal lamina, (3) connective tissue interposed between the trophoblast and the fetal blood vessel, (4) the basal lamina surrounding the blood vessel, and finally (5) the endothelial lining of the blood vessel itself. Through the placental barrier must pass, in a two-way stream, waste substances which must be eliminated by the fetus and substances from the mother which are needed for respiration, growth, fluid balance, and immunological defense of the fetus (Fig. 7-25). Understanding the nature of the placental barrier has become increasingly important in recent decades because of the recognition that a number of substances, e.g., certain medications or alcohol, that can be present in the blood of some pregnant women are capable of producing birth defects if they pass through the placental barrier and enter the embryonic circulation.

The basis of placental function is the pattern of circulation of maternal and fetal blood in relation to the villi and the placental barrier. On the maternal side, blood enters into the intervillous space through the open ends of about 30 spiral arteries of the uterus at a pressure of about 70 to 80 mmHg (Ramsay, 1965; Wilkin, 1965). The arterial blood, which is rich in oxygen and nutrients, passes over the villi in small fountainlike streams and then under reduced pressure settles back to the maternal base of the placental compartment, where it is removed via open-ended uterine veins (Fig. 7-22). The intervillous space occupied by blood in the mature placenta is about 150 ml, and near term this volume of blood is replaced about three times per minute. On the fetal side, blood enters the placental villi through branches of the umbilical arteries. Despite the fact that anatomically the blood is arterial, it is the physiological equivalent of venous blood—poor in oxygen and high in CO_2 and waste products. In the terminal branches of the villi the fetal vessels are in the form of capillary networks, and in this region the bulk of placental exchange occurs. The now replenished blood returns to the fetus through the drainage system of the umbilical vein.

The major functions of the placenta consist of transport and synthesis. Transport functions in both directions across the placental barrier. The surface area available for exchange (about $10m^2$) is greatly increased not only by the branches of the chorionic villi but also by vast numbers of microvilli which

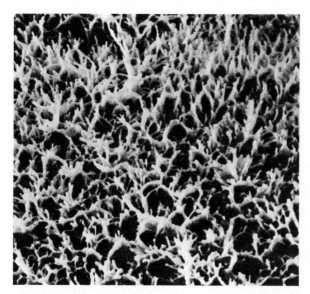

FIGURE 7-24
Scanning electron micrograph of the syncytial trophoblastic surface of the human placenta in the twelfth week of pregnancy. Numerous microvilli increase the absorptive surface of the placenta. ×9000. (Courtesy of Dr. Staffan Bergström.)

project from the surface of the syntrophoblast (Fig. 7-24). From the maternal side several classes of substances are transported (Fig. 7-25). One consists of easily diffusible substances, such as oxygen, water, and inorganic ions. Another class is composed of low-molecular-weight compounds, such as sugars, amino acids, and lipids, which serve as substrates for anabolic processes within the embryo. Their transfer requires active transport across the membranes of the placenta. Larger molecules, such as protein hormones and antibodies, require a means of transport involving pinocytosis as well as diffusion. An important class of transported macromolecules is maternal antibodies, which protect the newborn infant from disease until the immune system of the infant becomes functional. In many mammals, such as cattle, the placental barrier does not allow the passage of maternal antibodies into the fetus. The newborn calf must obtain antibodies from its mother's milk during the first 36 hours after birth (while its intestinal villi are still capable of absorbing undigested proteins) or it will become an immunological cripple. From the fetal side, the principal substances transported across the placenta are CO_2, water, electrolytes, urea, and other waste products of fetal metabolism.

The placenta is known to synthesize four hormones in the syntrophoblastic layer. Two are protein hormones (*human chorionic gonadotropin* and *human placental lactogen*[3]) and two are steroids (*progesterone* and *estrogens*). Chorionic gonadotropin, the hormone responsible for maintaining the corpus luteum, is produced early by the trophoblastic tissues, even before implanta-

---

[3]Also called *chorionic somatomammotropin*, this poorly understood hormone has both somatotropic and prolactinlike activity.

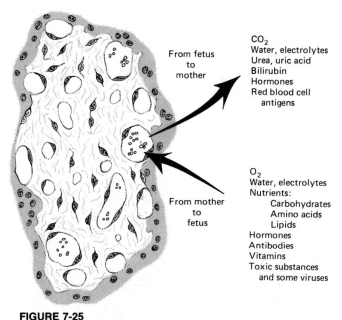

From fetus
to
mother

CO_2
Water, electrolytes
Urea, uric acid
Bilirubin
Hormones
Red blood cell
    antigens

From mother
to
fetus

O_2
Water, electrolytes
Nutrients:
    Carbohydrates
    Amino acids
    Lipids
Hormones
Antibodies
Vitamins
Toxic substances
    and some viruses

FIGURE 7-25
Diagram illustrating the major forms of exchange between a fetus
and its mother across a placental villus.

tion. The presence of this hormone in maternal urine has been the basis for many of the common tests for pregnancy. Under the sustaining influence of chorionic gonadotropin, the corpus luteum continues to produce progesterone and estrogens, which in turn act on the endometrium so that it continues to provide adequate support for the growth of the embryo. Within a couple of months the placenta synthesizes estrogens and progesterone in such quantities that pregnancy can be maintained even if the corpus luteum is surgically removed.

The part of the chorion not involved in the formation of the placenta also undergoes interesting changes. During the last half of pregnancy, the chorion laeve is pushed by the growing embryo tight against the uterine walls and actually fuses with the opposite decidua parietalis, thus obliterating much of the uterine cavity (Fig. 7-28). Adherent to the inner face of the chorion laeve is the amnion, which during the third month expands to fill the entire chorionic sac (Fig. 7-26) and soon thereafter becomes loosely attached to its inner face. At term, the amniotic sac contains almost a liter of amniotic fluid. A section passing through the tissue between the amniotic cavity and the muscular layer of the wall of the uterus, in a region clear of the placenta, will show a merging of the three originally separate structures. From the embryo toward the uterus these are, in order, the amnion, the chorion laeve, and the decidua parietalis (Fig. 7-28).

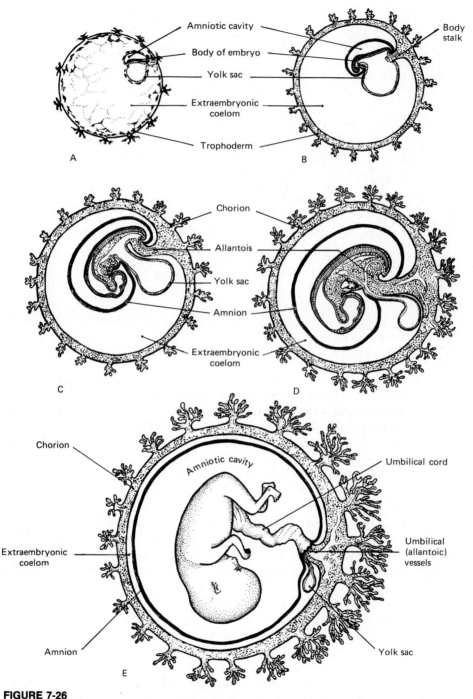

FIGURE 7-26
Early changes in interrelations of embryo and extraembryonic membranes. (See colored insert.)

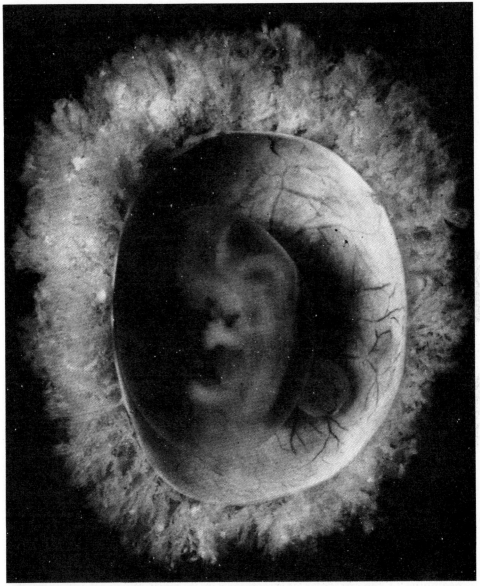

FIGURE 7-27
Chorionic vesicle of seventh week opened to expose the embryo within its intact
amnion. The small sphere to the right of the amnion is the yolk sac.
(Photograph of Chester Reather of Carnegie. Embryo No. 8537A.)

Birth and the Afterbirth

The placental attachment normally occurs relatively high up in the body of the
uterus. This results in the much-thinned decidua capsularis, the chorion laeve,

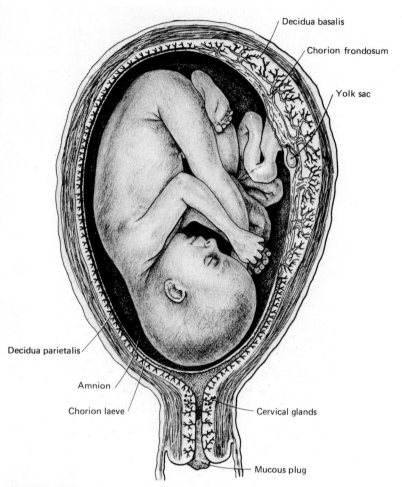

Decidua basalis

Chorion frondosum

Yolk sac

Decidua parietalis

Amnion

Chorion laeve

Cervical glands

Mucous plug

FIGURE 7-28
Diagram showing the relations to the uterus of a 5-month fetus and its membranes. The amniotic cavity is darkly stippled; the uterine cavity is lightly stippled.

and the adherent amnion being the only structures lying over the cervical outlet (Fig. 7-28). Together they form one composite fibrous membrane. With the beginning of the muscular contractions which mark the onset of labor, the amniotic fluid is squeezed into this thin part of the chorionic sac and the sac acts as a preliminary dilator of the cervical canal. As the periodic contractions become more frequent and more powerful, the investing membranes rupture at this region, freeing the embryo from its fetal envelopes but leaving the placenta still attached within the uterus. The retention of the placenta is of vital importance, for the process of birth ordinarily extends over several hours, and if the fetus were prematurely cut off from its uterine associations it would not survive the resulting interruption of its oxygen supply.

Continued uterine contractions, aided by voluntary contractions of the abdominal muscles, force the fetus into the slowly enlarging cervical canal until it is dilated sufficiently to permit the fetus to begin to move out of the uterus. When this has been accomplished, the obstetrician speaks of the first stage of labor as having passed. The second stage of labor is much briefer than the first. Once the fetus passes the cervical canal, it moves promptly through the vagina to "present itself" at the perineum. Dilation of the vulval orifice of the vagina progresses much more rapidly than did dilation of the cervix, and once the presenting part of the body—usually the head—passes this outlet, the rest of the body emerges rapidly. With the tying off and cutting of the umbilical cord, the relations with the uterus and placenta are ended and the newborn infant is for the first time an independently living individual.

In the usual course of events, some 15 to 20 minutes after the delivery of the fetus the uterus begins again to go into a series of contractions which loosen the placenta and the decidua from its walls and finally expel them. This is called the third stage of labor. Associated with the placenta are the torn remnants of the ruptured amnion, chorion laeve, and umbilical cord. This entire mass constitutes the *afterbirth*.

BASIC BODY PLAN OF YOUNG MAMMALIAN EMBRYOS

As the embryo gets past the stage of gastrulation, the period of organogenesis begins. So many important developmental events occur simultaneously that it is very difficult to study entire embryos. It is more common to study the development of individual organ systems or specific processes. The main purpose of this chapter is to provide an overview of the entire embryo during the period of early organogenesis.

It is important to recognize that the basic body plan of essentially all vertebrate embryos during the period of early organogenesis is essentially a carryover of that seen in the aquatic, gill-breathing ancestral forms of vertebrates. These characteristics include a circulatory system based on a simple tubular heart emptying into a ventral aorta, the breaking up of the ventral aorta into branches that supply the gills as they pass around the pharynx and the collecting of the dorsal branches of these arteries into a dorsal aorta (Fig. 8-17). The overall subdivision of the pharyngeal region into a system of gill (branchial) arches is retained, although with numerous modifications, throughout embryos of the vertebrate classes. Likewise, the segmental organization of the body (an adaptation for sinusoidal swimming movements) is prominent in the somites and spinal nerves and persists into adult life in the jointed vertebral column. Even the strung-out kidneys, so characteristic as an adaptation to the hypoosmotic environment of freshwater fishes, persist in the early embryonic stages of higher vertebrates. These common fundamental features of the organization of early embryos provide the basis for von Baer's law, which was introduced on p. 3.

There is a striking similarity in the general organization and external appearance of various amniote embryos (cf. Figs. 8-1, 8-2 and A-11 for early

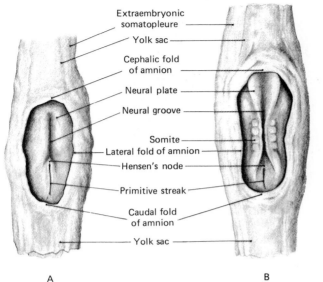

Extraembryonic
somatopleure

Yolk sac

Cephalic fold
of amnion

Neural plate

Neural groove

Somite

Lateral fold of amnion

Hensen's node

Primitive streak

Caudal fold
of amnion

Yolk sac

A   B

FIGURE 8-1
Drawings (×15) of pig embryos at the first appearance of the neural groove (*left*, from Carnegie Collection, C160-68) and at the time of the first somites (*right*, from Carnegie Collection, C190-2 and C196-1).

embryos and Fig. 1-23 for later embryos). As if to anticipate its ultimate cerebral dominance, the human embryo, however, soon begins to show a larger neural plate in its future forebrain region. The cephalic part of the neural plate

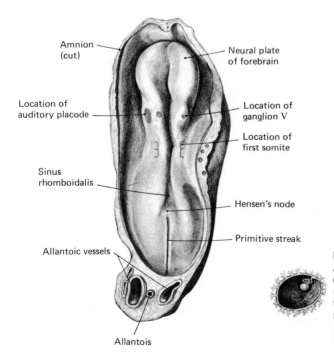

Amnion
(cut)

Neural plate
of forebrain

Location of
auditory placode

Location of
ganglion V

Location of
first somite

Sinus
rhomboidalis

Hensen's node

Allantoic vessels

Primitive streak

Allantois

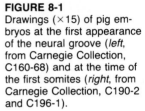

FIGURE 8-2
A human embryo at the beginning of somite formation (×45). (Sketch at the lower right shows the embryo and surrounding membranes at natural size.) (After Ingalls, 1920. *Carnegie Cont. to Emb.*, vol. 11.)

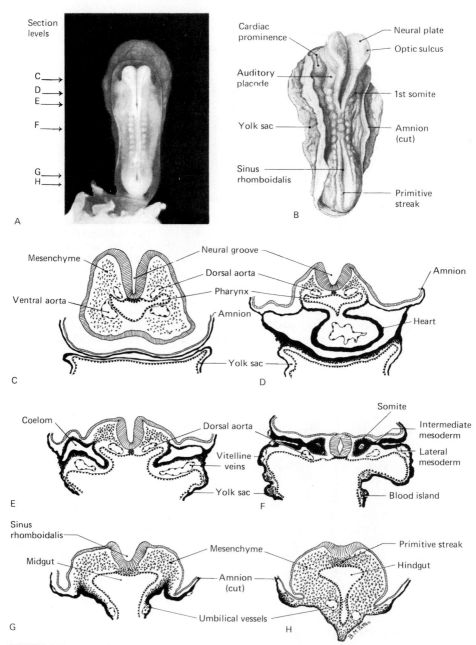

FIGURE 8-3
Structure of human embryo in the 7- to 8-somite stage, probable fertilization age of 18 to 19 days. (A) Bartelmez 8-somite embryo (University of Chicago, H 1404), photographed (×12½) before sectioning. (B) Reconstruction of the Payne 7-somite embryo (×22). C–H Sections of Bartelmez embryo at levels indicated in (A). Projection outlines (×60) schematically represented with ectoderm hatched; endoderm, a beaded line; mesenchyme, angular stippling; and the more solid parts of the mesenchyme in black.

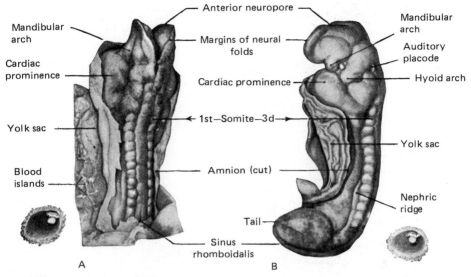

Mandibular arch

Cardiac prominence

Yolk sac

Blood islands

Anterior neuropore

Margins of neural folds

Cardiac prominence

← 1st—Somite—3d →

Amnion (cut)

Tail

Sinus rhomboidalis

Mandibular arch

Auditory placode

Hyoid arch

Yolk sac

Nephric ridge

A   B

FIGURE 8-4
Two human embryos of about 3 weeks' fertilization age. (A) Corner 10-somite embryo; probable age about 20 days (×25). (After Corner, 1919. *Carnegie Cont. to Emb.*, vol. 20.) (B) Heuser 14-somite embryo; probable age about 22 days (×30). (After Heuser, 1930. *Carnegie Cont. to Emb.*, vol. 22.) Sketches in lower corners show actual size of respective embryos and their chorionic vesicles.

of the young human embryo is so large and is expanding laterally so rapidly that its closure is delayed compared with the early closure in other embryos. In the midbody region, however, the general arrangement of structures and layers is essentially similar among amniote embryos. Between the third week and the fourth week the human embryo loses its originally straight body axis and begins to show a marked flexure in the cranial region (Figs. 8-4 and 8-5).

There is a basic similarity in the body plan of 4- to 4½-day chick embryos (Fig. A-36), 5-mm pig embryos (Fig. 8-6) and month-old human embryos (Fig. 8-7). Minor differences consist of the larger size of the eyes and optic regions of avian embryos and the relatively more advanced heart, liver, and mesonephros of mammalian embryos. The remainder of this chapter will summarize the main structural features of 4- to 6-mm pig embryos and 1-month human embryos.

EXTERNAL FEATURES

In the cephalic region the primordia of the developing sensory organs are prominent features. The nasal (olfactory) placodes appear as local ectodermal thickenings on either side of the head (Fig. 8-14E), and in more advanced embryos the nasal placodes become depressed to form the nasal pits (Fig. 8-7). The primordia of the eyes, located far laterally in the head (Fig. 8-7), are

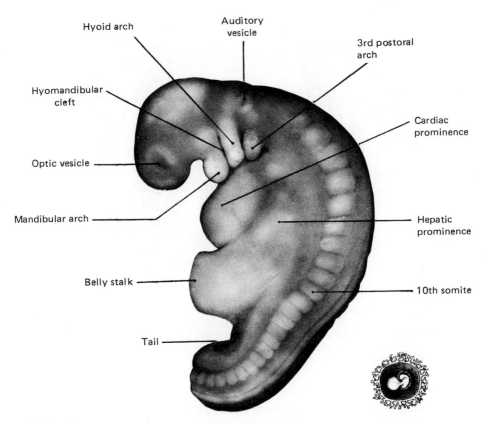

FIGURE 8-5
Human embryo toward the end of the fourth week. Retouched photograph (×26) of embryo 6097 in the Carnegie Collection; crown-rump length 3.6 mm; 25 pairs of somites. Sketch, lower right, shows actual size of embryo and its chorionic vesicle.

prominent landmarks, and the auditory vesicles and contours of the brain walls are sharply outlined in cleared embryos (Fig. 8-9). The face does not exist as a recognizable region.

The region of the throat is dominated by a system of branchial arches and clefts, a region homologous to the gills of primitive fishes. The branchial arches are best viewed from the ventral aspect (Figs. 8-12 and 14-1A and B). The first arch on either side is subdivided into two components: the *maxillary processes*, which form the lateral parts of the upper jaw, and the mandibular elevations, which merge with each other in the midventral line to form the arch of the lower jaw (*mandibular arch*). Posterior to the mandibular arch are three similar arches, the *hyoid* and the unnamed third and fourth postoral arches, all of which appear clearly in lateral views (Figs. 8-6 and 8-7). The hyoid arch is homologous to the *operculum* (gill cover) of fishes, and even in the human embryo it slightly overhangs the third and fourth arches.

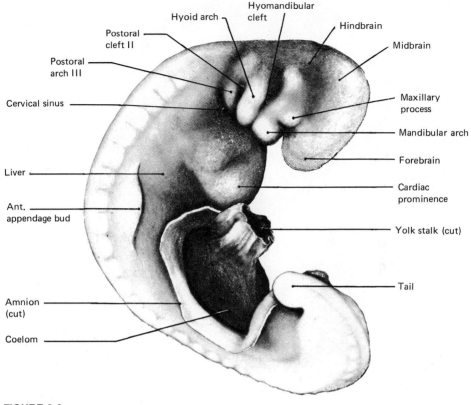

FIGURE 8-6
Photograph (×20) of a pig embryo of 5 mm.

Between the branchial arches are deep furrows which mark the position of the ancestral gill clefts. Although in mammalian embryos these furrows do not ordinarily break through into the pharynx, they are commonly called clefts because of their phylogenetic significance. Only the most cephalic of these clefts is named (*hyomandibular cleft*; Fig. 8-6); the others are designated by their postoral numbers. The entire region about the third and fourth postoral clefts becomes especially deeply depressed and is known as the *cervical sinus* (Fig. 8-7).

The upper body of the early embryo is dominated by the precociously large heart, which forms a surface bulge known as the *cardiac prominence* (Fig. 8-7). Along the dorsal side of the trunk the paired somites extend from just caudal to the auditory vesicle into the tail. Conspicuous in the midventral region is the body stalk, which in time becomes more discrete and elongated as the umbilical cord (Fig. 7-26).

The appendage buds first appear as flangelike projections from the body wall (Fig. 8-6). The arm bud slightly leads the leg bud in developmental progress.

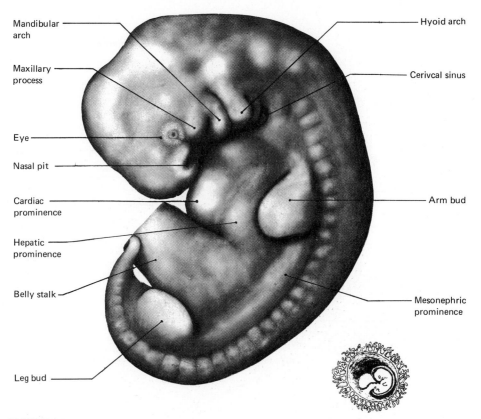

Mandibular arch

Maxillary process

Eye

Nasal pit

Cardiac prominence

Hepatic prominence

Belly stalk

Leg bud

Hyoid arch

Cerivcal sinus

Arm bud

Mesonephric prominence

FIGURE 8-7
Human embryo late in fifth week after fertilization; C-R 6.5 mm. Retouched photograph (×17) of embryo 6502 in the Carnegie Collection. Small sketch, lower right, shows actual size of embryo and its chorionic vesicle.

At this stage the human embryo has every bit as well developed a tail as does the pig embryo (cf. Figs. 8-6 and 8-7). Later in development the human tail normally undergoes regressive changes (Fig. 17-31) that leave the human with only a symbolic coccyx. Regression of the human tail is apparently brought about by intrinsic cell death (Fallon and Simandl, 1978). Occasionally this regression fails to occur, and a human infant is born with a sizable and unmistakable tail.

THE NERVOUS SYSTEM

The central nervous system has progressed from a simple neural tube to one in which the fundamental subdivisions of the brain are taking shape. At first, three subdivisions are apparent. These are the forebrain (*prosencephalon*), the midbrain (*mesencephalon*), and the hindbrain (*rhombencephalon*) (Fig. 12-23A). At the stage under consideration the brain is in the phase of transition

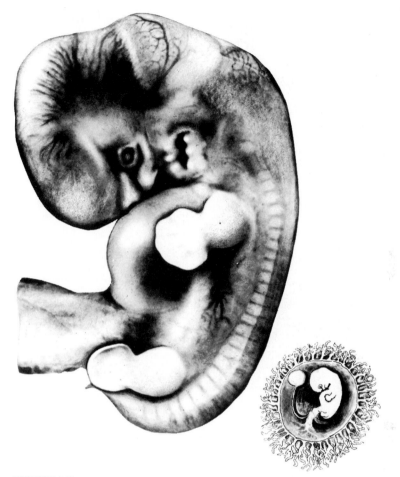

FIGURE 8-8
Human embryo a little over 6 weeks after fertilization, C-R 14 mm.
(Retouched photograph (×8) of embryo No. 1267A in the Carnegie
Collection.)

from the three- to the five-vesicle condition. There is a clear indication of the
separation of the forebrain into a rostral *telencephalon* and a *diencephalon*,and
the thickening of the walls of the more rostral part of the hindbrain presages the
differentiation of the *metencephalon* from the *myelencephalon* (Fig. 12-23C).
The mesencephalon remains undivided.

Projecting from the walls of the diencephalon are the *optic vesicles*, which
are beginning to invaginate to form the optic cups (Fig. 8-14D). In response to
an inductive signal from the optic vesicle, the ectoderm overlying the optic cup
has thickened and has begun to invaginate to form the *lens vesicle* (Figs. 8-14D
and 13-2D).

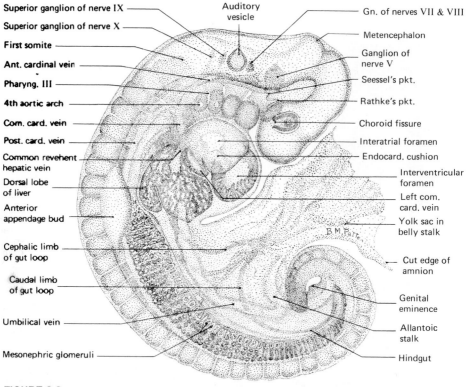

Superior ganglion of nerve IX
Superior ganglion of nerve X
First somite
Ant. cardinal vein
Pharyng. III
4th aortic arch
Com. card. vein
Post. card. vein
Common revehent hepatic vein
Dorsal lobe of liver
Anterior appendage bud
Cephalic limb of gut loop
Caudal limb of gut loop
Umbilical vein
Mesonephric glomeruli

Auditory vesicle

Gn. of nerves VII & VIII
Metencephalon
Ganglion of nerve V
Seessel's pkt.
Rathke's pkt.
Choroid fissure
Interatrial foramen
Endocard. cushion
Interventricular foramen
Left com. card. vein
Yolk sac in belly stalk
Cut edge of amnion
Genital eminence
Allantoic stalk
Hindgut

B. M. Patte.

FIGURE 8-9
Projection drawing (×17) of a lightly stained and cleared 5-mm pig embryo.

From 4 to 5 weeks in the human embryo many of the 12 cranial nerves begin to appear (Fig. 12-23B and C). Details of the development of the individual cranial nerves will be covered in Chap. 12.

Caudal to the myelencephalon the neural tube is more slender and gives rise the spinal cord. During these early stages it is a relatively simple appearing tube with a slitlike central canal and walls of closely packed cells of ectodermal origin. Alongside the spinal cord the ribbonlike neural crest has just broken up and formed the spinal (and also cranial) ganglia.

THE DIGESTIVE AND RESPIRATORY SYSTEMS

As was described in Chap. 7, the primitive gut tube is formed by the endoderm as the body itself folds into a tube. By the stage of the 5-mm pig and the 1-month human embryo, it has been delimited into a tubular *foregut* and *hindgut* and a *midgut*, which still has an open ventral region leading into the yolk stalk.

The intraembryonic gut tract of young mammalian embryos at first ends blindly at both its cephalic and caudal ends (Fig. 8-10B and C). An external

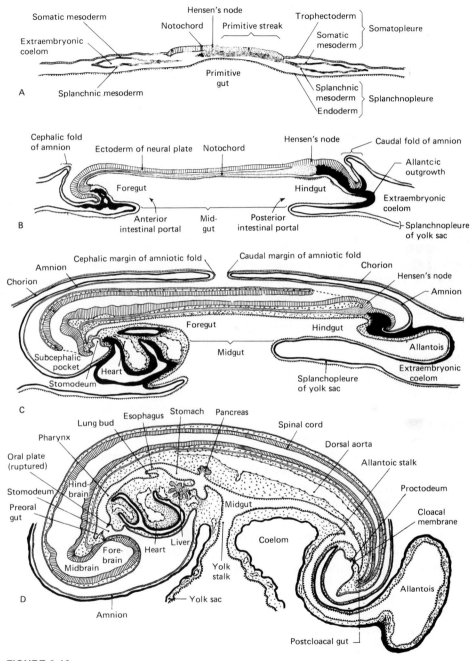

FIGURE 8-10
Sagittal sections of pig embryos to show establishment and early regional differentiation of the gut. The drawings indicate (A) the primitive-streak stage; (B) the beginning of somite formation; (C) embryos having about 15 somites; (D) embryos having about 25 somites, or about 5 mm.

depression called the *stomodeum* marks the future oral opening. As the stomodeum deepens, the ectoderm of its floor comes into direct contact with the endoderm of the foregut to form the *stomodeal,* or *oral, plate.* As is common when ectoderm and endoderm abut directly upon each other, the oral plate breaks through and establishes the oral opening into the foregut (Figs. 8-10D, 8-11C, and 8-12).

Arriving medially as a slender ectodermal diverticulum from the ventral part of the stomodeum is *Rathke's* pocket (Figs. 8-13 and 16-4). From its first appearance, Rathke's pocket is in close relationship to the *infundibular process* from the floor of the diencephalon. Together, these two structures will form the *hypophysis.*

On the endodermal side of the oral plate, the extreme cephalic end of the foregut remains as the *preoral gut* (Fig. 8-10D), or *Seessel's pocket* (Fig. 8-13). This structure serves only as an anatomical landmark and gives rise to no adult structure.

The cephalic end of the foregut is mainly involved in the formation of the pharynx, from which four pouches extend on either side toward the corresponding external gill furrow (Fig. 8-14B). Although the thin membrane that remains between the pharyngeal pouch and the external gill furrow in mammals does not ordinarily break down to form an open gill cleft such as that found in our water-living ancestors, the similarity of relationships is obvious. The significance of the structural relations in this region is further emphasized by the location of the *aortic arches,* which lie in closely packed mesenchymal tissue between the gill clefts. In the embryos of birds and mammals, the aortic arches do not form capillary beds as they do in gill-breathing animals. Nevertheless, the basic ancestral pattern can be seen in the way the aortic arches pass from the ventrally located heart to the dorsal aorta by way of the gill arches flanking the pharynx (Fig. 8-17).

In older embryos, a number of important structures arise from the endodermal lining of the pharynx (see Chap. 16). Of these, only the small cluster of cells which constitute the primordium of the *thyroid gland* has emerged from the floor of the pharynx at this stage.

Also from the floor of the posterior pharynx, a median ventral groove is rapidly converted into a tubular outgrowth parallel to the digestive tract. This groove is the *laryngeotracheal groove* (Fig. 8-14D), and the tubular outgrowth is the future *trachea,* the caudal end of which is just beginning to bifurcate to form the *bronchial* or *lung buds* (Figs. 8-14E and F and 15-7).

Caudal to the pharynx, the digestive tube passes from a short, narrow esophagus to a slight dilation that represents the future stomach. Immediately caudal to the stomach are the outgrowths of the gut that constitute the

---

FIGURE 8-11
Sagittal plans of human embryos in the third and fourth weeks to show establishment of the digestive system. (A) At the beginning of somite formation, about 16 days. (B) Seven somites, about 18 days. (C) Fourteen somites, about 22 days. (D) Toward the end of the first month.

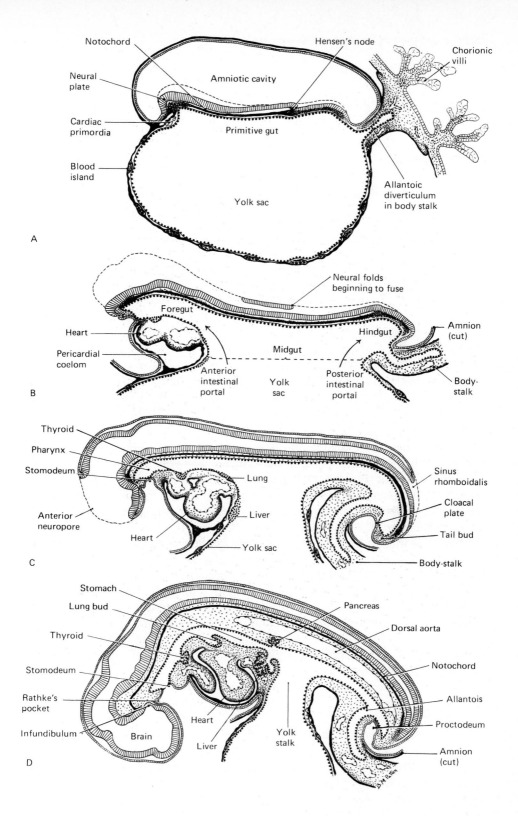

A

Notochord

Neural plate

Cardiac primordia

Blood island

Hensen's node

Amniotic cavity

Primitive gut

Chorionic villi

Allantoic diverticulum in body stalk

Yolk sac

B

Heart

Pericardial coelom

Foregut

Neural folds beginning to fuse

Hindgut

Amnion (cut)

Midgut

Anterior intestinal portal

Yolk sac

Posterior intestinal portal

Body-stalk

C

Thyroid

Pharynx

Stomodeum

Anterior neuropore

Heart

Lung

Liver

Yolk sac

Sinus rhomboidalis

Cloacal plate

Tail bud

Body-stalk

D

Stomach

Lung bud

Thyroid

Stomodeum

Rathke's pocket

Infundibulum

Brain

Heart

Liver

Yolk stalk

Pancreas

Dorsal aorta

Notochord

Allantois

Proctodeum

Amnion (cut)

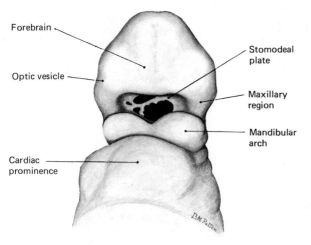

Forebrain

Optic vesicle

Cardiac
prominence

Stomodeal
plate

Maxillary
region

Mandibular
arch

FIGURE 8-12
Breaking through of the stomodeal plate to establish oral opening into foregut as seen in a face
view of a human embryo of the
fourth week. (Drawn ×30, from
stereophotographs of embryo
6097 in the Carnegie Collection.)

primordia of the *pancreas, liver* and *gallbladder.* The pancreas at this stage consists of two independent parts, a well-defined dorsal bud and a smaller ventral bud (Fig. 8-13). The original hepatic diverticulum arises ventrally from the gut as a mass of branching epithelial cords almost directly opposite the dorsal pancreatic bud (Fig. 8-13).

At this stage, the intestines are represented by a straight tube which runs parallel to the midsagittal plane of the embryo. The yolk stalk, arising from the floor of the midgut (Fig. 8-11D), serves as a useful landmark for indicating what part of the developing digestive tract came from the foregut and what part was derived from hindgut. In contrast to the foregut, the hindgut gives rise to only one diverticulum, the allantoic stalk which arises near its posterior end (Fig. 8-10D). Posterior to the allantoic stalk, the hindgut dilates slightly to form the *cloaca.*

Near the caudal end of the cloaca is a depression in the ventral body wall called the *proctodeum.* It deepens in a manner similar to that exhibited by a stomodeum at the anterior end of the foregut, leaving only a thin separating membrane of proctodeal ectoderm and cloacal endoderm. This membrane is known as the *cloacal plate,* or *cloacal membrane* (Fig. 8-10D). With its rupture, which occurs considerably later than the rupture of the oral plate, the originally blind hindgut establishes an outlet.

THE MESODERM

From the simple condition in which it was divided into somitic (paraxial), intermediate, and lateral plate components, the mesoderm has begun to undergo some specialization. The somites have developed past the purely epithelial stage (Fig. 6-23B). The cells of the sclerotomal region have broken away from the main body of the somite, leaving the dermatomal and myotomal layers (Fig. 6-23D).

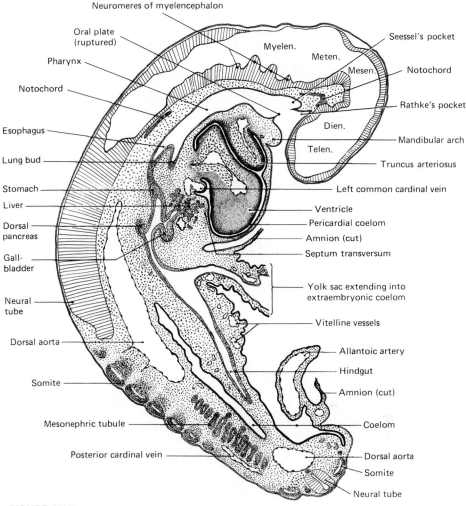

FIGURE 8-13
Longitudinal section of 5-mm pig embryo (×25). The caudal end of an embryo in this stage of development is usually somewhat twisted to one side (see Fig. 8-6). For this reason, sections which cut the cephalic region in the sagittal plane pass diagonally through the posterior part of the body. For a schematic plan of a completely sagittal section of an embryo of about this age see figure 8-10, D.

The intermediate mesoderm has differentiated into the tubules and ducts of pronephric and mesonephric kidneys (Fig. 8-14G and H), which are transient stages in the ontogenesis of the definitive urogenital system. By looking ahead to the schematic drawings of Figs. 17-6 and 17-7, one can better appreciate the relationship between the mesonephros and the metanephros, the permanent kidney. The urogenital (intermediate) mesoderm is concentrated in a pair of

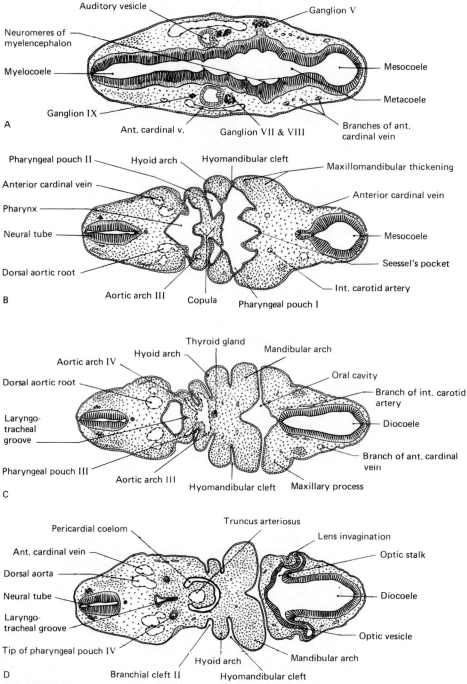

FIGURE 8-14
Transverse sections of 5-mm pig embryo (×18). Compare with entire embryo (Fig. 8-6) and with longitudinal section of embryo of same age (Fig. 8-13).

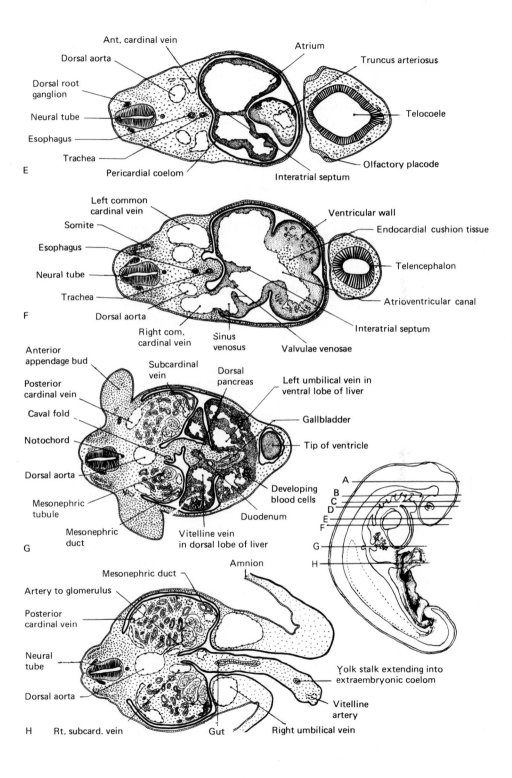

E
Ant. cardinal vein · Dorsal aorta · Dorsal root ganglion · Neural tube · Esophagus · Trachea · Pericardial coelom · Atrium · Truncus arteriosus · Telocoele · Olfactory placode · Interatrial septum

F
Left common cardinal vein · Somite · Esophagus · Neural tube · Trachea · Dorsal aorta · Right com. cardinal vein · Sinus venosus · Valvulae venosae · Ventricular wall · Endocardial cushion tissue · Telencephalon · Atrioventricular canal · Interatrial septum

G
Anterior appendage bud · Posterior cardinal vein · Caval fold · Notochord · Dorsal aorta · Mesonephric tubule · Mesonephric duct · Subcardinal vein · Dorsal pancreas · Developing blood cells · Duodenum · Vitelline vein in dorsal lobe of liver · Left umbilical vein in ventral lobe of liver · Gallbladder · Tip of ventricle

H
Mesonephric duct · Artery to glomerulus · Posterior cardinal vein · Neural tube · Dorsal aorta · Rt. subcard. vein · Amnion · Gut · Right umbilical vein · Yolk stalk extending into extraembryonic coelom · Vitelline artery

ridges which run on the dorsal mesentery (Figs. 8-14G and 17-17). At this stage there is still no trace of gonadal tissue.

At a very early stage, the lateral plate mesoderm splits into two layers, one associated with ectoderm and one associated with endoderm. By convention, a layer of mesoderm plus ectoderm is called a *somatopleure* and a layer of mesoderm plus endoderm is called a *splanchnopleure*.

THE CIRCULATORY SYSTEM

Basic Plan of the Embryonic Circulatory System

The embryonic circulatory system can be analyzed in terms of three major circulatory arcs, with the heart as the common center and pumping station. One arc is entirely intraembryonic in its distribution. Its vessels bring food materials and oxygen to all parts of the growing body and return waste materials from them. The other two circulatory arcs have both intra- and extraembryonic components. The vitelline arc carries blood to and from the yolk sac. The other arc carries blood to and from the allantois for gaseous interchange (Fig. A-42). As the blood from the three arcs is returned to the heart for recirculation, it is constantly mixed so that its food material, oxygen, and accumulated waste products are maintained at serviceable levels.

In placental mammals radical changes in the source of food supply and basic living conditions, compared with reptiles and birds, alter the way the two extraembryonic circulatory arcs operate. The yolk sac of higher mammalian embryos, although formed in characteristic relations to other structures, is small and empty; its circulatory arc, therefore, has lost its significance as a purveyor of food. Nevertheless, the vessels of this arc still form and are for a time quite conspicuous. In what may be called a "phylogenetic hangover," the vitelline vessels, although deprived of their primary function, still bring the first blood cells into the embryonic circulation from their place of formation in the yolk-sac splanchnopleure, just as they did in ancestral forms with food-laden yolk sacs.

In mammalian embryos the allantoic arc takes over the functions abandoned by the vitelline arc as well as continuing to carry out its own earlier responsibilities. The allantois is either closely applied to the uterine lining, as in the pig (Fig. 8-17A), or sends little rootlets of its own tissue (called *chorionic villi*) into the uterine mucosa, as in the human (Fig. 8-17B). In either situation, maternal blood and fetal blood are brought close together so that the fetal blood can absorb food and oxygen from the maternal blood and pass its own waste materials back to the maternal circulatory system. Thus, the allantoic circulation of a mammalian embryo serves as a provisional mechanism for food getting, respiration, and excretion.

The Establishing of the Heart

In mammalian embryos, the heart arises from paired mesodermal primordia situated ventrolaterally beneath the pharynx (Fig. 8-15). The cardiac primordia

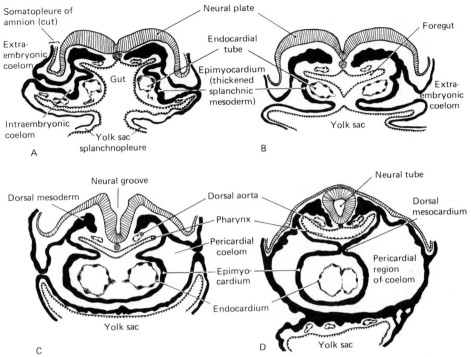

FIGURE 8-15
Sections cut transversely through the cardiac region of pig embryos of various ages to show the origin of the heart from paired primordia. (Projection diagrams ×65, from series in the Carnegie Collection.) (A) 5-somite embryo; (B) 7-somite embryo; (A) 10-somite embryo; (D) 13-somite embryo.

are composed of two layers as well as being paired right and left. The inner layer is called the *endocardium*, because it is destined to form the internal lining of the heart. The outer layer, derived from thickened splanchnic mesoderm, is known as the *epimyocardium*, because it will give rise to both the heavy muscular layer of the heart wall (*myocardium*) and its outer covering (*epicardium*).

While these changes have been occurring in the heart, folding off of the embryonic body has been going on with concomitant progress in the closure of the foregut at the level of the heart (cf. Fig. 8-15A and B). As a result the paired endocardial tubes are brought progressively closer together. Finally they are approximated to each other and fused into a single tube lying in the midline (Figs. 8-15 and 8-16).

In the same process the epimyocardial layers are bent toward the midline enwrapping the endocardium. Ventral to the endocardial tubes the epimyocardial layers of opposite sides come into contact with each other. Where this contact occurs, the limbs of the mesodermal folds next to the endocardium fuse with each other, forming an outer layer of the heart that no longer is interrupted ventrally (Fig. 8-15C). Thus the originally paired right and left coelomic

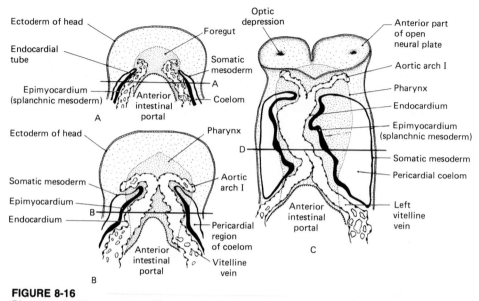

FIGURE 8-16
Diagrams showing progress of fusion of cardiac primordia in the pig as seen in ventral views. (A) 5-somite embryo; (B) 7-somite embryo; (C) 13-somite embryo. The embryos are supposed to be viewed as transparent objects with the outlines of the cardiac primordia showing through. The lines A, B, and D indicate the levels of sections A, B, and D in Figure 8-15.

chambers become confluent to form a median unpaired pericardial cavity in the same process which establishes the heart as a median structure. Dorsally the right and left epimyocardial layers become contiguous, but here they do not fuse immediately as happens ventral to the heart. They persist for a time as a double-layered supporting membrane called the *dorsal mesocardium.* In this manner the heart is established as a nearly straight double-walled tube suspended mesially in the most cephalic part of the coelom.

The early heart soon undergoes a change in shape from a straight tubular form to an S-shaped configuration (Fig. A-46). During this phase the heart functions like a simple tubular peristaltic pump. The veins converging to enter the heart become confluent in a thin-walled chamber called the *sinus venosus* (Figs. 8-14F and 18-19C and D). The sinus venosus opens through a slitlike orifice into the atrial portion of the heart. Backflow of blood is prevented by well-developed flaps known as the *valvulae venosae* (Figs. 8-14F and 8-19). From the atrial region, which is just beginning to bulge out into right and left chambers, blood enters the muscular ventricles. Internal subdivision of the early atrial and ventricular regions is just beginning. Details of cardiac partitioning will be covered in Chap. 18. From the ventricle, blood passes into the *truncus arteriosus* and then on to the body by way of the ventral aortic roots. Often the transitional area between ventricle and truncus is called the *conus* (Fig. 18-18). Despite the early modifications of the structure of the heart,

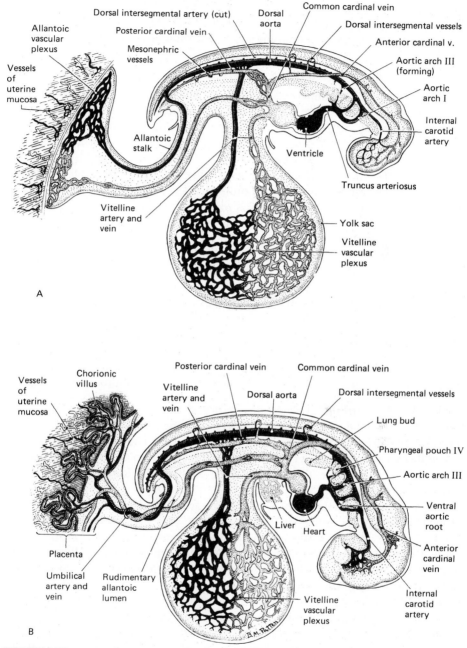

FIGURE 8-17
Semischematic diagrams showing the basic plan of the circulatory system of young mammalian embryos. At this stage all the major blood vessels are paired right and left. For the sake of simplicity only the vessels on the side toward the observer are shown. (A) Pig embryo of about 4 mm (age about 16 to 17 days); (B) human embryo of about 4.5 mm (fertilization age about 4 weeks).

the blood entering the caudal end of the heart through the sinus venosus is still pumped out of the heart into the truncus as an undivided stream, just as was the case in younger embryos containing a straight tubular heart.

The Formation of Blood and Blood Vessels

While the heart is becoming established, the main vascular channels characteristic of young embryos are also making their appearance. In a manner similar to the genesis of the endocardial tubes themselves, cords and knots of mesodermal cells become aggregated along the future course of a developing vessel. These strands of cells then form hollowed-out tubes lined by a layer of thin, flattened *endothelial cells*. In later stages some vessels also become extended by the formation of budlike outgrowths from their walls. Typically, a meshwork of small vascular channels is formed first. Gradually, some of these primitive channels become enlarged to form the main vessels.

Blood cells and early blood vessels form as aggregates of splanchnic mesodermal cells lining the yolk sac (Figs. 18-2 and 18-3). These aggregates are called *blood islands*. The differentiation of the central cells of the blood islands into blood cells is described in Chap. 18. The cells around the early blood cells become flattened and differentiate into vascular endothelial cells. As the blood islands develop, they coalesce and become incorporated into the vitelline circulatory arc. These vitelline channels feed blood cells into the early beating heart, which distributes them throughout the vascular channels (Fig. 8-17). In human embryos the first blood cells are drawn from the yolk sac into the vascular system toward the end of the third week of development.

The Arterial System Vertebrate embryos pump their blood from the ventrally located heart around the pharynx to the dorsal aorta by way of a series of six paired blood vessels called *aortic arches*. The aortic arches appear in a cephalocaudal sequence (Fig. 8-17) and then undergo an irregular sequence of regression or preservation. The fate of each of the aortic arches will be described in Chap. 18.

The aortic arches empty into the dorsal aorta, which initially is a paired vessel throughout its entire length (Fig. A-20). This paired condition persists cephalic to the arm buds, where the vessels are called the *dorsal aortic roots*. Caudal to the arm buds the paired aorta fuse to form a single midline vessel.

Three major arterial trunks lead off the dorsal aorta in young embryos (Fig. 8-17). At the level of the first aortic arches, the paired *internal carotid arteries* grow to the developing brain. The main *vitelline artery* arises at midbody level and extends ventrally to supply the vitelline vascular plexus on the yolk sac. Toward the caudal end of the aorta, the large *allantoic arteries* grow out along either side of the allantoic stalk. These vessels ultimately become known as the *umbilical arteries*.

The Venous System The main return channels for the intraembryonic circulatory arc are the cardinal veins (Fig. 8-17). The paired *anterior cardinal*

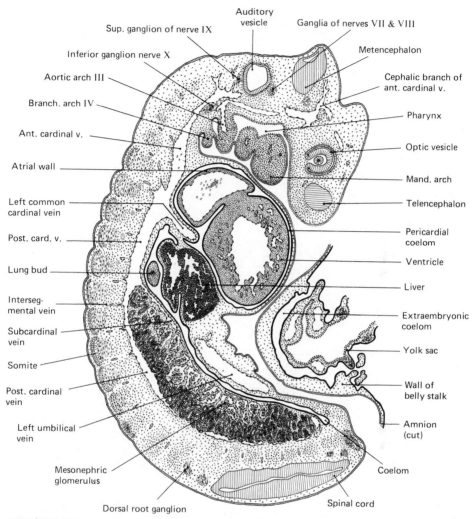

FIGURE 8-18
Drawing of a parasagittal section of a 5.5-mm pig embryo. (Projection outlines ×22.) The section was taken from the series to the left of the midline, at a plane especially favorable for showing the cardinal veins.

veins, which collect blood from the head, and the *posterior cardinal veins*, which are the primary drainage channels of the caudal half of the body, converge to form the *common cardinal veins*, which empty into the sinus venous of the heart. Major pairs of veins also bring blood from the two extraembryonic circulatory arcs into the body. Arising as collecting channels in the vitelline vascular plexus over the yolk sac, the main *vitelline veins* (sometimes called the *omphalomesenteric veins*) pass into the body through the substance of the liver to empty into the sinus venosus. The growing cell cords

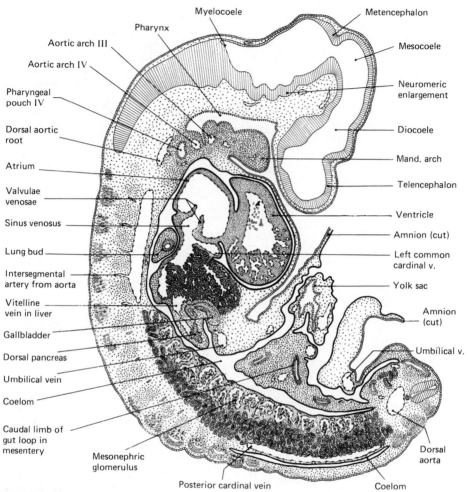

FIGURE 8-19
Drawing of parasagittal section of 5.5-mm pig embryo. (Projection outlines ×22.) The section was taken from the series to the right of the midline, at a plane favorable for showing the entrance of the sinus venosus into the right atrium.

of the liver soon break up the proximal portion of the vitelline veins into irregular channels called *hepatic sinusoids*. The return channels from the allantoic arc are initially referred to as the *allantoic veins*, but in time as the placenta forms and the body stalk is converted into the umbilical cord, they are known as the *umbilical veins*. Although paired when they are first formed, the umbilical veins undergo a complex series of fusions and obliterations (Fig. 18-14) during later development.

CELLULAR DIFFERENTIATION AND THE DEVELOPMENT OF MUSCULAR AND SKELETAL TISSUES

Up to this point we have seen the embryo develop from a single-celled zygote through cleavage, gastrulation, and the period of neurulation; in the process a number of distinctly recognizable cell types as different as mesenchyme, neural epithelium, and syntrophoblast have appeared. The generation of cell diversity (there are roughly 200 distinctly different types of cells in an adult mammal) is one of the fundamental processes of development; this process is often referred to as *cytodifferentiation*. Cytodifferentiation has been formally defined in a number of different ways, many of which imply specific states of being or mechanisms that are the source of considerable disagreement among embryologists and developmental biologists.

This chapter will begin with a general discussion of cytodifferentiation and will continue with the development of muscular and skeletal tissues. Since the cytodifferentiation of skeletal and muscle cells does not occur in a vacuum, cytodifferentiation in these tissues will be dealt with in the context of the development of the skeleton and musculature as functional organ systems, not just as isolated cells. Other well-studied systems of cytodifferentiation (e.g., lens fibers, epidermal cells, and red blood cells) will be described in other chapters.

THE GENERATION OF CELL DIVERSITY

All of the many cell types in the adult body are ultimately descended from a single cell: the fertilized egg. Implicit in this statement is the recognition that all the information required to generate these different cells is present in the zygote. The information for the production of specific adult proteins is encoded in the nuclear or mitochondrial DNA, but maternally derived components of the cytoplasm or membranes may be important regulatory influences, especially in early development. As was discussed in Chap. 4, maternally derived mRNAs are also the basis for the formation of many proteins during cleavage in a number of species.

Constancy of the Genome and Differential Gene Expression

Contemporary molecular genetics and developmental biology are based on the premise that, with only a few exceptions, all the cells of an organism at all stages of its life history possess a complete complement of genetic information like that of the zygote. As development proceeds and different types of cells appear, different portions of the genome are expressed in the various cell types.

This assumption was not always generally accepted. The German biologist August Weismann, one of the giants of nineteenth-century biology, in 1892 proposed an elaborate theory, according to which, cells progressively lose genetic information as development proceeds. For example, as the arm takes shape, the cells of the shoulder region contain all the information required to guide the formation of the entire arm. However, cells of the forearm have lost the information required for formation of the upper arm, although they can still guide the formation of the hand and fingers.

A number of classical experiments, ranging from Spemann's hair loop constriction of the amphibian zygote (Fig. 4-22) to nuclear transplantation (Fig. 1-27), have shown that single nuclei from cells taken from the period of cleavage up to the adult are capable of directing the development of a complete individual from a single cell. In addition, more recent cell and molecular hybridization studies have demonstrated the presence of a wide variety of genes in cells that normally do not express them. For example, if a human liver cell is fused with a mouse muscle cell (*somatic hybridization*), the human liver can be shown to produce human muscle-specific proteins which would normally never be expressed in a liver cell (Blau et al., 1985).

Molecular hybridization experiments have also demonstrated the presence in chromosomes of genes that are not active in the cell. In one experiment, Barnett et al. (1980) produced a cDNA corresponding to a yolk protein in *Drosophila*. When the radioactive cDNA was applied to the polytene chromosomes of the larval salivary gland, the cDNA probe bound to a well-defined band on a chromosome, demonstrating that a cell that does not make yolk proteins still possesses the gene.

There are, however, some exceptions to the rule of genomic constancy in all the cells of an organism. The most obvious are those cases (e.g., erythrocytes and platelets in mammalian blood) in which the entire nucleus is lost from the mature cell. Less extreme but still obvious on simple microscopic examination is the phenomenon of *chromosomal diminution* in *Ascaris* (Fig. 9-1). In this case, as the zygote undergoes early cleavage, a specialized region of germ plasm is segregated in certain of the blastomeres (future germ cells). In all the

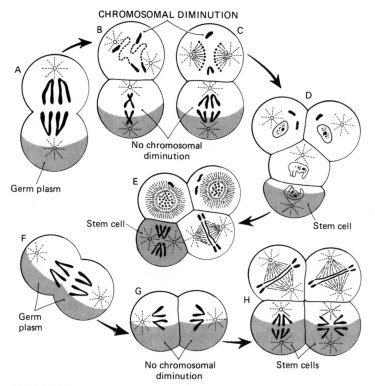

FIGURE 9-1
Relationship between the distribution of germ plasm (gray) and
chromosomal diminution in *Ascaris*. In normal development (A–E) only the
cells that contain germ plasm retain the full complement of chromosomal
material. In cells not containing germ plasm, chromosomal diminution occurs. In centrifuged embryos (F–H) the germ plasm is abnormally distributed.
As in the normal embryo, chromosomal diminution does not occur in cells
containing germ plasm. (*After C. H. Waddington, 1966, Principles of Development and Differentiation, Macmillan, New York.*)

blastomeres that do not contain germ plasm, portions of the chromosomes lose
their integrity and disappear, leaving only incomplete chromosomes in these
somatic cells. A more recently recognized exception to the rule of genomic
constancy involves antibody-producing cells, in which rearrangements of
genetic material result in the loss of some potential antibody-creating genes and
the creation of others (Hozumi and Tonegawa, 1976).

Differential gene expression means that only a small fraction of the total
genome is expressed in any given cell and that different parts of the genome are
expressed in different cells. Although this may seem intuitively obvious, it is
difficult to demonstrate directly. One method of demonstrating differential gene
expression is simply to look at the products made by different types of mature
cells. It is more difficult, however, to show that other gene products are not
being made in extremely small amounts by the same cell. In fact, there is some

evidence that very small numbers of protein molecules, representing a variety of genes, can be produced in single cells.

One of the early breakthroughs in demonstrating differential gene expression came as a result of studies of the banding patterns and puffing of the giant *polytene chromosomes* of insects (Fig. 1-16; Beermann, 1952). When these studies were first done, it was assumed that each band on the chromosomes represented a gene and that the puffs were indicative of an activity (which we would now call transcriptional) of these genes. A number of studies involving the incorporation of isotopic precursors of RNA, the use of specific inhibitors of RNA synthesis, and hybridization techniques have shown that the puffs are indeed regions of RNA synthesis. Different patterns of puffing on these chromosomes have been shown both at different developmental stages and in different types of tissues. Experimental studies have shown that with the addition of *ecdysone*, the insect molting hormone, new puffs appear on the polytene chromosomes and old ones regress.

Among the vertebrates, the loops on the lampbrush chromosomes (Fig. 3-15) in oocytes at the diplotene stage of meiosis have similarly been considered to reflect regions where specific gene expression is taking place. In situ hybridization techniques (Fig. 9-2) have been used to demonstrate specific locations on lampbrush chromosome loops where histone mRNAs are produced (Old et al., 1977).

A number of other molecular hybridization studies have shown that a given type of mRNA is found in some cell types but not in others or that a given cell type does not contain demonstrable amounts of all the mRNAs that can be extracted from the organism. Such molecule-specific studies have confirmed at a different level what had been intuitively obvious to morphologists and biochemists for years, namely, that there are many different cell types in the body and that the basis for these differences is differential gene expression.

Isoforms and the Progressive Specialization of Cells During Development

Differentiation has often been considered to be the progressive specialization of a cell during development. According to this concept, the fertilized egg is said to be a highly unspecialized (or undifferentiated) cell, and successive generations of progeny are assumed to become increasingly specialized as they approach the mature condition. This view of cellular development uses the stable adult state of a cell as the basis for comparison. It is becoming increasingly apparent, however, that at any stage of development cells are uniquely specialized for the situation in which they find themselves. The amphibian oocyte or zygote, for instance, with its intrinsic supply of yolk and its diverse store of maternal RNAs, ribosomes, and enzymes, abounds in subcellular adaptations that are required for the initiation of development. At a slightly later stage of development, the ubiquitous mesenchyme is composed of cells that are specialized for moving, detecting changes in the substrate, and

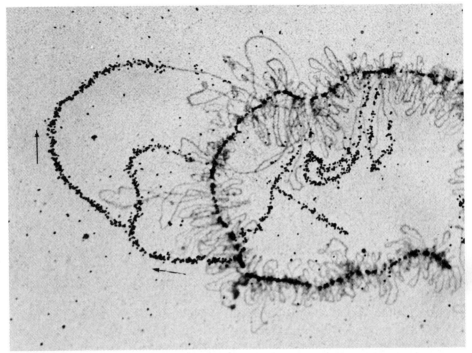

FIGURE 9-2
An early attempt to use in situ hybridization to localize histone mRNAs on chromosomal loops of
newt lampbrush chromosomes. H³-labelled histone gene DNA from the sea urchin was applied to
the chromosomal preparations. The rows of black dots (silver grains) deposited in an autoradiograph
were reported to show areas where the labelled histone DNA bound to corresponding histone
mRNAs being formed on the chromosomal loops, although some non-specific labelling also
occurred. (*Courtesy of H. G. Callan, from Old et al., 1977.*)

engaging in morphogenetic activities related to pattern formation. Even in an adult individual, the different stages in the maturation of a given cell type have special properties that are uniquely appropriate for them and not for later or earlier stages in the sequence of maturation. For example, within the series of cells in the differentiation pathway of a mature red blood cell (see Chap. 18), the early stages that are still present in the bone marrow are capable of sensing the mitogenic influences of erythropoietin and then responding by undergoing mitosis. Such properties are important in maintaining physiological levels of red blood cells. In contrast, mature red blood cells have lost the nucleus (Fig. 18-4), and their small size and biconcave shape maximize the surface-to-volume ratio, an important characteristic in a cell whose primary function is gas transport and exchange.

In following the pathway of differentiation of almost any cell type, one sees the orderly replacement of one cell type by that of the next stage of development. Commonly the cell of the next generation has many of the properties of its predecessor, but it also has new ones that are somewhat different. The members of such a series of successive replacement events are called *isoforms* (Caplan et al., 1983). Although isoforms were introduced in this text in the context of cellular replacements, there are also many examples of molecular isoforms. Molecular isoforms are typically proteins, often enzymes, which have a common general function but exhibit slightly different properties. Because molecular isoforms typically have a different primary structure, they are products of different genes.

Concrete examples of both molecular and cellular isoform transitions are encountered in the development of mammalian red blood cells (see Chap. 18). The first circulating red blood cells are derived from the yolk sac and are nucleated. Later in development, these nucleated cells are replaced by nonnucleated cells that arise from precursors in the liver, spleen, or bone marrow. The terminal nucleated and nonnucleated red blood cells are cellular isoforms. The hemoglobin molecule serves as a good example of a series of molecular isoforms. Hemoglobin consists of four polypeptide chains attached to a heme molecule. At different stages of development new polypeptide (globin) chains are substituted for those which previously made up the hemoglobin molecule (Fig. 18-6). Although the overall function of hemoglobin is always to carry oxygen, the affinity of the molecule for oxygen varies depending on the structure of the globin molecules of which it is composed. Even at the level of gross structure, the replacement of baby (deciduous) teeth by their adult counterparts can be looked upon as an isoform transition. Later in this chapter, a series of isoform transitions in developing muscle will be discussed in detail. Although much remains to be learned about isoforms, it appears that each stage in an isoform series is adapted for the conditions found in the embryo at that particular time.

There are other circumstances when cells are replaced by progeny with a distinctly different phenotype. In Chap. 6 transitions from epithelial to mesenchymal phenotypes and vice versa were discussed. Such abrupt changes are as much a part of the overall sequence of differentiation in a cell line as are the more gradual isoform transitions.

Patterns in the Generation of Cell Diversity

When one studies the lineage of almost any cell type throughout embryogenesis, one is struck by the observation that there are various patterns by which cells become different from one another. At some periods, the progeny of cells do not seem to differ from their progenitors. After other cell divisions the daughter cells have properties that are noticeably different from those of their predecessors. Holtzer (1970) has used the term *proliferative mitosis* to describe a cell division in which both daughter cells are identical to the precursor cell (Fig. 9-3). In contrast, a mitosis resulting in at least one daughter cell with a different phenotype from that of its precursor is called a *quantal mitosis*. (For

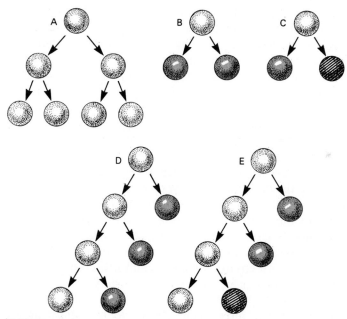

FIGURE 9-3
Possible interrelationships between mitosis and cytodifferentiation. (A) Proliferative mitosis. There is no change in the cell type of the daughter cells. (B) Mitosis followed by differentiation of the daughter cells into the same cell type. (C) Mitosis followed by differentiation of the daughter cells into different cell types. (D) A series of mitoses in which one daughter cell maintains the stem cell and the other differentiates into a single cell type throughout the series. (E) A series of mitoses in which the stem-cell line is maintained but the other daughter cell changes its phenotype after repeated stem-cell divisions.
(*After D. E. S. Truman, 1974, The Biochemistry of Cytodifferentiation, Halsted Press.*)

varieties of quantal mitosis, see Fig. 9-3.) Examples of proliferative mitosis are the cell divisions of mesenchymal cells in the early limb bud and epithelial cells in the early neural plate. When the daughter cells of a quantal mitosis can be recognized, they are usually given a new name. For example, when a cartilage precursor cell arises from a population of general mesenchymal cells, it is called a *chondroblast*. If, instead, the cell is a progenitor of bone, it is called an *osteoblast*. One of the most controversial questions in the field of cytodifferentiation is, What causes the switch from a proliferative to a quantal mitosis? Does a quantal mitosis arise as the result of a genetic program that is entirely inherent in a cell, or is it the result of some type of environmental influence that is received and acted upon by the cell? Could both mechanisms play a role?

Lineages versus Environment in the Generation of Cell Diversity

Speculation concerning the mechanisms by which cells in an embryo undergo diversification has tended to fall into two theoretical camps (Holtzer et al., 1985). One stresses the importance of cell lineages and intrinsic information as the basis for generating cell diversity. According to this viewpoint, there are never any truly undifferentiated or uncommitted cells. At any time in an organism's life history, all cells are specialized for some function. Cells change from one type to the next solely on the basis of information contained in the cells that give rise to them, and mitosis (quantal mitosis) is required for the progression of a given cell into the next developmental stage. There is a strong resemblance to the classical descriptions of mosaic development, as exemplified by the embryos of mollusks or the nematode, *Caenorhabditis elegans*, in which development according to strict cell lineages is the rule.

At the opposite extreme is the viewpoint that the cells of an early embryo are truly uncommitted to a future fate and are undifferentiated in the sense that they are not specialized for any function. On such an empty developmental slate, environmental events (e.g., inductions) impose instructions that in essence tell the cell what to become. In the extreme case an undifferentiated cell, if exposed to the appropriate environmental conditions, could be transformed into any of a large number of cell types, possibly without even undergoing mitosis.

As new and improved methods have been applied to the study of the diversification of cells in an embryo, it has become increasingly apparent that elements of both of these hypotheses can play a role. With improved intracellular markers, it has been shown that well-defined cell lineages play a more prominent role in normal development, even in higher vertebrates, than had been previously recognized. On the other side of the coin, systems once thought to be strictly mosaic have been shown to respond to environmental influences.

Embryonic inductions, certain types of exposure to hormones, and some types of positional influences result in a change in phenotype by a group of cells. Commonly such changes in phenotype are preceded or accompanied by a burst of mitosis that would probably qualify as a quantal mitosis. It is still not completely resolved whether an induction, for instance, merely facilitates a change in phenotype that is programmed in the cell (*permissive induction*) and could have occurred as the result of some other influence as well or whether the induction actually provides some type of specific information without which the change in phenotype would not occur (*instructive induction*). Some investigators feel that in response to a given environmental influence or intracellular event, a given cell has at most two options for change (binary choice theory). Others feel that especially after an instructive induction a given undifferentiated cell can become transformed into any of a variety of cell types.

At certain late stages in their life history, many cells seem to be channeled into a slot (determined) that allows them to become only one specialized type of cell. The cell might still have the option of remaining a germinative or stem cell of the same type or it could go on to produce progeny that have a specialized structure and function. One of the narrowest definitions of differentiation would be the process by which such a "channeled" cell is transformed from a functionally immature to a functionally mature cell (see Fig. 15-8).

The Influence of the Cytoplasm on Gene Expression and the Generation of Cell Diversity

One of the oldest hypotheses in developmental biology states that different cell types ultimately arise from the fertilized egg because, as the zygote is subdivided by cleavage divisions, the nuclei of the blastomeres are exposed to different types and amounts of cytoplasmic components (*determinants*). This idea was generated in the late 1800s, when embryologists studying the early development of certain marine invertebrates with areas of different colored cytoplasm in the eggs noticed that the colors were unequally distributed among the blastomeres. There have been classic demonstrations of the effect of cytoplasmic influences on determining the type of cell that will ultimately form.

In many cases cells containing vegetal pole cytoplasm will become germ cells. In examples already covered in this text, the blastomeres of *Ascaris* embryos that contain a certain type of cytoplasm will not undergo chromosomal diminution but will become germ cells (Fig. 9-1). Even after centrifugation, additional cells which contain vegetal cytoplasm will retain their full complement of chromosomes. Similarly, blastomeres of anuran embryos containing germ plasm (Fig. 3-1) and *Drosophila* embryos containing a similar *pole plasm* (Illmensee and Mahowald, 1974) eventually become germ cells.

Cytoplasmic determinants play a role in the specification of cell types other than those of the germ line. In ascidian embryos there are colored bands of cytoplasm that can be followed during early cleavage. It was known

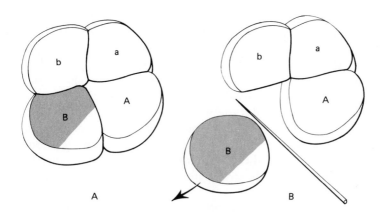

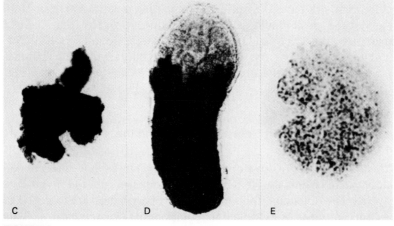

FIGURE 9-4
Experiment showing the importance of the localization of cytoplasmic determinants in ascidian (*Ciona intestinalis*) development. If the B4.1 pair of blastomeres (containing a muscle determinant) is removed during early cleavage (B), the determinant-bearing B blastomeres develop acetylcholinesterase activity (D), whereas the remaining blastomeres (E) do not produce the enzyme. The histochemical reaction for acetylcholinesterase activity leaves the cells black. A control embryo (D) shows acetylcholinesterase activity in the developing tail (black) but not in the rest of the embryo. (*After Whittaker et al., 1977.*)

to the early embryologists (Conklin, 1905) that cells containing material from what is called the yellow crescent ultimately differentiate into muscle. Other distinctive cytoplasmic regions become segregated into cells that become specific structures of the embryo. More recently, Whittaker et al. (1977), working on another ascidian species, removed from an eight-cell embryo the blastomeres that normally would have been the progenitors of the muscle and mesenchyme of the embryo (Fig. 9-4A). They stained cellular descendants both from the removed blastomeres and from the remainder of the embryo

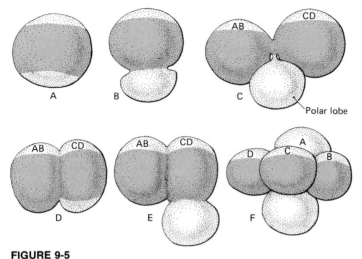

FIGURE 9-5
Cleavage in the mollusk *Dentalium*, showing the polar lobe. (*After E. B. Wilson, 1904, J. Exp. Zool.* 1:1.)

for the activity of *acetylcholinesterase*, an enzymatic marker of muscle differentiation. The cells descended from the removed blastomeres showed intense acetylcholinesterase activity, whereas those derived from cells that did not contain the cytoplasmic determinants for muscle failed to stain (Fig. 9-4B).

Another classic example of the role of specific cytoplasmic regions in generating cell diversity is seen in embryos of the mollusk, *Dentalium*. Early in cleavage a prominent cytoplasmic *polar lobe* appears and then disappears, only to appear again (Fig. 9-5). One can either remove the polar lobe from the two-cell stage or separate the blastomere containing the polar lobe from the four-cell embryo. In the first experiment, embryos growing in the absence of polar lobe cytoplasm did not form mesoderm. The later cell-separation experiments showed that only cells containing some polar lobe cytoplasm could give rise to mesodermal derivatives. Despite the number of years that have elapsed since this clear-cut experiment was performed, surprisingly little is known about the chemical nature of this cytoplasmic influence of the polar lobe on gene expression.

The nuclear transplantation and cell hybridization experiments already mentioned provide specific molecular evidence that a changed cytoplasmic environment can dramatically alter the course of gene expression both by shutting down previously active genes and by activating previously dormant genes. For example, DeRobertis and Gurdon (1979) transplanted *Xenopus* kidney cell nuclei into oocytes of *Pleurodeles*, a salamander. After several days proteins characteristic of *Xenopus* oocytes were detected in the *Pleurodeles* oocyte. In addition, the transplanted kidney nuclei ceased producing detectable amounts of a number of kidney-specific proteins.

Molecular Mechanisms in the Control of Differential Gene Expression in Differentiating Cells

It is generally accepted that the basis for generating cell diversity and cytodifferentiation is the formation of different gene products in different cells at different times. There are many levels of control of gene expression, all of which can play a role in the determination and differentiation of cell lines. Specific molecular mechanisms of gene expression and its control are beyond the scope of this text and constitute major portions of some fine recent textbooks on molecular genetics and cell biology. However, we shall briefly examine the pathways leading to the synthesis of protein molecules to gain some appreciation of the numerous ways in which the final expression of a gene product can be influenced (Fig. 9-6).

At the level of the DNA molecule the molecular structure of the gene itself can be changed by adverse environmental influences, such as x-rays and chemical mutagens, but in antibody-producing cells certain rearrangements of the genetic message occur in the normal course of events. More important,

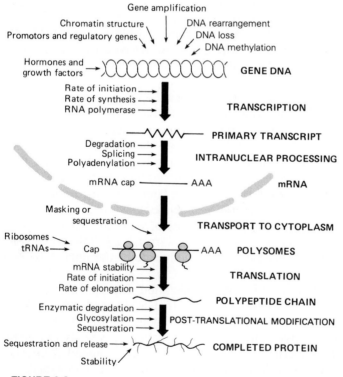

FIGURE 9-6
Gene expression in a typical eukaryotic cell. Large type on right shows sequence of gene expression leading to protein synthesis. Smaller type on left shows ways in which gene expression can be influenced.

however, are the controls that determine which genes will be transcribed into RNA molecules. Inactivation of an X chromosome in female mammals is a case in which a significant portion of the genome is automatically prevented from being expressed (Fig. 3-41). A fundamental problem is access to the coding region of a specific gene. In the normal configuration, DNA chains are tightly complexed to histones (basic proteins) and acidic nuclear proteins to form nuclear chromatin. At regular intervals, like beads on a necklace, the DNA (about 140 base pairs) is wound around a complex of histone molecules to form a structure called a *nucleosome* (Fig. 9-7). The nucleosomes are separated from one another by "spacer" stretches about 60 base pairs long. For simple access to the nucleotide bases along a DNA strand, the nucleosomes and other types of binding between the DNA and its associated proteins must be dealt with.

Exposure of a gene segment alone does not ensure that its encoded information can be transcribed into a corresponding RNA molecule. Somewhat removed from the gene itself (about 30 and 80 base pairs away) are two promoter regions—short sequences of nucleotides that are important in the binding of *RNA polymerase II*, which is essential for the formation of mRNAs. Other recognition sites exist for *RNA polymerase I* (for large ribosomal RNAs) and *RNA polymerase III* (for small RNAs, e.g., tRNAs and 5S ribosomal

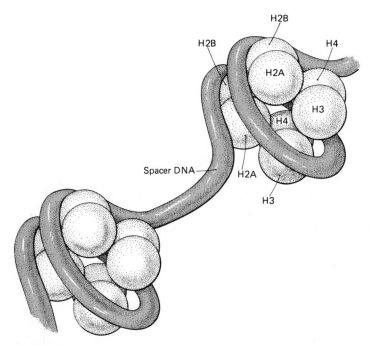

FIGURE 9-7
Model of nucleosome structure. The DNA coils around a histone octomer containing histones H2A, H2B, H3, and H4. (*Redrawn from S. F. Gilbert, 1985, Development Biology, Sinauer, after Wolfe.*)

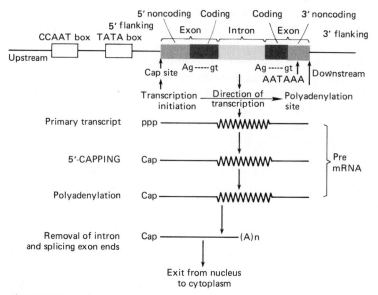

FIGURE 9-8
The structure of a typical gene, with RNA formation and intranuclear processing indicated. (*Adapted from C. F. Graham and P. F. Wareing, 1984, Developmental Control in Animals and Plants, Blackwell.*)

RNA). In some cases methylation of certain bases on DNA can influence the level of transcription of specific genes.

The molecule of mRNA that is just formed from the DNA template (the *primary transcript*) is not yet in a functional form. It is subjected to a number of *processing* steps before it leaves the nucleus (Figs. 1-6 and 9-8). Recent studies have demonstrated a surprising degree of similarity in the unprocessed nuclear RNAs between early blastomeres and cells at later stages of development. Yet substantial differences are seen in the processed cytoplasmic mRNAs. The interested reader is referred to a review by Davidson and Britten (1979) for an influential hypothesis on how processing events can be involved in differential gene expression.

After processed mRNA molecules pass from the nucleus into the cytoplasm, the expression of the genetic information contained in them is subject to further controls. At the level of the mRNA molecule itself, certain molecules can be temporarily or permanently rendered ineffective as a basis for protein synthesis by chemical inactivation (e.g., masked messages; see p. 102) or by being sequestered in certain regions of the cell where they cannot engage in protein synthesis. The length of time that a population of mRNA molecules remains intact before being degraded can be an important factor in determining the number of protein molecules produced.

At the level of translation of protein molecules, any of a number of factors that influence the efficiency of polypeptide synthesis (e.g., initiation, rate of elongation) can play an important role in controlling the expression of a particular gene product. Translational regulation is employed in a number of developmental systems. It seems to be particularly important in cases where a rapid burst of protein synthesis is required after some type of developmental or metabolic signal. In this case the mRNA can already be present in the cytoplasm, ready to act within a few minutes' notice. Good examples of such systems are the stored maternal mRNAs in sea urchin and amphibian oocytes that can be activated within minutes after fertilization and the mRNAs that produce proteins in many hormonally stimulated cells.

Even after synthesis of a polypeptide chain the regulation of gene expression is not complete. Many proteins are synthesized in an inactive form and become functional only when certain groups of amino acids have been cleaved off; most proteins require the addition of carbohydrate side chains (usually in the Golgi apparatus) to become fully functional; and still others must combine with other proteins (e.g., the subunits of enzymes) in order to become active. The collagen molecule (Fig. 1-9) is a classic example of a protein that undergoes extensive posttranslational regulation.

The examples of regulation of gene expression mentioned in this section have just scratched the surface of this complex but important field. At many institutions advanced courses in genetics are available to students who wish to pursue this aspect of development in greater detail.

THE FORMATION OF MUSCLE

The ability to change shape, one of the fundamental properties of animal cells, is due to the presence of varying amounts of contractile proteins in the cytoplasm. Muscle cells are uniquely specialized for contraction, but their contractions are designed to facilitate massive movements of gross structures in the body. About 45 percent of the mass of the body is made up of muscle: skeletal, cardiac, and smooth. Each type of muscle has a unique mode of embryogenesis which is adapted to its structural and functional roles in the body. For instance, cardiac muscle must beat continuously once the heart begins to function. Therefore, in the growing heart cardiac muscle cells must be able to divide while the heart is beating. Skeletal muscle, by contrast, is not required to contract during the early stages of its formation, but in order to produce the massive but intricately coordinated movements of body parts that are required in postnatal life, skeletal muscle cells must form long muscle fibers which are highly oriented, and connected with a well-defined anatomical origin and insertion. In addition, each muscle fiber must be precisely integrated into a functional neural circuit or coordinated action of the muscle will be impossible. Skeletal muscle is one of the most intensely studied tissues with respect to the regulation of gene expression for contractile proteins. This tissue serves as a good example of the succession of both molecular and cellular isoforms in development.

Skeletal Muscle

The Embryological Origin of Muscle Cells   With only a few minor exceptions (e.g., the striated muscle of the iris and the ciliary muscle of the eye, which originate from neural crest ectoderm), skeletal muscle is derived from somatic mesoderm. There is some evidence that in amniotes certain cells of the epiblast layer are determined to become muscle (myogenic) cells as they are invaginating through or leaving the primitive streak during gastrulation. The mechanism of this determination is totally unknown. After gastrulation, the vast majority of the myogenic cells take up temporary positions in the *paraxial mesoderm* (Fig. 5-16). The paraxial mesoderm is the precursor of the somites in the trunk and the somitomeres in the head. Wachtler and Jacob (1986) have demonstrated the presence of myogenic cells in the prechordal mesoderm of the head (Fig. 5-16) shortly after its formation.

For many years the origin of the skeletal muscle cells, especially of the limbs and branchial arches, was debated. Some investigators believed that the origin was the lateral plate mesoderm, and others believe that it was the somites or paraxial mesoderm. This controversy has been decided in favor of the latter viewpoint by the results of cellular marking experiments involving the removal of somites in chick embryos and their replacement with a graft of somites from the Japanese quail (Christ et al., 1977; Chevallier et al., 1977; Fig. 9-9). Quail cells can be easily distinguished from cells of the host chick embryo because the nuclei of quail cells contain large masses of dense chromatin associated with the nucleus. The grafted quail tissue becomes readily integrated with the body of the host, and quail cells undergo surprisingly normal patterns of migration within the chick tissues. In these experiments cells from the quail somites were shown to form the muscles in question.

During this early stage in their life history, the premuscle cells have already undergone several changes in shape, from epiblastic epithelial cells to mesenchymal cells leaving the primitive streak and back to epithelial cells in the early somites. (They remain mesenchymal in the paraxial mesoderm of the head.) The development of these premuscle cells into muscle fibers and entire muscles will be described in subsequent sections.

The Cytodifferentiation of Muscle   At one level the differentiation of a muscle fiber can be viewed as a straightforward succession of cell types, starting with mesenchymal cells (Fig. 9-10). With passing time and a number of cell divisions, these cells are converted into spindle-shaped *myoblasts* which are beginning to get together the cytoplasmic machinery required for the production of the specialized contractile proteins. The next stage of myogenesis consists of the fusion of myoblasts into long syncytial *myotubes*, which possess centrally located chains of nuclei and which are beginning to assemble organized bundles of contractile filaments. Finally, as the central nuclei of the myotube migrate toward the periphery and large amounts of contractile proteins have been laid down, the myotubes become connected into *muscle fibers*.

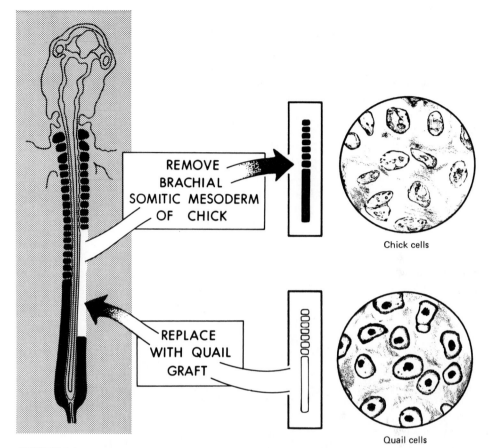

FIGURE 9-9
Diagram of an operation in which a row of somites is removed from a chick embryo and replaced by somites from a quail embryo. The dense masses of chromatin in the nuclei of quail cells serve as natural cellular markers. (*Adapted from Christ et al., 1977, Anat. Embryol.* **150:***171.*)

When one delves into the mechanisms of myogenesis, it soon becomes apparent that (1) the cytodifferentiation of muscle is an extremely complex process and (2) despite intensive efforts directed toward understanding how muscle fibers form, there is still disagreement regarding even some of the fundamental mechanisms in myogenesis. Most schemes purporting to illustrate the cytodifferentiation of muscle begin with a mesenchymal cell that has the potential to form a number of cell types, including cartilage cells and connective-tissue cells. At some time a restriction event occurs, and the cell or its descendants, although looking the same as before, is committed to the muscle-forming lineage. Such cells are often called *presumptive myoblasts*. On the basis of recent evidence, however, it is possible that restriction into a purely myogenic cell line occurs while the cells are still part of the epiblast or in the primitive streak of the early embryo.

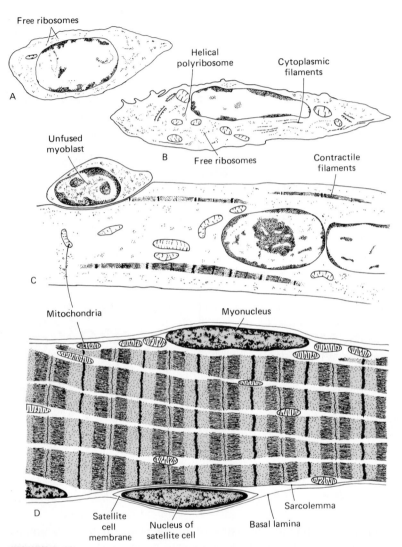

FIGURE 9-10
Major steps in the embryonic differentiation of a skeletal muscle. (A) Unspecialized mesenchymal cell. (B) Myoblast. This spindle-shaped cell possesses large numbers of free ribosomes, including helical polyribosomes upon which the myosin molecules are formed. Cytoplasmic filaments and microtubules are present, but identifiable contractile proteins are not found. (C) Myotube. This is a long multinucleated cell formed by the fusion of mononucleated myoblasts. The nuclei are arranged in long central chains. Myofilamentogenesis is actively occurring, and bundles of well-ordered contractile filaments are present in the periphery. The large numbers of ribosomes attest to the continued protein synthetic activity. The small mononucleated cell alongside the myotube will eventually fuse with the myotube during further maturation. (D) Cross-striated muscle fiber. The nuclei have moved to a peripheral location, and the bulk of the cytoplasm (sarcoplasm) is filled with bundles of contractile filaments demonstrating the characteristic banding pattern of skeletal muscle.

There is considerable controversy regarding the mechanism underlying restriction events in myogenesis. Holtzer et al. (1986) view muscle development as starting with a small number of founder cells, the progeny of which expand as clones within a very restricted and predictable lineage of myogenic cells. According to this viewpoint, a given line of cells is expressing an internally controlled genetic program of developmental change, with outside influences acting only in a permissive manner to facilitate the change. A number of other investigators place greater emphasis on extrinsic influences (e.g., inductions, hormonal or positional effects) as means of influencing the course of differentiation.

Many investigators feel that after a certain number of divisions, presumptive myoblasts are transformed into committed myoblasts which will ultimately go through a terminal mitotic division to become *postmitotic myoblasts*. There is less agreement about how this step affects the cells' synthetic capabilities. Myoblasts are beginning to generate the RNA transcripts that will be needed to form massive amounts of contractile proteins, but how many of these proteins are or can be produced by single myoblasts is an open question.

The next major step in myogenesis is the fusion of individual myoblasts. The biological events surrounding the fusion of myoblasts and the developmental significance of fusion have been the object of intense inquiry. Myoblasts may be programmed to undergo fusion as the result of an internal program that leads them to pass through a quantal mitosis and withdraw from the cell cycle. According to another interpretation (Merrifield et al., 1984), environmental conditions that result in a prolonged G_1 phase of the cell cycle can lead to the initiation of fusion. Fusion involves a precise alignment of myoblasts involving Ca^{2+}-mediated recognition, adhesion, and union of the plasma membranes (rev. by Wakelam, 1985).

The formation of multinucleated muscle fibers by the fusion of separate cells is one of two competing hypotheses. According to another interpretation, repeated division of the nuclei of myoblasts without cytokinesis is the mechanism of multinucleation. Several lines of experimentation have determined that the fusion mechanism is the correct one. One of the most convincing experiments involved the use of allophenic mouse embryos (Mintz and Baker, 1967). In this experiment (Fig. 9-11) mouse embryos homozygous for different *isozymes* (different molecular forms of the same enzyme) of the enzyme isocitrate dehydrogenase were fused (see Fig. 4-24). These isozymes (aa and bb) have different rates of electrophoretic migration. According to the internal division model, only enzyme forms aa and bb would be expected to be found, but if myoblasts bearing different genotypes fused, an intermediate ab form of the enzyme would also be formed. The latter case held true, thus confirming the validity of the fusion model.

Shortly after fusion, the *myotube* is extremely actively engaged in the synthesis of *actin* and *myosin*, the principal contracile proteins of muscle. Myosin, a large filamentous protein composed of several subunits, is synthesized on characteristic helical polyribosomes containing 50 to 60 ribosomes.

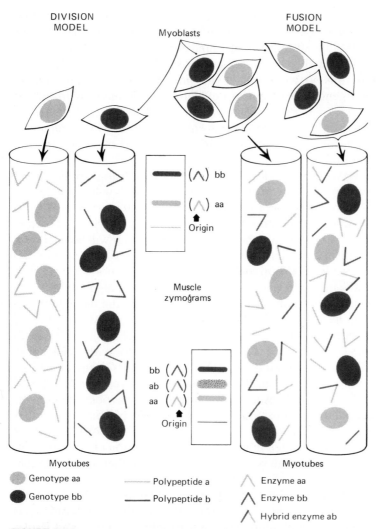

FIGURE 9-11
The use of isozyme differences and allophenic mice in testing the "division"
vs. the "fusion" model of myogenesis. The enzyme isocitrate dehydrogenase
is a dimer composed of two polypeptide chains, aa or bb, depending on the
isozyme type. In allophenic embryos made from embryos of genotypes aa
and bb, the internal division model would produce muscle fibers having only
aa or bb isozyme bands after electrophoresis (upper zymogram). According to
the fusion model, myoblasts of different genetic background would fuse into
hybrid forms (ab) of the enzyme (lower zymogram). The latter result was obtained. (*Adapted from Mintz and Baker, 1967, Proc. Nat. Acad. Sci. 58:592.*)

The newly synthesized contractile proteins undergo *self-assembly* into thick
and thin filaments, which in turn are assembled into functional units around the
peripheral regions of the myotube. As the myotube matures, the contractile

filaments occupy a greater share of the cytoplasm and the nuclei migrate from their central location to positions just beneath the plasma membrane. By this stage the structure is properly known as a *skeletal muscle fiber*. Further growth of the myotube, or the muscle fiber, is accomplished by means of the cytoplasmic fusion of additional myoblasts with the muscle fiber. Some myoblasts do not immediately fuse with the muscle fiber but instead remain as unspecialized mononucleated cells in a position between the muscle fiber and its surrounding basal lamina. These cells are known as *satellite cells* (Fig. 9-10D). Later in postnatal life satellite cells or their progeny can fuse with the growing muscle fibers. After damage to a muscle, satellite cells are activated and become the source of new muscle fibers during regeneration (Mauro, 1979).

Differentiation of a skeletal muscle fiber is not completed when the fiber is filled with myofilaments and peripheral nuclei. Another major step, requiring interaction with a motor nerve, results in the final enzymatic and functional differentiation of the immature muscle fiber into one of several types (fast, slow, and intermediate). Information supplied by the nerve causes changes in the mitochondria and the contractile proteins themselves and results in a muscle fiber's being fast or slow contracting, or fatigable or fatigue-resistant. This last stage of neurally induced differentiation occurs after birth and represents a constant interaction throughout life.

Isoform Transitions in Developing Muscle The cytodifferentiation of muscle has a dimension different from the version presented in the previous section. In early avian and human limb buds there appear to be at least two distinct populations of myoblasts (White et al., 1975), which could be considered to be cellular isoforms. Defined largely by their reactions to in vitro culture, these myoblasts constitute early and later populations of myoblasts, each of which gives rise to myotubes of different morphologies. These populations of myogenic cells arise in the limb bud at different times, and the myoblasts of the "early" population do not seem to be precursors of the "late" population. Rather, they may arise from a common precursor population at a time before myogenic cells enter the limb bud. Seed and Hauschka (1984) postulated that the early myoblasts may respond to morphogenetic cues provided by the connective-tissue cells of the limb bud and that the myotubes descended from these cells may serve as a template upon which mature myotubes are organized to form the definitive muscles. Bonner and Adams (1982) have provided evidence in favor of a third population of myoblasts that require some type of interaction with motor nerves in the limb bud for the establishment of a normal population of cells.

At the molecular level of organization muscle is characterized by the presence of a succession of isoforms of both contractile proteins and enzymes during development and also by the presence of different isoforms of contractile proteins in the different types of muscle fibers found in the adult. Of these proteins, the subunits of the myosin molecule have received the most attention. The myosin molecule is a large protein composed of several subunits (Fig.

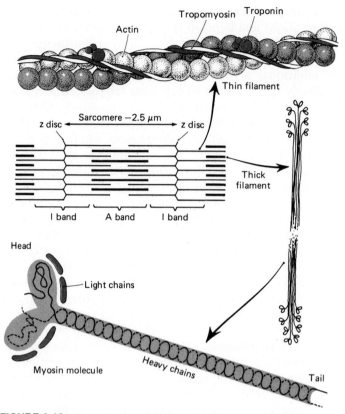

FIGURE 9-12
A sarcomere from vertebrate striated muscle, showing details of the
molecular structure of the thick and thin filaments.

9-12). Each molecule contains two heavy chains (MHC) and a series of light
chains (LC). Fast-contracting muscle fibers have one LC_1, two LC_2 and one
LC_3 light chain subunits, whereas slow muscle fibers contain two LC_1 and two
LC_2 chains. One of the functions of myosin is ATPase activity during
contraction. (Energy production, derived from splitting ATP, is important in
the movement of cross bridges between actin and myosin during muscle
contraction.) It has been postulated that differences in ATPase activity may
account at least in part for the difference in the speed of contraction between
fast and slow muscle fibers.

The myosin molecule has been shown to undergo a series of isoform
transitions during ontogenesis (Fig. 9-13). In developing fast muscle there is a
succession of three myosin heavy chains (embryonic MHC_{emb}, neonatal
MHC_{neo}, and adult fast MHC_f) and, in embryonic muscle, a transition between
an embryonic (LC_{1emb}) and a mature form ($LC1_f$) of light chain one. Interestingly, although the transition between the neonatal (MHC_{neo}) and the adult
(MHC_f) isoform of the myosin heavy chains occurs during a period when major

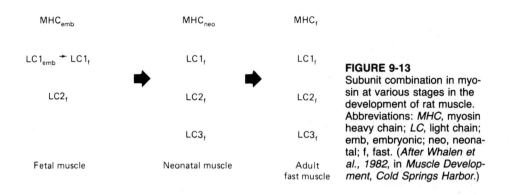

FIGURE 9-13
Subunit combination in myosin at various stages in the development of rat muscle. Abbreviations: *MHC*, myosin heavy chain; *LC*, light chain; emb, embryonic; neo, neonatal; f, fast. (*After Whalen et al., 1982*, in *Muscle Development, Cold Springs Harbor*.)

readjustments are being made in the innervation of neonatal muscles, it appears to be independent of innervation (Whalen et al., 1982).

Other proteins of muscle undergo similar isoform transitions during embryonic development and regeneration in the adult. In fact, during regeneration there is a recapitulation of the major patterns of isoform transitions that occur during normal development. At the genetic level, the DNA segments that code for the different myosin isoforms are considered to be part of a large *multigene family* in which there seems to be an overall mechanism for coordinating the expression of the various isoforms and subunits.

The Development of an Entire Muscle   Regardless of its specific type, almost every muscle goes through a series of common developmental steps. From a common origin in the paraxial mesoderm, spindle-shaped myogenic cells (presumptive myoblasts) migrate toward the region which will be occupied by the mature muscle. These cells migrate through a fibrillar extracellular matrix, but in contrast to the neural crest, factors leading to or guiding the migration of myogenic cells have received very little study. Similarly, it is not known what causes the migration of myogenic cells to cease.

It has been well established that the pathway of migration and the final resting site are not controlled by information inherent within the myogenic cells at any given body level. This has been determined by grafting somites from one craniocaudal level in quail embryos in place of somites or somitomeres from another level in the chick. Regardless of the origin of the myogenic cells of the graft, they migrate along a pathway appropriate to the site in which they were placed and form normal muscles appropriate for that site. For example, if somites that normally supply muscle cells to the hindlimb are grafted into the forelimb area, the cells derived from them form perfectly normal forelimb muscles (Jacob et al., 1982). The migrating myogenic cells collect in a peripheral staging area that is called a *muscle blastema* or *premuscle mass*. If cells from several somites enter a single premuscle mass, there seems to be an almost random mixing within the mass. These observations about the lack of specificity of the myogenic cells themselves have led most investigators to suspect that the morphogenesis of muscles is controlled by the connective tissue associated with the muscle.

The next stage in the formation of a muscle is the splitting of the common muscle mass into the primordia of individual muscles. The basis for the splitting of the premuscle mass is not known. Although early nerve fibers may be associated with the premuscle mass at this time, it is well established that the presence of nerves is not required for splitting. By this stage in muscle development early myotubes have begun to form from the myoblasts, and commonly the orientation of the myotubes already corresponds to that of the muscle fibers of the mature muscle. The overall shape of the muscle is established before the tendons appear. The connective tissue and tendons of the muscle arise from local connective tissue (usually from the lateral plate), and only secondarily do the tendons attach to the muscle.

As soon as the developing muscle fibers construct the rudiments of an organized contractile apparatus, the muscles become capable of weak contractions. Functional innervation occurs early, and the first functional innervation of a muscle is purely motor. Later in prenatal development sensory nerve fibers interact with certain groups of muscle fibers and cause them to transform into the specialized *intrafusal muscle fibers* of the *muscle spindles*, the stretch receptors of a muscle. Zelená (1957) found that if sensory nerve fibers are prevented from interacting with muscle fibers at a critical stage of prenatal development in the rat, muscle spindles do not form.

When skeletal muscles first become functional, they contract very slowly, and neonatal limb muscles, at least, are very similar in their contractile speeds. Only later, as the locomotor functions of the body as a whole progress, do the gross contractile properties of the muscle mature, resulting in the establishment of fast and slow muscle fibers. These changes in contractile function reflect underlying changes in neural signaling to individual muscle fibers and subsequent changes in the isoforms of the major contractile proteins.

Most muscles do not add significant numbers of muscle fibers after the neonatal period, but individual muscle fibers grow by adding *sarcomeres* (fundamental units of contractile proteins) at either end and by increasing their cross-sectional area. As the muscle fibers grow, progeny of satellite cells are added to the muscle fiber syncytium in order to maintain a manageable nucleocytoplasmic ratio. Muscle fibers are quite stable in adult life, but if they are injured, they are able to regenerate (Carlson, 1973; Mauro, 1979). Individual muscle fibers can regenerate completely, but at the level of an entire muscle the extent of repair can range from minimal to complete functional return, depending on other conditions, such as the blood supply and the extent of reinnervation. In mammals regenerating muscle fibers arise from satellite cells (Snow, 1977), and at the cellular level a regenerating muscle fiber recapitulates many of the morphological and molecular events that it went through during embryogenesis.

Development of the Trunk and Limb Musculature Recent studies have shown that the muscles of the trunk and limbs all arise from the somites (Jacob et al., 1986). Three different modes of cellular behavior are involved in formation of the muscle of the abdomen, the back, and the limbs from the somites. The limb muscles arise from individual cells that break free from the ventrolateral margins of the somites (as early as the HH stage 14 in the chick wing bud; see inside of back cover) and migrate as individuals into the early limb bud. Further stages in the development of limb muscles will be described in detail in Chap. 11.

The precursor cells of the abdominal muscles also leave the somite, but they do so in the form of discrete somite buds. The emigrating cells in the flank form large premuscle masses, which soon split into four muscle masses, corresponding to the three oblique muscles of the abdominal wall, as well as the rectus abdominis muscle. Final shiftings and growth processes complete the molding of the ventral abdominal muscles.

The precursors of the intrinsic back muscles do not leave the region of the somite; rather, they remain in the myotomal region of the developing somite (Fig. 6-23D). In the deep areas, the metameric arrangement of the somites is retained, and short muscle fibers connect processes of one vertebra to the next. More superficially, muscle fibers from adjacent vertebrae fuse to form slips of muscle of varying lengths that connect more distant regions of the vertebral column.

Development of Cranial and Cervical Musculature   Any discussion of the embryogenesis of the cranial musculature must take into account the debate regarding the existence of somitomeres and the limited information about the myogenic potential of the prechordal plate (Noden, 1983; Wachtler and Jacob, 1986). There is now little doubt that the paraxial mesoderm constitutes the main source of cranial musculature and that at least some of the cells constituting the extraocular musculature have passed through the prechordal plate. These findings seem to put to rest a great deal of speculation in the literature of comparative anatomy and embryology that the branchial-arch musculature has a special "visceral" origin. Many investigators now feel that there is little difference between the formation of head and trunk musculature. Others, however, feel that despite the lack of evidence for a visceral formation of the branchial-arch musculature, there are still fundamental differences in the mechanisms of muscle formation between the head and the trunk.

Myogenic cells migrate out from the paraxial mesoderm in a manner similar to that of prospective limb muscle cells. Depending on the location, these cells migrate through either mesoderm-derived or neural crest mesenchyme. Like the rest of the body, there seems to be no level-specificity for cranial myogenic cells. Morphogenesis of the cranial muscles seems to be determined by the connective tissue in which the myogenic cells arise. In the case of muscles of the face and ventral neck, the neural crest origin of the muscle-associated mesenchyme means that the cranial neural crest is likely to possess considerable morphogenetic information.

Wachtler and Jacob (1986) have presented evidence suggesting that competent myogenic cells are present in the prechordal plate but are absent from the

TABLE 9-1
EMBRYOLOGIC ORIGINS OF THE MAJOR CLASSES OF MUSCLE

Embryologic origin	Derived muscle	Innervation
Somitomeres 1 through 3 and/or prechordal plate	Most extrinsic eye muscles	Cranial nerves III, IV
Somitomere 4	Jaw-closing muscles	Cranial nerve V (mandibular branch)
Somitomere 5	Lateral rectus of eye	Cranial nerve VI
Somitomere 6	Jaw-opening and other second-arch muscles	Cranial nerve VII
Somitomere 7	Third-arch branchial muscles	Cranial nerve IX
Somites 1 and 2	Intrinsic laryngeal muscles and pharyngeal muscles	Cranial nerve X
Occipital somites (1 through 7)	Muscles of tongue, larynx, and neck	Cranial nerves XI and XII Cranial cervical nerves
Trunk somites	Trunk muscles Diaphragm Limb muscles	Spinal nerves
Splanchnic mesoderm	Cardiac muscle	Autonomic
Splanchnic mesoderm	Smooth muscle of gut and respiratory tracts	Autonomic
Local mesenchyme	Other smooth muscle Vascular Arrector pili muscles	Autonomic

paraxial mesoderm of the head in early embryos. Whether myogenic cells later pass into the paraxial mesoderm from the prechordal plate mesoderm has not been determined.

The origins and neural relations of the major groups of cranial and cervical muscles are shown in Table 9-1. Some uncertainty still surrounds the origin of the extraocular muscle fibers. The classical view, based mainly on descriptive studies, is that they arise from special preotic myotomes. Their exact origin has not been determined, but some line of passage from the prechordal plate and through the paraxial mesoderm seems likely. Many of the muscles of the neck arise from the occipital somites, as do the muscles of the tongue, which secondarily migrate (along with their nerve XII) into the oral cavity.

Cardiac Muscle

The heart is formed from the splanchnic mesoderm of the early embryo (Figs. A-21 and 9-14). The cardiac muscle cells themselves arise from mesenchymal cells located in the inner layer of the epimyocardium. Early cardiac myoblasts, like those of skeletal muscle, are spindle-shaped mononuclear cells, but they

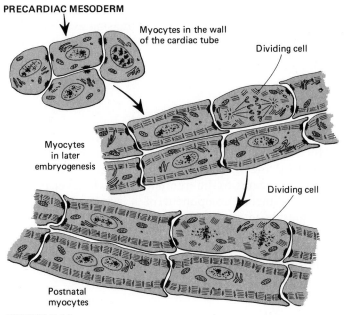

PRECARDIAC MESODERM

Myocytes in the wall
of the cardiac tube

Dividing cell

Myocytes
in later
embryogenesis

Dividing cell

Postnatal
myocytes

FIGURE 9-14
Stages in the histogenesis of cardiac muscle. Note that during mitosis
there is a partial disassembly of the contractile filaments. [*Adapted
from P. Rumyantsev, 1982*, Cardiomyocytes (*Russian*).]

exhibit certain unique features. Perhaps foremost among these is the presence
of comparatively large numbers of myofibrils in the cytoplasm and the
consequent ability of cardiac myoblasts to undergo pronounced contractions.
Another feature which distinguishes the developing cardiac muscle cell from its
counterpart in skeletal muscle is the ability of the cardiac muscle cell to
undergo mitotic divisions even though the cytoplasm contains numerous
bundles of contractile filaments (Rumyantsev, 1982). Many types of cells in the
body lose the ability to divide if they have already produced specialized
cytoplasmic structures characteristic of their fully differentiated state. In view
of the requirement for early and continuous functioning of the heart during
embryonic development, it is not surprising that the cells of the heart deviate
from this general rule by producing contractile filaments while they increase in
number. Cardiac muscle cells do not fuse to form a syncytium as do skeletal
muscle myoblasts. Instead, adjoining cells develop specialized intercellular
connections that in mature muscle account for the appearance of the *interca-
lated disks* which are found between the cells.

Cardiac muscle cells also contain actin and myosin isoforms. In the adult
there are distinct differences between myosin isoforms found in atrial and
ventricular muscle and in skeletal muscle as well. However, one light chain,
($LC1_{emb}$), which is found in skeletal muscle as well as in atrial and ventricular

muscle of the embryo, continues to be expressed in muscle fibers of the atrium and the conducting system (Purkinje fibers). There is increasing evidence that a number of other muscle proteins are commonly expressed by both atrial and embryonic or regenerating skeletal muscle fibers.

Smooth Musculature

The smooth musculature arises from mesoderm that in the course of development applies itself as an outer coat around the primary epithelial lining of hollow internal organs. Much of the smooth muscle, for example that surrounding the digestive and respiratory tracts, arises from splanchnic mesoderm, but as a general rule smooth muscle seems to differentiate from whatever type of mesenchyme surrounds the epithelial component of a structure. There is evidence that the smooth muscle in many of the blood vessels arises from somatic mesoderm and that the smooth muscle of the iris (the *sphincter pupillae* muscle) is of ectodermal origin. Very little is known about the developmental mechanisms underlying the cytodifferentiation and histogenesis of smooth muscle.

FORMATION OF THE SKELETON

All the components of the skeleton are derived from mesenchyme—of mesodermal origin in the limb, trunk, and part of the head and of neural crest origin in the face and branchial-arch region. Mesenchyme can be converted into skeletal elements by forming bone directly (*intramembranous ossification*) or by first forming a cartilaginous model which is subsequently replaced by true bone (*endochondral bone formation*).

There are two major divisions of the skeleton. The *axial skeleton* includes the bones of the head, the vertebral column, and the ribs. The bones of the axial skeleton surround important soft tissues, principally the brain and spinal cord. So intimate is their relationship to the structures they protect that their initial formation is dependent on inductive influences from the central nervous system, and their final size and morphology are the result of intrinsic potential combined with growth pressures from the underlying soft tissues (Hall, 1978). The *appendicular skeleton* consists of the bones of the limbs and the limb girdles. The bones of the limb differ in several respects from those of the axial skeleton. In contrast to the bones of the head and vertebral column, they are the central structures and are surrounded by the soft tissues with which they are associated. The evidence for the induction of limb bones by ectodermal structures is much more tenuous than that for the induction of axial bones.

Contrary to the static image conjured up by a skeletal preparation made of dried bones, skeletal tissue is highly reactive in both prenatal development and postnatal life. In postnatal life the morphology of the skeleton adapts to changing patterns of mechanical function by some of the same cellular processes of bone formation and destruction that occur in the embryo.

Cartilage Formation

Cartilage forms from mesenchyme in many areas of the embryo, such as the limbs, vertebral column, respiratory tract, and skull (Fig. 9-15). In some cases, for example the cartilaginous precursors of vertebrae, chondrogenesis is known to be initiated by an inductive process. In other cases, for example, limb bones, the chondrogenic stimulus has not been definitely determined. Regardless of the initiatory mechanism, the sequence of chondrogenesis is remarkably similar wherever it occurs.

The most prominent feature of the differentiation of cartilage is a change in the character and amount of the extracellular matrix that surrounds the differentiating cartilage cells. The extracellular matrix surrounding precartilaginous mesenchymal cells is rich in hyaluronic acid and also contains small amounts of type I collagen. In the earliest stages of chondrogenesis, *chondro-*

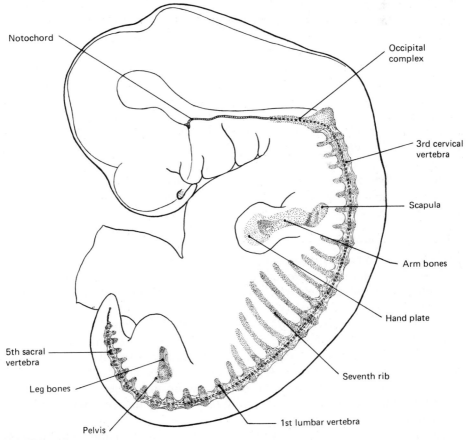

FIGURE 9-15
Diagram of precartilage primordia in a 9-mm human embryo. (*Adapted from several sources, chiefly the work of Bardeen.*)

blasts (precursor cells of cartilage) are still associated with high levels of hyaluronic acid. Levels of both *hyaluronidase* and *chondroitin sulfate*, a characteristic matrix component of mature cartilage, begin to increase. Hyaluronic acid is associated with the migration and proliferation of early embryonic cells, and the removal of hyaluronic acid by hyaluronidase often coincides with the onset of overt differentiation. Chondroitin sulfate is secreted in minute amounts by very early chondrogenic cells, but a specific inductive stimulus, e.g., by the notochord and/or neural tube of cells of the sclerotome, results in a dramatic increase in the synthesis of chondroitin sulfate.

A switch in the production of collagen from type I to the cartilage-specific type II, along with changes in the *proteoglycan* associated with the rapidly increasing amounts of extracellular matrix, marks a definite threshold in the differentiation of cartilage. Like the contractile proteins of muscle, the *cartilage proteoglycans* (Fig. 1-12) pass through several different isoform states as cartilage differentiates and matures. The core protein of the molecule appears to stay the same, but the carbohydrate side chains, including chondroitin sulfate and keratan sulfate, attached to it change over time. This is a good example of *posttranslational modification* of gene expression.

As the cartilage matrix, which binds large amounts of water, increases in amount, the cartilage cells (*chondrocytes*) embedded in it become more widely separated from one another. They also continue to multiply by mitosis, expanding the nonrigid matrix as they do so. The growth of cartilage by internal expansion due to both mitosis and continued secretion of matrix is called *interstitial growth*. Cartilage and other hard materials can also expand by *appositional growth*, or the laying down of matrix by chondrogenic cells on the outer edge of the mass of cartilage. In time, some of the embryonic cartilage is replaced by true bone in endochondral bone formation. In other areas, such as the trachea, the cartilage remains as such throughout life.

Histogenesis of Bone

Much remains to be learned about the nature of the cells that give rise to bone. From the early stages of osteogenesis several kinds of cells are involved in the formation of bone. *Osteoblasts*, which lay down the matrix and ultimately become entrapped in the matrix as *osteocytes*, differentiate from mesenchyme, which, as in cartilage, may originate from mesoderm or neural crest ectoderm, depending on the location of bone formation. Closely associated with bone formation is localized bone removal. This is accomplished by multinucleated *osteoclasts*, which arise from a separate hematogeneous (blood-derived) population of cells. A third population of cells associated with developing bone constitutes the *marrow*, a complex group of blood-forming cells which will not be dealt with in this chapter.

The differentiation of bone involves the production of an abundant intercellular matrix of specialized collagenous fibers and ground substance which has a strong tendency to calcify rather than take up water as cartilage matrix does.

Throughout both embryogenesis and postnatal life the growth and maintenance of bone are based on a delicate balance between the deposition of new bone and the resorption of previously deposited bone. These two opposed processes often take place within a few hundred microns of one another. Particularly in later growth and remodeling the internal architecture of a bone is highly responsive to changes in its mechanical environment. An exciting hypothesis regarding growth control in bone states that through the piezoelectric[1] properties of the osseous matrix, mechanical deformation is translated into differences in electrical potential. These differences are then sensed by bone-forming cells, which react in predictable ways. According to Bassett (1971), a net negative charge stimulates osteoblastic activity and the deposition of new bone, whereas osteoclastic activity and bone resorption occur in areas with a new positive charge. This concept has recently been put to medical use in the stimulation of fracture healing in bones. Some idea of the state of this field can be gained by reading the reports given in a conference on bioelectricity (Liboff and Rinaldi, 1974).

Modes of Bone Formation

Some bones, such as the flat bones of the skull, form directly from mesenchyme without an intervening cartilaginous phase (*intramembranous bone formation*). In areas of well-vascularized mesenchyme, bone-forming cells (*osteoblasts*) secrete a delicate axis of type I collagen fibers along with other mucopolysaccharide matrix molecules. Osteoblasts line up along such strands and continue to secrete matrix until they are surrounded by it (Fig. 9-16). Owing to special properties of the matrix and the associated enzyme, *alkaline phosphatase*, crystals of calcium phosphate in the form of *hydroxyapatite* harden the matrix into a rigid shell around the bone cells (*osteocytes*). Only by maintaining a communications network of cell processes from one osteocyte to the next in the various layers (lamellae) of bone that form can deeply embedded osteocytes receive the required oxygen and nutrients from nearby capillaries. Newly forming membranous bones consist of an irregular scattering of *trabeculae* (from the Latin, meaning *little beam*), which are the calcified form of the original connective-tissue matrix secreted by the osteoblasts. As the bone grows, the trabeculae interconnect to form a spongy meshwork of bone. The spaces among the trabeculae become occupied by *bone marrow*.

In *endochondral bone formation* a model of the bone is first formed from hyaline cartilage. Later in development, the cartilage is removed and bone is deposited in its place. The actual process of bone formation appears to differ little from that seen in intramembranous bone formation.

---

[1]*Piezoelectricity* is a term borrowed from the physical sciences and refers to electrical potential produced by the mechanical deformation of certain types of nonconducting crystals. Some biological structures possess similar properties. In the case of bone, the collagen fibers in the matrix are piezoelectric.

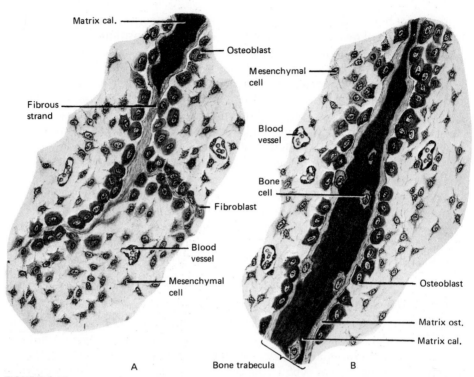

FIGURE 9-16
Formation of trabecula of membrane bone. Projection drawings from the mandible of a pig embryo 130 mm in length. The two parts of this illustration cover a single trabecula which was growing from either end. The areas shown in A and B were directly continuous in the actual material, the bottom of B fitting on top of A. The areas were separated in drawing merely as a matter of economy in the space occupied by the illustration.

When a mass of cartilage is about to be replaced by bone, very striking changes take place in its structure. The cells which have been secreting cartilage matrix begin to hypertrophy. Almost concurrently, calcium salts are deposited in the matrix surrounding them. The calcified matrix does not permit an adequate exchange of oxygen and metabolites between blood vessels and the cartilage cells. Consequently the cells die and are not able to maintain the integrity of the matrix surrounding them. The matrix becomes eroded, and this process of destruction continues until the cartilage is extensively honeycombed. Meanwhile, the tissue of the perichondrium overlying the area of cartilage erosion becomes active. There is rapid cell proliferation, and the new cells, carrying blood vessels and young connective tissue with them, begin to invade the honeycombed cartilage (Fig. 9-17).

Development of Characteristic Skeletal Elements

Details of the formation of many of the characteristic skeletal elements vary from region to region and even from bone to bone. It is beyond the scope of this

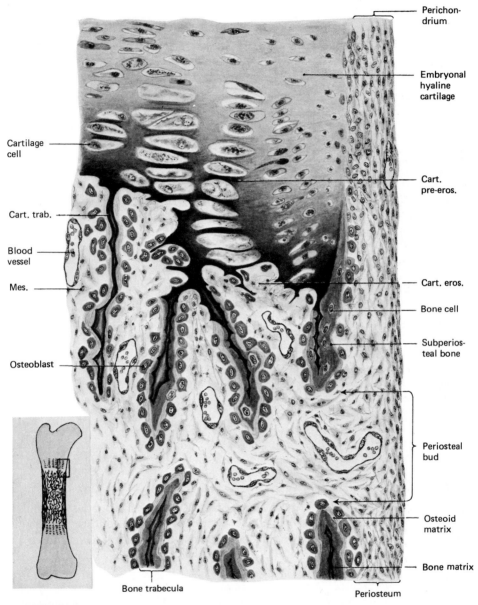

Perichon-
drium

Embryonal
hyaline
cartilage

Cartilage
cell

Cart.
pre-eros.

Cart. trab.

Blood
vessel

Mes.

Cart. eros.

Bone cell

Subperios-
teal bone

Osteoblast

Periosteal
bud

Osteoid
matrix

Bone matrix

Bone trabecula

Periosteum

FIGURE 9-17
Drawing showing periosteal bud and an area of endochondral bone formation from the radius of
a 125-mm sheep embryo. The small sketch indicates the location of the area drawn in detail.

book to deal with the formation of all the different bones or groups of bones.
For a review of the mechanisms involved in the development of a number of
specific bones, the reader is referred to the book by Hall (1978).

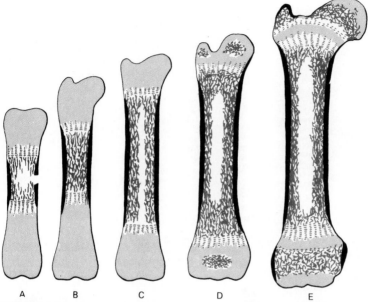

A   B   C   D   E

FIGURE 9-18
Diagrams showing ossification in a long bone. Light gray areas represent cartilage; dark gray and black areas indicate bone. (A) Primary ossification center in
shaft; (B) primary center plus shell of a subperiosteal bone; (C) entire shaft ossified; (D) ossification centers appearing in the epiphyses; (E) entire bone ossified except for epiphyseal cartilage plates and articular surfaces.

Long Bones Long bones begin as purely cartilaginous miniatures of their
adult counterpart. Through endochondral ossification the cartilaginous model
is converted into bone. The primary site of ossification occurs in the center of
the shaft, or *diaphysis* (Fig. 9-18A). By a mechanism like that illustrated in Fig.
9-17, ossification proceeds toward either end of the bone. Secondary ossification centers appear at either end of the bone (Fig. 9-18D), and the cartilage
remaining between the primary and secondary ossification centers is known as
the *epiphyseal plate*.

Formation of Joints An early indication of the formation of a freely
movable joint (*diarthrosis*) is an area where mesenchyme is less concentrated
between the precartilaginous mesenchyme of two skeletal elements (Fig.
9-19A). As the perichondrium takes shape, an area of loose connective tissue
caps the ends of the bones (Fig. 9-19B). With ossification starting in the bones,
the ends of the bones remain cartilaginous as the *articular surfaces*, and the
connective tissue at the ends of the bones disappears (Fig. 9-19C and D).
Meanwhile, the original perichondrium is converted into the tough, fibrous
joint capsule.

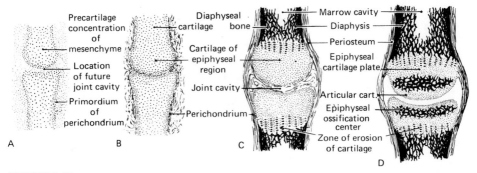

FIGURE 9-19
Schematic diagrams showing four stages in the development of a joint.

The initial development of a joint is accomplished on the basis of intrinsic factors, but maintenance of a joint requires some form of mechanical function. If an embryonic joint is subjected to prolonged culture in vitro or if the limb is paralyzed, fusion of the tissues of the joint is the rule.

Formation of Vertebrae and Ribs The segmental nature of the mature vertebral column reflects its origin from the somites of the early embryo. The vertebral column arises from mesodermal cells originating in the sclerotomal portion of the somite (Fig. 6-23D). Although the cells of the sclerotome possess a chondrogenic bias and produce small numbers of molecules characteristic of the skeletal matrix, their participation in the formation of axial skeletal elements depends on inductive influences from the notochord and the ventral half of the spinal cord (Holtzer and Detwiler, 1953). Specific details of the inductions vary among the classes of vertebrates, but there is increasing evidence that elements of the extracellular matrix produced by the notochord and spinal cord are the effective inductive agents when they come in contact with migrating cells of the sclerotome. In birds the centra of the vertebrae are induced by the notochord, and the neural arches by the spinal cord.

The earliest morphological step in the formation of the centrum of a vertebra is a migration of cells from the sclerotomal portions of the somites on either side toward the midline, where they become clustered about the notochord (Fig. 9-20A). The sclerotomal cells from each somite pair are densely packed in the caudal part and loosely packed in the cranial part (Fig. 9-20B). In the human embryo some cells from the condensed caudal portion migrate craniad and begin to differentiate into the intervertebral disk. The remainder of the condensed portion of the sclerotome then migrates caudad, whereas the loosely packed cells of the following somite migrate craniad (Fig. 9-20B). These migrating masses of cells derived from two somites then join to form the primordium of the centrum of a vertebra in a position interdigitating between two myotomes (Fig. 9-20C). Soon, paired concentrations of mesenchymal cells

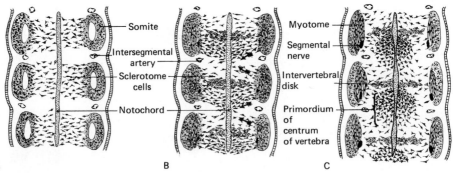

FIGURE 9-20
Semischematic coronal sections through dorsal region of young embryos to show how vertebrae become intermyotomal in position. Note that the primordium of a centrum is formed by cells originating from sclerotomes of both adjacent pairs of somites.

extend dorsally and laterally from the center to establish the primordia of the neural arches and the ribs.

The derivation of the vertebrae from mesenchymal contributions of adjacent somites places them between somites rather than opposite them. This means that when muscle tissue develops from the myotomal part of the somites, it lies across the intervertebral joints, a location which effectively sets it up in proper mechanical relation to the skeletal units which it will move.

In the chick, both the spinal ganglia and the notochord are separately involved in the morphogenesis of certain segments of the vertebral column. If the spinal ganglia are removed, neural-arch cartilage forms, but it forms an unsegmented rod even though individual centra form (Fig. 9-21B). Conversely, removal of the notochord results in the absence of segmented centra (Fig. 9-21C). If both notochord and spinal ganglia are removed, an unsegmented cylinder of cartilage forms around the spinal cord (Fig. 9-21D). The effect of the spinal ganglia on segmentation of the neural arches may be largely mechanical, for they are present before the vertebrae begin to form and may act as barriers to the migration of sclerotomal cells (Hall, 1977). Little is known about the role of the notochord in the morphogenesis of the centra of the vertebrae.

By the time ossification begins, the rib cartilages become separated from the vertebra (Fig. 9-22A), but the cartilaginous primordium of the vertebra itself remains in one piece. Figure 9-22B to 9-22E shows the homologous components in vertebrae from different levels. One can see that all the vertebrae contain components that are developmentally homologous with the ribs.

During the formation of the vertebral column the regions of the notochord that are within the developing vertebrae themselves eventually disappear. Between vertebral bodies mesenchymal cells surrounding the notochord form the intervertebral disks. Within the *intervertebral disks* the notochord persists as a mucoid structure known as the *nucleus pulposus*.

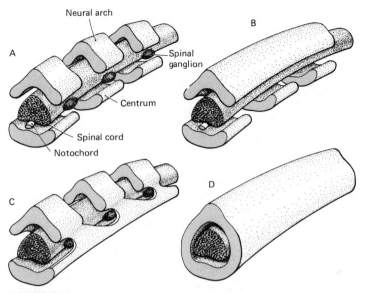

FIGURE 9-21
Results of removal of the spinal cord and/or notochord on the morphogenesis of the vertebral column of the chick embryo. (A) Normal embryo showing segmental arrangement of the neural arches, centra, and spinal ganglia. (B) Excision of the spinal ganglia results in unsegmented neural arches but does not affect segmentation of the centra. (C) Excision of the notochord results in unsegmented centra but does not influence the morphogenesis of the neural arches. (D) Excision of both notochord and spinal ganglia results in a totally unsegmented vertebral column. (*After Hall, 1977, Adv. Anat. Embryol. Cell Biol.* **53***(IV):1.*)

Development of the Skull   The skull consists of two major subdivisions: the *neurocranium*, which surrounds the brain, and the *viscerocranium*, which surrounds the oral cavity, pharynx, and upper respiratory tract. Both of these subdivisions have components which arise as cartilaginous models and are later replaced by endochondral ossification. Both also have membrane bones which arise by direct ossification from mesenchyme. The vertebrate skull is so complex from both the ontogenetic standpoint and the phylogenetic standpoint that only a brief outline of the important aspects of its development will be given here. Those interested in greater detail are referred to the monograph by DeBeer (1937).

The bones of skull arise from mesenchyme, but in contrast to the rest of the skeleton, many of the mesenchymal precursor cells appear to have originated from the neural crest. Like the vertebral column, virtually all the bones of the skull are formed as the result of inductions between epithelial structures and skeletogenic mesenchyme. The *chondrocranium* (the cartilaginous base of the neurocranium) is induced by the notochord, whereas the membranous bones of the neurocranium are induced by the parts of the brain which they ultimately protect (Fig. 9-23). The elements of the viscerocranium (Fig. 16-5D), by

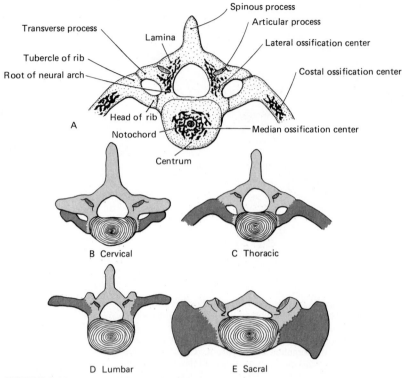

FIGURE 9-22
The component parts and the locations of the ossification centers in developing vertebrae. (A) Diagram showing locations of various ossification centers in a thoracic vertebra and the associated ribs. (B–E) Drawings showing characteristic components of
vertebrae from different levels. The neural-arch component is represented by light
gray, the costal component by darker gray, and the centrum by concentric lines.

contrast, require an inductive stimulus from the pharyngeal endoderm in order
to take shape.

The *cartilaginous neurocranium*, which forms the base of the skull, consists
of several masses of cartilage. Around the cranial end of the notochord is a
cartilaginous plate called the *parachordal cartilage*, which is derived from the
sclerotomal portions of the four occipital somites (Figs. 9-24A). This plate
forms the base of the occipital bone at the base of the skull. Rostral to the
parachordal cartilage are the *prechordal cartilages* (constituting the paired
hypophyseal cartilages), which form the bone surrounding the pituitary gland,
and the *trabeculae cranii*, which form the ethmoid bone in the nasal region.
Lateral to this axis are other pairs of cartilaginous elements which are
associated with the sense organs (Fig. 9-24A). These cartilaginous structures
are eventually replaced by bone.

The membranous neurocranium includes the large flat platelike bones of the
cranial vault. After induction by specific parts of the brain, these bones remain

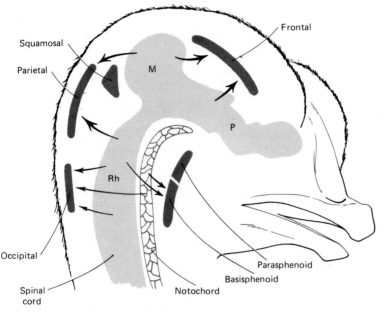

FIGURE 9-23
Formation of bones in the developing avian skull as the result of inductive interactions between specific regions of the central nervous system and/or notochord and the surrounding ectomesenchyme of the head. (*After Schowing, 1968, J. Embryol. Exp. Morph. 19:83.*)

separated by fibrous areas called cranial sutures as well as by larger soft areas called fontanelles. Throughout the period of fetal and postnatal growth, these bones adapt to the changing size and growth patterns of the brain.

The *cartilaginous viscerocranium*, which is principally of neural crest origin (Fig. 6-15), is an integral part of the branchial-arch system. Details of its further development will be presented in Chap. 16. The bones of the *membranous viscerocranium* form in association with the cartilaginous core of the first branchial arch. They ultimately form the adult bones of the upper and lower jaws as well as part of the temporal bone, which becomes incorporated into the neurocranium.

Progress of Ossification in the Skeleton as a Whole

Each of the more than 200 bones of the body has its own developmental history involving the formation of the connective tissue or the cartilaginous mass which precedes it; appearance of local erosion centers if the bone is preformed in cartilage; number, location, and time of appearance of ossification centers; growth in length and diameter; development of epiphyses; time of fusion of epiphyses and diaphysis; and finally the development of muscle ridges and articular facets. It would be neither possible nor desirable in a book of this sort to attempt systematic survey of the development of all, or even the majority, of

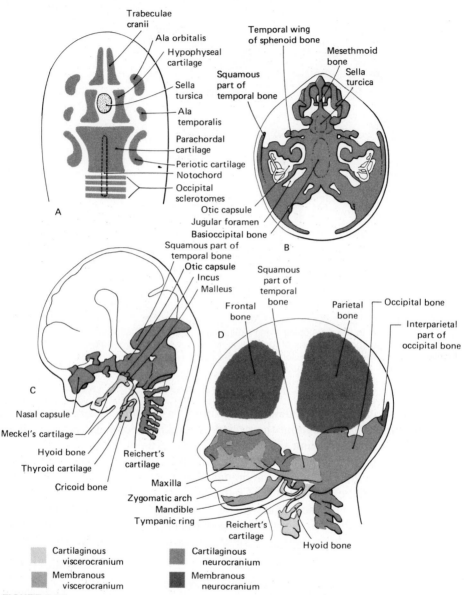

FIGURE 9-24
Diagrams illustrating the origins and development of the major bones of the skull. (A) Embryo of about 6 weeks (viewed from above), showing the primordial cartilages that will form the chondro-cranium. (B) Embryo of about 8 weeks (viewed from above), showing the chondrocranium. (C) Lateral view of the embryo illustrated in B. (D) Skull of 3-month embryo. (See color insert.)

the bones. The human embryos in Figs. 9-25, 9-26, and 9-27, stained with the bone-seeking dye *alizarin red*, will provide a good visual summary of early ossification in the skeleton as a whole.

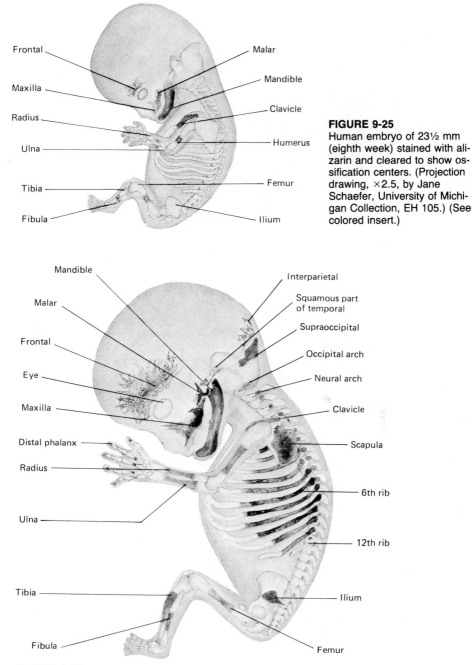

FIGURE 9-25
Human embryo of 23½ mm (eighth week) stained with alizarin and cleared to show ossification centers. (Projection drawing, ×2.5, by Jane Schaefer, University of Michigan Collection, EH 105.) (See colored insert.)

FIGURE 9-26
Human embryo of 39 mm (9 weeks) stained with alizarin and cleared to show the progress of ossification. (Projection drawing, ×2.5, by Jane Schaefer, University of Michigan Collection, EH 149.) (See colored insert.)

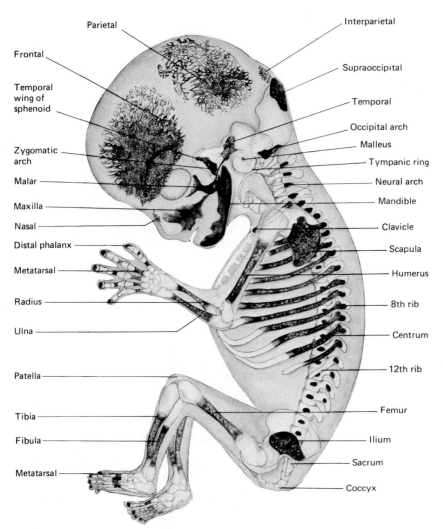

FIGURE 9-27
Human embryo of 49 mm (tenth week) stained with alizarin and cleared to show the developing skeletal system. (Projection drawing, ×2.5, by Jane Schaefer, University of Michigan Collection, EH 70.) (See colored insert.)

CHAPTER 10

THE SKIN AND ITS DERIVATIVES

As the interface between the body and the external environment, the skin is faced with physiological and mechanical challenges as diverse as retaining body fluids, reducing friction in water, maintaining body temperature, and protecting against the bites and scratches of predators. In addition, pigmentation patterns and odor from skin glands are important in social and sexual communication. In order to accommodate these many functions, the integument of vertebrates has evolved a wide variety of appendages. Some of the major classes of skin appendages are scales, feathers, hairs, horns, teeth, nails and claws, sweat and sebaceous glands, and mammary glands. Despite the bewildering phenotypic diversity even among the vertebrates, the organization and development of the skin and its appendages follow a well-defined and regular pattern.

The vertebrate integument consists of a multilayered ectodermally derived *epidermis* resting upon a layer of mesodermal connective tissue called the *dermis*. As they mature, developing epidermal cells produce increasing amounts of the specialized intracellular protein *keratin*. A typical integumentary appendage often contains components derived from both epidermis and dermis. Both development and maintenance of the skin and its appendages involve a series of communications between the ectoderm and the mesoderm.

EARLY DEVELOPMENT OF THE INTEGUMENT

The early embryo is covered by a single layer of ectodermal cells which are initially not closely associated with the underlying mesenchymal cells. Rather, the ectoderm rests upon a loose layer of extracellular matrix. Two major

355

changes occur in the early ectoderm. First, it begins to stratify into two layers: a deep layer of basal ectodermal cells and a newly forming superficial layer, the *periderm*, which covers the surface of amniote embryos before the ectoderm is transformed into a well-differentiated epidermis. The second change is the appearance of a well-defined basal lamina beneath the basal layer of ectodermal cells.

Formation of a recognizable dermis lags behind differentiation of the epidermis. Throughout most of the body the cells of the dorsal dermis arise from the dermatome segments of the maturing somites, and those of the ventral dermis come from the somatic mesoderm of the lateral plate (Fig. 6-23). In contrast, the dermis of the face and part of the ventral neck contains cells that originate in the neural crest (Fig. 6-17).

Despite the seeming retardation in morphological development of the dermis, most experimental evidence indicates that the dermis or its precursor strongly influences the development of the ectoderm into a definitive epidermis. Later in this chapter, considerable attention will be given to the series of continuing interactions between ectoderm and dermal mesenchyme that result in the formation of the mature skin and its derivatives.

Two specializations are seen in the ectoderm of early amphibian embryos. Shortly after closure of the neural folds, large numbers of ciliated cells rapidly appear over the entire ectodermal surface of the embryo (Fig. 11-6C). Possibly because of inhibition of neighboring cells by ciliated cells, no two adjacent cells are ciliated. Although the existence of ectodermal cilia has been known for almost a century (Assheton, 1896), their physiological role is still poorly understood. The ciliary beat causes currents of water to flow over the embryo, and this may facilitate respiration or cleansing of the surface of the embryo.

The postneurula amphibian ectoderm also possesses the property of electrical conductivity. This was demonstrated by Chuang and Dai (1961), who extirpated the neural plates of amphibian embryos and connected several embryos in series in head-to-tail fashion (*telobiosis*) (Fig. 10-1). When the ectoderm of one embryo in the chain was stimulated, the other embryos in the chain responded by movement. The period of ectodermal conductivity is correlated with the presence of large gap junctions between adjacent ectodermal cells (Chuang-Tseng et al., 1982).

HISTOGENESIS OF THE SKIN AND ITS DERIVATIVES

During embryonic development, as in postnatal life, the developing skin serves as the direct interface between the embryo and its external environment. In amniote embryos, the immediate environment consists of amniotic fluid. There is indirect evidence suggesting a certain degree of fluid and metabolic exchange between the skin of the early embryo and the amniotic fluid. As the developing skin approches its normal postnatal structure, the barrier function of the epidermis becomes more prominent and exchange with the external environment is increasingly mediated through the glands that form in the skin. Late

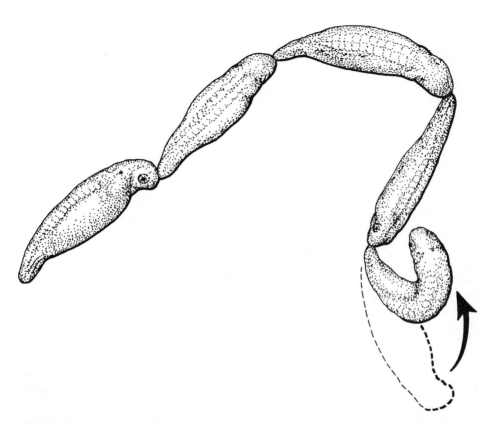

FIGURE 10-1
An early experiment demonstrating ectodermal communication in the amphibian embryo. When five embryos are linked together in a head-to-tail fashion, stimulation of the ectoderm in a posterior embryo causes a twitching of the anteriormost embryo. (*Adapted from Chuang and Dai, 1961, Scientia, 12:41.*)

mammalian features are covered with a whitish secretion called the *vernix caseosa*, which is assumed to serve as a means of insulating the skin from the amniotic fluid.

Epidermis

The histogenesis of the human epidermis has been described in great detail and will serve as the basis for this section. The normal adult epidermis is a multilayered structure, the thickness of which varies in different parts of the body (Fig. 10-15). The basal layer rests upon a basal lamina which separates it from the dermis. As its name implies, the cells of this layer are mitotically active. The progeny of the basal cells become progressively displaced to more superficial layers as additional cells are produced by the dividing stem cells of the basal layer.

As a given cell rises toward the surface of the epidermis it undergoes a series of inexorable changes that ultimately result in its transformation to a flat, nonliving shell full of keratin, the protein that marks the terminal differentiation of the epidermal cell (Fig. 10-15). As the epidermal cells differentiate during their passage to the surface, their distinctive features produce a layering effect. Immediately above the basal layer is a relatively thick layer, which is called the *stratum spinosum* because the desmosomes which bind adjacent cells together give the cells a spiny appearance in histological preparations. The cells of this layer have begun the active synthesis of the precursors of the keratin proteins. To the electron microscopist, these appear as filaments of intermediate thickness (6 to 9 nm) arranged in loose bundles throughout the cytoplasm. The filaments, which are aggregates of individual keratin polypeptide subunits, undergo a major change in conformation and take on a granular appearance. The granules have long been called *keratohyalin*, and because of their prominence, the layer of cells that contains them is called the *stratum granulosum*. As the cells in this layer age, their nuclei disappear and the cytoplasm becomes filled with tightly packed aggregates of further changed keratin filaments. These dead scalelike cells of the outer *stratum corneum* are eventually shed from the body.

In all mammals studied, the epidermis undergoes a characteristic sequence of differentiation during embryonic life (Holbrook, 1983). The single layered ectoderm (Fig. 10-2A) becomes a bilayered epidermal structure as a layer of flattened peridermal cells forms on its surface (Fig. 10-2B and C). The *periderm* is found on the developing epidermis of all amniote embryos, including reptiles. It is most prominent during the period of early epidermal differentiation before the definitive layers of the epidermis have formed, and it has long been considered to be an embryo's adaptation designed to provide a protective covering for the fetus. However, in recent years there has been a gradual accumulation of evidence suggesting that the cells of the periderm may be actively involved in the exchange of water, Na^+, and possibly glucose between the skin and the amniotic fluid.

The next stage is that of the three-layered epidermis. Progeny of the basal layer, which is in a period of intense proliferation, form an *intermediate layer* (Fig. 10-2D). The term *intermediate* can be applied to more than its location, because the cells of this layer also represent an intermediate stage in the differentiation of epidermal cells with the definitive pattern of keratinization. Starting in the fifth month, cells of the lower part of the intermediate layer begin to form typical keratohyalin granules, thus becoming recognizable on the stratum granulosum. The older intermediate cells, which are located superficially to the newly formed granular layer, have also undergone a certain degree of keratinization, but the pattern of keratin filaments is not characteristic of mature epidermis.

As these events are occurring in the intermediate layer, the cells of the periderm are also changing. In the 3-month human fetus, prominent glycogen-

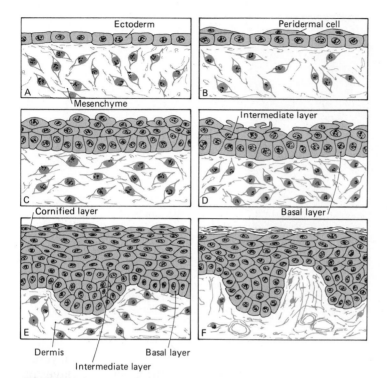

FIGURE 10-2
Stages in the histogenesis of human skin. (A) 1 month; (B) almost 2 months; (C) 2½ months; (D) 4 months; (E) 6 months; (F) adult skin.

filled blebs arise on the surface of the peridermal cells (Figs. 10-2D and 10-3). By the fifth month the blebs have begun to collapse, and shortly thereafter the periderm layer begins to break up.

The gradual breaking up of the periderm coincides with the differentiation of a *stratum corneum* from the first-formed granular cells (Fig. 10-2E and F). Even in the late fetus, the sequence of events in the formation of the stratum corneum is remarkably similar to that which is seen in postnatal life. It is in the stratum corneum that regional differences in the histogenesis of the epidermis are most apparent. For instance, the ectoderm is much thicker over the palms and soles than it is over the rest of the body. Since this occurs in utero, where no greater mechanical pressure is applied to these regions than to other parts of the body, it points to fundamental regional differences in the developmental control of epidermal morphogenesis.

Immigrant Cells in the Epidermis

Despite its apparent isolation from the rest of the body by its underlying basal lamina, the epidermis is not a homogeneous tissue derived solely from surface

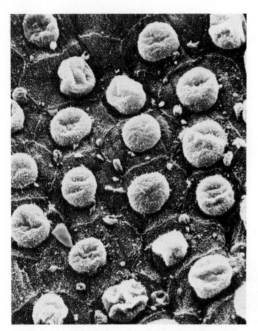

FIGURE 10-3
Scanning electron micrograph showing surface blebs on the peridermal cells of a 3-month human fetus. (*Courtesy of K. A Holbrook, from Holbrook, 1983.*)

ectodermal cells. Three types of foreign cells invade the embryo's epidermis and remain there during adult life.

Melanoblasts, migrating out from the neural crest (see Chap. 6), reach the dermis in human embryos during the second month and penetrate the epidermis early in the third month. The differentiation of melanoblasts into *melanocytes* (definitive pigment cells) is associated with the formation of pigment granules (*melanosomes*) from more immature forms (*premelanosomes*). Not until the melanocytes have matured is pigmentation present in the skin. There are large racial differences in the extent of pigmentation in the skin, but the number of melanocytes in the skin does not differ much from race to race. Instead, there is more pigment per cell in the melanocytes of individuals with dark skin. The skin of albinos usually contains a normal number of melanocytes, but they fail to accumulate pigment because they lack the enzyme *tyrosinase*, which converts the amino acid tyrosine to *melanin*. In heavily pigmented races melanocytes invade the embryonic epidermis earlier than is the case in embryos of nonpigmented races.

Toward the end of the first trimester of pregnancy, the fetal epidermis is invaded by a population of *Langerhans cells* (Wolff and Stingl, 1983). Langerhans cells are not readily distinguishable from ordinary epidermal cells (*keratinocytes*), but they contain distinctive cytoplasmic granules and are readily identified histochemically by their high membrane-bound ATPase or by certain surface antigens. For years the origin and function of these cells were enigmatic, but they are now known to arise from precursors in the bone marrow and to penetrate all layers of the epidermis. Although the story of their function is still unfolding, these cells are recognized as the most peripheral outposts of the immune system. They process antigens that penetrate the epidermis and then cooperate with T lymphocytes in the epidermis to initiate a cell-mediated response against the antigen.

A third immigrant cell type, the *Merkel cell*, arrives in the fetal epidermis after the first two foreign cell types (about 16 weeks in the human embryo). Suspected of arising from the neural crest, Merkel cells become associated with free nerve terminals and serve as slow-adapting mechanoreceptors for the skin.

Hair

Hairs are very diverse structures when one considers the varieties of sizes and degrees of coarseness of hairs on the body of a single individual mammal, yet they all go through a characteristic sequence of developmental stages. In the human, hair formation first becomes recognizable over the eyebrows, scalp, lips, and chin of embryos and spreads throughout the body in a craniocaudal direction.

Hair formation is first recognized when a cluster of basal epidermal cells begins to project as a bud downward into the dermis (Fig. 10-4A). As the epidermal bud continues to extend downward, a condensation of dermal mesenchymal cells known as the *dermal papilla* begins to indent the tip (Fig. 10-4B). The epidermal *hair bulb* that partially surrounds the dermal papilla like an inverted cap is the source of the hair itself.

In the next stage of development, primordium of the hair and its associated structures take shape (Fig. 10-4C). The incipient hair is first represented by a cone of rapidly proliferating epidermal cells from the inner wall of the epidermal hair bulb. The growing hair continues to push upward through the center of the hair follicle until it reaches the surface of the fetal epidermis. At that point it emerges from the level of the epidermis, sometimes bearing a small cap of peridermal cells. Fully formed hairs continue to push upward through the canal that they have created within the epidermal follicle.

The differentiation of an embryonic hair is a relatively leisurely process, taking several months in the human. As the hair bulb begins to mature, it is infiltrated by melanocytes, which provide the pigment that colors dark hair. The melanocytes later become more tightly localized at the root of the hair bulb. The epidermal cells that constitute the hair shaft begin to undergo keratinization during the fifth fetal month. Keratinization of the hair is marked by the formation of hard granules of keratin complex called *trichohyalin*, which imparts hardness to the hair.

Two other structures also form in conjunction with the developing hair follicle. One is the *sebaceous gland* (Fig. 10-4C and D), which begins as a bulge of epidermal cells midway along the length of the follicle. The cells of the sebaceous gland differentiate to form an oily secretion called *sebum*, which is discharged onto the surface of the skin via the sheath of the hair follicle. The

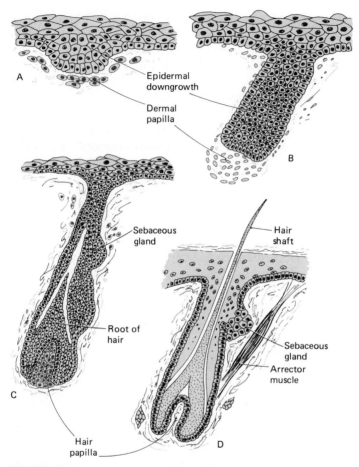

FIGURE 10-4
Major stages in the formation of a human hair. (A) Hair primordium (12 weeks); (B) early hair peg (15–16 weeks); (C) bulbous hair follicle (18 weeks); (D) adult hair.

mammalian fetus secretes a whitish material which, when combined with desquamated epidermal cells, forms a presumably protective coating known as the *vernix caseosa*.

The other major structure associated with the hair follicle is a thin slip of smooth muscle forming from dermal mesenchyme. On one end it attaches to another bulge of epidermal cells along the hair follicle, and the other end is embedded in the dermis near its junction with the epidermis. An *arrector pili muscle* (Fig. 10-4D) is attached to each hair in mammals, and the contraction of these muscles is responsible for raising the fur when an animal is angry or cold. In humans, contraction of the arrector pili muscles accounts for ''goose bumps'' when we get cold.

The first hairs to emerge over the fetal body are close together and very fine. Known as *lanugo*, they are prominent during the seventh and eighth months of human pregnancy. They are normally shed just before birth and are replaced by coarser definitive hairs, which are thought to arise from new follicles. This can be looked upon as an extension of the developmental isoform strategy from the level of molecules and cells to that of more complex tissues.

Feathers

Like hair, a feather begins as a concentration of dermal cells beneath an epidermal placode (Fig. 10-5A). Macroscopically, *feather rudiments* at this

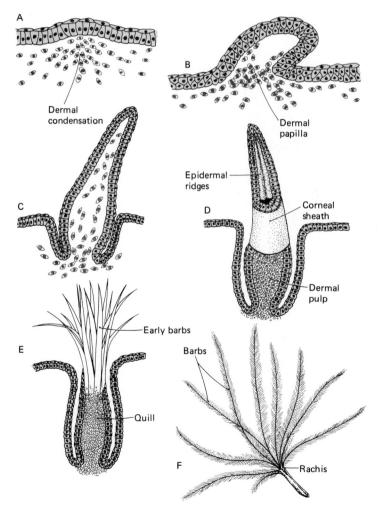

FIGURE 10-5
Stages in the formation of a down feather.

stage appear as small, whitish spots on the skin of 7- to 8-day chick embryos. Owing to cell proliferation in both epidermal and dermal components, the originally flat feather rudiment rises above the surface of the skin as a *feather bud* (Fig. 10-5B). The epidermis at its base sinks down into the dermis to form a pitlike structure called the *feather follicle* (Fig. 10-5C), and the rapidly growing feather bud assumes a conical shape.

Further development depends on the type of feather to be formed. The first feathers to appear in the embryo are *down feathers*, in which the barbs all arise in a circle at the same level from a short shaft (Fig. 10-5F). The most prominent feather type in the mature bird is the familiar *contour feather* (Fig. 10-6D).

In the development of a down feather, the thickened epidermis of the elongating feather bud forms a series of roughly parallel columns, called *barb*

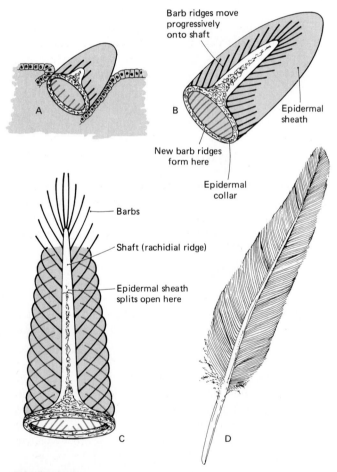

FIGURE 10-6
Stages in the formation of a contour feather.

ridges, beneath a cornified sheath of surface cells (Fig. 10-5E). As the down feather matures, the cells of the dermal pulp retract from the feather bud and the epidermal components harden as a result of keratinization. Final eruption of the down feather occurs when the cornified outer epidermal sheath splits open and allows the barb ridges or columns to spread out in a plumelike fashion to form the definitive down *feather barbs* (Fig. 10-5E). The barbs possess regular branching structures called *barbules*, which are responsible for the insulating properties of feathers. The structure of the barbules differs from one species to the next and serves as a valuable species-specific marker for experimental embryological studies.

Up to the stage of the conical feather bud, the development of contour feathers is morphologically similar to that of down feathers. Thereafter, the two types of feathers go their separate ways during their development. At the base of the cone-shaped contour feather bud an epidermal collar produces a series of parallel barb ridges (Fig. 10-6A), but soon the dorsal part of the epidermal collar begins to elongate to form the long shaft of the feather (Fig. 10-6B). Early elongation of the feather shaft occurs entirely within the cone of the feather bud, and the barb ridges bend beneath the cornified outer epidermal sheath until their ends almost touch on the side opposite the shaft. As the outer epidermal sheath of the feather bud begins to split, the apical part of the shaft and its associated barbs are freed from their conical restraints. The freed barbs unroll and flatten out in a typical feather form, while at the base the barbs are still encased in the intact epidermal hull (Fig. 10-6C). From this description, it should be apparent that the apex of a feather is the oldest region and that new parts are added proximally.

Like hair, feathers develop in association with dermal smooth-muscle elements. However, contour feathers are moved by two main muscles, an erector muscle that causes the fluffing of feathers for display purposes or warmth and a depressor muscle that allows the feathers to be flattened upon the body during flying.

Scales

As examples of scale development, we shall consider those which cover the toes and tarsometatarsal region of the chicken leg (Sawyer, 1972). The initial placodal stages of scale development (Fig. 10-7A and B) differ from those of the feather in that there is not a pronounced condensation of dermal cells beneath the placode. However, as the placodes elevate into ridges (Fig. 10-7C and D), dermal cells aggregate and proliferate at the apical end of the scale that is now beginning to take shape. The apical end of the scale continues to grow until it overlaps the basal portion of the next scale in line (Fig. 10-7E). By this point, both the epidermal and peridermal cells covering the inner and outer sufaces of the scale have undergone complex series of differentiative changes and have begun to form keratins specific for each surface of the scale.

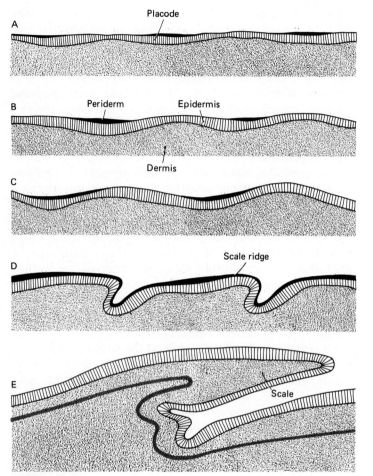

FIGURE 10-7
Stages in the formation of a scale in the bird. (A) Placode stage; (B) asymmetrical placode stage; (C) hump stage; (D) definitive scale ridge stage; (E) mature scale. (*After Sawyer, 1972, J. Exp. Zool.* **181**:*367*.)

TISSUE INTERACTIONS IN INTEGUMENTARY DEVELOPMENT

Almost all aspects of integumentary development depend on a continuing series of reciprocal communications between the ectoderm and its underlying mesenchyme. The availability of both a wide variety of skin appendages and some valuable genetic mutants has provided embryologists with powerful tools for the experimental analysis of the factors that lead to both the morphogenesis and the differentiation of the skin and its derivatives. Most of the experimental strategies have involved separating the ectoderm from its underlying mesenchyme and then allowing these components to develop alone or in combination with mesenchyme or ectoderm from other regions, species, or stages.

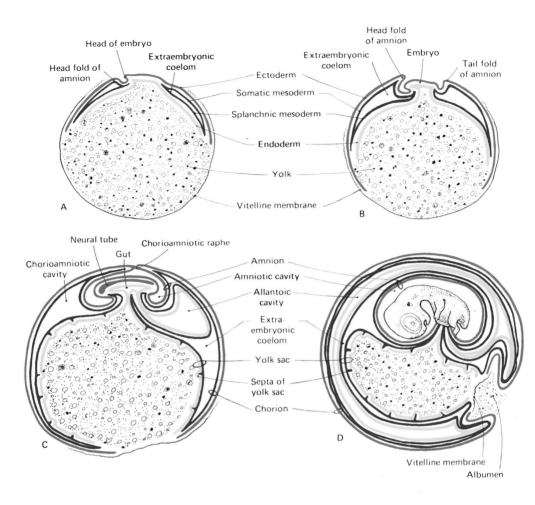

Fig. 7-4 Schematic diagrams to show the extraembryonic membranes of the chick.(D, after Lillie.) The embryo is cut longitudinally. The albumen, shell membranes, and shell are not shown; for their relations, see Fig. 7-1. (A) Embryo early in second day of incubation. (B) Embryo early in third day of incubation. (C) Embryo of 5 days. (D) Embryo of 9 days.

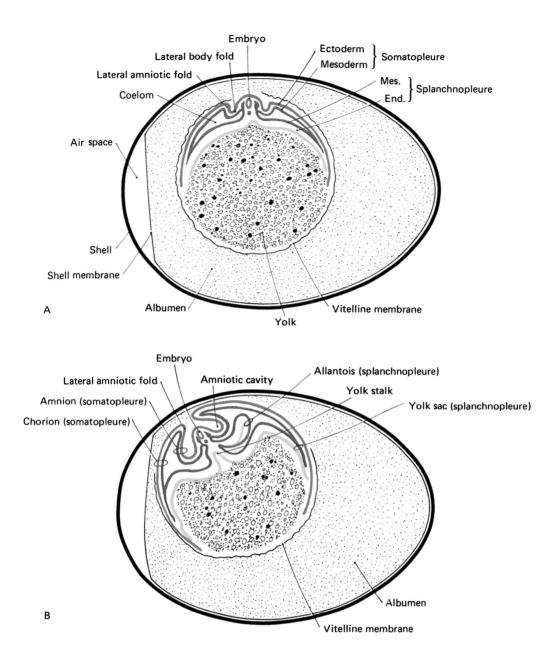

Fig. 7-1 Schematic diagrams to show the extraembryonic membranes of the chick. (After Duval.) The diagrams represent longitudinal sections through the entire egg. The body of the embryo, being oriented approximately at right angles to the long axis of the egg, is cut transversely. (A) Embryo of about 2 days' incubation. (B) Embryo of about 3 days' incubation. (C) Embryo of about 5 days' incubation. (D) Embryo of about 14 days' incubation.

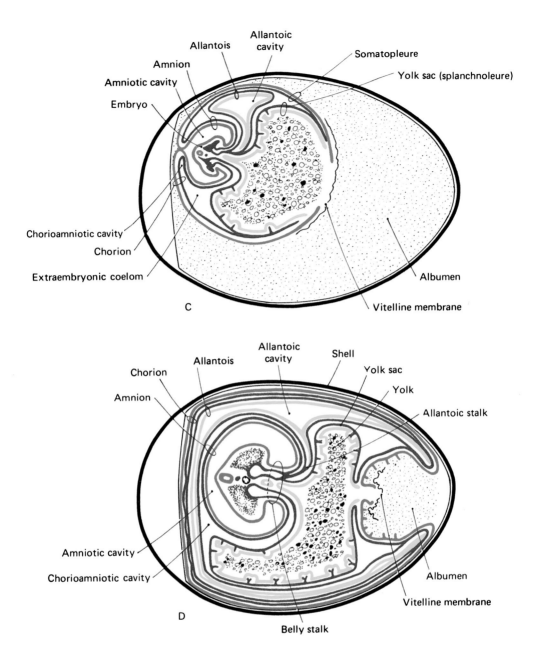

Allantois
Allantoic
cavity
Amnion
Somatopleure
Amniotic cavity
Yolk sac (splanchnoleure)
Embryo

Chorioamniotic cavity
Chorion
Extraembryonic coelom
Albumen

Vitelline membrane

C

Allantoic
cavity
Allantois
Shell
Chorion
Yolk sac
Amnion
Yolk
Allantoic stalk

Amniotic cavity
Chorioamniotic cavity
Albumen

Vitelline membrane

D

Belly stalk

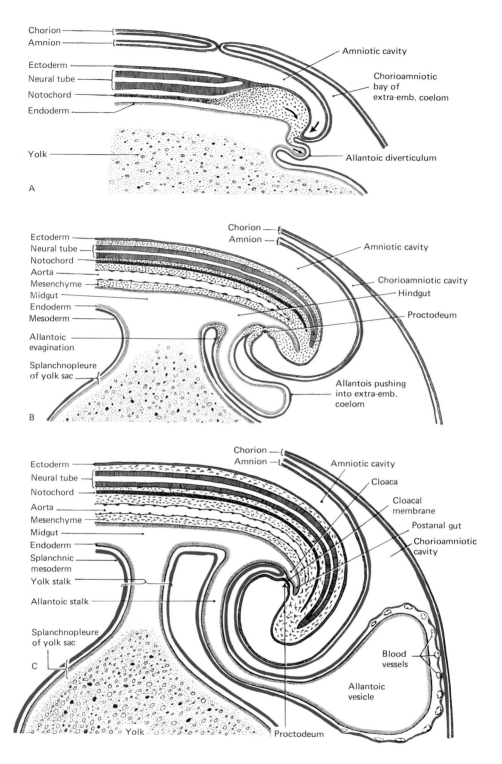

Fig. 7-6 Schematic longitudinal-section diagrams of the caudal regions of a series of chick embryos to show the formation of the allantois. (A) At about 2½ days of incubation; (B) at about 3 days; (C) at 4 days.

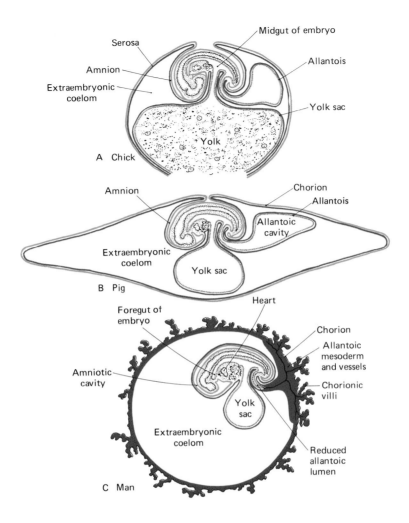

Fig. 7-13 Diagrams showing interrelations of embryo and extraembryonic membranes characteristic of higher vertebrates. Neither the absence of yolk from its yolk sac nor the reduction of its allantoic lumen radically changes the human embryo's basic architectural scheme from that of more primitive types.

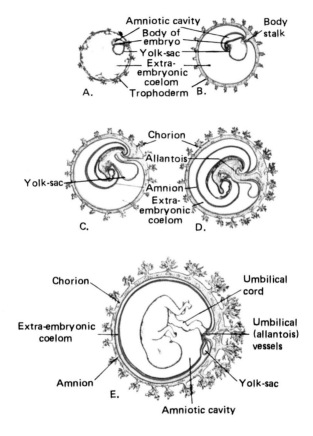

Fig. 7-26 Early changes in interrelations of embryo and extraembryonic membranes.

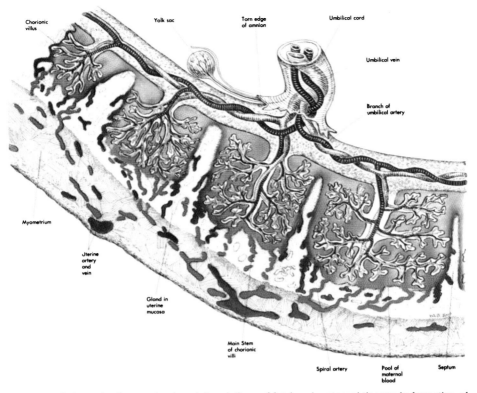

Fig. 7-22 Schematic diagram to show interrelations of fetal and maternal tissues in formation of placenta. Chorionic villi are represented as becoming progressively further developed from left to right across the illustration.

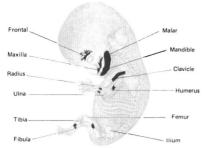

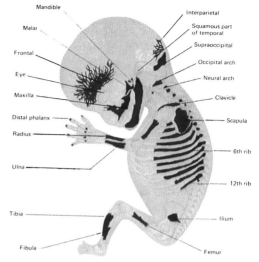

Fig. 9-25 Human embryo of 23½ mm (eighth week) stained with alizarin and cleared to show ossification centers. *(Projection drawing, X1.5, by Jane Schaefer. University of Michigan Collection, EH 105.)*

Fig. 9-26 Human embryo of 39 mm (nine weeks) stained with alizarin and cleared to show the progress of ossification. *(Projection drawing, X2.5, by Jane Schaefer. University of Michigan Collection, EH 149.)*

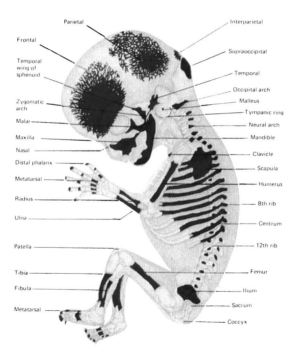

Fig. 9-27 Human embryo of 49 mm (tenth week) stained with alizarin and cleared to show the developing skeletal system. *(Projection drawing, X2.7, by Jane Schaefer, University of Michigan Collection, EH 70.)*

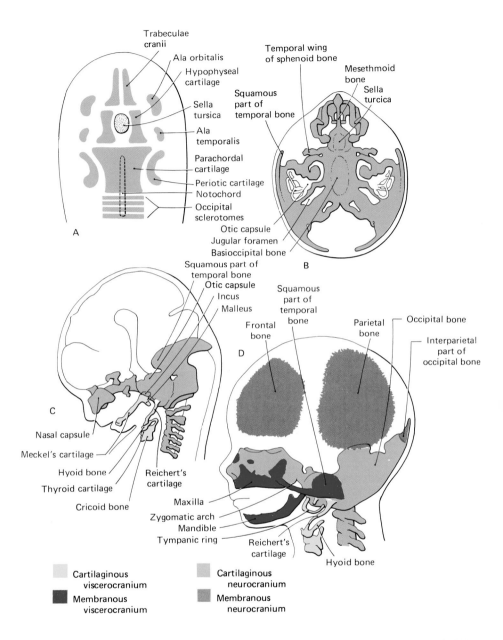

Fig. 9-24 Diagrams illustrating the origins and development of the major bones of the skull. (A) Embryo of about 6 weeks (viewed from above), showing the primordial cartilages that will form the chondrocranium. (B) Embryo of about 8 weeks (viewed from above), showing the chondrocranium. (C) Lateral view of the embryo illustrated in B. (D) Skull of 3-month embryo.

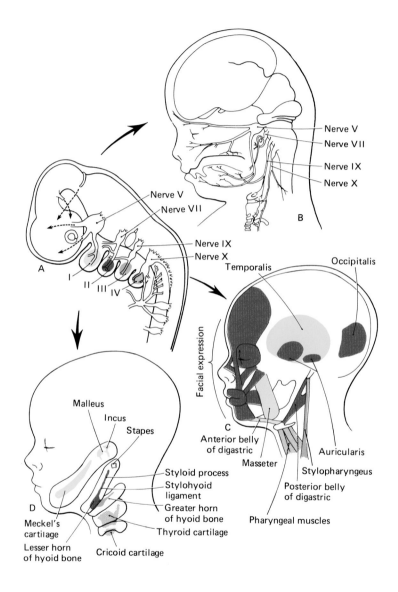

Fig. 16-5 Schematic diagrams showing major derivatives of structures that constitute the branchial arches. (A) 5-week embryo; (B–D) 4- to 5-month fetuses. The gray tones of structures in C and D correspond to those of the branchial arches depicted in A.

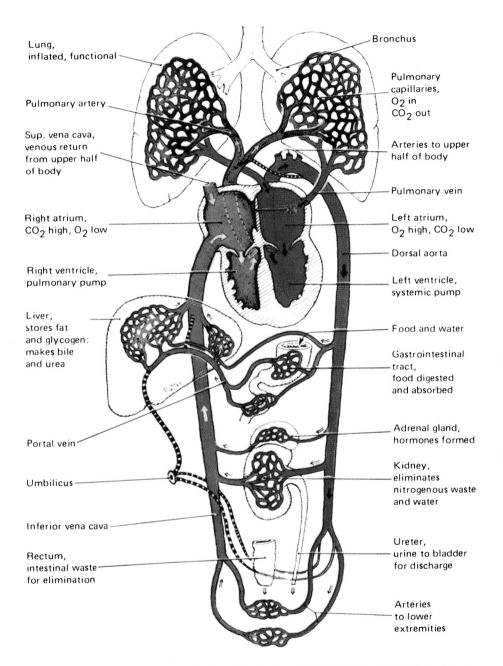

Fig. 18-1 Plan of the postnatal circulation. *(After Patten, 1963, in Fishbein,* Birth Defects. *Courtesy of the National Foundation and the J.B. Lippincott Company, Philadelphia.)* The heavily cross-banded structures were important fetal vessels (cf. Fig. 18-29) which after birth ceased to carry blood and gradually became reduced to fibrous cords. The asterisk indicates the valvula foraminis ovalis in the closed position characteristic for postnatal life.

Labels, clockwise from upper left:

Lung, inflated, functional

Pulmonary artery

Sup. vena cava, venous return from upper half of body

Right atrium, CO_2 high, O_2 low

Right ventricle, pulmonary pump

Liver, stores fat and glycogen: makes bile and urea

Portal vein

Umbilicus

Inferior vena cava

Rectum, intestinal waste for elimination

Bronchus

Pulmonary capillaries, O_2 in CO_2 out

Arteries to upper half of body

Pulmonary vein

Left atrium, O_2 high, CO_2 low

Dorsal aorta

Left ventricle, systemic pump

Food and water

Gastrointestinal tract, food digested and absorbed

Adrenal gland, hormones formed

Kidney, eliminates nitrogenous waste and water

Ureter, urine to bladder for discharge

Arteries to lower extremities

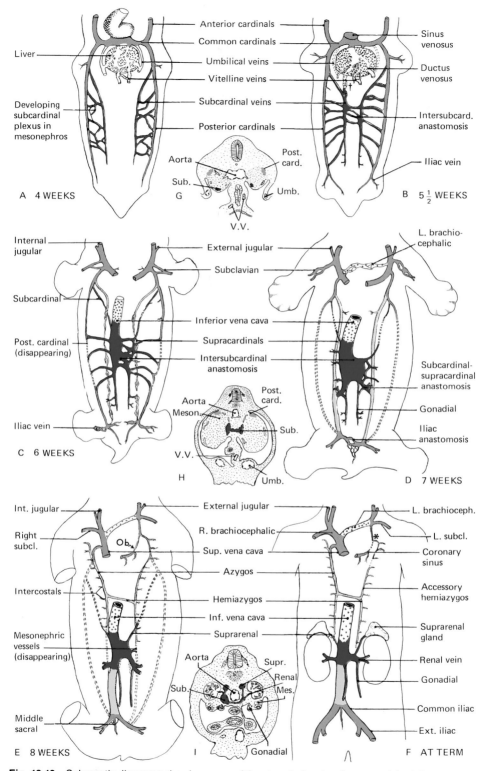

Fig. 18-12 Schematic diagrams showing some of the steps in the development of the inferior vena cava. Cardinal veins are shown in black; subcardinals are stippled; supracardinals are horizontally hatched. Vessels arising independently of these three systems are indicated by small crosses. *(Based on the work of McClure and Butler.)* Abbreviations: *Ob.,* oblique vein of left atrium; *,* left superior intercostal; †, mesenteric portion of inferior vena cava; *subl.,* subclavian vein.

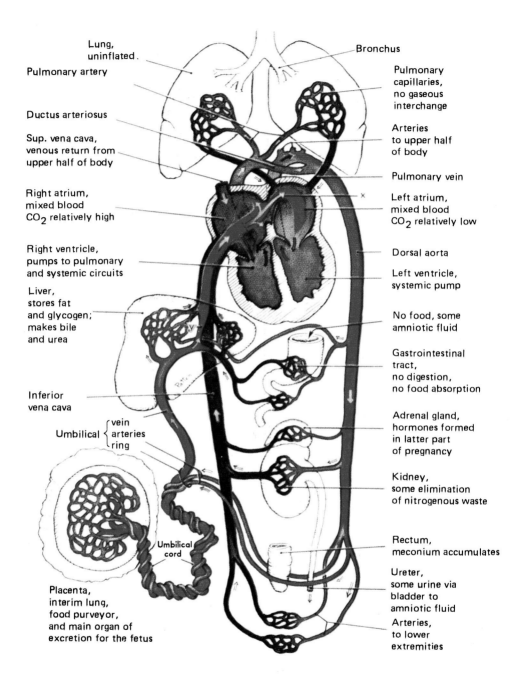

Lung, uninflated.

Pulmonary artery

Ductus arteriosus

Sup. vena cava, venous return from upper half of body

Right atrium, mixed blood CO_2 relatively high

Right ventricle, pumps to pulmonary and systemic circuits

Liver, stores fat and glycogen; makes bile and urea

Inferior vena cava

Umbilical { vein arteries ring

Placenta, interim lung, food purveyor, and main organ of excretion for the fetus

Umbilical cord

Bronchus

Pulmonary capillaries, no gaseous interchange

Arteries to upper half of body

Pulmonary vein

Left atrium, mixed blood CO_2 relatively low

Dorsal aorta

Left ventricle, systemic pump

No food, some amniotic fluid

Gastrointestinal tract, no digestion, no food absorption

Adrenal gland, hormones formed in latter part of pregnancy

Kidney, some elimination of nitrogenous waste

Rectum, meconium accumulates

Ureter, some urine via bladder to amniotic fluid

Arteries, to lower extremities

Fig. 18-29 Plan of the fetal circulation at term. The ductus venosus in the liver is marked by †. The arrow in the heart marked by * indicates the passage of blood from the right atrium to the left as it occurs during atrial diastole. This flow pushes the valvula into the open positions here represented. When the atria contract, the valvula moves back against the septum, closing the foramen ovale against the return flow and thus forcing all the blood in the left atrium to enter the left ventricle. *(After Patten, in Fishbein, 1963,* Birth Defects, *Courtesy of the National Foundation and the J.B. Lippincott Company, Philadelphia.)*

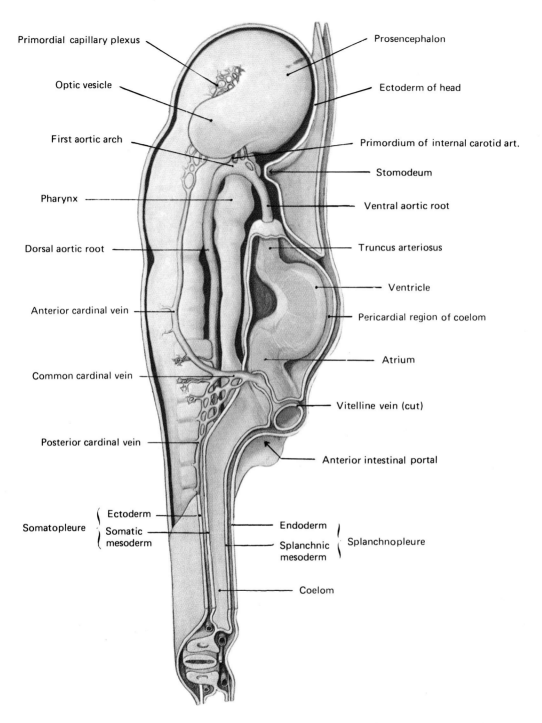

Fig. A-25 Diagrammatic lateral view of dissection of a 38-hour chick. The lateral body wall of the right side has been removed to show the internal structures. Note especially the relations of the pericardial region to that part of the coelom which lies farther caudally and the small anastomosing channels of the developing posterior cardinal vein from which a single main vessel is later derived.

Primordial capillary plexus

Optic vesicle

First aortic arch

Pharynx

Dorsal aortic root

Anterior cardinal vein

Common cardinal vein

Posterior cardinal vein

Somatopleure { Ectoderm
Somatic mesoderm }

Prosencephalon

Ectoderm of head

Primordium of internal carotid art.

Stomodeum

Ventral aortic root

Truncus arteriosus

Ventricle

Pericardial region of coelom

Atrium

Vitelline vein (cut)

Anterior intestinal portal

Endoderm }
Splanchnic mesoderm } Splanchnopleure

Coelom

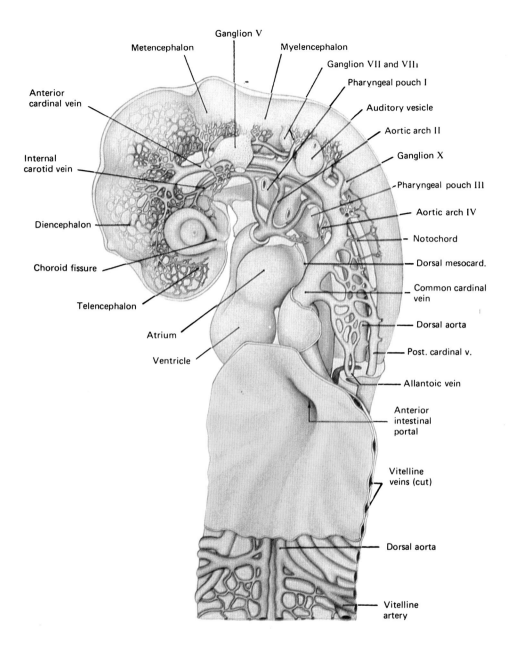

Fig. A-32 Drawing to show the deeper structures of the cephalothoracic region of a 60-hour chick, exposed from the left. The basis of the illustration was a wax-plate reconstruction made from serial sections; the smaller vessels and the primordial capillary plexuses were added from injected specimens. Note especially the relations of the aortic arches to the pharyngeal pouches and the way in which the large veins enter the sinus venosus. Although the vitelline veins are not fully exposed by this dissection, the bulges they cause in the endoderm on either side of the anterior intestinal portal clearly suggest the way the main right and left veins become confluent with each other to enter the sinus venosus as a short median trunk.

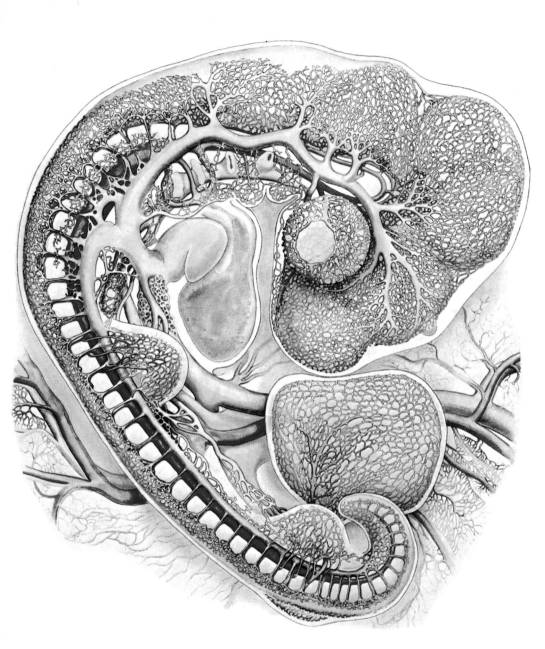

Fig. A-47 The blood vessels of a 4-day chick. The basis of the illustration was the same wax-plate reconstruction from which Figure A-44 was drawn. The smaller vessels and the primordial capillary plexuses were added from injected specimens. The labeled diagram of Figure A-44 will serve as a means of identifying the main vessels.

Details of many of the hundreds of such experiments that have been performed can be found in reviews by Sengel (1976, 1983).

One of the simplest ways to test the importance of tissue interactions in skin development is to separate the ectodermal and mesenchymal components and raise them separately. When grown in isolation, ectoderm remains ectodermal in character, without differentiating into an epidermis. No epidermal append-ages are formed. Similarly, isolated dermal mesenchchyme fails to develop into a normal dermis. These straightforward tests show the importance of ecto-dermal-mesenchymal communication in the normal development of skin.

Further experimentation has revealed much more about the nature of the communication between developing dermis and epidermis. The strategy behind such experiments is first to separate a piece of developing skin into its epidermal and dermal components, usually by incubating the piece in a solution of trypsin for a few minutes. Then the epidermis can be recombined with a foreign dermis or vice versa. The new piece of "recombinant" skin is then allowed to develop in vitro or as a chorioallantoic graft. Analysis consists of examining the types of appendages that develop from the pieces of hybrid test skin. The power of this approach can be better appreciated when one realizes the tremendous variety of skin appendages that can serve as useful experimen-tal endpoints. In the bird, for instance, the predominant integumentary appendage is the feather, but there are different types of feathers in different areas of the body. Not all regions of the avian skin are feathered. Some are smooth (glabrous); other areas in the legs are covered with scales (scutate and reticulate are the main forms of scales). Other integumentary specializations in the chicken are the beak, the comb, the spurs, the cornea, and the claws.

Many early recombination experiments consisted of combining dermis from one region of a chick embryo with epidermis from another region (*heterotopic recombination*). The results of many such experiments led to one general conclusion: The location from which the piece of dermis is taken determines the nature of the integumentary appendages that are formed. For example, epidermis from a wide variety of locations forms back feathers when combined with back dermis (Fig. 10-8A). By the same token, feather-forming ectoderm, if combined with pieces of dermis from various regions, will differentiate differently depending upon the source of the dermis. Experiments like those just summarized illustrate the ability of the dermis to direct the course of gene expression of the ectoderm with which it is combined. This type of direction is often called *instructive induction*.

Another relatively simple experimental strategy—combining a piece of dermis with a piece of ectoderm from another species—has proved very important in understanding the nature of dermal-epidermal interactions. Such experiments have shown that the ectoderm does not passively follow the dictates of the dermis without exhibiting any individuality of its own. One of the first demonstrations of this was a classic experiment by Spemann and Schotté (1932), who grafted flank ectoderm from the gastrula of a newt to the prospective oral region of a frog gastrula. They also performed the reciprocal experiment of grafting ectoderm from

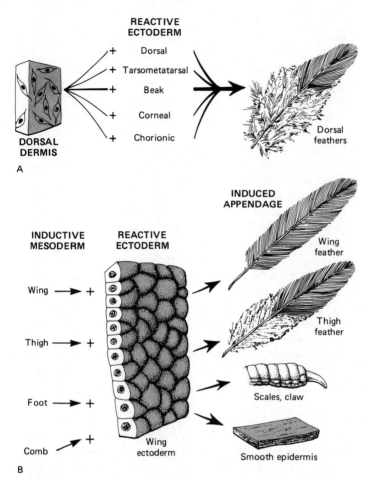

FIGURE 10-8
Experiments showing the regional reactivity of the dermis in inducing ectodermal derivatives. (A) Ectoderm from almost any source responds to a dorsal dermis by forming dorsal feathers. (B) When wing ectoderm is combined with dermis from a variety of locations, it forms derivatives appropriate to the region from which the dermis was derived. (*B adapted from J. W. Saunders, 1980, Developmental Biology, Macmillan.*)

a frog flank to the prospective oral region of a newt. The mouth of the frog tadpole is toothless but has a horny jaw for nipping off bits of vegetation; caudal to the mouth is a pair of suckers which are used to stabilize the newly hatched tadpole. The newt, by contrast, has teeth in the jaw and a fingerlike balancing structure near each corner of the mouth. The recombinant embryos showed interesting hybrid characteristics. In each case the transplanted flank ectoderm had undergone differentiation into mouth parts, but the nature of the mouth part was characteristic of the donor species. Thus, the frog larva was adorned with newt teeth and balancers, and the newt larva was equipped with tadpole suckers and mouth parts (Fig. 10-9).

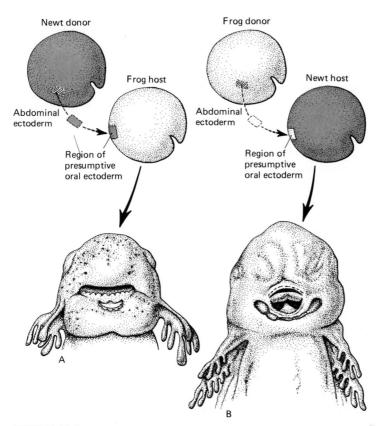

FIGURE 10-9
Experiments showing the importance of both site and intrinsic genetic information as determinants of the result of grafting experiments. (A) Newt abdominal ectoderm grafted into the presumptive oral region of a frog embryo host forms mouth parts (including balancers) that are characteristic of newts. (B) Frog abdominal ectoderm grafted into the presumptive oral regions of a newt embryo host forms mouth parts (with suckers) that are characteristic of frogs. (*Adapted in part from H. Spemann, 1938, Embryonic Induction, Yale Univ. Press.*)

Subsequent recombination experiments have had the same results. In almost all cases of *heterospecific recombinations* the location from which the piece of dermis was taken determines the region-specific nature of the integumentary appendages, but the species-specific nature of the appendages is appropriate to that of the ectoderm, from which the bulk of the appendage is formed. For example, if a piece of back ectoderm from a chick embryo is combined with back mesoderm from a mouse embryo, appendages appropriate to the back are formed. These appendages, however, are not hairs but are back feathers because the chick ectoderm can make only the back-type appendages within its genetic repertoire, namely, back feathers. Mouse mesoderm cannot instruct avian ectoderm to form hairs. Another example should suffice to establish the general principles learned from heterospecific recombinations. When corneal ectoderm from a chick embryo was combined with mouse dermis, the dermis induced the corneal ectoderm to form appendages appropriate to the site of origin of the dermis, but since the ectoderm had to react according to its own genetic heritage, feathers rather than hairs formed (Coulombre and Coulombre, 1971).

The types of experiments just summarized make dermal-epidermal interactions appear to be simple one-way avenues of communication, with the dermis sending out a location-specific message and the epidermis replying with a species-specific response. However, the story is often more complex and more subtle. Sengel (1976) proposed the following sequence of inductive events leading to the formation of typical skin appendages, such as feathers, scales, and hairs: The initial message emanates from the predermal mesenchyme and instructs the overlying ectoderm to form thickened placodes in a region-specific pattern. This initial instruction, however, does not specify which kind of appendage is to be constructed. Next, the ectodermal placode sends back a message causing the predermal mesenchymal cells to condense beneath the placodes. The condensed dermis then sends a second, class-specific inductive message that is transmitted back to the thickened ectodermal placode. This second dermal inductive message specifies the formation of a specific kind of appendage, for example, a contour feather. In a manner that is not well understood, the second dermal inductive signal transmits information that can be translated into both the specific form of the appendage and the synthesis of specific keratin proteins by the epidermis.

The experimental analysis of several mutations that give rise to integumentary abnormalities has begun to establish the connection between dermal induction and gene expression in the epidermis. As an example, we shall concentrate on a mutant called *scaleless* (sc/sc), an autosomal recessive gene in chickens. Chickens homozygous for the scaleless gene lack the large scutate scales which normally cover the shank of the leg; they are also missing most of their feathers and have demonstrable abnormalities in other types of scales, the scleral ossicles of the eye, and other structures. Defective scale formation has received the most attention (rev. by Sawyer, 1983). In scale formation, the predermal mesenchyme is normal but the overlying ectoderm fails to form a normal placode. Various recombinations between normal and scaleless ectoderm and mesenchyme have shown that "scaleless" dermis does not acquire its late scale-inducing properties because it has failed to obtain some necessary information from the epidermis. As a consequence, the late scale-forming dermis is unable to induce the formation of proper scale, and in the epidermis the genes for β-keratin, which is a major component of feathers and the outer layer of scales, remain unexpressed (Fig. 10-10). However, further studies have shown that scaleless epidermis is indeed capable of forming the normal complement of β-keratins. McAleese and Sawyer (1982) combined scaleless epidermis (10- to 16-day) with 14- to 16-day normal scutate scale dermis (Fig. 10-11) and found that the scaleless epidermis formed all the β-keratin components found in normal scales.

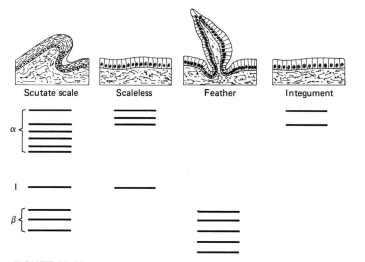

FIGURE 10-10
The spectrum of keratin polypeptides obtained from several regions of
the integument of the normal chick embryo and from all regions of the
scaleless mutant that in normal embryos would have formed scales.
In the scaleless mutant, proteins of the β-keratin family are not expressed. (*Adapted from Sawyer, 1983.*)

The nature of the defect in the scaleless mutant has not been characterized
exactly, but evidence points to the extracellular matrix that forms the interface
between the developing epidermis and dermis. The collagen fibers in the

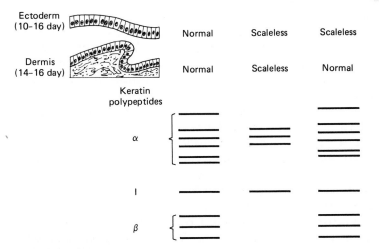

FIGURE 10-11
Results of recombination experiments between ectoderm and dermis between *scaleless* and normal embryos. *Scaleless* ectoderm does not produce β-keratins, but if it is combined with normal
dermis, it is then capable of forming all the β-keratins that are formed by normal epidermis.
(*Adapted from Sawyer, 1983.*)

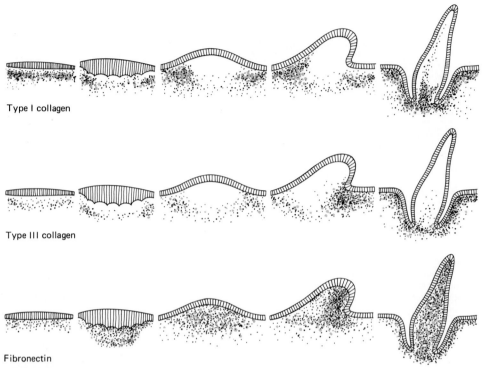

Type I collagen

Type III collagen

Fibronectin

FIGURE 10-12
Distribution of extracellular matrix components (indicated by gray dots) during feather morphogenesis. (*After Mauger et al., 1984, in Matrices and Cell Division, Alan R. Liss.*)

basement membrane are less regularly distributed than normal, and several
other components of the matrix (fibronectin, proteogylcans) are also abnormal.
How such abnormalities may relate to the defective transmission of inductive
messages from one time to the next remains to be elucidated.

In recent years, increasing attention has been focused on the nature of the
interface between the epidermis and dermis in areas where inductive interactions
are occurring. Before the start of feather formation and in areas of glabrous
(nonfeathered) skin, a regular basal lamina intervenes between the epidermis and
dermis and the distribution of collagen (types I and III) and fibronectin is uniform
along the dermal-epidermal interface. This changes considerably as the feather
rudiments take shape (Fig. 10-12). In areas where feather papillae begin to take
shape, long epidermal cell processes penetrate the underlying basal lamina and
come in close contact with dermal-cell processes, which become aligned close to
and parallel to the deep side of the basal lamina. Initially, types I and III collagens
become concentrated in areas between feather germs and are almost absent in the
early feather germs themselves. In contrast, fibronectin becomes concentrated in
the region of feather formation (Fig. 10-12).

As the feather bud begins to grow out, fibronectin remains abundant in the dermal core of the feather bud. During early outgrowth collagen becomes concentrated around the base of the feather bud, with type I more prominent on the cranial side (obtuse angle) and type III more prominent on the caudal side (acute angle). Presumably, these collagen deposits help stabilize the outgrowing feather. As is the case with other structures that develop through a continuing series of epithelial-mesenchymal interactions (e.g., salivary glands; see Chap. 14), collagen is diminished or removed in areas of inductive or morphogenetic activity, whereas it accumulates around the resulting structures once morphogenetic stability has set in. During the entire process of feather formation, laminin and type IV collagen remain evenly localized to the basal lamina, supporting the notion that these molecules are more heavily involved in the attachment of the epidermal cells to the basal lamina than as morphogenetic agents.

PATTERN FORMATION IN THE INTEGUMENT

Feathers

Epidermal appendages such as feathers are not just randomly distributed over the skin. Rather, they form in a highly ordered pattern. In avian embryos feathers form as specific patches of the skin called *feather tracts* (*pterylae*) (Fig. 10-13). Feather tracts are separated from one another by featherless areas called *apteria*.

The feather papillae which populate a feather tract arise in a well-defined and well-controlled sequence. In the tracts that cover the back and neck (Fig. 10-13), the nearly simultaneous appearance of feather primordia in a single row at the midline (in the case of the back) marks the start of the feather tract called the primary row. Shortly thereafter, a secondary row of feather papillae arises on either side of the primary row. In a regular geometric pattern, the feather germs of the secondary row arise in staggered fashion laterally to those of the primary row (Fig. 10-13A). Feather germs of the tertiary row next form laterally to those of the secondary row. This sequence of additional rows continues until the feather tract is filled out.

Because of their alternation of positions from row to row, the feather papillae are arranged in a highly regular hexagonal lattice (Fig. 10-14). The cellular mechanisms underlying the formation of the lattice are incompletely understood, but a tentative picture is beginning to emerge. Each feather papilla is the site of the local inductions between epidermis and dermis that were described earlier in this chapter. Early work suggested that the spacing and pattern of the feather papillae are determined by an oriented matrix of collagen. This conclusion was supported when treatment of the skin with collagenase disrupted the hexagonal lattice and interfered with the early steps of feather papilla formation. Studies on scaleless mutants have shown major disruptions in feather pattern despite a normal rate of collagen synthesis (Goetinck and Sekellick, 1972). However, in this mutant the hexagonal lattice does not form normally. Recombination studies have shown that the lack of a normal dermal

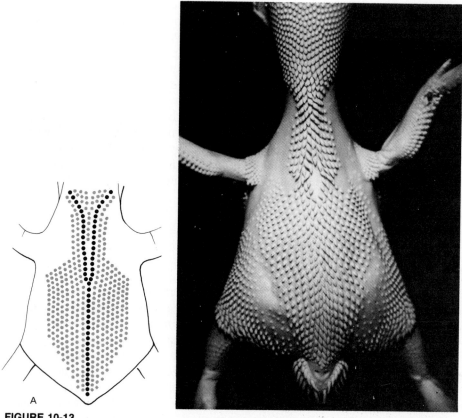

FIGURE 10-13
Early pattern of feather tract formation in the chick embryo. (A) Drawing of back and saddle tracts. Starting at the midline (black circles), each more lateral row of feather germs (gray circles) arises successively. (B) Dorsal views of 10½-day chick embryo showing the feather papillae arranged in regular rows in feather tracts. (*Courtesy of P. Sengel, photograph by A. Mauger in Sengel, 1976.*)

lattice occurs when either normal or scaleless dermis is combined with scaleless epidermis. Therefore, it appears that a genetic defect in scaleless epidermis is associated with the disruption of a dermal pattern that serves as the basis for the distribution of feather germs within the feather tracts.

The major unanswered question is how an overall morphogenetic blueprint is translated at the local level in a manner that results in the exact specification of the site of each feather primordium. A series of experiments by Sengel (1976) suggests an orderly sequence of morphogenetic signal calling in the feather tract that covers the back (Figs. 10-13 and 10-14). Once the feather germs of the middorsal row are established (how this occurs is not known), they control the lateral extension of the hexagonal lattice in the dermis, which specifies the locations of the feather germs in the next most lateral row on either side. When the secondary feather rows on each side are established, they in turn specify

FIGURE 10-14
The regular hexagonal arrangement of the precursors of the dermal papillae of the feather germs in the chick embryo. (*From J. Saunders, 1983, Developmental Biology, Macmillan.*)

the tertiary rows by a similar mechanism. This process of serial signaling for the next most lateral row continues until the edge of the feather tract is reached. Why feather rows stop being formed at this point is not known, but discovering the nature of the stop signal may be as important to our understanding of pattern formation in the skin as discovering how feather rows are initiated.

EPIDERMAL DIFFERENTIATION

When associated with prospective dermis, the embryonic ectoderm undergoes a series of changes that result in its differentation into a multilayered epidermis.

Throughout life, the epidermis is under the continuing influence of the dermis, and its regional characteristics, for example, the thick epidermis of the palm and sole and the thin epidermis of the ear, are determined to a large extent by signals emanating from the underlying dermis. Recombination experiments with adult epidermis and dermis have shown that the thickness of the epidermis is controlled to a large extent by the dermis. For example, if the thick epidermis from the sole is combined with the dermis from the ear, the epidermis thins out and even develops fine hair follicles.

Regardless of regional differences, the epidermis all over the body has the same fundamental organization and function. It is a multilayered structure which is almost totally geared toward providing the body with an impermeable protective covering. Epidermal cells arise from the mitotic activity of germinative cells in the basal layer (Fig. 10-15) of the epidermis. The newly produced epidermal cells are steadily displaced toward the surface as additional new cells

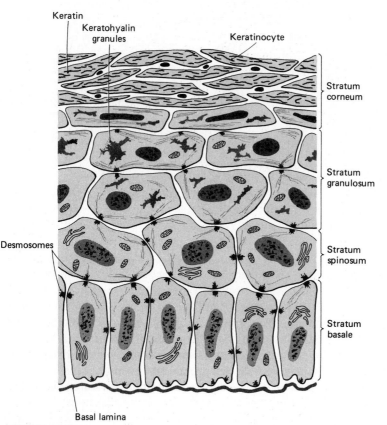

FIGURE 10-15
Layering and differentiation of the epidermis. Cells arising in the basal layer (stratum basale) progress toward the surface, undergoing terminal differentiation into keratinocytes as they move out.

are produced beneath them, and as they mature they undergo a characteristic set of changes that are closely associated with the production of keratin proteins which accumulate within the epidermal cells. It is now recognized that "keratin" represents a large family of proteins. Historically, keratins have been divided into A and B types,and, like the collagens, each general type of keratin is an aggregate which can be composed of many varieties of polypeptide subunits.

Cells of the basal layer (Fig. 10-15) contain the constellation of cytoplasmic organelles associated with protein synthetic activity. In addition, they also contain scattered bundles of 6- to 8-nm-thick intermediate filaments which represent aggregates of subunits of keratin and are characteristic of the differentiating epidermal cell.

As the epidermal cells (*keratinocytes*) are pushed outward with the next layer (*stratum spinosum*), they develop extensive networks of keratin filaments which converge on the *desmosomes*, the small patchlike structures that bind one epidermal cell to its neighbor. The outer cells of the spinous layer begin to accumulate another marker of epidermal differentiation, *keratohyalin granules*.

Keratohyalin granules are prominent components of the *stratum granulosum* of the epidermis. These granules, which are easily seen with the light microscope, are composed of a number of proteins, and their function is just beginning to be understood. By the time the keratinocytes have moved out to the granular layer, the nuclei develop the classical signs of terminal differentiation: They become compressed, the nuclear chromatin becomes dense, and the nuclear membrane shows signs of breaking up. Not only do the keratin filaments become more prominent, different keratin subunits are synthesized on different types of RNAs (Fuchs and Green, 1980). In general, the keratin subunits that are made in more mature keratinocytes have higher molecular weights than do those synthesized in the basal cells.

The keratinocytes next pass through a thin transitional layer, where the nuclei are lost and the cells become noticeably flattened. At this point the cells become little more than flattened bags of keratin filaments which constitute the outer *stratum corneum*. The cells of the stratum corneum are held together by a histidine-rich protein, *filaggrin*, which is derived from a component of the keratohyalin granules that is secreted into the intercellular spaces. The cells of the stratum corneum typically accumulate to about 15 to 20 layers, but the thickness of the layers varies considerably over the body. Ultimately, the desmosomes and the intercellular material that hold these cells together become degraded and the surface cells are shed. Acccording to Marks et al. (1983), 1309 cells/cm^2/hour are shed from the surface of the forearm. In buildings, much of the shed epidermis accumulates as house dust.

Recent work (Dale et al., 1985) has shown a close correlation between the morphology of the developing human epidermis and the production of specific keratin proteins (Fig. 10-16). In the embryonic period (<9 weeks), the two-layered epidermis, which consists of a basal layer and the periderm, contains three keratins (40-KD, 45-KD, and 52-KD) which are found in simple epithelia and two keratins (50-KD and 58-KD) which are markers of stratified

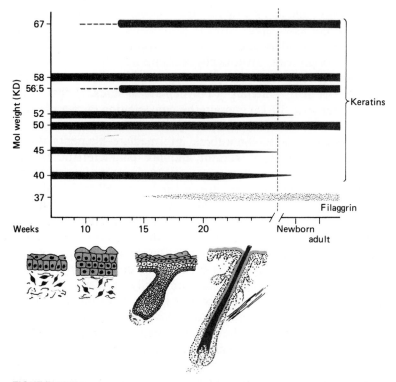

FIGURE 10-16
Diagram showing the expression of keratins and filaggrin during human fetal skin development. The lower figures show approximate stages in hair development at the times indicated. (*Graph adapted from Dale et al., 1985*, J. Cell Biol. **101**:*1266*.)

epithelia. During the period of epidermal stratification (9 to 12 weeks), when the intermediate layer appears and the epidermis becomes truly stratified, traces of 56.5-KD and 67-KD keratins, which are characteristic of keratinized epidermis, are detectable in the outer cells of the intermediate layer. The next period of epidermal differentiation, the period of follicular keratinization (13 to 32 weeks), is characterized by a major increase in the amounts of 56.5-KD and 67-KD keratins in the cells that constitute the developing hair follicles. Filaggrin, which is an important component of keratinized epidermis, also appears at this time. During the remainder of the fetal period, when the epidermal cells in the interfollicular areas become keratinized, the simple epithelial keratins (40-KD, 45-KD, and 52-KD) decline and then disappear around the time of birth.

The order of appearance of keratin types, i.e., those of simple vs. stratified vs. keratinized epithelia, correlates well with the morphological differentiation of the epidermis in both the embryo and the adult. The fact that the keratins characteristic of stratified and keratinized epithelia appear just before the appearance of the corresponding epithelial types suggests a close relationship between their synthesis and structural changes in the epidermal cells. Interestingly, all the keratins found in prenatal epidermis can be found in adult epithelia, suggesting that unique embryonic keratin isoforms may not exist. This is in marked contrast to many other tissues, such as skeletal muscle (see Chap. 9).

In normal epidermis there is a stable equilibrium between the production and the loss of epidermal cells, and recently it has been recognized that typical mammalian epidermis consists of small cellular domains called *epidermal proliferative units*. An epidermal proliferative unit consists of a hexagonal array of mitotically dividing basal cells and is capped by a highly regular column of flattened cornified cells (Fig. 10-17). Within a proliferative unit basal cells divide, and as the daughter cells move into the spinous layer, they flatten to form the base of the cellular column that covers the basal cells of the unit (Potten, 1974).

Considerable attention has been directed toward factors that control the proliferation of epidermal cells. Two distinctly different strategies for regulating the mitotic rate have been proposed. One requires a positive stimulus to stimulate mitosis in the local epidermal cells. Such a stimulatory agent has been identified. While investigating the production of *nerve growth factor* (see Chap. 12), Cohen (Cohen and Elliott, 1963) isolated from the mouse submaxillary gland a protein (*epidermal growth factor*) which stimulates epidermal mitosis and also accelerates the keratinization of embryonic epidermis.

The other main strategy of mitotic control in the epidermis involves the principle of negative feedback. It is assumed that the basal epidermal cells have an intrinsic tendency to multiply until they are told not to do so. According to the *chalone theory* (Bullough, 1972), differentiating epidermal cells produce a tissue-specific but not species-specific molecule (chalone) that diffuses to the basal cells and inhibits their proliferation if its concentration is high enough. According to this theory, if the superficial epidermal cells are scraped off, the

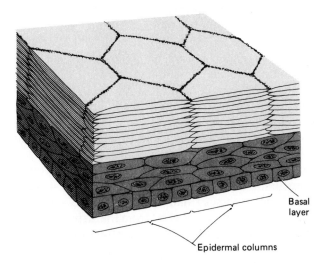

Basal
layer

Epidermal columns

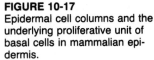

FIGURE 10-17
Epidermal cell columns and the underlying proliferative unit of basal cells in mammalian epidermis.

concentration of chalone will decrease, allowing the basal cells to divide more rapidly than normally. This quickly replaces the lost cells. As the number of new cells approaches normal, the chalone that they produce will reach normal concentrations and restrain further proliferation of the basal cells. The validity of the chalone theory (and even the existence of chalones) has been vigorously debated in recent years, but the concept of regulation of growth of an organ by products produced by its own cells (autoregulation of growth) is an important one.

Normally the life span of a human epidermal cell from generation to its being shed from the surface of the skin is about 4 weeks. However, some skin diseases stem from a lack of control of epidermal proliferation. For example, in *psoriasis* epidermal cells are shed less than a week after they are generated.

PIGMENTATION PATTERNS IN THE SKIN

Biological diversity is nowhere more evident than in the types and patterns of pigmentation of animals. The myriad colors and patterns one finds in animals of a typical rain forest or coral reef are attributable to a class of cells called *chromatophores* which originate in the neural crest and from there migrate in well-defined, species-specific patterns to their final peripheral destinations (see Chap. 6). There are three major types of chromatophores:

1 The *melanophores* give yellowish-brown, brown, or black coloration because of their content of *melanin* or a derivative of melanin.

2 *Xanthophores*, containing carotenoid or pteridine pigments, are responsible for the yellow to red colors.

3 *Iridophores* produce metallic effects (silvery or bluish) because structural elements within the cells reflect light.

Other colors, such as green, are reflected blue light interacting with yellow xanthophores. The white light one sees in some hairs and feathers is caused by the scattering of light by air spaces in the epidermal appendages.

By far the majority of developmental studies of pigmentation have been concerned with black pigment. The remainder of this section will deal with studies that have shed some light on the mechanisms involved in the genesis of well-defined patterns of pigmentation in amphibians and birds. In all these studies, two fundamental questions have underlain much of the experimentation. The first is how much of a pigment pattern is attributable to genetic information that is contained in individual pigment cells, and the second is the role of the environment surrounding the pigment cells in determining their distribution and expression of color.

An early approach to the problem of the genesis of pigment patterns involved transplantations of pigment precursor cells (actually segments of neural crest) between embryos of striped (*Taricha torosa*) and non-striped (*T. rivularis*) salamanders (Twitty, 1949). When a piece of neural crest from the striped species is grafted onto an early embryo of the nonstriped form, the larva develops a stripe on the flank skin beneath the graft (Fig. 10-18). Conversely, a neural crest

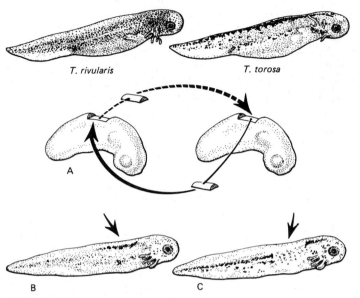

FIGURE 10-18
Interspecific neural crest grafts between *Taricha rivularis* (left) and *T. torosa* (right). The grafted neural crest cells form pigment patterns characteristic of the donor (scattered distribution in *T. rivularis* and a stripe in *T. torosa*). (*Adapted from V. Twitty, 1966, Of Scientists and Salamanders, Freeman.*)

graft from the nonstriped species into an embryo of the striped one results in a larva with a patch of flank skin containing evenly distributed melanophores. Since the pigment cells from the neural crest grafts migrate through host tissues, this experiment shows that much of the pattern of pigmentation can be attributed to intrinsic properties of the migrating pigment cells.

An additional experiment provided further evidence suggesting that the intrinsic properties of migrating pigment cells are very important. Normally, pigment cells spread ventrally from both sides of the neural crest (Fig. 10-19A). Twitty removed the neural crest and grafted it halfway down the flank (Fig. 10-19B). He found that pigment cells spread out from both sides of the graft. Those on one side of the graft migrated dorsally (opposite to their normal direction of migration) over the top of the embryo and then down the other side.

An in vitro experiment suggested that mutual repulsion of pigment cells may be responsible for part of their behavior. Individual pigment cells were sucked into fine-bore capillary tubes and remained almost motionless (Fig. 10-20A). However, when two or three cells were sucked into a tube, they began to migrate away from one another.

A more difficult question is why the pigment cells in *T. torosa* form a dorsolateral stripe, whereas those of other species (e.g., *T. rivularis*) do not. A secondary question is why the stripe is located dorsolaterally and not ventrally.

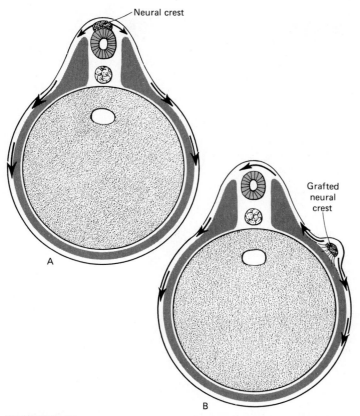

FIGURE 10-19
(A) In normal development of the salamander, neural crest cells migrate
from the neural crest ventrally just beneath the ectoderm. (B) When a
segment of neural crest is grafted to the flank of a host, some neural
crest cells migrate dorsally and then ventrally down the other side of the
host (arrows). This experiment shows that the normal neural crest pathway allows migration in either direction. (*Adapted from V. Twitty, 1966,
Of Scientists and Salamanders, Freeman.*)

When the melanophores of *T. torosa* first migrate away from the neural crest, they
become dispersed evenly over the flank, but then they secondarily reaggregate to
form a stripe. The reaggregation behavior is closely correlated with the full
differentiation of the early pigment cells into melanophores and the adhesion and
retraction of other melanophores toward the most differentiated areas. Such
cellular behavior ultimately results in the concentration of most of the melanophores into a well-defined stripe that is located near the dorsal margin of the somite
(Fig. 10-21). That some feature of the dorsal margin of the somite is important in
determining the location of the stripe has been strongly suggested by an experiment in which Twitty first removed the neural crest of a donor embryo and then
removed a segment containing somite, neural tube, and notochord (Fig. 10-22).

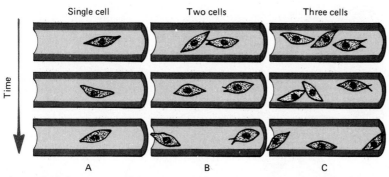

FIGURE 10-20
Experiment on migration of newt pigment cells in a capillary tube. (A) A single
cell does not migrate. (B) and (C) Two or three cells placed in a tube migrate
away from one another, suggesting that they may respond to diffusible substances that they secrete. (*Adapted from V. Twitty, 1966, Of Scientists and Salamanders, Freeman.*)

This segment was turned and grafted into a host embryo so that the neural tube and
dorsal margin of the somite were located in the midflank region. The pigment cells
migrated out from the neural crest of the host and over the graft. When the phase
of secondary reaggregation of melanophores took place, they were localized along
the dorsal margin of the somite of the graft, in the midflank of the embryo, showing

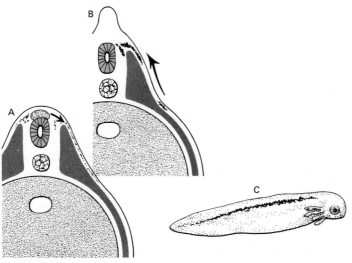

FIGURE 10-21
The sequence of events leading to the formation of a dorsal band in *Taricha
torosa*. (A) Early, pigment cells (dark dots) are distributed along the flank.
(B) Later, they migrate dorsally and aggregate into a dorsolateral stripe (C).
(*After V. Twitty, 1966, Of Scientists and Salamanders, Freeman.*)

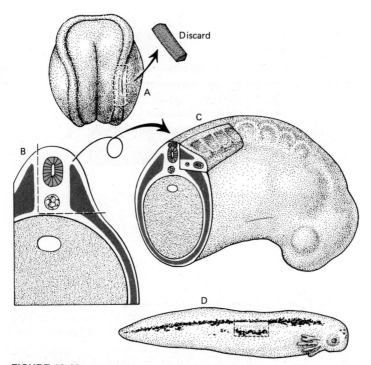

FIGURE 10-22
An experiment showing the role of local substrate in determining the pattern of pigment-cell migration from the neural crest. (A) The neural folds (source of neural crest cells) of a donor *T. torosa* embryo is removed. (B) A block of tissue containing neural tube, notochord, and somites is removed and turned 90° (as well as reversed along the anteroposterior axis) and placed into a normal host (C). Pigment cells from the neural crest of the host migrate into the graft and form a dorsolateral stripe in a position appropriate for the grafted, not the host, tissues (rectangle in D). (*Adapted from V. Twitty, 1966, Of Scientists and Salamanders, Freeman.*)

that properties of the environment as well as of the pigment cells themselves are important in determining the pattern of pigmentation.

These experiments were performed long before the era of increased awareness of the extracellular matrix, and the nature of the environment surrounding the pigment cells received little attention. Most recent work (see also the section on the neural crest in Chap. 6) has suggested a more prominent role of the environment in the establishment and maintenance of pigment patterns than was suspected by earlier investigators. For example, the belly skin of *Xenopus* is white. Yet Ohsugi and Ide (1983) have shown that incompletely differentiated melanophores (*melanoblasts*) are present in the white belly skin. This was done by incubating ventral skin in dopa (3,4-dihydroxyphenylalanine), a precursor of melanin. The treated cells become dark. Thus in *Xenopus*, an environmentally influenced persistence of an immature stage of differentiation of melanophores provides the basis for the dorsoventral pattern of pigmentation.

Studies of pigmentation patterns in feathers have reinforced some of the conclusions derived from the studies of amphibians (Rawles, 1948). Feathers become pigmented by means of the migration of melanoblasts into the feather papillae and their formation of melanin granules at a later time. When neural crest–derived melanoblasts from a Barred Rock chicken are grafted to the base of the wing bud of a White Leghorn host, they spread throughout the developing wing and become incorporated into the developing feathers. As the melanocytes differentiate and produce visible pigment, the pattern of pigmentation is that of the donor from which the melanocytes were derived, in this case a barred wing (Fig. 10-23). The conclusion that information intrinsic to pigment cells determines pattern was strengthened by an experiment involving the transplantation of melanoblasts derived from strains of chickens with sex-linked differences in plumage pattern (Willier and Rawles, 1944). Melanoblasts derived from male donors form a male pattern of pigmentation and those from female donors form a female pattern in a white host regardless of the sex of the host. That the body of the host provides information that determines the regional characteristics of the pigment pattern is readily seen when pigment cells from a striped bird are introduced into breast feathers of a white host. The feathers become striped, but the stripes are narrower than those of wing feathers.

How banding in feathers occurs is not completely understood because the white regions between colored bands also contain melanocytes which do not express pigment characteristics. Nickerson (1944) suggested that the periodic production of bars could be due to the production of inhibitory substances by the melanocytes of a pigmented band in the feather papilla. The net effect would be to inhibit the expression of melanin by pigment cells within a certain range of the inhibitor. As the feather bud grows out, the concentration of inhibitor at its base is reduced, and after a certain width of white, the more proximal melanocytes are again able to produce melanin granules. This is not the only possible explanation. Another possibility is that an essential precursor of melanin is used up by the melanocytes of a pigment band, thus depriving the melanocytes nearby of the opportunity to form pigment. Classic systems such as this are awaiting reinvestigation with modern techniques and thought patterns as clear as those of the investigators who first defined these systems of analysis.

MAMMARY GLANDS

A number of glandular structures are derived from epidermal downgrowths and are the result of inductive interactions between the epidermis and underlying mesenchyme. Among these are sweat glands, some scent glands, sebaceous glands, mammary glands, and in a broad sense, teeth. Of these structures, teeth and mammary glands have received the most experimental attention. Teeth are dealt with in Chap. 14. This section will concentrate on mammary glands,

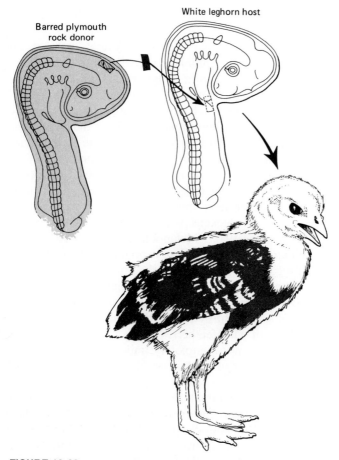

FIGURE 10-23
Results of grafting melanoblasts from a Barred Rock chick embryo
donor into a White Leghorn host embryo. The grafted cells spread
out into the wing and formed a pigment pattern characteristic of
the donor. (*Based on the experiments of Rawles, 1948.*)

which have been extensively investigated with respect to inductive interactions, the extracellular matrix, and hormonal influences on their development.

The first indication of mammary gland development is the appearance of a
pair of bandlike ectodermal thickenings (*milk lines*) that run from the region of
the armpit (*axilla*) to the groin (*inguinal region*) (Fig. 10-24A). The actual
number and location of individual mammary glands vary considerably from one
species to another, but they are all located somewhere along the length of the
milk lines. Some mammals, such as pigs and dogs, develop a series of pairs of
mammary glands spread out along the length of the milk lines. In others,
mammary tissue is confined to a given region along the milk line, for example,
in the pectoral region in humans or near the posterior ends of the milk line in

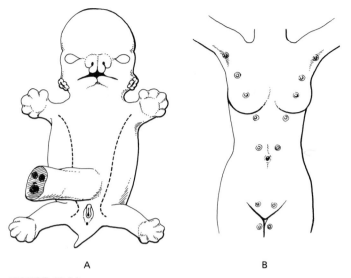

FIGURE 10-24
(A) The "milk line" (dashed lines) in a generalized mammalian embryo.
Mammary glands form along this line. (B) Common sites of formation of
supernumerary nipples along the course of the milk line in the human.

cattle and whales. In humans, supernumerary nipples can develop anywhere
along the original milk line (Fig. 10-24B).

In humans, the original milk line is a raised ridge of ectodermal cells underlain
by a slight concentration of mesodermal cells (Fig. 10-25). As mammary gland
development continues, a solid plug of ectodermal cells at the site of the future
nipple pushes its way into the subjacent mesenchyme. Soon a cluster of branching
ectodermal cords foreshadows the formation of the hollow mammary ducts. Hogg
et al. (1983) have stressed the consistency with which developing glandular and
other tubular structures begin as solid cords of epithelium invading an underlying
connective tissue, before a system of hollow tubules forms. Other examples of
invasive epithelia in the form of solid cords penetrating into mesenchyme are the
liver, pancreas, salivary and thyroid glands, vascular endothelium, and hair
follicles. Similarly, invasive epithelial tumors assume the form of solid clumps of
cords of cells as they invade connective tissue.

The initial ingrowth of mammary duct primordia into the underlying mesenchyme is the result of an ectodermal-mesenchymal inductive interaction of
the type that is so common in the formation of many glandular structures in the
body. The mammary mesoderm is the inductor, and the ectoderm is the
responding tissue. There are two types of mammary mesoderm, a fibroblastic
mammary mesenchyme that closely surrounds the epithelial duct primordia
and a precursor of the mammary fat pad. Recent work suggests that interaction
with tissue of the mammary fat pad rather than the fibroblastic mesoderm is the
major factor that results in the shaping of the characteristic mammary duct

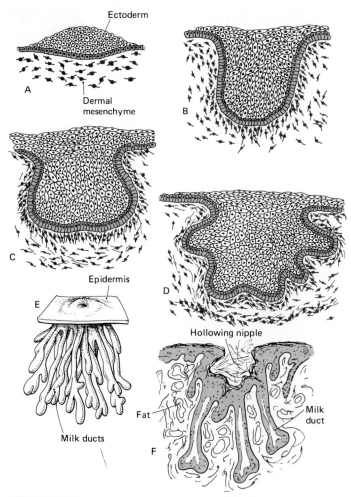

FIGURE 10-25
Stages in the histogenesis of the human mammary gland. (A) Sixth week; (B) seventh week; (C) tenth week; (D) fourth month; (E) sixth month; (F) eighth month. (*After Patten.*)

system (Sakakura et al., 1982). As is the case with other glandular tissue, it is becoming increasingly apparent that the extracellular matrix is a major mediator of the inductive message. The mesoderm seems to control the pattern of branching of the mammary ducts, but the nature of the ductal epithelium is an intrinsic property of mammary ectoderm. This is illustrated by an experiment in which mouse mammary ectoderm was combined with salivary gland mesenchyme (Sakakura et al., 1976). Although the expanding mammary ducts assumed the branching pattern of salivary gland epithelium, the epithelial cells were shown to synthesize α-lactalbumin, one of the milk proteins.

Not surprisingly for a structure that is a prominent secondary sexual characteristic, the developing mammary gland is exquisitely sensitive to its hormonal environment. One of the earliest hormonal effects is a negative one. Although rudimentary ductal tissue persists in the human male, the mammary primordia of male mice disappear early in development as a result of testosterone-induced cell death. Experimental analysis has shown (1) that male and female mammary epithelial primordia are equally sensitive to the effects of testosterone and (2) that testosterone does not act directly upon the ductal epithelium; instead, its effect is mediated through the mammary mesenchyme.

The first point was established by culturing female mammary primordia in the presence of testosterone (Kratochwil, 1971). The early mammary bud is cut off at the neck, where it joined the surface ectoderm, just as is the case in the normal male mammary primordium. Conversely, if a male gland rudiment is cultured in the absence of testosterone, it develops like a female gland (Fig. 10-26).

The role of the mesenchyme in mediating testosterone-induced repression of mammary ductal epithelium was demonstrated with the help of a genetic mutant called *androgen insensitivity syndrome* (*testicular feminization*), which is found in both mice and humans. In this condition functional testosterone receptors are missing from XY males. Although the testes produce a great deal of testosterone, the tissues cannot respond to it (see Chap. 17). One of the results is that males have typical female breast development. Recombinants have been made with normal and mutant tissues (Kratochwil and Schwartz, 1976). In the presence of testosterone, mutant ectoderm combined with normal mesenchyme regresses, whereas normal ectoderm combined with androgen insensitive mesoderm goes on to develop normal ducts regardless of the genetic sex of the ectoderm (Fig. 10-26). The ability of testosterone to influence mesoderm to cause the death of the epithelium in mice is confined to mammary tissue. In the presence of testosterone, mammary mesoderm will not result in the death of nonmammary ectoderm, nor will nonmammary mesoderm kill mammary ectoderm (Durnberger and Kratochwil, 1980).

Further development of the female mammary gland is closely tied to its hormonal environs (Topper and Freeman, 1980). After the initial simple duct system has been laid down in the embryo, the hormonally unstimulated mammary gland remains in an infantile condition until puberty (Fig. 10-27A). Under the influence of estrogens and against a background of a number of other hormones, the ducts proliferate and branch and the pad of fatty tissue beneath the ducts begins to enlarge (Fig. 10-27B). The mammary gland remains in this mature but resting condition until pregnancy, when increased amounts of progesterone, as well as prolactin and placental lactogen, stimulate the development of secretory alveoli at the ends of the branched ducts (Fig. 10-27C). As the alveoli develop, the individual epithelial cells produce increasing amounts of the cytoplasmic organelles associated with protein synthesis and secretion, and large numbers of microvilli appear on their apical surfaces.

Active lactation, which involves the synthesis of milk proteins (*casein* and α-*lactalbumin*) and lipids, depends on the release of prolactin from the anterior

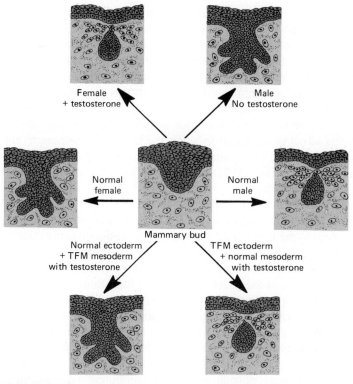

FIGURE 10-26
Roles of genetic specificity and testosterone in the development of mouse mammary gland tissue. With normal tissues, addition of testosterone to the female mammary rudiment (top left) causes prospective duct tissue to detach and regress, as in normal male development. Conversely (top right), male rudiments assume a female configuration in the absence of testosterone. In the testicular feminization (TFM) mutant, if normal mammary ectoderm is cultured with TFM mammary ectoderm in the presence of testosterone, mammary duct epithelium proceeds to develop (lower left). In contrast, if normal male mammary mesoderm is combined with TFM ectoderm in the presence of testosterone, the normal male pattern of separation and regression of mammary ductal epithelium occurs. This shows that the genetic defect is expressed in TFM mesoderm. (*After the experiments of Kratochwil, 1971.*)

pituitary (Fig. 10-27C). Prolactin release is controlled by a complex set of interactions that begins with the tactile stimulation of sensory nerves in the nipples by the nursing infant. This stimulus is carried to the hypothalamus and suppresses the release of *prolactin-inhibiting factor* by the hypothalamus. The absence of negative feedback inhibition by the hypothalamus allows the anterior pituitary gland to secrete prolactin. The actual ejection of milk is triggered by the release of *oxytocin* by the posterior pituitary gland. This is a rapid effect of the suckling stimulus. The immediate cause of milk ejection is the contraction of smooth-muscle-like *myoepithelial cells*, which ring the

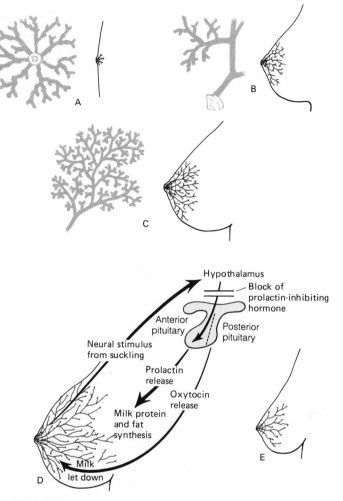

FIGURE 10-27
Diagrams showing development of the duct system and hormonal control of the human mammary gland. (A) newborn; (B) young adult; (C)
adult; (D) lactating adult; (E) postlactation.

mammary alveoli and are sensitive to oxytocin. Another effect of suckling is the
inhibition of release of LH-RH (*luteinizing hormone–releasing hormone*) by the
hypothalamus. This inhibits ovulation and serves as a natural form of birth
control.

After nursing ceases, a reduced release of prolactin, along with the presence
of nonejected milk within the mammary gland, results in the cessation of milk
production. In time, the mammary alveoli regress and the duct system of the
mammary gland involutes to the state that it was in before pregnancy (Fig
10-27E).

CHAPTER 11

LIMB DEVELOPMENT

The appendages of vertebrates have undergone a remarkable degree of diversification during the unfolding of the vertebrate classes. Paleontological evidence suggests that the earliest appendages consisted of a pair of fin folds that ran along the entire length of the body of some of the most primitive fishes. Soon this arrangement gave way to discrete outgrowths, the pelvic and pectoral fins, which may represent the localized persistence of portions of the original fin folds. With the spreading of vertebrates into terrestrial habitats, their limbs took on a considerably different configuration from that of their aquatic forebears. Since then, the fundamental arrangement and the pattern of development of the tetrapod limb have been preserved despite the evolutionary radiation that has led to the appearance of structures as diverse as flippers, wings, hooves, and the human hand.

INITIATION OF LIMB DEVELOPMENT

The limb arises as a condensation of cells from the lateral plate mesoderm and its ectodermal covering (Fig. 11-1). The limb primordia in amniotes appear along the *Wolffian ridges* (Fig. 11-2), which run along the lateral surface of the body. The limb does not begin to form until the primordia of most other organs are laid down and the musculature of the body wall beneath the undifferentiated limb bud is already well differentiated. This is reminiscent of the phylogenesis of the chordates, in which appendages arose only after the chordate line was well established. The cephalocaudal rule of development applies to the limbs, and at any given time the anterior limbs are more advanced than the posterior ones.

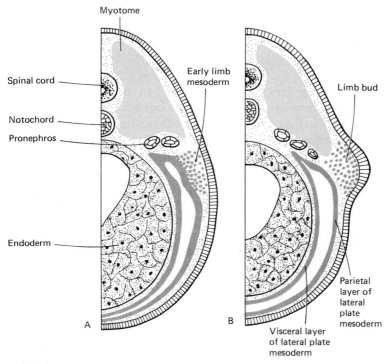

FIGURE 11-1
Diagrams illustrating the origin of the amphibian limb bud. (*After Balinsky,
1975, An introduction to Embryology, 4th ed. W. B. Saunders, Philadelphia*)

Some form of activation of the lateral mesodermal cells seems to be required
for the initiation of limb development, but the nature of the stimulus is poorly
defined. The early limb primordium is located close to the somites, and there is
evidence that the presence of somites is necessary for limb development to
occur. The proximity of the pronephros to the limb primordium is intriguing,
but definitive evidence for a major causal relationship between pronephros and
limb is lacking.

After the mesodermal cells of the early limb primordium are activated, they
next act upon the overlying ectoderm. In most classes of vertebrates the
ectoderm responds to the influence of the future limb mesoderm by thickening.
This early action of the limb mesoderm on the ectoderm occurs before the limb
primordium begins to project beyond the surface of the body wall as the limb
bud. At this early stage the flat-surfaced limb primordium is known as the *limb
disk*.

Experiments have demonstrated that the mesoderm rather than the ectoderm is the prime mover in early limb development. If limb mesoderm is
combined with the early flank ectoderm, the ectoderm thickens and partici-
pates in the formation of a limb. In contrast, a combination of limb ectoderm

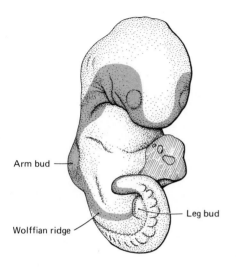

FIGURE 11-2
Ventrolateral view of a 30-somite (4.2-mm) human embryo, showing the thickened ectodermal ring (gray). The portion of the ring between the upper and lower limb buds is known as the Wolffian ridge. (*After O'Rahilly and Gardner, 1975, Anat. Embryol.* **148**:*1*.)

with flank mesoderm does not result in limb development. These experiments not only illustrate the primacy of limb mesoderm, they also show that at this early stage the ectoderm of the body is still not completely determined, because ectoderm that would normally form flank skin can still assume the specializations that are involved in promoting outgrowth of the limb.

REGULATIVE PROPERTIES OF THE EARLY LIMB PRIMORDIUM

The limb primordium is a *self-differentiating system*; i.e., once established, it contains all the information required to attain its normal form even if it is removed from the body of the embryo. The early *limb disk* is also a highly regulative system. This has been shown by some simple but highly instructive experiments (Fig. 11-3):

1 When part of a limb disk is removed, the part that is left continues to develop into a perfectly normal limb.

2 When a limb disk is split into two halves which are prevented from fusing, each half gives rise to a complete limb.

3 When two equivalent halves of a limb disk are placed together, a single limb forms.

4 When two harmonious limb disks are superimposed, the cells become reorganized so that they form only a single limb.

5 Disaggregated limb mesoderm that is grafted to the flank reorganizes and often forms a normal limb.

These experiments demonstrate that the limb primordium possesses regulative properties very similar to those of entire early embryos, such as the sea urchin (on which Hans Driesch's experiments defining regulative characteristics were

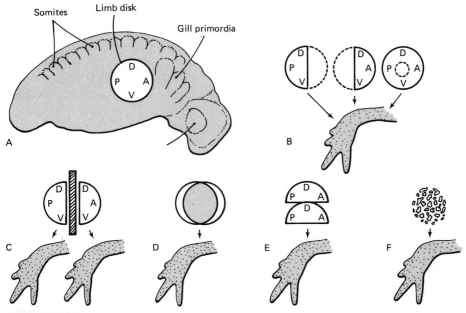

FIGURE 11-3
Experiments demonstrating the regulative properties of the limb disk of the early amphibian embryo. (A) Normal embryo. (B) Normal limb development after removal of either half or the central region of the limb disk. (C) When halves of the limb disk are separated by a barrier, each half develops into a complete limb of the same polarity. (D) Superimposition of two limb disks of the same polarity results in the formation of one normal limb. (E) Combining two identical halves of a limb disk results in a single limb. (F) Mechanical disruption of the limb disk followed by its reaggregation is followed by normal development. (*After Harrison and Swett.*)

first made) and the mammalian blastocyst, which has received careful study only in recent years.

AXIAL DETERMINATION OF THE LIMBS

The limb is an asymmetric structure which is commonly defined in terms of three axes--the proximodistal (P-D), the anteroposterior (A-P), and the dorso-ventral (D-V). Experiments conducted early by Ross Harrison in this century demonstrated clearly that the individual axes of the limb are fixed (determined) at different times of development. This was done by rotating the limb disks of *Ambystoma* embryos (Fig. 11-4). When a right forelimb disk was rotated 180° around its center (Fig. 11-4A), the limb that formed looked like a normal left arm growing from the right side of the body. A left limb disk grafted into the place of the right disk with only its A-P axis reversed with respect to that of the body of the host also developed into a normal-looking left arm, whereas a left arm disk grafted to the right side with only the D-V axis reversed developed into a normal right arm. In all three of these experiments development along the

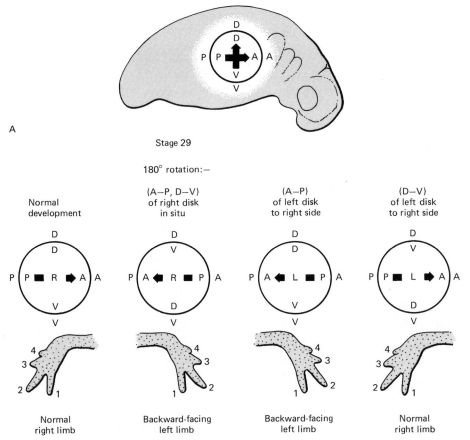

FIGURE 11-4
Diagrams of experiments demonstrating axial determination in the limb disk of the amphibian (*Ambystoma*) embryo by means of grafting and rotation of the limb disk with respect to the surrounding tissues. (A) Experiments performed on early embryos showing that only the anteroposterior axis is determined. (B) Experiments performed on older embryos, showing that both the anteroposterior and dorsoventral axes are determined. For details, see the text. (*After Harrison and Swett.*) Abbreviations: *A*, anterior; *D*, dorsal; *P*, posterior; *V*, ventral; *L*, left; *R*, right.

A-P axis of the early limb disk proceeded on the basis of information intrinsic to the rotated limb disk. On the other hand, development along the D-V axis was dictated by the body of the host, not by axial information residing within the disk. Relations along the P-D axis were not altered in these experiments. These results showed that the A-P axis of the early limb disk was determined but that the D-V axis was not. Other work showed that the P-D axis was also undetermined at this time.

At a later stage of embryonic development the same operations as those outlined here were performed (Fig. 11-4B). The types of limbs that grew out from these rotated primordia differed from those in the previously described

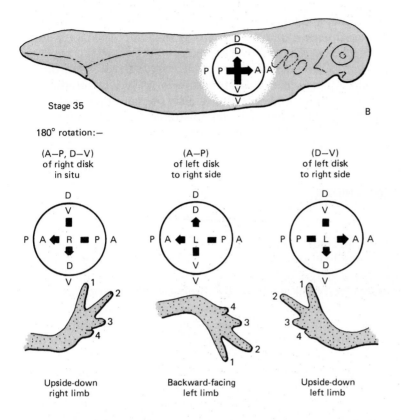

Stage 35

B

180° rotation:−

(A−P, D−V) of right disk in situ	(A−P) of left disk to right side	(D−V) of left disk to right side

Upside-down right limb

Backward-facing left limb

Upside-down left limb

series. Now both the A-P and the D-V axes of the outgrowing limbs corresponded entirely with the original axes of the rotated limb primordia and were independent of any influence by the body.

Additional experiments of a similar nature showed that the P-D axis was determined last. Thus it was established that in the limb the sequence of axial determination is A-P→D-V→P-D (Swett, 1937). This sequence in the limb may represent a general one in the major external organs, for a similar sequence of axial determination occurs in the developing ear and in the retina of the eye.

OUTGROWTH OF THE LIMB BUD

As soon as the limb primordium begins to project from the surface of the embryo, it may be properly called a *limb bud*. Structurally, the *limb bud* consists of a mesodermal core of homogeneous-appearing mesenchymal cells, which are supplied with a capillary network. Nerve fibers have not yet grown into the early limb bud. Covering the mesodermal core of the limb bud is a sheet of ectoderm, which in most vertebrates is thickened at the apex of the limb bud.

In 1948, Saunders conducted an illuminating study that opened the way for a vast amount of subsequent experimentation on the mechanism of limb development. He was interested in discovering the role, if any, of the band of thickened ectoderm (*apical ectodermal ridge*) that covers the convex outer margin of the wing bud in chick embryos. He removed the apical ectodermal ridge from the wing buds of chick embryos, and shortly thereafter outgrowth of the limb ceased (Fig. 11-5). This experiment demonstrated a critical interaction between the apical ectodermal ridge and the underlying mesodermal core during growth of the wing bud.

APICAL ECTODERMAL RIDGE

Structure of the Ridge

The appendage buds of most vertebrates possess a terminal thickening. Typically, the thickening is in the form of an apical ectodermal ridge running along the anteroposterior margin of the bud. There is remarkable uniformity in the shape and appearance of the apical ridge throughout the vertebrates. In birds and mammals it commonly stands out as a nipple-shaped structure in cross section (Figs. 11-5 and 11-6), whereas in fishes and some reptiles the apical epidermis appears to be thrown into a tight fold. The amphibians show an unusual dichotomy in the morphology of the limb ectoderm. Anuran amphibian embryos have apical ectodermal ridges that are of the same general form as those of birds and mammals. Urodeles, on the other hand, do not have apical thickenings on the limb buds (Fig. 11-6C). Why this latter group does not have apical ridges and how limb outgrowth occurs in their absence has not been explained.

The apical ridge begins to take shape shortly after the limb bud begins to project from the lateral surface of the embryo. It is oriented along the anteroposterior (pre- to postaxial) plane of the limb bud and is somewhat symmetrical in shape; i.e., it is thicker along its posterior portion in birds. The apical ectodermal ridge typically attains its greatest degree of development when the limb is at the paddle-shaped stage of development. Thereafter, as the digits begin to differentiate, the ridge begins to regress.

There are two principal histological configurations of the apical ectodermal ridge in higher vertebrates. In birds it is arranged as a pseudostratified columnar epithelium (Fig. 11-5B), whereas in mammals it is a stratified cuboidal or squamous epithelium. Kelley and Fallon (1976) have described in detail the structure of the human apical ectodermal ridge during its major stages of development and decline (Fig. 11-7). One of the major structural characteristics of the apical ectodermal ridge is the presence of numerous gap junctions connecting the cells of the ridge with one another (Fig. 1-19). Fallon and Kelley (1977) have postulated that the cells of the apical ectodermal ridge are electrically and metabolically coupled through the gap junctions. Histochemical studies (Milaire, 1969) have shown that the cells of the apical ridge contain

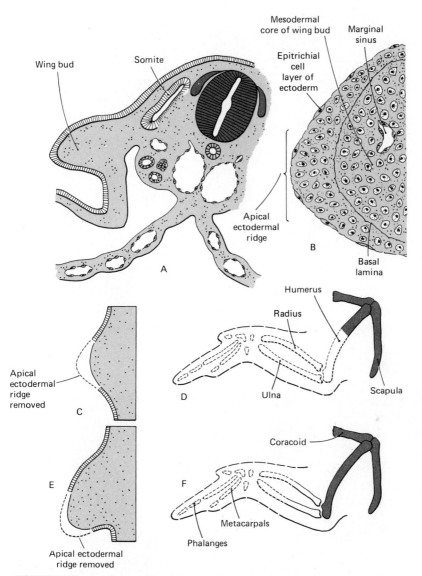

FIGURE 11-5
The effect of the removal of the apical ectodermal ridge on the development of the wing bud. (A) Diagram showing location of the center of mesodermal proliferation involved in forming the core of the wing bud. (B) Cell detail drawing of tip of wing bud of 3-day chick to show the apical ridge.
(C) Ridge removal in 3-day chick. (D) Skeletal deficiency resulting from third day removal of apical ridge. (E) Ridge removal at 4 days. (F) Skeletal deficiencies resulting from fourth-day removal of apical ridge. In D and F the parts of the skeleton which develop are stippled; the parts failing to develop are shown in outline only (*C–F, showing the results of apical ridge removal on skeletal development, based on the work of Saunders, 1948, J. Exp. Zool., 108.*)

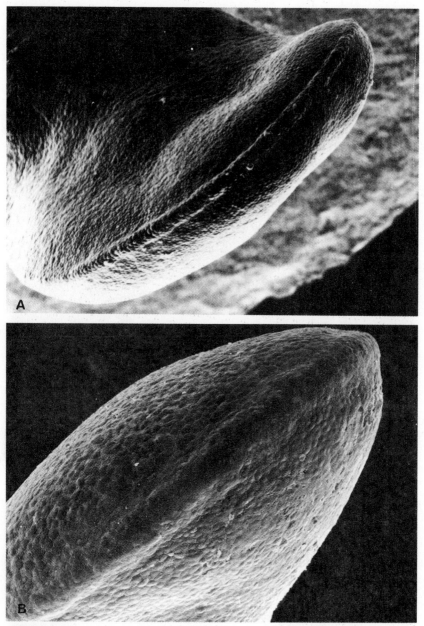

FIGURE 11-6
Scanning electron micrographs of limb buds in the embryos of (A) the chick (stage 26); (B) the human (stage 15); (C) the axolotl (stage 39). Note the pronounced apical ectodermal ridge in the limb buds of the chick and human embryos and the absence of a ridge in the axolotl. (*A, courtesy of Dr. John Fallon; B, from Kelley and Fallon, 1976, Devel. Biol. **51**:241; courtesy of the authors; C, from Tank et al., 1977, J. Exp. Zool. **291**:417.*)

C

large amounts of glycogen and RNA as well as high concentrations of both alkaline phosphatase and acid phosphatase. These chemical properties are characteristic of groups of embryonic cells that are actively involved in growth and differentation.

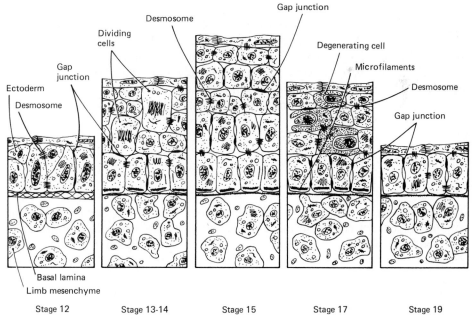

FIGURE 11-7
Schematic diagram illustrating the changes in structure during the building up and reduction of human apical ectodermal ridge. At stage 12, the basal lamina between ectoderm and mesoderm is double-layered, with cross-links. At later stages it is a single-layered structure. (*Adapted from Kelley and Fallon, 1976, Devel. Biol,* **51***:241.*)

Properties of the Apical Ectodermal Ridge

Although there is still considerable discussion concerning its exact role in development of the limb, the apical ectodermal ridge is considered by most researchers to act as a stimulator of outgrowth of the limb bud. Shortly after surgical removal of the apical ectodermal ridge, outgrowth of the limb bud ceases (Saunders, 1948). A genetic wingless mutation of chickens provides a natural experiment that produces similar results (Zwilling, 1949). In this mutant, early wing development is normal but then comes to a halt. Correlated in time are the regression and disappearance of the apical ectodermal ridge. Conversely, additional apical ectodermal ridges, whether applied surgically in the laboratory or provided through the action of a genetic mutant (*eudiplopodia*), result in the formation of supernumerary wing tips in chick embryos.

Whatever its exact role in limb development, the apical ectodermal ridge seems to be remarkably nonspecific in its actions. Surgical exchange experiments have shown that the apical ridge of a chick embryo can promote outgrowth of limb mesoderm not only from other species of birds but also from mouse embryos. The apical ectodermal ridge does not specify the proximodistal level of morphogenesis of the limb but rather acts as a nonspecific stimulator of outgrowth. Rubin and Saunders (1972) grafted apical ridges from older chick leg buds (and consequently from a more distal level) onto young leg buds and found that normal development ensued, despite the difference in age and proximodistal level between the apical ectodermal ridge and the underlying mesoderm. When the apical ridge begins to regress, however, it gradually loses its ability to foster growth of the limb bud. Even when the apical ectodermal ridge is dissociated or turned inside out it can reorganize and fulfill its usual role. Although there is some indirect evidence that the cells of the apical ectodermal ridge secrete diffusible substances into the underlying mesoderm of the limb bud, further evidence must be accumulated before this function can be considered to be established.

MESODERM OF THE EARLY LIMB BUD

Structure

The mesoderm of the early limb bud consists of a homogeneous-appearing mound of mesenchymal cells embedded in a loose matrix of collagen fibers and mucopolysaccharide ground substance. The cells are undifferentiated in appearance, with large nuclei, prominent nucleoli, and scanty basophilic cytoplasm. Mitotic activity of these cells is prominent throughout the limb bud. Beneath the apical ectodermal ridge is a zone, several hundred microns thick, which is rich in ground substance and in which the cells have a high rate of division. This region has been called the *progress zone* (Summerbell et al., 1973), and it may play an important role in growth and morphogenesis of the limb. The limb bud is well vascularized and contains a prominent marginal sinus beneath the apical ectodermal ridge. Nerves are not present within the early limb bud, but nerve fibers grow in as development progresses. Differentiation of the mesoderm will be dealt with later in this chapter.

Role of the Mesoderm

The interaction between ectoderm and mesoderm in the limb bud is not entirely a one-way process. There is evidence that the mesoderm of the early limb bud stimulates the early ectoderm to form the apical ectodermal ridge and that once the apical ridge is established a continuing influence of the mesoderm is required for maintenance of its integrity.

Mutants for winglessness and polydactyly in chickens have furthered our understanding of the important role of the mesoderm. In the wingless mutant early development of the wing bud is normal, but the apical ridge soon degenerates and further outgrowth of the wing bud ceases. Recombination experiments have shown that if "wingless" mesoderm is covered with normal wing ectoderm, the "normal" apical ridge undergoes degeneration and development of the wing is arrested (Zwilling, 1956). Similarly, if limb ectoderm is placed over nonlimb mesoderm, the apical ectodermal ridge regresses. In polydactyly, combinations of normal mesoderm and mutant ectoderm give rise to normal limbs, whereas mutant mesoderm plus normal ectoderm produces polydactylous appendages. Such experiments indicate that the mesoderm continuously stimulates the overlying ectoderm to maintain an active apical ectodermal ridge which in turn stimulates further outgrowth of the mesoderm of the limb bud. Zwilling (1961) has proposed that the mesodermal influence is due to the existence of a mesodermal "maintenance factor," but this factor has not been isolated or defined. Polydactyly in humans is often due to a genetic recessive condition, and it is not uncommon in certain human populations (e.g., certain American Amish communities) where the total genetic pool is relatively restricted.

It has been repeatedly demonstrated that the character of the morphogenesis of the appendage is governed by the mesodermal component rather than by the ectoderm. If wing mesoderm is combined with ectoderm of the foot, a wing covered with feathers develops, whereas if distal hindlimb mesoderm is covered with wing ectoderm, a leg and a foot covered with scales are formed. The mutual interactions between limb mesoderm and ectoderm operate even across species barriers. Limb mesoderm of a duck combined with ectoderm of a chick gives rise to a webbed foot, and early limb bud mesoderm of a mouse forms toes when it is placed in approximation to ectoderm of the wing bud of a chick (Cairns, 1965).

Cell Death in the Developing Limb

Among the many processes involved in the development of the limb bud, cell death is prominent (Saunders et al., 1962). Appearing in the forelimb first in the region of the future axilla (armpit), patches of dying cells are later seen in the

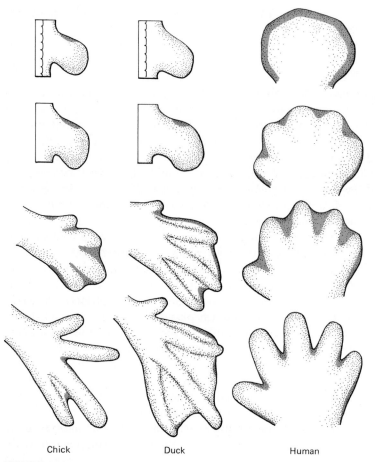

Chick   Duck   Human

FIGURE 11-8
Zones of cell death at various stages of embryonic development of the
chick and duck foot and the human hand. (*Chick and duck after Saunders
and Fallon, 1966, in* Major Problems in Developmental Biology, *ed. M.
Locke, Academic Press, New York; human after Menkes et al., 1965, Rev.
Roumaine Embryol. Cytol.* **2**:*161*.)

elbow region and between the developing digits; in the hindlimb a similar
pattern is seen (Fig. 11-8). The area of cell death in the future axilla of the chick
is particularly well defined in the chick embryo; it is called the *posterior
necrotic zone.* More circumscribed areas of cell death are associated with
differentiation of the skeleton, particularly the joints, and with the histogenesis
of certain groups of muscles. Even in the development of mammalian digits, the
abbreviation of the first digit, which contains only two phalangeal segments as
opposed to three for the other digits, may be due to the establishment of a
localized zone of cell necrosis at the end of the first digital primordium (Milaire,
1977). In cases of webbed digits, whether normal for the species (e.g., the duck)

or as a developmental anomaly (*zygodactyly*), the persistence of webs of soft tissue between the digits is associated with reduced amounts of cell death.

Prominent areas of cell death, particularly in the interdigital areas, have been seen in all major groups of amniote embryos (reptiles, birds, and mammals, including humans), but interdigital cell death has not been reported in amphibian embryos. Cameron and Fallon (1977) speculated that cell necrosis as a developmental mechanism in limb development may have originated phylogenetically in conjunction with the branching of the amniotes from the anamniote line of vertebrates.

The cells that constitute the necrotic areas of the limb buds appear to be genetically determined to die at a particular stage of development. At a certain stage of development, what Saunders calls a *death clock* is set, and the affected cells, although apparently healthy for some time afterward, are committed to dying at a critical stage of development (Saunders, 1969). The visible onset of cell death is preceded by a reduction in protein and nucleic acid synthesis and an increase in the activity of hydrolytic enzymes. Grafting experiments involving cells of the posterior necrotic zone of chick embryos have shown that although the death clock is set at approximately stage 17 of development, reversibility, or a "reprieve of the death sentence," is possible up to stage 22 if the cells of the posterior necrotic zone are grafted to the dorsal side of the wing bud. After stage 22, however, the commitment to death is irreversible, even though obvious signs of cell necrosis are not visible until stage 24.

The Zone of Polarizing Activity

Another discrete area of limb mesenchyme with characteristic properties is located near the posterior edge of the limb bud near its junction to the body wall. The cells of this area, known as the *zone of polarizing activity*, stimulate the formation of a supernumerary appendage if they are grafted to the anterior (preaxial) part of the limb bud (Fig. 11-9). The zone of polarizing activity was first identified in the wing bud of the chick embryo, very close to the posterior necrotic zone of cell death (Saunders and Gasseling, 1968). Since then a polarizing zone has been identified in the hind limb of birds and in the limb buds of urodele and anuran amphibians, reptiles, and mammals, including human beings.

In the wing bud of the chick, on which the vast majority of the experimental work has been performed, polarizing activity, as assayed by grafting techniques, is only weakly detectable at the earliest stages of limb outgrowth (stages 15 and 16). Thereafter it is quite strongly developed until the late stages, when early digital differentiation begins (Fig. 11-10). MacCabe et al. (1977) developed an in vitro assay system showing that the zone of polarizing activity is the source of a diffusible factor (apparently not identical to Zwilling's maintenance factor) which can cause thickening of the cells of the apical ectodermal ridge.

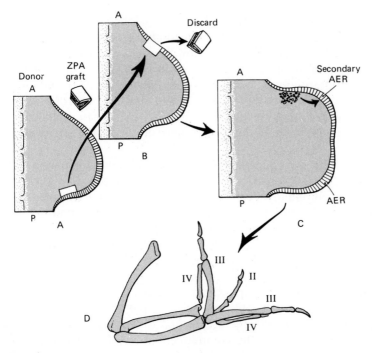

FIGURE 11-9
Scheme illustrating the grafting of a zone of polarizing activity (ZPA) from one wing bud of a chick embryo into the anterior part of another wing bud. A secondary apical ectodermal ridge forms distal to the graft, and a duplicated wing tip develops. The figure on the right shows the skeleton of such a wing and illustrates the mirror image symmetry of the two wing tips. (*After Saunders, 1971, Ann. N.Y. Acad. Sci.* **193**:*29.*) Abbreviations: *A*, anterior; *P*, posterior.

Polarizing activity is not confined to the zone of polarizing activity. Recent work has shown that cells of the somites, Hensen's node, and even the mesonephros of the chick possess polarizing activity. In addition to the tissues mentioned previously, the application of retinoic acid (a vitamin A derivative) to the anterior part of the avian wing bud results in duplications like those caused by grafts of polarizing tissue (Tickle et al., 1982). Despite the unquestioned activity of the cells of the zone of polarizing activity, their role in normal limb outgrowth and morphogenesis is uncertain, for if this zone is excised, development of the appendage proceeds entirely normally. According to one suggestion (Fallon and Crosby, 1975), polarizing activity may play a role in normal development at the earliest inductive stages but not during subsequent stages of limb development. Other opinions either assign a much more important ongoing function during limb morphogenesis or suggest that this zone may not have any function in normal limb development.

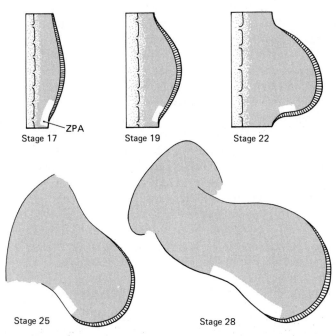

FIGURE 11-10
Schematic drawings (at the same scale) showing the changing location of the zone of polarizing activity (ZPA) in the chick wing at different stages of development. The stages are those of Hamburger
and Hamilton (1951). (*Adapted from Saunders, 1972, Ann. N.Y.
Acad. Sci.* **193**:*29.*)

Morphogenetic Control of the Developing Limb

Despite the considerable experimental progress that has been made in the analysis
of certain aspects of limb development, our understanding of the overall control of
the process is still far from complete. This has led investigators to propose a
number of widely differing hypotheses regarding limb morphogenesis. Detailed
consideration of these theories is beyond the scope of this book, but some of the
more influential ideas will be summarized here. Typically, investigators have
viewed morphogenetic control of the developing limb in terms of three axes, as
defined by the Cartesian coordinate system, and most hypothetical mechanisms
have been confined to explaining development along a particular axis.

Outgrowth along the proximodistal axis has received by far the most
attention. According to the Saunders-Zwilling hypothesis (Zwilling, 1961),
which centers on the interactions between the apical ectodermal ridge and the
underlying mesoderm, the apical ectodermal ridge exerts an inductive action on
the mesoderm, causing it to increase in mass by means of cell proliferation. The
continued activity of the apical ridge is in turn controlled by the influence of the
mesoderm on the apical ridge, in the form of an undefined mesodermal
maintenance factor.

The role of the ectoderm was viewed in a completely different light by Amprino (1965), who postulated that the epithelial hull of the limb bud exerts its morphogenetic effect by mechanical means. He felt that the growth of the ectoderm results in a proximodistal shifting of that layer with respect to the mesoderm of the limb bud and that the ectodermal shifting creates new space that is quickly filled in by the proliferating mesodermal cells.

A more recent hypothesis is known as the *progress zone model* (Summerbell et al., 1973; Wolpert et al., 1975). The progress zone is the distal 300 μm of mesoderm beneath the apical ectoderm of the wing bud in the chick. Under the apparent influence of the apical ectodermal ridge, the cells in the progress zone are said to remain both undifferentiated and in a continuous state of mitosis. As these cells divide, some of the daughter cells leave the progress zone and are then free to differentiate into discrete components of the appendage. What the cells will become (i.e., their proximodistal level of differentation) depends on how long the cells have resided in the progress zone. Cells that leave the progress zone early during the course of limb development are supposed to be endowed with positional information enabling them to develop into the humerus or femur, for instance, whereas cells that leave the progress zone at a later time would contain positional information compatible with the development of tissues of the next segment of the limb (i.e., radius or ulna). The last of the cells to emerge from the progress zone would form the terminal phalangeal bones.

There are currently two major hypotheses regarding morphogenetic control along the anteroposterior axis. One hypothesis (Summerbell and Honig, 1981) states that the cells of the zone of polarizing activity produce a chemical signal (morphogen) that diffuses throughout the limb bud, setting up a gradient of decreasing concentration from the posterior to the anterior part of the limb bud. Mesenchymal cells along the limb bud are able to sense the gradient, and they set up a pattern based upon their reading of the gradient. According to this hypothesis, the formation of supernumerary limb parts after the grafting of polarizing tissue into the anterior margin of the wing bud (Fig. 11-9) is due to the setting up of a secondary high point in the morphogen gradient. When anterior cells, formerly bathed in a low concentration of the morphogen, are exposed to a high concentration, they become specified to form supernumerary posterior wing parts.

The other hypothesis (Iten, 1982) is not based on gradients but rather assumes that around the circumference of the limb bud the cells have a continuous array of different positional values. Local interactions among cells with different positional values are postulated to result in the establishment of the limb pattern. This hypothesis explains the formation of supernumerary limbs after certain grafts by interactions between cells of widely disparate positional values. If cells from opposite ends of the limb are placed in direct contact with one another, local interactions are said to fill in the intervening positional values and, in the process, set up the pattern for a secondary limb.

Morphology of Later Limb Development

Up to this point, our consideration of limb development has been confined to the early stages, in which the limb bud consists of a mass of homogeneous-appearing mesodermal cells covered by a sheet of ectoderm. Despite our inability to detect differences among cells or groups of cells of the limb bud even with the most sophisticated morphological and chemical techniques currently available, experimental analysis has revealed that during these early developmental stages axial determination and the pattern for the morphogenesis of specific components of the limb have been set up. On the basis of this morphogenetic blueprint the differentiation of individual tissue components and cell types and later growth of the limb occur. This sequence of events is becoming so clear that it is almost axiomatic to say that "morphogenesis precedes differentiation."

When viewed grossly, limb development proceeds from a flattened disk of determined cells to an early mound-shaped outgrowth, the limb bud (Fig. 11-11). As the limb bud grows out, it flattens distally, forming a paddle-shaped structure. Early digital primordia appear like rays within the thin paddle, and the elbow or knee becomes readily recognizable. By this time essentially all the major components of the limb have been laid down, and further development consists principally of growth and the rotation of the limb to its final definitive configuration.

The differentiation of limb components follows a pronounced proximodistal gradient and a lesser gradient along the anteroposterior axis, with the direction of the latter gradient varying among the classes of vertebrates. This means that even though all the prospective parts of the limb are represented in the early limb bud (Fig. 11-12), the *stylopodium* (upper arm or thigh) differentiates first, followed by the *zeugopodium* (forearm or leg) and finally the *autopodium* (hand or foot).

Differentiation of the Skeleton

The skeletal tissues arise from cells originally present within the limb bud. The first morphological indication of formation of the skeleton is the condensation of mesenchymal cells in the central core of the proximal part of the limb bud. These cells begin to secrete matrix material, and soon hyaline cartilage becomes recognizable. In the mammalian forelimb, the scapula and humerus form first (Forsthoefel, 1963). Then the postaxial components of the zeugopodium and autopodium (ulna and digits IV and V) appear. Finally, the preaxial skeletal elements (radius and digits III, II, and I) become established (Fig. 11-13). In amphibian limbs, the gradient of differentiation along the anteroposterior axis is reversed and the radius is established before the ulna. The skeleton of the limb remains cartilaginous until the gross form of the limb is well established. Only at a later period in development does the replacement of cartilage by bone begin. The formation of joints is discussed in Chap. 9.

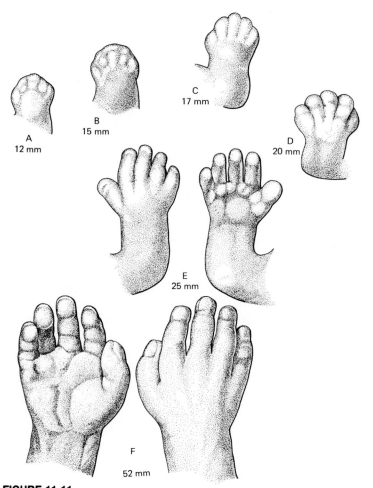

FIGURE 11-11
Stages in the development of the human arm and hand. The lengths indicated refer to the crown-rump length. (*After Scammon, from Retzius.*)

Differentiation of Muscles

It is now well established that the cellular precursors of limb muscles migrate into the limb buds from the somites (see Chap. 9). This was shown by the technique of grafting quail somites into early chick embryos and determining which structures of the limb were derived from chick or quail cells. The muscle fibers (and associated satellite cells) are somite-derived, whereas the tendons and other connective tissue of the muscle arise from local limb bud tissue. In fact, if "muscleless" limbs are produced by removing the somites in early embryos, the appropriate tendons still form in the absence of the muscles.

Quail grafting experiments have shown that in addition to muscle cells, pigment cells, Schwann cells, osteoclasts and chondroclasts, and endothelial

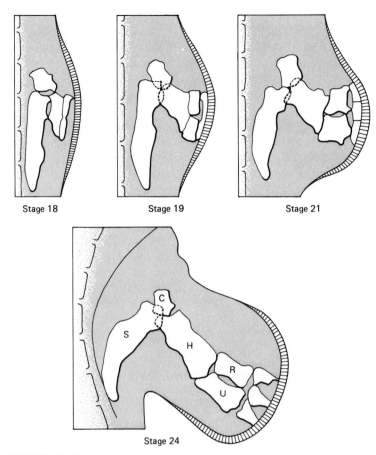

Stage 18   Stage 19   Stage 21

Stage 24

FIGURE 11-12
Maps of prospective bone-forming areas of the developing wing bud. Based
upon reconstructions of limb buds bearing radioisotopically labeled grafts of
tissues. Stages based on Hamburger and Hamilton (1951). (*Adapted from
Stark and Searls, 1973, Devel. Biol. **33**:138.*) Abbreviations: *C*, coracoid; *H*,
humerus; *R*, radius; *S*, scapula; *U*, ulna.

cells of blood vessels enter the limb bud secondarily. In contrast, fibroblasts,
osteoblasts and chondroblasts, fat cells, and smooth muscle cells arise from the
limb bud itself. The extent of migration of precursor cells of muscle at various
stages of limb development is shown in Fig. 11-14.

The formation of muscular elements in the limb takes place shortly after the
skeletal elements begin to take shape. Čihák (1972) described four fundamental
phases in the ontogenesis of muscle pattern:

1 The condensation of mesenchymatous cells to form common flexor and
extensor muscle masses. In these masses, layers corresponding to the fundamental muscular layers of the adult limbs take shape.

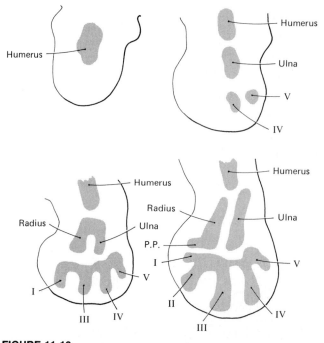

FIGURE 11-13
Development of skeletal primordia in embryonic forelimbs of the mouse. (*Drawings based upon photomicrographs in Forsthoefel, 1963, Anat. Rec. 147:29.*) Abbrev: *p.p.*, prepollex.

2 The formation of individual muscle primordia within the common muscle masses. Within the wing bud of the chick embryo, Shellswell and Wolpert (1977) described a binary pattern of splitting of the common muscle masses, first into groups and then into individual muscles. By the time the muscle primordia have taken shape, the muscle cells themselves are in the myoblastic or early myotube stage. Almost nothing is known about the mechanisms involved in splitting of the common muscle masses into individual primordia except that nerves are not essential.

3 The phase of reconstruction of the muscle primordia. When first formed, the muscle primordia are in a phylogenetically primitive arrangement. During the phase of reconstruction, many of the primordia become rearranged to a configuration characteristic of the species. For example, in the human hand, some muscle primordia from different layers fuse to form single muscles (e.g., dorsal interosseous muscles). In contrast, other primordia (e.g., parts of the contrahentes muscle layer of the palm) disappear through cell death, despite the fact that the cells within them have differentiated to the point of containing myofilaments (Grim, 1972).

4 The development of the definitive form of the muscle. Prominent in this phase is the formation of connective-tissue elements and their integration with

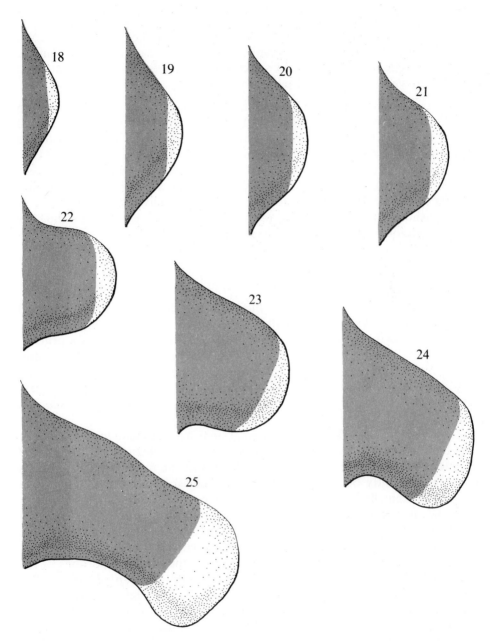

FIGURE 11-14
Distribution of myogenic cells (shaded areas) in the leg buds of quail embryos. The numbers refer to Hamburger and Hamilton stages of development. (For stages, see inside of back cover.) (*After Jacob et al., 1986.*)

the muscle fibers. It is now known that the bodies of the muscles first take shape apart from the tendons and that the muscles and their tendons are joined together only at a later time. Because of their highly independent modes of origin, it appears likely that when the primordia of the skeletal elements, muscles, and tendons are being established, the cells that constitute them are taking their cues from a common set of morphogenetic guidelines that were established when their original pattern was set up.

After the muscles in the limb have been fully formed, they must still undergo considerable growth to keep pace with the rest of the body both during the fetal period and after birth. In many species new muscle fibers are added to the muscles up to about the time of birth. Shortly after birth the formation of new muscle fibers ceases, and the number of muscle fibers remains more or less constant throughout postnatal life. Muscle fibers are syncytial structures, and their nuclei can no longer divide. Growth of the individual muscle fiber is accomplished by the incorporation of satellite cells (see page 333) into the fiber and the addition of new units of contractile proteins (sarcomeres), usually at the ends of the muscle fibers.

Development of the Vascular Supply

The earliest limb bud is supplied by a fine capillary network arising from several segmental branches of the aorta (Fig. 11-15). Soon, some channels within the network are favored over others. One of the first to take shape is a primary central artery, which distributes blood to the limb bud. Blood from the central artery travels through a capillary network and then collects in a *marginal sinus* that extends around the apex of the limb bud, just beneath the apical ectodermal ridge, and returns the blood to the body through venous channels coursing along both the pre- and postaxial borders of the limb.

The capillary networks and the marginal sinus itself respond to growth of the limb bud by sending out new vascular sprouts. A careful study by Seichert and Rychter (1971) showed that the marginal sinus in the developing wing of the chick is constantly being re-formed by the coalescence of capillary sprouts which grow apically from the already existing marginal sinus. Blood first drains from the marginal venous channels into superficial venous plexuses of the body, but as development continues, a progressively greater proportion of the blood is shunted into deeper channels. As the digital rays become established, the apical marginal sinus begins to break up, but the proximal portions of the marginal channels persist in mammals as the *basilic* and *cephalic veins*. The fundamental arrangement of deeply situated arteries and superficial veins persists in the adult limb, but some deep veins exist as well.

The formation of the major arteries of the extremities is a complex process in which several major arterial trunks successively take ascendency and then regress, to be replaced by still other major vessels. As an example of this process, the major steps in the development of the definitive arterial pattern of the human arm are illustrated in Fig. 11-16.

The role of the vasculature in limb development has been viewed in several different ways over the years. Some of the earliest opinions suggested that vascular patterns at specific stages of development are phylogenetically fixed

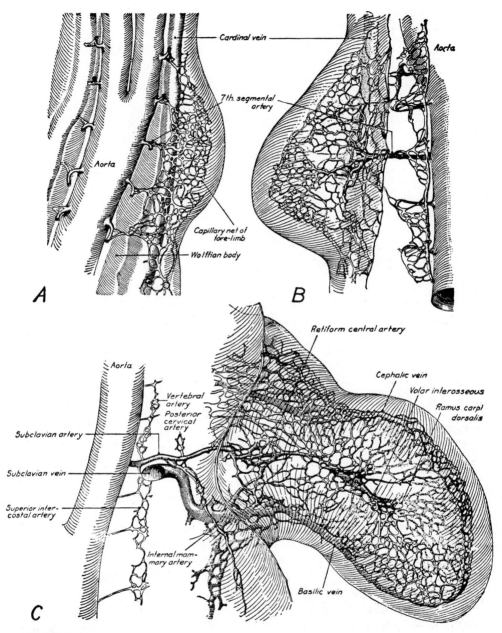

FIGURE 11-15
Diagrams from injected pig embryos showing three stages in the early development of the anterior limb bud. (A) Embryo of 4.5 mm (equivalent to 4-week human embryo). (B) Embryo of 7.5 mm (equivalent to 5-week human embryo). (C) Embryo of 12 mm (equivalent of 6-week human embryo). (*After Wollard, 1922, Carnegie Cont. Embryol.* **14**:*139.*)

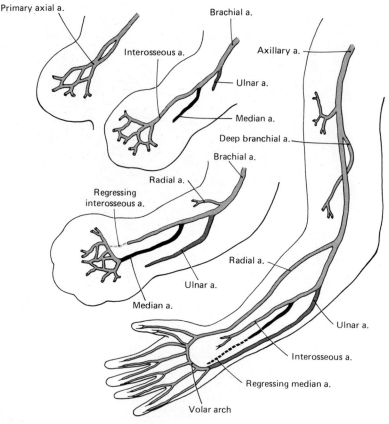

FIGURE 11-16
Development of the arteries of the human arm.

and are in a sense almost independent of the tissues developing alongside them. A more widespread viewpoint is that the vascular pattern to a large extent reponds to the growth and development of other tissues in the limb bud and that morphogenesis of the vascular system is almost entirely a secondary phenomenon that is dependent on the differentiation of other tissues within the limb. In recent years increasing attention has been paid to the possibility that the pattern of the vasculature may of itself act as a primary agent in morphogenesis and differentiation. The close relationship between the marginal sinus and the apical ectodermal ridge is looked upon by some as a functionally important association instead of as a mere anatomical juxtaposition. According to others (e.g., Caplan and Kautroupas, 1973), local differences in the metabolic environment brought about by the patterns of the embryonic arteries and veins may be a determining factor in the course of differentiation followed by some groups of cells. In contrast, Wilson and Orr-Urtereger (1986) have shown that the differentiation of arteries and veins is a relatively late process, occurring after the onset of cartilage and muscle differentiation. They are skeptical about the role of blood vessels as determiners of pattern or specific tissue differentiation. More information is required before the role of blood vessels in limb development can be firmly established, but it is likely that the blood vessels will be shown to be prime movers in some aspects of development and secondary responders in other aspects.

Innervation of the Embryonic Limb

The early limb bud forms in the absence of nerves, and it remains uninnervated during the early period of outgrowth. In higher vertebrates nerves grow into the limb slightly later than the first appearance of the skeletal primordia and roughly at the time when the common muscle masses take shape (Fig. 11-17).

The interrelationships between nerves and many components of the limb have stimulated a number of questions concerning morphogenetic influences of nerves on the limb and factors that determine the pattern of innervation of the limb. It is quite well established that overall morphogenesis of limbs is independent of the nerve supply. This has been shown in a number of grafting experiments in which limb buds, placed in ectopic locations either on the body or, in birds, on the chorioallantoic membrane, underwent normal development.

In another type of experiment, *aneurogenic limbs* have been created by surgically removing the neural tube of early embryos. In these circumstances the limbs grow in the absence of nerves. Both the gross morphology of the limbs and the patterns of internal tissues are nearly normal. The nerves do, however, appear to influence the rate of mitosis of mesenchymal and epithelial tissues, and some minor deviations from normal morphology in aneurogenic limbs are probably due to deficiencies in the number of cells available to form the primordia of certain structures. In this way, nerves may indirectly affect morphogenesis.

The factors that determine the pattern of nerves within a limb are still being debated. Extreme views are (1) that the pattern of innervation is determined entirely by the configuration of the other tissue components of the limb, with ingrowing nerves passively following environmental cues, and (2) that the ingrowing nerves themselves are uniquely specified, allowing them to home in upon specific predetermined structures within the limb. Both postulates have been subjected to experimental analysis, and it now appears that the answer to this question may lie somewhere between these two extremes. If limbs are grafted to ectopic sites or if pieces of the embryonic nervous system are grafted so that the limb is innervated by foreign nerves, the main pattern of nerve trunks is often recognizable, but the overall pattern of innervation is usually abnormal (Piatt, 1956, 1957). Yet other work on fishes and amphibians has strongly suggested that motor nerve fibers have a preference for a particular muscle, even to the point of functionally displacing other nerve fibers that have been allowed secondarily to innervate the muscle. These questions will be examined in greater detail in Chap. 12.

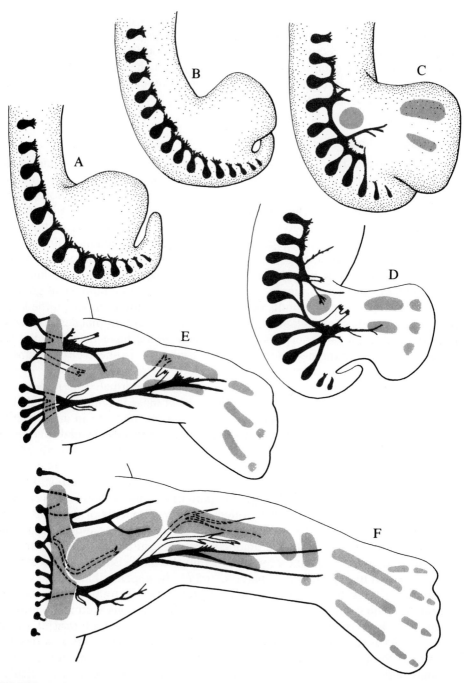

FIGURE 11-17
Stages in the innervation of the hind limb of the chick. (A) Stage 23; (B) stage 24; (C) stage 25; (D) stages 26–27; (E) stages 28–29; (F) stage 30. (*Adapted from B. Fouvet, 1973, Arch. Anat. Microsc. Morphol. Exp.* **62**:*269.*)

Rotation of the Limbs

Both the anterior and posterior limbs undergo pronounced changes of position during embryonic life. It is easiest to use the human limb as a model for describing these changes. At the stage of early differentiation of the digits the arm bud projects outward at right angles to the body. (This is equivalent to holding one's arm horizontally, with the thumb upward and the palm forward when standing.) As the elbow joint forms, the forearm and hand bend ventrally so that the palm faces the trunk. (Without dropping the upper arm at the shoulder, flex your elbows, bringing the palms of your hands against your chest.) Then the arm undergoes a 90° rotation about its long axis so that the elbow points in a caudal direction. (Drop your elbow to your side.)

The leg undergoes a similar set of axial changes. Early in development the leg bud projects straight out from the body parallel to the arm, with the future great toe cranial (like the thumb) and the sole facing in the same direction as the palm. As the knee forms, the lower leg and foot bend so that the sole faces the trunk. The leg also undergoes a 90° rotation, but in an opposite direction to that of the arm. This brings the knee to point in a cranial direction.

Despite the changes in configuration of the limbs, their homologous structures and early embryonic relations can be determined by the segmental pattern of their innervation. In the arm, the thumb and the radial side of the forearm (preaxial structures) receive the most cranial innervation of the limb, as do the great toe and the tibial side of the leg.

LIMB REGENERATION

The limbs of some vertebrates (e.g., salamanders) are endowed with the capacity to regenerate after they have been amputated. Amputation is followed by epidermal wound healing, which covers the amputation surface, and by the removal of internal debris created by the original wound. Within a few days, the limb stump enters the period of *dedifferentiation*, during which the differentiated tissues of the distal part of the limb break up and are replaced by a population of primitive-appearing cells. These cells aggregate and proliferate at the end of the limb stump to form a regeneration blastema (Fig. 11-18), which is similar in many respects to an embryonic limb bud. As the blastema grows, it forms a new limb in a manner that parallels very closely the sequence of events that occurs in the embryonic limb bud (Fig. 1-25).

Although there are many parallels between the embryonic development and regeneration of a limb, there are also some differences, reflecting the tighter integration of the regenerating limb to the body. At least three conditions—a wound epidermis, mesodermal damage, and an adequate nerve supply—are normally required for the initiation of limb regeneration. The role of other factors, such as hormones and bioelectric currents, is unclear.

Both the dedifferentiative phase and subsequent morphogenesis involve the reading of positional information and response to it by the cells involved in regeneration. In contrast to the embryonic limb bud, which must set up a

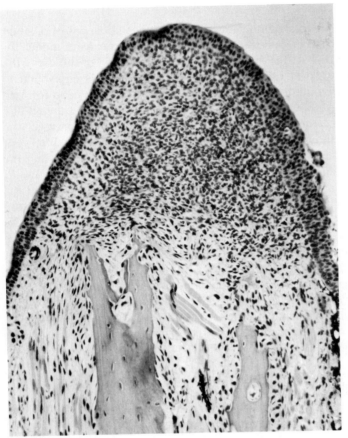

FIGURE 11-18
Regeneration blastema in an adult newt. Twenty-five days previously
the forearm was amputated through the radius and ulna. The
blastema is the dense aggregation of cells in the top half of the figure.
In the lower half one can see the ends of the amputated radius and
ulna and some of the forearm muscles. (*From B. M. Carlson, 1974, in
G. V. Sherbet, ed. Neoplasia and Cell Differentiation. S. Karger,
Basel.*)

system of positional values, the regenerating limb develops in association with
mature cells of the limb stump which have retained a record of their position
and which, under appropriate conditions, can express this information.

The regeneration of extremities is most successful in fish, salamanders, and
anuran tadpoles. During metamorphosis, most frogs and toads lose the ability
to regenerate limbs. Lizards are renowned for their ability to regenerate tails,
but their limb regenerative capacity is poor. The limbs of birds and mammals
do not regenerate. A major question in regeneration is whether the limbs of
higher forms have lost their capacity to regenerate entirely or whether they
retain a latent ability to regenerate if missing factors are supplied. Numerous

attempts have been made to stimulate the regeneration of limbs in frogs and higher vertebrates. Some degree of success has been reported using such means as changes in the hormonal environment, an increase in the nerve supply, and the application of electrical currents to the amputated limb. A number of factors are likely to be involved in the loss of regenerative power, and the stimulation of partial regeneration of limbs in higher vertebrates suggests that the loss of the ability to regenerate may not be irreversible.

CHAPTER 12

DEVELOPMENT OF
THE NERVOUS SYSTEM

The nervous system is the chief coordinating system of the body. Information about the environment within the body and surrounding the body is collected by sense organs and brought to the central nervous system—the brain and spinal cord—through the *sensory (afferent) nerves*. Within the central nervous system the sensory signals are distributed to a variety of integrating centers, some very simple and some very complex, which generate messages leading to an appropriate response. Many responses involve coordinated muscular movements, which represent the sum of a large number of messages passing through tracts of nerve fibers within the central nervous system and ultimately leaving the brain or spinal cord through *motor (efferent) nerves*, which ultimately terminate on individual muscle fibers. Some responses are mediated via the *autonomic (sympathetic or parasympathetic) nerves* and can result in changes in glandular secretions, heart rate, or blood pressure. Other responses appear to be confined to the brain and are involved in functions such as thinking and memory.

A number of fundamental developmental processes are involved in the embryogenesis of the nervous system. Some dominate the embryo during their occurrence; others are limited to certain parts of the nervous system. The major processes are (1) induction, both the primary induction of the neural plate and the secondary inductions emanating from the early brain and spinal cord, (2) proliferation, both as a response to primary induction and as a prelude to the morphogenesis and growth of specific portions of the nervous system, (3) cellular communication and the adhesion of like cells, (4) cell migration, of which there are many striking examples at many stages and in many regions, (5) the differentiation of neurons and glial cells, including both structural and functional maturation, (6) the formation of specific connections between groups of neurons, (7) the stabilization or elimination of interneural connections, resulting in the death of unconnected cells, and (8) the development of integrated neural function, allowing the embryo and newborn to undertake coordinated movements.

EARLY STAGES IN THE ESTABLISHMENT OF THE NERVOUS SYSTEM

Clones and Cell Lineages

Experiments conducted on amphibian embryos (Hirose and Jacobson, 1979) have suggested to some investigators that even in the early cleavage stages individual blastomeres give rise to consistent populations of cells within the nervous system and within other areas of the body as well. Each blastomere is said to serve as the source of a clone of cells which ultimately constitute a *compartment*, or *clonal domain*. This is determined by injecting a single blastomere with horseradish peroxidase as an intracellular label. After varying numbers of cell divisions, the embryos are fixed and histochemically stained for peroxidase activity. Only the cells which are the progeny of the originally labeled blastomere are stained. In this way maps of clonal domains derived from identifiable blastomeres can be constructed. An example of a map of a clonal domain in the central nervous system of a *Xenopus* embryo is shown in Fig. 12-1.

These experiments show a remarkable precision in the distribution of the progeny of a single labeled cell, but the interpretation of the results is controversial. At one extreme, the results could be interpreted from the purely mosaic point of view, which would say that within a given cell lineage, development is autonomous of any external influences. This viewpoint was challenged by Gimlich and Cooke (1983), who produced secondary embryos in *Xenopus* by grafting the dorsal lip of the blastopore to the ventral part of the embryo and noting that cells which would normally have become belly ectoderm were transformed into neural tissue. Jacobson (1985) has pointed out ways in which determined lineages and environmental factors might work at different times to generate certain groups of cells in the nervous system and elsewhere in the body. Although it is surprisingly difficult to sort out the variables in this version of the "nature vs. nurture" argument, there is little doubt that large populations of cells in the nervous system can arise from only a handful of founder cells.

Induction of the Nervous System and Neurulation

The primary induction of the nervous system by the chordamesoderm acting on the overlying ectoderm has been covered in Chap. 6. One of the striking changes that occur during the period of primary induction is the distribution of *cell adhesion molecules* (CAMs) (see page 15).

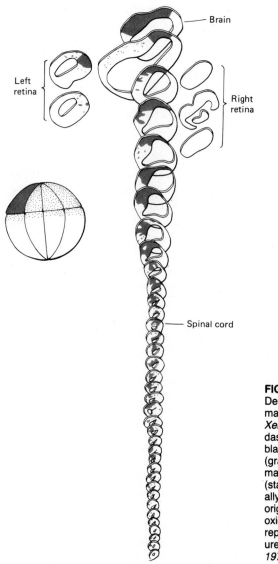

FIGURE 12-1
Demonstration of cell lineages in the formation of the central nervous system in *Xenopus* (Anura). Horseradish peroxidase was injected as a marker into one blastomere (LD 1.2) of a cleaving embryo (gray cell in sketch of embryo). The animal was allowed to develop into a larva (stage 32) and was then fixed and serially sectioned. Cells descended from the original labeled blastomere retain the peroxidase label and are stained gray on the representative sections shown in this figure. (*Adapted from Hirose and Jacobson, 1979, Devel. Biol.* **71**:*191*.)

The early epiblast of the chick embryo contains both N-CAM and L-CAM, but as the neural plate takes shape, the cells that constitute it possess only N-CAM. Conversely, the nonneural ectodermal cells lose their N-CAM and express only L-CAM on their surfaces.

In addition to bringing about a change in the shape of the cells of the future neural plate and neural tube (Figs. 6-9 and 6-12), the primary inductive event stimulates the proliferation of the neuroepithelial cells. Increased proliferation by the responding tissue is a common consequence of induction in other systems as well. We have also seen the origin of the neural groove by the infolding of the thickened neural plate ectoderm, the closure of the neural groove to form the neural tube, and the separation of the tube from the overlying parent ectoderm (Figs. 8-1, 8-2, and 8-3). Coincident with the separation of the neural tube from the cutaneous ectoderm, the cells of the neural crest migrate out from the junction between neural and cutaneous ectoderm and either become widely distributed throughout the body or remain near the spinal cord as the segmentally arranged spinal ganglia (Fig. 6-16).

Shortly after the neural plate is established, the new neural primordium itself acts as a secondary inductor. Prominent examples are inductions of sensory structures, such as the lens of the eye (Fig. 13-4), the otic vesicle (inner ear), and the series of ectodermal placodes (Fig. 13-1). Later, the brain and spinal cord also induce the bony structures that protect them (Fig. 9-23).

Early Morphogenesis of the Nervous System

Almost as soon as it is independently established, the neural tube becomes markedly enlarged cephalically (Fig. 12-2). This expanded portion is the primordium of the brain. Caudally the neural tube (the forerunner of the spinal cord) remains of relatively uniform diameter.

In its enlargement the brain at first exhibits three regional divisions: the primary forebrain, midbrain, and hindbrain, or, to use their more technical synonyms, the *prosencephalon*, *mesencephalon*, and *rhombencephalon* (Fig. 12-12). The three-vesicle stage of the brain is short-lived. The prosencephalon becomes subdivided into two regions, the *telencephalon* and *diencephalon;* the mesencephalon remains undivided; and the rhombencephalic regions become differentiated into the *metencephalon* and *myelencephalon*. Thus in place of three vesicles, five are established. Using this organization as a basis, one can trace the later differentiation of some of the more important parts of the nervous system.

HISTOGENESIS OF THE CENTRAL NERVOUS SYSTEM

Establishment of the Neuroepithelium

The ectoderm of the open neural groove and early neural tube is a thick pseudostratified epithelium. An epithelium of this type appears to contain several layers of cells because the nuclei are found in several different planes, but in reality the nuclei are located within very long and somewhat irregularly shaped cells, the cytoplasm of which extends throughout the entire thickness of the epithelium. The *neuroepithelial cells* possess a high degree of mitotic activity, and in confirmation of the early work of Sauer (1935), autoradiographic research with tritiated thymidine used as a marker has indicated that during different phases of their life history the nuclei of the neuroepithelial cells occupy different positions within the neural epithelium (Langman et al., 1966).

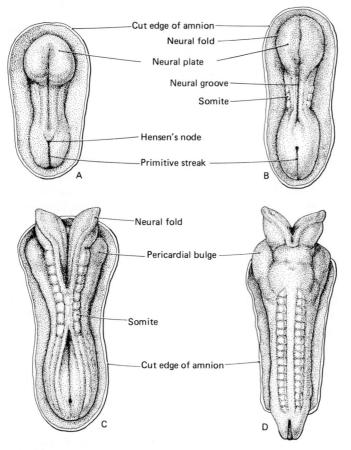

FIGURE 12-2
Early stages in formation of the human central nervous system. (A)
18 days; (B) 20 days; (C) 22 days; (D) 23 days. (*After Sadler, 1985
Langman's Medical Embryology, 5th ed., Williams & Wilkins.*)

The synthesis of DNA occurs in nuclei located close to the external limiting
membrane (Fig. 12-3). Then the nuclei move toward the lumen of the neural
tube, and the cells undergo mitotic division.

Before closure of the neural tube, the nuclei of the daughter cells again
move toward the external limiting membrane, where they may again synthesize DNA and repeat the germinative cycle. In this manner the number of
neuroepithelial cells is greatly increased. Following closure of the neural
tube, some of the neuroepithelial daughter cells migrate away from the lumen
past the DNA-synthesizing neuroepithelial cells to positions immediately
beneath the external limiting membrane (Fig. 12-3). These cells, called
neuroblasts, begin to produce processes which are the forerunners of axons
and dendrites.

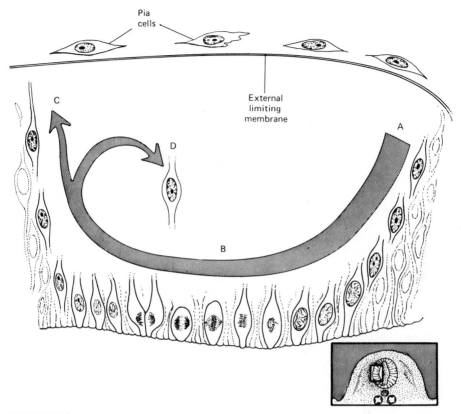

FIGURE 12-3
Diagrammatic representation of cellular events occurring in the early neural tube. When it first forms, the neural tube is a pseudostratified epithelium with the external limiting membrane serving as its base and the boundary of the central canal serving as its apex. The nuclei of neuroepithelial cells begin synthesizing DNA in the basal areas (A) and then migrate within the cytoplasm toward the apical part of the cells, where they undergo mitosis (B). Nuclei of the daughter cells then migrate back toward the external limiting membrane, where they either begin to differentiate as neuroblasts (C) or return to the pool of proliferating neuroepithelial cells (D). (*Adapted from several sources, chiefly Sauer and Langman.*)

As the primitive neuroepithelium matures, it can be subdivided into layers (Fig. 12-4).[1] The innermost layer, called the *ventricular*, or *ependymal layer,* contains cells which are still in the mitotic cycle. Ultimately this layer becomes the *ependyma*, a columnar epithelium which lines the central canal and the ventricular system of the central nervous system. The peripheral part of the

---

[1]The terminology of the layers of the developing neural tube is in a state of flux. Many textbooks still refer to three layers, the ependymal layer, the mantle layer, and the marginal layer. The Boulder Committee (1970) has advocated replacing the names with *ventricular layer*, *intermediate layer*, and *marginal layer*. In addition, this committee designates a fourth layer, the *subventricular layer*, between the ventricular layer and the intermediate layer. The latter terminology is becoming increasingly common in the research literature.

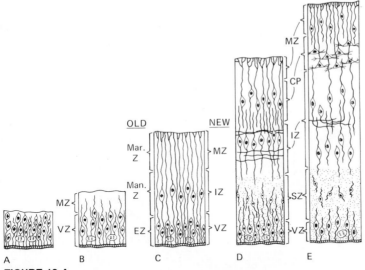

FIGURE 12-4
Diagram illustrating successive stages in the layering of the wall of the developing brain. (*After Rakic.*) The designation of the layers is that recommended by the Boulder Committee (*Anat. Rec.* **166**:*257*). At the earliest stage (A) the wall of the brain is a pseudostratified epithelium. Section "C" compares the traditional with the Boulder terminology. Abbreviations: *CP*, cortical plate; *EZ*, ependymal zone; *IZ*, intermediate zone; *MZ* (*Mar. Z*), marginal zone; *Man. Z*, mantle zone; *SZ*, subventricular zone; *VZ*, ventricular zone. Cell proliferation occurs in the ventricular zone and, to a lesser extent, the subventricular zone. As the brain matures, the layers of the cerebral cortex develop in the region of the cortical plate.

neural tube now consists of an expanding zone containing numerous cell processes but few cell bodies. This outer, cell-poor layer is called the *marginal layer*; it ultimately becomes the *white matter* of the central nervous system. The postmitotic neuroblasts which leave the inner ventricular layer form an intermediate layer of densely packed cells called the *mantle layer*. This layer will become the *gray matter*.

Neuroblasts

Some of the daughter cells generated in the neuroepithelium lose their ability to undergo mitosis and migrate toward the outer wall of the neuroepithelium (Fig. 12-3). These cells, called *neuroblasts*, are initially bipolar, having slender processes which connect with both the central luminal border and the external limiting membrane of the neural tube. The *bipolar cells* soon are detached from the inner luminal border by a retraction of the inner process, thus becoming *unipolar cells*. The unipolar cells then accumulate large amounts of rough endoplasmic reticulum (*Nissl substance*) in their cytoplasm and send out several cytoplasmic processes. At this stage the *multipolar neuroblasts* (Fig.

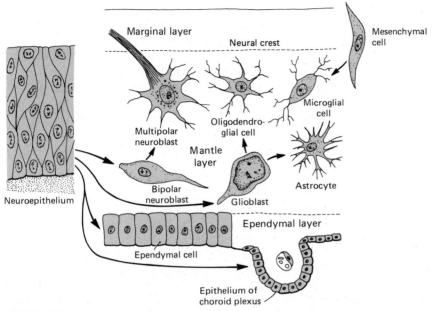

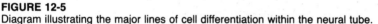

FIGURE 12-5
Diagram illustrating the major lines of cell differentiation within the neural tube.

12-5) are expending much of their developmental energy toward the production of the processes (axons and dendrites) which will make connections with other parts of the nervous system. There are many specific patterns of differentiation of individual neurons throughout the nervous system, ranging from the early formation of large macroneurons to the later differentiation of microneurons. The morphology of both the nerve cell bodies and their processes varies greatly from region to region within the nervous system. Details can be found in more specialized texts.

Neuroglia

After the first wave of neuroblast formation, other neuroepithelial cells follow a course of differentiation leading to the formation of neuroglial cells. Surprisingly little is known about the functions of mature neuroglial cells or their development in the embryo. Of the three classes of neuroglial cells, the *astrocytes* and the *oligodendroglial cells* arise from ectoderm, the former from the neural tube and the latter probably from the neural crest. The third type, the *microglia*, are active in phagocytosis after damage to the nervous system and are presumably of mesodermal origin (possibly a form of macrophage), although some researchers feel that microglia arise from neurectoderm.

The astrocytes and oligodendroglial cells may arise from a common intermediate cell type known as a *glioblast* or from separate precursor cells (Fig.

12-5). Some glioblasts develop long, fine processes and differentiate into astrocytes. The processes of astrocytes become closely associated with capillaries, and it has been suspected that among other functions they may serve as a transit system for metabolites within the substance of the central nervous system. The oligodendroglial cells are smaller and structurally more simple than astrocytes and become recognizable later in development. They appear as satellites around the cell bodies of neurons and are also involved in the myelination of nerve fibers within the white matter of the central nervous system. In contrast to neuroblasts and neurons, neuroglial cells retain the ability to divide, even in the adult.

Ependymal Cells

Other cells of the original neuroepithelium retain their epithelial character and differentiate into the columnar *ependymal cells* that line the central canal of the brain and spinal cord (Fig. 12-5). Much remains to be learned about the functions and potentialities of ependymal cells. In lower vertebrates the ependymal cells retain the potential for differentiating into neurons during regeneration of the central nervous system.

Development of Gray and White Matter in the Spinal Cord and Brain

During the period of development when the neurons differentiate, the appearance of the spinal cord undergoes marked changes. Some of the neuroblasts in the mantle layer of the cord send out processes very early in development. Others remain undifferentiated and continue to proliferate for a time, causing continued growth in the mantle layer. As it grows, the mantle layer takes on a very characteristic configuration, ultimately becoming butterfly-shaped in cross section. With this change in shape and with the transformation of its glioblasts into neuroglia and its neuroblasts into characteristic nerve cells, the mantle layer becomes the so-called *gray matter* of the spinal cord (Figs. 12-6 and 12-7).

During the growth of the mantle layer the originally extensive lumen of the neural tube is reduced by obliteration of its dorsal portion to form the small central canal characteristic of the adult cord (Fig. 12-7). The central canal is now lined by the epithelioid ependymal layer.

Meanwhile the outer (marginal) layer of the cord has been expanding extensively. This expansion is due to the secondary ingrowth of longitudinally disposed neuronal processes which constitute the conduction paths (tracts) between the various levels of the spinal cord and the brain. Because each of these fibers is enveloped in a sheath rich in myelin, the region of the cord in which they lie has a characteristic whitish appearance which contrasts strongly with the gray color of the richly cellular portion of the cord derived from the mantle layer. For this reason the fibers in the marginal layer of the cord are said to constitute its *white matter*. The main groups of these fibers are more or less

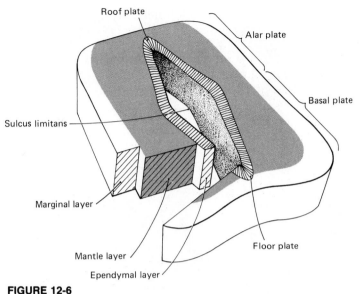

Roof plate

Alar plate

Basal plate

Sulcus limitans

Marginal layer

Mantle layer

Ependymal layer

Floor plate

FIGURE 12-6
Diagram illustrating the major subdivisions of the neural tube.

marked off from each other by the dorsal and ventral horns of the gray matter. They are known as the *dorsal*, *lateral*, and *ventral columns* of the white matter of the cord (Fig. 12-7). The dorsal columns contain the main sensory (afferent) paths to the brain, the ventral columns are primarily motor (efferent), and the lateral columns contain important ascending fiber tracts to the brain and also some of the main motor paths from brain to cord.

In dealing with the topography of the neural tube in cross section, it is customary to designate its thickened side walls as the *lateral plates*, its thin dorsal wall as the *roof plate*, and its thin ventral wall as the *floor plate* (Fig. 12-6). Extending along the inner surface of each lateral plate is a longitudinal sulcus (*sulcus limitans*), which divides the lateral plate into a dorsal afferent (carrying signals to the brain) part (*alar plate*) and a ventral efferent (carrying signals away from the brain) part (*basal plate*).

Histogenesis of the developing brain involves unique problems and unique solutions. A readily apparent problem involves the location of the gray and white matter. As we have just seen for the spinal cord and brainstem, the gray matter, containing the nerve-cell bodies, remains in the center and becomes surrounded by the myelinated neural processes which constitute the white matter. In the forebrain and cerebellum, by contrast, the gray matter is located at the periphery, often in several discrete layers, and the white matter is on the inside.

One cannot properly speak of the brain as a homogeneous entity because almost every part has a unique pattern of histogenesis. Isolated examples will be presented here for illustrative purposes. Developmental differences not only

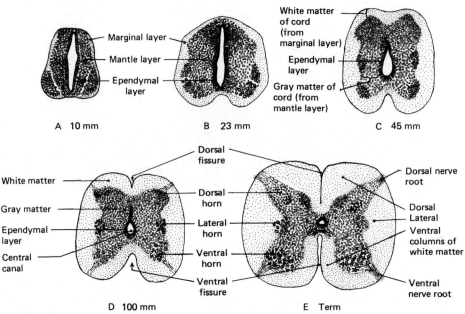

FIGURE 12-7
Transverse sections through spinal cord of the pig at various ages. Note especially the parts of the adult cord derived from the ependymal, mantle, and marginal layers of the embryonic neural tube.

are present from one region to another in the embryonic brain, they also are related to time. This can be illustrated by patterns of cellular proliferation. In almost all areas of the spinal cord or brain the proliferation of cells that will become neurons occurs near the central canal. Except for a few specialized areas, such as the cerebellum, the young neurons are postmitotic cells; that is, they will not divide again, although they may undertake relatively extensive migrations. If tritiated thymidine is injected into an embryo and the animal is not killed until it is mature, the radioactive label, which is incorporated into the DNA, remains concentrated in the neuroblasts that were undergoing their last round of DNA synthesis at the time the isotope was available. Older, postmitotic nerve cells do not incorporate the isotope at all, and other cells which remain in the mitotic cycle gradually lose the label by progressive dilution with each cell division. Radioautographs of brains prepared from mature mice which were exposed to [³H]thymidine at various times during pregnancy reveal strikingly different patterns of proliferation (Fig. 12-8).

One of the most important processes in the histogenesis of the brain is cell migration. From their site of origin near the ventricles of the brain, young neurons migrate out toward the periphery, where they often settle down in several discrete layers. According to Rakic (1975), young postmitotic neurons, usually fairly simple bipolar cells, use long processes of specialized cells called

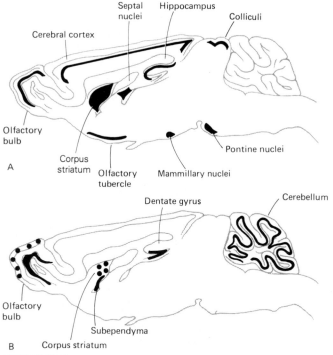

FIGURE 12-8
Sagittal sections through the brains of adult mice which were exposed to [³H]thymidine at 15 days (A) and 19 days (B) of gestation. Autoradiographic preparations were made from the sectioned brains. The location of labeled cells (black) illustrates the differing patterns of cell division during development. (*Modified from Langman, 1975, Birth Defects* **XI**(7):85.)

radial glial cells as guides in their journey to the periphery (Fig. 12-9). The radial glial cells, which ultimately become a form of astrocyte, are very prominent during the period of active neuronal cell migration. Their cell bodies are located close to the ventricular lumen, but a long process from each cell extends nearly to the surface of the brain; the migrating neurons are guided along their way by these radial processes. In multilayered areas of brain cortex, large neurons, constituting the innermost layer, migrate out first. Subsequent layers of gray matter are formed by smaller neurons migrating through the first and other previously formed layers to form a new layer of neurons at the periphery, so that whatever the number of layers of cortical gray matter, the innermost layer is the first formed and the outermost layer is the last (Angevine and Sidman, 1961). A mouse mutant called *weaver* is characterized by certain behavioral deficits related to abnormal function of the cerebellum. One of the defects in this mutant is an abnormality in the radial glial cells and a corresponding abnormal migration of the granule cells,

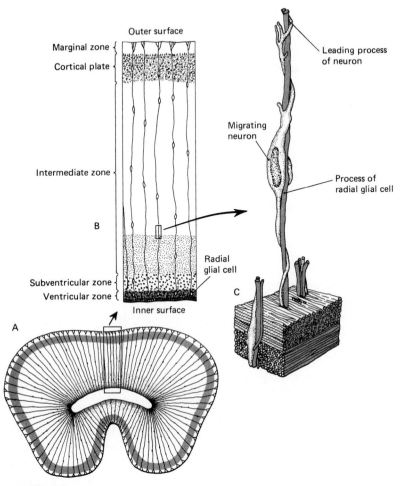

FIGURE 12-9
Schematic drawing illustrating the position of radial glial cells and their association with peripherally migrating neurons during development of the brain. (*After Rakic, 1975, Birth Defects* **XI***(7):100.*)

which form a characteristic layer in the cerebellar cortex (Rakic and Sidman, 1973).

The Meninges

The early neural tube is surrounded by a loose mesenchymal membrane. In the early fetal period two layers appear: a relatively thick outer layer of mesodermal origin which forms the *dura mater*, and a thin inner layer of presumed neural crest origin which ultimately becomes further subdivided into the thin *pia mater* and the middle *arachnoid layer*. Soon spaces which form within the pia-arachnoid layer become filled with the specialized *cerebrospinal fluid*, which bathes the spinal cord.

REGIONAL DIFFERENTIATION OF THE SPINAL CORD AND BRAIN

Spinal Cord

After the earliest period of histogenic activity, in which precursor cells begin to differentiate into neuronal and glial cells, development of the spinal cord is dominated externally by the development of the spinal nerves, one pair for each body segment.

During the period of development, when the neurons are differentiating and acquiring their sheaths, the spinal cord undergoes marked changes in its relations within the body. In young embryos the neural tube extends the entire length of the body and into the tail (Fig. 12-10). As the spinal column is formed, the growth of the neural arches encloses the spinal cord in the neural canal. Up to about 3 months the neural canal and the spinal cord are coextensive, and the segmentally arranged nerves pass outward through the intervertebral spaces directly opposite their point of origin. After this period, differential growth is such that neither the vertebral column nor the neural tube keeps pace with the expansion of the caudal part of the body (Fig. 12-11). The spinal cord lags much farther behind than does the vertebral column. Because the nerves are already established before these changes in relations occur, they appear to be dragged out caudally and pass back through the neural canal until they arrive at the intervertebral space which was originally opposite their point of origin (cf. Fig. 12-11A to 12-11D). Since the cephalic parts of the two systems are fixed with reference to each other, the extent of displacement is progressively greater in the more caudal regions. In a human fetus at term the spinal cord ends at about the level of the third lumbar vertebra, except for a small vestigial strand (*filum terminale*) representing the regressing terminal portion of the primitive neural tube (Fig. 12-11D). Postnatally, this differential growth continues until adulthood, when the end of the cord usually lies near the level of the first lumbar vertebra. Thus the sacral and coccygeal nerves emerging from the cord course almost directly downward for a considerable distance. The group of nerves thus pulled out in the lower portion of the spinal canal constitutes the *cauda equina*, so called because of its fancied resemblance to a horse's tail.

Brain

The development of the major adult divisions of the brain is shown in Fig. 12-12 and 12-13. Transcending several of the named regional areas of the brain is the *brainstem* (Fig. 12-15), a region which preserves quite closely the basic arrangement of the neural tube. It is a region through which bundles, or tracts, of nerve fibers pass to and from the brain. Through the basal plate, efferent

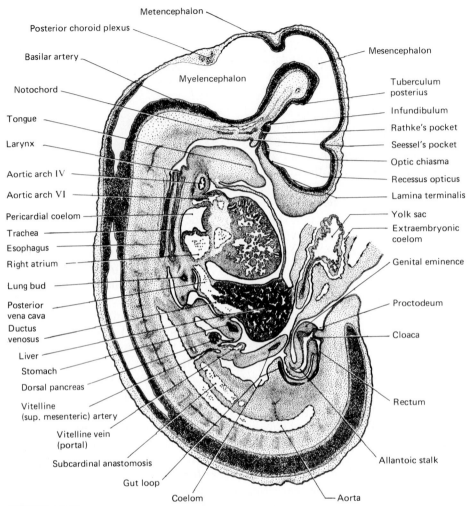

FIGURE 12-10
Sagittal section of 10-mm pig embryo. The general arrangement of internal structures is essentially similar to that in human embryos of the sixth week.

motor nerve fibers originating farther up in the brain descend toward their destinations in the spinal cord; through the region of the alar plate, bundles of afferent sensory nerve fibers head toward some of the integrating regions of the brain.

Superimposed on the fundamental tract-bearing portions of the brainstem are specializations that set the brain up as different from the spinal cord. Throughout the length of the brainstem are *nuclei* (aggregations of nerve cells), which serve as relay and integration centers for signals relating to specific bodily functions. For example, nuclei controlling basic respiratory functions

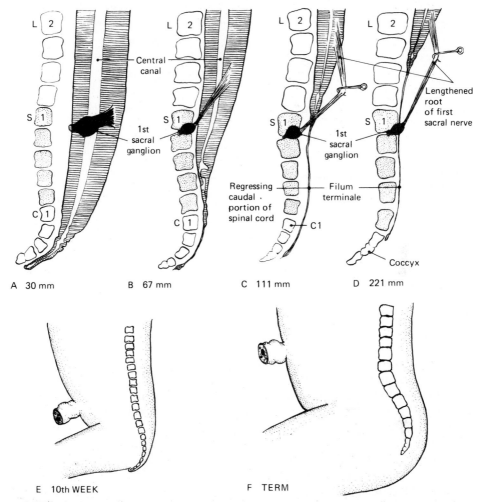

FIGURE 12-11

Diagrams showing changes in relations of caudal end of spinal column and spinal cord due to differential growth. (A–D) Relations of the first sacral nerve and ganglion at different ages used as an indicator of the changing position of the spinal cord within the spinal canal. (*After Streeter, 1919, Am. J. Anat., vol. 25.*) (E) Silhouette of embryo of 10 weeks; (F) at term, showing the shift cephalad of caudal part of spinal column. (*Redrawn, with some modification, after Schultz.*)

are located in the medulla. Direct input into a motor control from the brainstem is provided by the *cranial nerves*, which relate to the brainstem in much the same manner as the spinal nerves relate to the spinal cord.

Other important structures form secondarily off major regions of the brainstem. Late in development the *cerebellum*, the main motor coordinating center of the body, arises as a very complex outgrowth on the dorsal side of the *metecephalon*. On the dorsal surface of the *mesencephalon* appear the less

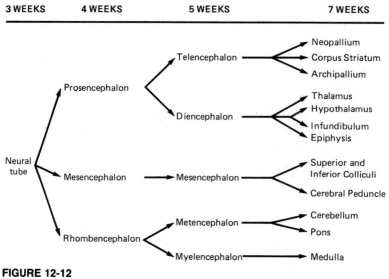

FIGURE 12-12
Summary of the major stages in development of the subdivisions of the human brain.

prominent but equally important *superior and inferior colliculi*, collectively called the *corpora quadrigemena* (Fig. 12-15), which are important processing centers for visual and auditory reflexes, respectively.

Higher in the brain (*diencephalon*) the original neural tube undergoes a number of striking local changes. Very early in development the optic vesicles arise as outpocketings from the ventrolateral walls of what will become the diencephalon. From the floor appears a median diverticulum called the *infundibulum*. As will be seen later, the infundibulum becomes enwrapped with *Rathke's pocket* (an extension of the stomodeal ectoderm) and will ultimately form the posterior pituitary gland.

The dorsal parts of the lateral walls of the diencephalon become greatly thickened by multiplication of the neuroblasts in the mantle layer and form the *thalamus*, a massive relay center that serves as the gateway of fibers passing from the brainstem and spinal cord to the cerebral hemispheres. Also in the diencephalon, above the pituitary gland, is the *hypothalamus*, another major integrating center that receives messages from many sources and converts them to hormonal signals mediated through the pituitary gland. The roof of the diencephalon gives rise to a small outgrowth, the *epiphysis* or *pineal body*, an endocrine gland which functions via the hypothalamus to modulate long-term reproductive cycles.

To some extent, the embryonic development of the telecephalon mirrors its increasing prominence during phylogeny. Of considerable prominence early in development is a region known as the *paleocortex* (*paleopallium*), which includes the structures involved in olfaction, starting with the first cranial nerve. The *archicortex* (*archipallium*) is another phylogenetically old system that is heavily involved in emotional and affective behavior. The archicortex, which includes the *hippocampus* and related structures, is relatively larger in lower mammals compared with higher mammals. Because of their complexity, the anatomical details of the older parts of the telencephalon will not be dealt with here. In humans, about 90 percent of the telencephalon is occupied by the *neocortex* (*neopallium*), the phylogenetically most recent part of the brain, which makes up the cerebral hemispheres. Although they appear relatively late in development, the cerebral hemispheres grow at a rapid rate and soon overshadow much of the upper part of the brainstem (Figs. 12-14 and 12-15).

Brain Vesicles and Cerebrospinal Fluid

As the brain takes shape, the simple central canal of the neural tube undergoes a number of changes to accommodate the various regions of the growing brain (Fig. 12-16). Expansions of the central canal become the four major *ventricles* of the brain (Fig. 12-13). The first two ventricles are found inside the expanding cerebral hemispheres. These lead into a median third ventricle that occupies both telencephalon and diencephalon. The central canal remains closest to its original form in the mesencephalon as the *aqueduct of Sylvius*, but it again expands to form the fourth ventricle in the myelencephalon.

In the roof of the third and fourth ventricles, small blood vessels from the pia mater press into the ependymal roof and push it ahead of them into the ventricles. These freely branching groups of vessels and their overlying ependymal epithelium are called the *anterior* and *posterior choroid plexuses*. These plexuses, plus others that form in the walls of the lateral ventricles (I and II) of the telencephalon, secrete *cerebrospinal fluid*, a clear fluid which bathes the tissues of the brain and spinal cord.

THE DEVELOPMENT OF A PERIPHERAL NERVE

The Components of a Peripheral Nerve

Regardless of their location in the body, most peripheral nerves have a similar organization (Fig. 12-17B). Cell bodies of *motor* (efferent) *neurons* in the ventral horn of the spinal cord send out *axons* that leave the cord through the *ventral root* and proceed in groups (*fascicles*) toward the periphery of the body, where groups of axons enter and innervate individual muscles. In higher vertebrates, the cell bodies of neurons innervating specific muscles are aggregated into fairly well circumscribed *motor pools* within the gray matter of the spinal cord. A given motor nerve fiber (axon) sends out several to several hundred branches, each of which terminates on an individual muscle fiber at the *neuromuscular junction* (*motor endplate*). This forms the basis for a *motor unit*.

The cell bodies of *sensory* (afferent) *neurons* are located in the *dorsal root ganglion* (Fig. 12-17B). These neurons send processes into the spinal cord through the dorsal root and long peripheral processes (*dendrites*) to the skin or

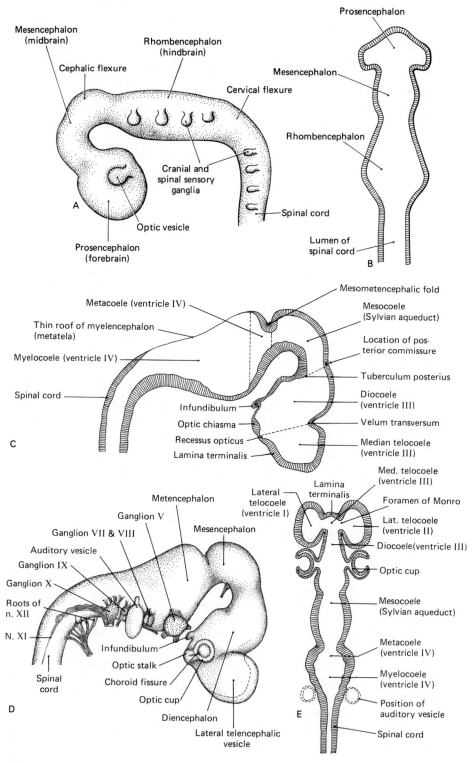

Mesencephalon (midbrain)

Cephalic flexure

Rhombencephalon (hindbrain)

Cervical flexure

Cranial and spinal sensory ganglia

Optic vesicle

Prosencephalon (forebrain)

A

Prosencephalon

Mesencephalon

Rhombencephalon

Spinal cord

Lumen of spinal cord

B

Metacoele (ventricle IV)

Thin roof of myelencephalon (metatela)

Myelocoele (ventricle IV)

Spinal cord

Infundibulum

Optic chiasma

Recessus opticus

Lamina terminalis

C

Mesometencephalic fold

Mesocoele (Sylvian aqueduct)

Location of posterior commissure

Tuberculum posterius

Diocoele (ventricle III)

Velum transversum

Median telocoele (ventricle III)

Metencephalon

Ganglion V

Ganglion VII & VIII

Auditory vesicle

Ganglion IX

Ganglion X

Roots of n. XII

N. XI

Spinal cord

D

Infundibulum

Optic stalk

Choroid fissure

Optic cup

Diencephalon

Lateral telencephalic vesicle

Mesencephalon

Lateral telocoele (ventricle I)

Lamina terminalis

Med. telocoele (ventricle III)

Foramen of Monro

Lat. telocoele (ventricle II)

Diocoele (ventricle III)

Optic cup

Mesocoele (Sylvian aqueduct)

Metacoele (ventricle IV)

Myelocoele (ventricle IV)

Position of auditory vesicle

Spinal cord

E

440

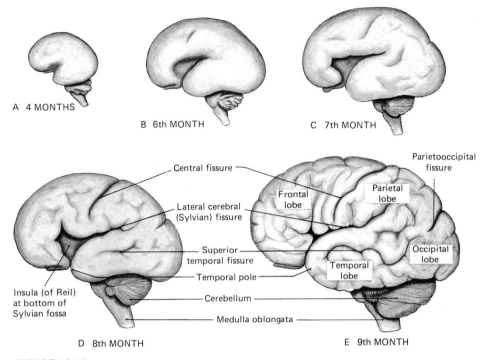

FIGURE 12-14
Lateral view of fetal brains at various stages in development. (*Modified from Retzius.*)

other specialized sensory structures, such as the *muscle spindles* (stretch receptors) in skeletal muscles. Near the spinal cord, the dorsal and ventral roots join and the afferent and efferent processes collect into a *peripheral nerve*.

In addition, components of *autonomic* (visceral) *nerves* pass through some parts of peripheral nerves. Cell bodies of efferent neurons arise in the *lateral horn* of the gray matter in the spinal cord (Fig. 12-7), and their axons leave the cord along with the motor axons in the ventral root. They soon branch out, however, into a *ramus communicans*, which can lead to a purely autonomic nerve or *ganglion*. Here a synapse with a second (second-order) autonomic neuron can occur.

Neuronal processes are either *myelinated* or *unmyelinated*. In the central nervous system, myelinated nerve processes constitute much of the white

FIGURE 12-13
Diagrams showing the topography of the brain in early human embryos. (A) Surface view of the 3-vesicle stage. (B) Schematic frontal plan of the brain at this stage. (C) Sagittal section through the five-vesicle stage. The conventional lines of demarcation between the adjacent brain vesicles are indicated by broken lines. (D) Surface view of brain with position of cranial ganglia and nerve roots indicated. (E) Schematic frontal plan of brain as it would appear if the flexeures had all been straightened out before cutting.

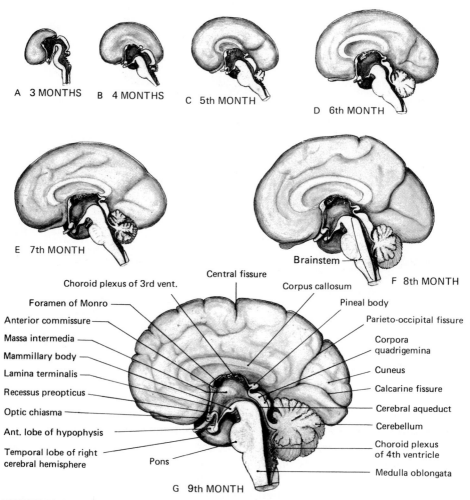

A 3 MONTHS   B 4 MONTHS   C 5th MONTH   D 6th MONTH

E 7th MONTH   F 8th MONTH

Brainstem

Central fissure
Choroid plexus of 3rd vent.
Corpus callosum
Foramen of Monro
Anterior commissure
Massa intermedia
Mammillary body
Lamina terminalis
Recessus preopticus
Optic chiasma
Ant. lobe of hypophysis
Temporal lobe of right
cerebral hemisphere
Pons

Pineal body
Parieto-occipital fissure
Corpora quadrigemina
Cuneus
Calcarine fissure
Cerebral aqueduct
Cerebellum
Choroid plexus of 4th ventricle
Medulla oblongata

G 9th MONTH

FIGURE 12-15
Sagittal sections of fetal brains at various stages of development (*Modified from Retzius.*)

matter and unmyelinated nerve fibers are found within the gray matter. *Myelin* is a multilayered sheath of phospholipid material which is wrapped around nerve processes like layers of a jelly roll (Fig. 12-19), and it serves as a form of insulation. In the central nervous system *oligodendroglial cells* are the myelinating agents, whereas in the peripheral nervous system the myelination of nerve fibers is accomplished by *Schwann cells.*

Axonal Outgrowth

The formation of a nerve begins with the outgrowth of axonal processes from some of the large motor neurons within the basal plate of the spinal cord or

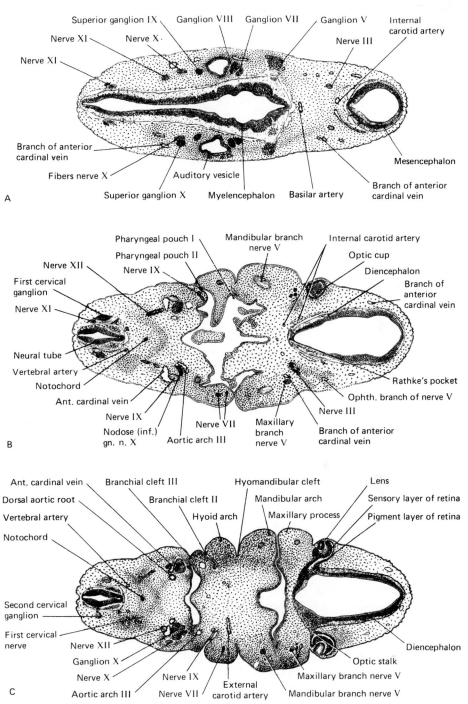

FIGURE 12-16
Three transverse sections from the series of the 9.4-mm pig embryo used in making the reconstruction illustrated in Fig. 18-9. (Projection drawings, ×15.) By laying a straightedge across either of these reconstructions at the level of the marginal line bearing the serial number of a cross section, its relations within the body as a whole are precisely indicated. (A) Section 96 passing through the myelencephalic region; (B) section 142 passing through the pharynx; (C) section 159 at the level of the optic cups.

443

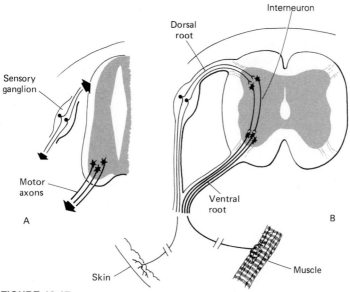

FIGURE 12-17
Diagrams illustrating the formation of a peripheral nerve. (A) Early stage, showing outgrowths of axons (arrows) from developing motor neurons in the basal plate of the spinal cord and the outgrowth of sensory processes toward both the spinal cord and the periphery. (B) Outgrowth completed, with nerve processes extending between peripheral end organs to the spinal cord.

brainstem (Fig. 12-17). The demonstration by Harrison (1910) that embryonic nerves grow by the progressive extension of processes from the nerve-cell body put to rest decades of debate regarding not only the mechanism of nerve growth but also the structural nature of the nervous system itself. In this experiment Harrison explanted cells from the neural tubes of frog embryos into serum in a hollow in a glass slide and watched the processes grow out (Fig. 12-18). An important problem was solved, and in the process a new technique—tissue culture—was established.

The first processes which grow out in the formation of a nerve are called *pioneering fibers*, and they typically arise from large neurons. The tip of the slender process of an outgrowing nerve fiber is expanded to form a *growth cone*, from which numerous fine pseudopodial processes (*filopodia*) actively extend and retract (Fig. 12-18). If this outgrowth is followed, it can be seen to advance by ameboid movements, tending to progress along anything that furnishes a favorable substratum along which it may move. The growth cone appears to pull out behind itself a long slender cytoplasmic process (the younger nerve fiber), but in reality much of the motive force for outgrowth lies in the cell body, which actively produces and sends down the growing nerve fibers the materials for further growth.

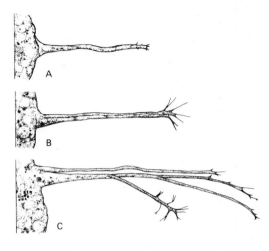

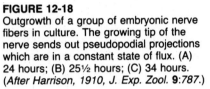

FIGURE 12-18
Outgrowth of a group of embryonic nerve fibers in culture. The growing tip of the nerve sends out pseudopodial projections which are in a constant state of flux. (A) 24 hours; (B) 25½ hours; (C) 34 hours. (*After Harrison, 1910, J. Exp. Zool.* **9**:*787.*)

The continued production and transport of cytoplasmic materials down the nerve fibers is called *axonal transport* (Weiss and Hiscoe, 1948), and it is a major means by which a neuron is able to maintain continued and adaptive contact with the periphery. Even in the adult, if a nerve is severed, the part of the nerve distal to the cut degenerates but new growth cones form at the ends of the proximal processes and renewed outgrowth, in this case called *regeneration*, brings the nerve back into contact with its end organ.

How does an outgrowing process of a peripheral nerve find its proper place in the body? Over the years many hypotheses have been put forth, but there is still no definitive answer. Most peripheral nerves seem to grow out by a continued testing of the microenvironment in the region of the growth cone by the filopodia. Direct observation has shown that the filopodia retract from unfavorable substrates but that they remain fixed to a favorable substratum and direct the entire nerve processes in that direction. In vitro experimentation (Weiss, 1934) has shown that the physical nature of the substratum is an important determinant of the direction and nature of outgrowth. This has been called *contact guidance*.

At a finer level than mechanical channeling, it is now well established that growth cones sample their microenvironment. Some chemical substrates are definitely preferred over others as surfaces over which nerve processes elongate. In one experiment conducted in vitro, growing nerve fibers confronted with a grid of polyornithine surrounding squares of palladium preferentially chose the former as a pathway (Letourneau, 1982). Nerve fibers show definite preferences for certain types of extracellular matrix, with *laminin* being a favored substrate. In fact, a currently popular hypothesis regarding the inability of nerve processes to regenerate in the adult brain or spinal cord is a deficiency of laminin as a substrate to support the regenerative elongation of axons and dendrites.

After languishing in disfavor for a number of decades, hypotheses regarding the response of growing nerve processes to electrical influences are now accorded greater respectability. A number of experiments have shown that nerve fibers do respond to electrical fields and that growing *neurites* (nerve processes) have a predilection for regions of electronegativity rather than regions with a greater positive charge.

Largely by elimination, many neurobiologists have been led to hypothesize that growing nerve fibers may be led along by or attracted to chemical gradients that act over short distances (perhaps a few hundred microns). However, very few concrete examples of chemical gradients or growth-inducing substances have been clearly identified. A notable exception is *nerve growth factor*, a well-characterized protein that stimulates the growth of sensory and sympathetic nerves (Levi-Montalcini, 1976).

A number of recent investigations have provided evidence that local factors, such as cell density, can be important determinants of axonal outgrowth. Tosney and Landmesser (1985) have proposed the following order of preference for the choice of substrate by outgrowing nerves: (1) other outgrowing neurites, (2) loosely packed cells or low concentration of glycosaminoglycans, and last, (3) densely packed cells or high concentrations of glycosaminoglycans. At the level of the individual axon, it has been difficult to rule in or rule out single factors that determine the paths of growth of a nerve in vertebrates, although in invertebrates there is evidence for highly specific chemical cues that determine branch points, for instance, as a nerve develops (Goodman and Bastiani, 1984).

Development of a Nerve to the Limb

This discussion will be based on studies of the innervation of the avian hindlimb (Fig. 11-17), a system that has received considerable attention in recent years (Landmesser, 1984). In all, about 20,000 motor neurons send axons into the hindlimb. As axons first grow out from the spinal cord, they soon come together in aggregates (spinal nerves) and preferentially pass through the anterior portions of the somites, possibly because the cells are less densely packed in the anterior than in the posterior part of the somite. When Keynes and Stern (1984) rotated somites, the nerve processes preferentially sought out the original anterior regions of the somites.

The neurites then pass through the loose mesenchyme that marks future openings in the pelvic girdle (they do not penetrate the precartilage mesenchyme of the girdle itself) and enter an anatomically complex *plexus* of nerve fibers at the base of the limb bud. The tips of the neurites remain in the plexus for nearly a day before outgrowth resumes. The reason for the delay is not clear, but by the time the neurites leave the plexus, a considerable degree of order has emerged and nerve fibers that will innervate specific end organs begin to travel together as discrete bundles (fascicles). The neurites leave the plexus in dorsal or ventral trunks, which lead to the dorsal and ventral muscle masses in the limb bud. The cues that lead the nerve fibers to choose dorsal or ventral roots seem to be local ones within the plexus, because correct choices are made even if the limb buds are experimentally altered, as in the creation of limb buds with two dorsal halves.

The major nerve trunks leading into the limb bud form along early "highways" through which the outgrowing nerve processes collect. Even neurites from totally foreign nerves (e.g, those to the wing) can follow the main nerve trunk highways in the leg. Why the neurites follow specific routes is not entirely clear. It is known that early muscle masses (see Chap. 11) are not required, for if they are prevented from forming, the main nerve trunks still grow out in a normal pattern (Lewis et al., 1981). However, some cues from individual muscles seem to be important in causing nerve fibers to branch out from the main trunks to supply the individual muscles. The nature of these cues is obscure, but displaced nerve fibers will deviate from their course to innervate the correct muscle.

Once within the early muscle primordia, the axons send branches to large numbers of early muscle fibers (Bennett, 1983). During late embryonic life and even in the early neonatal period, it is common for branches of more than one motor axon to innervate a single muscle fiber. Shortly after birth, however, a process of sorting out begins. This results in the retraction of extra axons from individual muscle fibers and their eventual innervation by only one motor axon.

Sensory nerve fibers also enter developing muscles, and late in the prenatal period they induce certain groups of immature muscle fibers to form *muscle spindles*. Other sensory nerve fibers form specialized terminals in the tendons and become *Golgi tendon organs*. Both these and the spindles act as *stretch receptors* (*proprioceptors*) of muscles.

In addition to the peripheral connections, the neuronal processes leading into the spinal cord must make the proper connections in order for coordinated movements of the limb to occur. At the simplest level, that of a *segmental reflex*, central processes (those in the cord) from the sensory neurons connect with those of the motor neurons, often through a short interneuron (Fig. 12-17B). Such simple connections enable simple reflexes, such as drawing one's hand away from a hot stove, to occur. Other central processes join with those from other segments of the spinal cord to form large tracts of nerve fibers leading to the brain through the white matter of the spinal cord. Here they go to the main motor and sensory centers of the brain, where complex integrative actions lead to the generation of volitional coordinated movements.

As well as making the proper connections, the developing nerve fibers must become insulated from one another. The myelination of peripheral nerve fibers is accomplished by the wrapping of portions of the Schwann cells many times around the nerve processes (Fig. 12-19). Some peripheral nerve fibers are considered to be unmyelinated, but even these are partially embedded within the cytoplasm of Schwann cells.

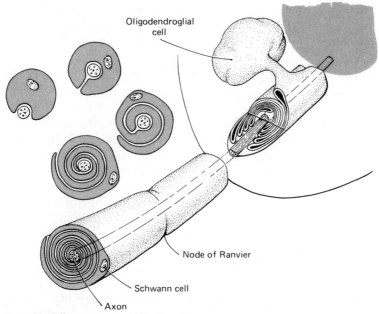

FIGURE 12-19
Myelination in the central and peripheral nervous systems. In a peripheral
nerve myelination is accomplished by Schwann cells wrapping around the
axon many times like a jelly roll (upper left). In the central nervous system oligodendroglial cells are the myelinating agents.

Autonomic Nerves

Accompanying the motor and sensory nerve fibers within the proximal parts of
spinal nerves are efferent and afferent fibers of the involuntary, or *autonomic*,
nerves. The efferent nerve fibers leave the spinal cord through the ventral roots
along with the motor axons, but then they leave the nerve through a separate
branch and make connection with the cell bodies of second-order neurons of
neural crest origin, located in chains of ganglia that run on each side of the
body ventral to the vertebral column or in other, isolated ganglia scattered
throughout the thorax and abdomen. These second-order neurons, which are
part of the *sympathetic* division of the autonomic nervous system, send out
processes to the visceral structures, including blood vessels and skin glands.
Afferent fibers return from these structures to cell bodies, which are located
in the dorsal root ganglia. Central connections are then made with the spinal
cord.

Studies with the chick-quail marker system (Le Douarin, 1986) have shown
that neural crest derived from levels caudal to somite 5 gives rise to components of the sympathetic nervous system, whereas precursors of the parasympathetic system are confined to the levels of the first seven somites and caudal
to somite 28 (Fig. 12-20).

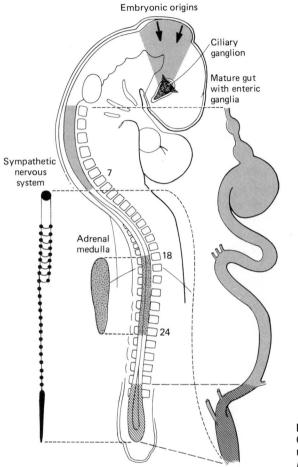

FIGURE 12-20
Origins of the avian autonomic nervous system. (*After N. Le Douarin, 1986.*)

The differentiation of autonomic neurons involves at least two major steps (Bunge et al., 1978). The first occurs early in the development of the neural crest, when some cells become determined to be components of the autonomic nervous system. These cells then migrate down the appropriate pathway and settle down in relation to a specific end organ. Despite the well-defined fates of cells originating at specific somitic levels, these appear not to be rigidly fixed. For example, if neural crest from the cephalic region (normally destined to become parasympathetic neurons) is transplanted to the level of somites 18 to 24, the cells migrate out and settle into the adrenal medulla, a functional component of the sympathetic nervous system (Le Douarin and Teillet, 1974). Conversely, if neural crest from the trunk (normally destined to become sympathetic neurons) is grafted into more cephalic levels, the cells migrate down the gut lining and differentiate into parasympathetic (enteric) neurons.

The second major stage in the differentiation of an autonomic nerve involves the choice of the neurotransmitter that the nerves will use (Black, 1982). Although it is becoming increasingly apparent that perhaps more than 10 different neurotransmitters are employed in various regions of the nervous system, the bulk of the autonomic nerves are either *cholinergic* (i.e., use acetylcholine as the transmitter) or *adrenergic* (noradrenergic) (i.e., use norepinephrine). The second-order (terminal) neurons of the parasympathetic division are cholinergic, whereas those of the sympathetic division are adrenergic. When, and even before, autonomic neurons first arrive at their final destination in the body, they are noradrenergic. Then they undergo a phase of definitive differentiation during which the neurons select the intrinsic neurotransmitter substance that characterizes their mature state. There is considerable evidence that early transmitter development proceeds independently of other differentiative events, such as the elongation of neurites and the innervation of target organs.

Even at a later stage of their development (after the last mitosis), autonomic neurons still have considerable liability with respect to the transmitter substance they produce (Patterson, 1978). Through the processing of microenvironmental signals from their end organs and possibly also the pattern of neural stimuli impinging on them from the central nervous system, autonomic neurons finally respond by committing themselves to becoming adrenergic or cholinergic.

Effect of the Periphery on Development of the Spinal Cord

Later development of the spinal cord is greatly influenced by its relationship with peripheral structures. Mere inspection of a mature spinal cord reveals swelling in the cervical and lumbar regions, from which the nerves to the upper and lower limbs arise. These differences are also reflected in the nerves themselves. The diameter of the nerves supplying the limbs is much greater than that of the nerves found in other regions of the trunk.

The relationship between peripheral load and the degree of development of the nervous system has been amply demonstrated in experimental studies. Reduction of the mass of peripheral tissue supplied by a given nerve is followed not only by a decrease in diameter of the nerve that would have supplied the tissue but also by a reduction in the size of the corresponding sensory ganglion and ventral (motor) horn. The reduction in size is largely due to a greater amount of neuronal cell death than is seen in controls. Even in normal development, large numbers of neurons die within the spinal cord (Fig. 12-21). Hughes (1961) found that in the lumbar region of the spinal cord in normal *Xenopus* almost 90 percent of the original motor neurons die by the time of metamorphosis. After removal of a limb bud, the effect is still more striking. Hamburger (1958) removed a limb bud in $2^{1}/_{2}$-day chick embryos and by 6 to 7 days found a massive wave of cell death which eliminated the entire population of 20,000 motor neurons normally destined to supply the limb. In contrast, the addition of peripheral tissue into the area supplied by a nerve (such as the

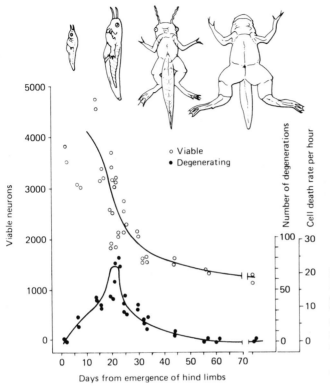

FIGURE 12-21
Numbers of viable and degenerating neurons in the ventral horn of the spinal cord of *Xenopus* at different stages of development. (*Adapted from Hughes, 1961, J. Embryol. Exp. Morph.* **9**:*269.*)

transplantation of a limb bud onto the flank of an early embryo) results in a dramatic increase in the size and number of nerve cells going to that area (Fig. 12-22; Detwiler, 1920).

The spinal cord is not the only part of the nervous system in which cell death is an important regulatory mechanism during development. Clarke (1985) has listed 15 cell groups in the developing vertebrate nervous system where from 18 to 100 percent of the neurons are lost as a result of cell death. The amount of cell death can be reduced by enlarging the target or reducing competition for a particular target of neurons. Conversely, neuronal death can be greatly increased by destroying the target or increasing the neuronal competition for the target.

In the rat certain muscles associated with sexual function in the male (e.g., the levator ani muscle) are extremely sensitive to testosterone. The same muscles form in the female embryo, but they soon regress because of a lack of hormonal stimulation. The regression is attended by the death of the motor neurons in the spinal cord that supply the muscles. If testosterone is given to females, the muscles persist and neuronal death is prevented. In addition to reacting to the status of the muscles (which are directly responsive to

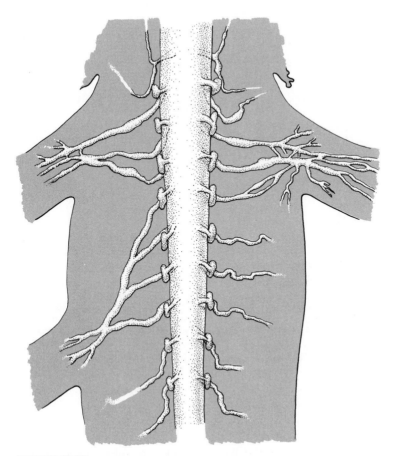

FIGURE 12-22
The spinal nerves of a normal salamander (*right*) and one to which an extra limb has been grafted (*lower limb on left*). The nerves (and corresponding sensory ganglia) to both normal and grafted limbs are larger than those supplying only the body wall. (*After Detwiler.*)

testosterone), the motor neurons also appear to be directly influenced by the hormone.

It has also been shown that specific chemical influences profoundly affect the development of some components of the nervous system. Nerve growth factor, an insulinlike protein that has been isolated from several normal and abnormal mammalian tissues, exerts a profound effect on the growth of sensory, and particularly sympathetic, neurons in a wide variety of vertebrate embryos (Levi-Montalcini, 1958). The demonstration of nerve growth factor has stimulated a search for other specific chemical factors which might exert an effect on the developing nervous system as well as on other components of the body.

THE CRANIAL NERVES

The spinal nerves are segmentally arranged, and all of them are built on the same general plan. The cranial nerves (Figs. 12-23 and 12-24), although most

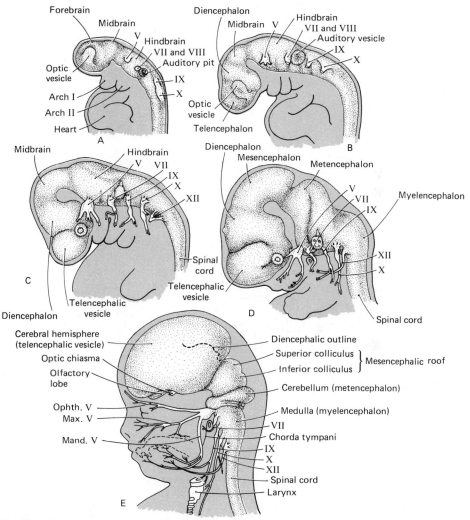

FIGURE 12-23
Five stages in early development of brain and cranial nerves. (*Adapted from various sources, primarily figures by Streeter and reconstructions in the Carnegie Collection.*) (A) At 20 somites; probable fertilization age of 3½ weeks. (*Based on the Davis embryo.*) (B) at 4 mm; fertilization age of about 4 weeks. (C) at 8 mm; fertilization age of about 5½ weeks. (D) At 17 mm; fertilization age of about 7 weeks. (E) At 50 to 60 mm; fertilization age of about 11 weeks. The cranial nerves shown are indicated by the appropriate roman numerals: V, trigeminal; VII, facial; VII, acoustic; IX, glossopharyngeal; X, vagus, XI, spinal accessory; XII, hypoglossal. Abbreviations: *Mand. V*, mandibular division of trigeminal nerve; *Max.V.*, maxillary division; *Ophth. V*, ophthalmic division of the fifth cranial nerve.

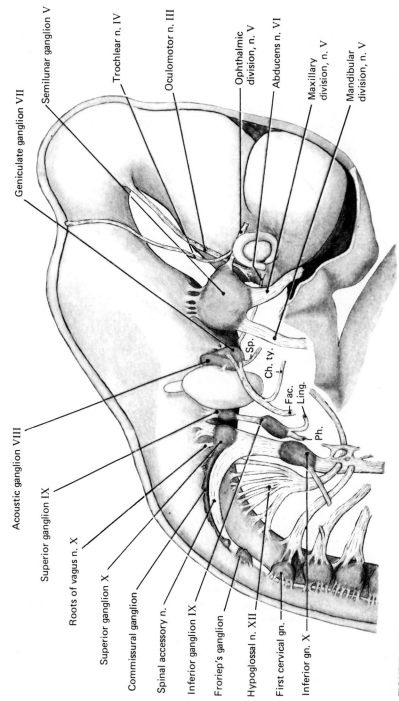

FIGURE 12-24
Reconstruction of the brain and cranial nerves of a 12-mm pig embryo. (*After F. T. Lewis, 1902, Am. Jour. Anat. vol. 2.*) Abbreviations: *Ch. ty.,* chorda tympani branch of the seventh (facial) nerve; *Fac.,* facial nerve; *Ling.,* lingual branch of ninth nerve; *Ph.,* pharyngeal branch of ninth nerve; *Sp.,* greater superficial petrosal nerve.

TABLE 12-1
SUMMARY OF CRANIAL NERVES

Cranial nerve	Associated component of central nervous system	Functional components	Distribution
Olfactory (I)	Telencephalon	Special sensory (olfaction)	Olfactory area of nose
Optic (II)	Diencephalon	Special sensory (vision)	Retina of eye
Oculomotor (III)	Mesencephalon	Motor Autonomic (minor)	Intraocular and four extraocular muscles
Trochlear (IV)	Mesencephalon	Motor	Superior oblique muscle of eye
Trigeminal (V)	Metencephalon	Sensory Motor (some)	Derivatives of branchial arch I*
Abducens (VI)	Myelencephalon	Motor	Lateral rectus muscle of eye
Facial (VII)	Myelencephalon	Motor Sensory (some) Autonomic (minor)	Derivatives of branchial arch II*
Auditory (VIII)	Myelencephalon	Special sensory (hearing, balance)	Inner ear
Glossopharyngeal (IX)	Myelencephalon	Sensory Motor (some) Autonomic (minor)	Derivatives of branchial arch III*
Vagus (X)	Myelencephalon	Sensory Motor Autonomic (major)	Derivatives of branchial arch IV*
Accessory (XI)	Myelencephalon Spinal cord	Motor Autonomic (minor)	Gut, heart, visceral organs Some neck muscles
Hypoglossal (XII)	Myelencephalon	Motor	Tongue muscles

*See Fig. 16-3.

likely derived from segmental structures in the phylogenetic past, have lost some of their regular segmental arrangement and have become very highly specialized (Table 12-1). One of the major phylogenetic changes has been the lack of union of the dorsal and ventral roots of the cranial nerves. Instead, the cranial nerves of mammals represent for the most part either the primitive dorsal or the primitive ventral roots.

The cranial nerves can be subdivided into several groups on the basis of their embryological origin and overall function. The first two cranial nerves (the olfactory and optic) can be considered to be extensions of brain tracts rather than true nerves. Several of the cranial nerves (III, IV, VI, and XII) appear to have evolved from primitive ventral (motor) roots. Nerves V, VII, IX, and X

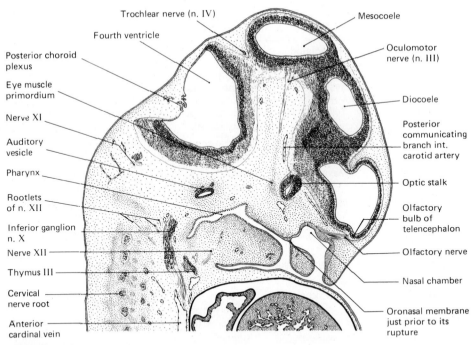

FIGURE 12-25
Projection drawing of parasagittal section of head of 15-mm pig embryo. The section is to the right of the midline, in a plane especially favorable for showing the relations of the nasal pits and the olfactory and oculomotor nerves.

are mixed motor and sensory nerves, and each nerve supplies the derivatives of a different branchial arch (Figs. 16-3 and 16-5). Traditionally, the motor components of these nerves have been placed into a separate functional category (special visceral efferent), but recent data showing that the branchial-arch muscles are derived from somitomeres suggest that these muscles do not differ greatly in origin or type from other muscles derived from somites.

The sensory components of nerves V, VII, IX, and X, as well as the auditory nerve (VIII) (Fig. 12-26), have a more complex origin. Neurons of some of the ganglia are derived from neural crest, whereas neurons of other ganglia or parts of ganglia are derived from ectodermal placodes (Figs. 12-27 and 12-28). Ectodermal placodes are more fully discussed in the introduction to Chap. 13.

THE DEVELOPMENT OF NEURAL FUNCTION IN THE EMBRYO

When dealing with the intricacies of morphology and mechanisms in the developing nervous system, one must not lose sight of the main reason for having a nervous system, namely, the generation and coordination of most of the functional activities of the body. In many developing systems form and function are closely linked. This is particularly true in the nervous system,

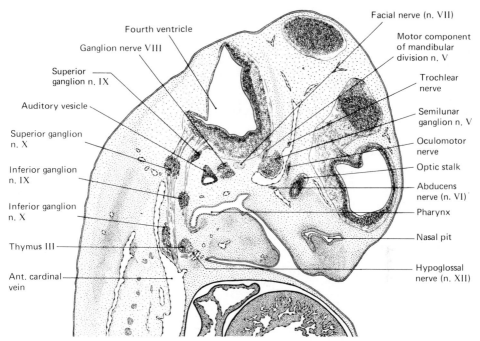

FIGURE 12-26
Projection drawing of parasagittal section of head of 15-mm pig embryo. The plane of section is slightly farther to the right than that shown in the preceding figure. It is particularly favorable for showing the position of origin and the ganglia of the trigeminal, glossopharyngeal, and vagus nerves.

where both the appropriate neural pathways and the end organs must be sufficiently mature to permit function. Early studies of neural function involved the gross examination of movement patterns and reflexes in embryos and fetuses. More recently, behavioral testing has been correlated with anatomical and physiological studies of the segments of the nervous system involved in the function under study.

The Development of Motor Function in the Human Embryo

Many aspects of development of the neuromuscular system are reflected in the movements of the embryo as a whole. Until about 6 weeks, the embryo merely rocks back and forth passively in its bath of amniotic fluid, with its only intrinsic movements being the regular beating of the heart. Toward the end of the second month the embryo makes its first spontaneous movements. These are simple side-to-side twitchings which serve to indicate the functional maturation of the musculature in the body wall. Actually, about a week earlier the embryo is capable of making weak twitches in the neck in response to striking the lips or nose with a fine bristle (Hooker, 1952). This behavioral

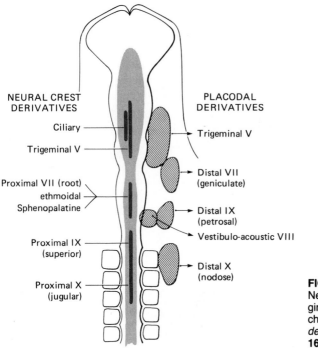

NEURAL CREST DERIVATIVES

Ciliary

Trigeminal V

Proximal VII (root)
ethmoidal
Sphenopalatine

Proximal IX
(superior)

Proximal X
(jugular)

PLACODAL DERIVATIVES

Trigeminal V

Distal VII
(geniculate)

Distal IX
(petrosal)

Vestibulo-acoustic VIII

Distal X
(nodose)

FIGURE 12-27
Neural crest and placodal origins of sensory ganglia in the chick. (*After D'Amico and Noden, 1983,* Am. J. Anat. **166**:*445.*)

pattern signifies that the first functional reflex arcs have been laid down. In succeeding days, the twitching movements and simple reflex capabilities extend progressively caudad, in conformance with the general cephalocaudal rule of develoment.

By the start of the third month reflex movements of the hands can be elicited; they are followed a week later by foot reflexes. The beginnings of limb function are accompanied by the rapid development of facial movements. At 10 or 11 weeks early swallowing and the first movements that foreshadow the rhythmical breathing motions of the chest are seen. The fetus seems to anticipate its major postnatal requirements by developing swallowing, breathing, and grasping movements early. By the end of the third month of pregnancy almost the entire skin of the fetus is sensitive to touch, and although its movements are still feeble, the fetus is very active at this time. However, only rarely can the mother detect these movements, mainly because the fetus is still so small that it could comfortably fit into the palm of one's hand.

Until the twelfth week, the movement patterns of the fetus are highly irregular and the magnitude of a response often far exceeds the intensity of the stimulus. Soon, however, the pattern of largely undirected responses to a stimulus ceases and more predictable reflex responses begin to occur.

Toward the end of the fourth month the mother can often feel fetal movements ("quickening"). The grip of the fetus becomes stronger, and weak

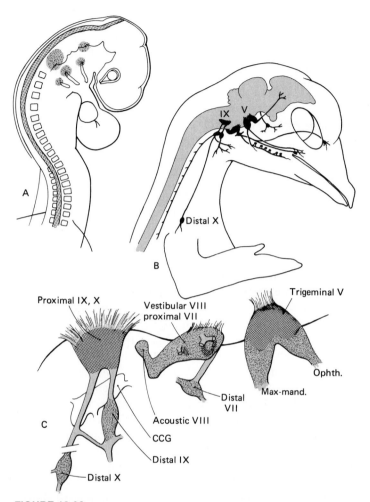

FIGURE 12-28
Contributions of neural crest and placodal cells to sensory ganglia in
the chick. The stipple and sand tones in the ganglia in (C) refer to the
primordia in (A). (*A, adapted from N. Le Douarin et al., 1986,* TINS
9:*175; B, adapted from D'Amico and Noden, 1983,* Am. J. Anat.
166:*445.*)

but nonsustained breathing movements are possible. The fetus soon begins to
alternate periods of rest with periods of activity. Very premature fetuses, born
late in the fifth month of pregnancy, are able to breathe spontaneously for short
periods of time, but these breathing movements cannot be sustained. The
remainder of pregnancy is occupied to a great extent by the maturation of reflex
and behavioral patterns which have already been set up. During this period the
development of sensory function proceeds rapidly and in a fairly regular
sequence. After the early appearance of general cutaneous sensation, the

functions of taste, balance (vestibular system), hearing, and vision mature, often in that sequence (Gottlieb, 1976; Bradley and Mistretta, 1975). The maturation of function continues for several years into postnatal life, and the sequence of functional changes often follows the morphological sequence of myelination of the nerves involved.

SEX HORMONES AND BRAIN DEVELOPMENT

Recent research has demonstrated clearly the existence of clear-cut differences between the brains of male and female laboratory mammals. These are reflected in a number of aspects of morphology, biochemistry, and behavior. As is the case with other sexually dimorphic structures in the body (see Chap. 17), the brain of mammals seems to have a baseline of structural and behavioral functions that is inherently female. Exposure of the central nervous system to circulating testosterone during a critical period of development (around or just after the time of birth in rats and before birth in primates) results in the embryo's being irreversibly imprinted with male characteristics.

If a male rat is castrated just before the critical period, it will exhibit typical female mating posture later in life, even if it is given testosterone. Conversely, a female rat given high doses of testosterone during the critical period does not show female sexual behavior (lordosis, or the arching of the back) but rather attempts to mount other females. The hormonal factors involved in sexual differentiation of the brain are complex. It is striking that the testosterone effect actually appears to be mediated by its conversion into estrogen in the brain. Estrogen levels in the early female brain are kept low because circulating estrogens outside the brain are avidly bound by α-*fetoprotein*, which prevents them from entering the brain and influencing its development at the critical period. One of the hormonal effects is a difference in the size of certain aggregations of neurons (nuclei) in the brain. In some cases, the differences can be seen with the naked eye (Gorski et al., 1980).

Birds have also been shown to exhibit striking effects of hormones on neural development. In certain finches, exposure to testosterone is correlated with singing behavior. In normal males, seasonal fluctuations in singing are directly correlated with changes in levels of testosterone, and females injected with testosterone break out into male song patterns (Nottebohm, 1980). These changes in behavior are related to significant changes in the growth of certain regions of the brain.

CHAPTER 13

THE SENSE ORGANS

The sense organs are an individual's windows to the surrounding world. It should not be surprising, therefore, that they are derived principally from the outer, ectodermal germ layer. The sense organs start out as thickened ectodermal placodes (Fig. 13-1), each one the result of a secondary inductive stimulus emanating from some part of the developing central nervous system.

The ectodermal placodes and their development show striking similarities to certain aspects of the neural crest. Gans and Northcutt (1983) have postulated that phylogenetically, placodes and neural crest are parallel derivatives of a single precursor, presumably the epidermal nerve plexus that controls many sensory, integrative, and even motor functions in hemichordates and proto-chordates. Although both neural crest and placodal cells possess the ability to migrate, placodes are confined to the head, in contrast to neural crest, which is distributed along the length of the body axis. The ectodermal placodes of vertebrates are arranged in two series. The *dorsolateral series* (Fig. 13-1) lies close to the neural crest and forms the special sense organs, whereas the ventrolateral, or *epibranchial, placodes* are more closely associated with the pharyngeal pouches. These latter contribute to the sensory ganglia of certain cranial nerves, particularly those that are associated with the taste buds (Fig. 12-28).

The vertebrate sensory organs that are functionally most important and structurally most complex are the eyes and ears. Their development will be described in detail in this chapter. The ears are composite structures in which much of the sound-collecting apparatus is derived from the branchial-arch system and the pharynx. The organs of smell and taste are morphologically so intimately bound with the overall development of the face and oral region that their embryogenesis will be treated together in Chap. 14. The cutaneous senses

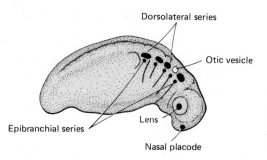

FIGURE 13-1
Ectodermal placodes in the head of an
amphibian (*Ambystoma*) larva. (*After
Yntema, 1937, J. Exp. Zool. 92:93.*)

are bound to the development of the peripheral nerves, which was described in
Chap. 12.

THE EYE

The vertebrate eye is a very complex organ, the constituents of which are
derived from several different primordial sources, both ectodermal and mesodermal, in the cephalic region of the embryo. Of necessity, its proper function
requires almost perfect alignment of the components directly involved in vision
as well as a high degree of transparency of those through which light rays pass.
The normal embryology of the eye demonstrates beautifully some of the
fundamental processes and concepts of development. Early development of
many of the components of the eye depends on inductive interactions between
one component and another. This is followed by a phase of intracellular
differentiation, starting with a burst of mitosis and then RNA synthesis leading
to the formation of specific classes of intracellular proteins in some cases and
extracellular fibers and matrix in others. In some aspects of eye development
the influence of extracellular materials and the migration of cells play an
important role. The development of the retina and its central connections
involves not only the differentiation of a very complex neural tissue but also
one of the best-documented cases of integration between one part of the
embryo and another. Finally, mechanical influences are important in obtaining
and maintaining proper alignment of the tissues in the visual pathway. Even
later in development, after many of the components of the eye have been laid
down, they may continue to exert sustained influences on one another.

Primary Optic Vesicle

After the initial induction of the neural ectoderm by the underlying *chordamesoderm*, the first morphological indication of eye formation is an outpocketing
of the wall of the diencephalon. In human embryos the evagination of the optic
vesicles begins very early. By the middle of the third week depressions appear
in the still open neural plate where it widens out in the future forebrain region.
These early depressions, called *optic sulci* (Fig. 13-2A), are the initial shallow

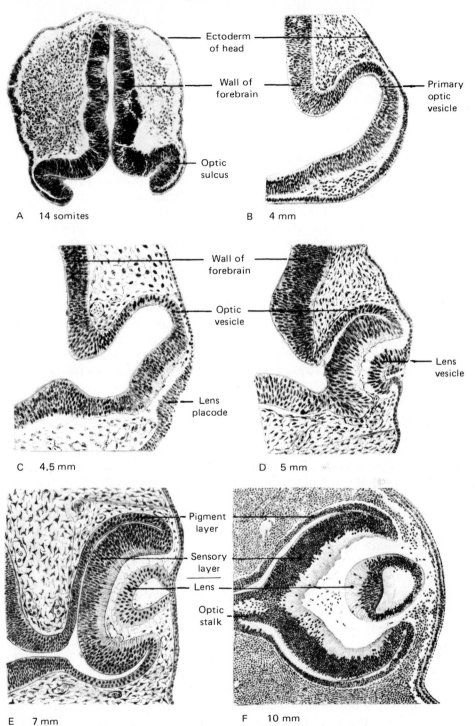

Ectoderm
of head

Wall of
forebrain

Optic
sulcus

A   14 somites

Primary
optic
vesicle

B   4 mm

Wall of
forebrain

Optic
vesicle

Lens
placode

C   4.5 mm

Lens
vesicle

D   5 mm

Pigment
layer

Sensory
layer

Lens

Optic
stalk

E   7 mm

F   10 mm

FIGURE 13-2
Early development of optic cup and lens in human embryos. Photomicrographs and drawings
from various sources placed in corresponding orientation and brought to same magnification
(×100). (*A, from Heuser; B, from Fischel; C–E, from Ida Mann; F, from Prentiss.*)

concavities that later become deepened to form the *primary optic vesicles*. After the neural plate in the brain region has been closed, the optic vesicles appear as rounded protuberances from the lateral walls of the forebrain (Fig. 12-23A).

Formation of the Optic Cup from the Primary Optic Vesicle

Toward the end of the fourth week, the lumen of the primary optic vesicle is still broadly continuous with the lumen of the forebrain and the walls of the vesicle differ little from the walls of the parent forebrain (Fig. 13-2B). At the start of the fifth week the distal portion of the optic vesicle begins to flatten (Fig. 13-2C), and very soon thereafter it invaginates so that the single-walled primary vesicle is transformed into the double-walled optic cup (Fig. 13-2D). As the invagination becomes more complete, the original lumen of the optic vesicle is reduced to a vestigial slit between the inner and outer layers of the newly formed cup (Fig. 13-2C to 13-2F). At the same time there is rapid differentiation of the two layers of the cup. The outer layer becomes much thinner, and by the sixth week in human embryos, it begins to show melanin granules, which foreshadow its ultimate conversion into the *pigment layer of the retina*. The inner "collapsed" layer of the cup becomes much thickened, an indication that it has begun the elaborate series of changes by which it will become the *sensory layer of the retina*. This is the layer that will receive the visual images and convert them into signals which will be transmitted to other regions of the brain via the optic nerve.

The invagination that forms the optic cup occurs not at the center of the optic vesicle but eccentrically, toward its ventral margin. This makes a gap in the continuity of the wall of the optic cup which is known as the *choroid fissure* (Fig. 13-3). As development of the eye progresses, the choroid fissure partially envelops the *hyaloid artery*, or *central artery of the retina* (Fig. 13-3E), a branch of the internal carotid artery which in the embryo supplies many of the structures within the eyeball. The *optic stalk* (Fig. 13-3B and C), along part of which the choroid fissure extends, is invaded by processes of the nerve cells which differentiate in the sensory layer of the retina. After these processes have grown down the wall of the optic stalk, on their way toward making synaptic connections in the brain, the optic stalk is known as the *optic nerve*.

Establishment of the Lens

Before it begins to invaginate, the primary optic vesicle comes into very close contact with the overlying head ectoderm (Fig. 13-2B). During this time the ectoderm, which has already been subjected to preparatory influences from other cephalic tissues, is induced by the optic vesicle to form lens (Fig. 13-4) (Spemann, 1938; Lewis, 1904). The effects of the inductive event are first made manifest at about the end of the fourth or the beginning of the fifth week (human embryos of 4 to 5 mm), when the superficial ectoderm immediately overlying

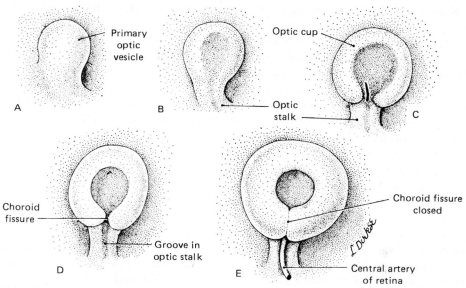

FIGURE 13-3
Stages in the formation of the optic cup and the choroid fissure. (*Adapted from several sources, primarily Streeter.*)

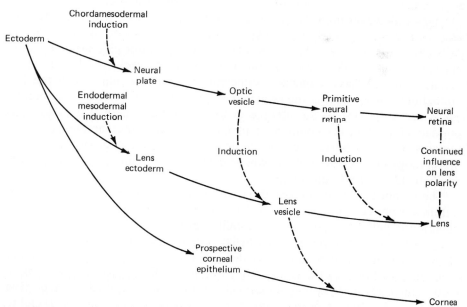

FIGURE 13-4
Scheme of major inductive events occurring in the embryonic eye. Inductive events or tissue interactions are indicated by broken arrows.

the optic cup begins to develop a local thickening known as the *lens placode* (Fig. 13-2C).

The importance of close contact between the optic cup and the overlying ectoderm in lens induction is demonstrated by numerous experiments in which these structures have been separated either by increased distance or by the interposition of foreign objects placed between them. In most cases there is an interference with lens formation. These experimental results are supported by natural experiments in the form of genetic mutants in which lens formation fails to occur. In the *eyeless* mutants in mice the optic cup does not come as close to the surface ectoderm as it should, and in the *fidget* mutant a reduction in mitosis in the optic cup interferes with proper contact with the ectoderm.

As the cavity in the optic cup deepens, the lens placode is invaginated into the cup to form an open *lens vesicle* (Fig. 13-2D). During the fifth week, the lens vesicle is closed (Fig. 13-2E) and then breaks away completely from the parent ectoderm to constitute a rounded epithelial body lying in the opening of the optic cup (Fig. 13-2F). Before the end of the sixth week, the cells on the deep pole of the lens begin to elongate, presaging their transformation into the long transparent elements known as *lens fibers* (Figs. 13-2F and 13-9A).

Differentiation of the Lens

The development of the lens into a transparent structure with the appropriate optical qualities involves a highly orchestrated sequence of intracellular differentiative events culminating in the synthesis of a specific class of lens proteins called *crystallins* (Piatigorsky, 1981). As we have just seen, the early lens vesicle soon undergoes an asymmetrical development, with the deeper cells elongating into lens fibers and the cells of the outer pole retaining a low epithelial configuration. By the end of the seventh week, the lens fibers in human embryos have elongated sufficiently to make contact with the lens epithelium, thus reducing the original cavity in the lens vesicle to a potential slit. During the rest of the lifetime of the individual, there is a progressive buildup of the fibrous component of the lens. The outer epithelial portion remains, but its relative prominence decreases greatly. A broad view of the morphology of lens development can be gained by examining in sequence Figs. 13-2C to 13-2F, 13-9, and 13-10.

The essence of cytodifferentiation within the lens is the transformation of mitotically active epithelial cells from the outer pole of the lens to the elongated postmitotic cells, containing the crystalline proteins, in the main body of the lens. The transformation from epithelial cells to lens fibers takes place in an equatorial region which surrounds the entire lens (Fig. 13-5). The low lens epithelium just outside the equatorial region is mitotically active throughout in embryonic eyes, but postnatally the cells in the central epithelial region (Fig. 13-5) cease dividing. A germinative region of dividing cells remains. Daughter cells from this region move into the equatorial zone, where they lose the ability to divide and begin to elongate. At the same time, the cytological correlates of

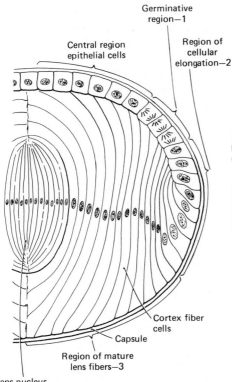

Germinative
region—1

Central region
epithelial cells

Region of
cellular
elongation—2

MORPHOLOGICAL
CHARACTERISTICS

1

Basophilic
Rough endoplasmic reticulum
Cells replicate

2

Cell volume increases
Nuclei enlarge
Nucleoli enlarge
Increase in ribosomal population
Cells no longer replicate

3

Acidophilic
Smooth endoplasmic reticulum
Nuclei decrease in size
Nucleoli decrease in size
Ribosomes break down

Cortex fiber
cells

Capsule

Region of mature
lens fibers—3

Lens nucleus
fiber cells

FIGURE 13-5
Organization of the vertebrate lens. During growth of the lens, cells from the
germinative region begin to elongate, stop dividing, and begin to synthesize
specific lens crystallin proteins. The elongated cells in the lens nucleus are
the oldest of the lens fiber cells. Toward the edge of the lens, the fiber cells
are successively younger. (*After Papaconstantinou, 1967, Science* **156**:*338.*)

RNA synthesis (nuclear and nucleolar enlargement and increased numbers of
ribosomes causing cytoplasmic basophilia) are becoming prominent. These
changes are preparatory to the formation of massive amounts of lens crystallins
within the elongated lens fibers.

From the equatorial zone the lens fibers continue to elongate. They lose the
cytological characteristics of protein-synthesizing cells and become eosinophilic cells, with shrunken and condensed nuclei and a continuing reduction
with age of the ribosomes and the rough endoplasmic reticulum on which the
crystallins and other proteins are made. The body of the lens contains layers of
lens fibers that are organized much like an onion. Those at the very center of
the lens (called the *nucleus*) are the first lens fibers that were formed in the
embryo. The tips of the lens fibers grow toward the external and internal poles
of the lens. At each pole of the lens there is a place, called the *lens suture*,

where fibers arising at opposite points on the equator meet one another. Farther toward the periphery, in the cortex of the lens, the lens fibers are successively younger. New lens fibers are continually fed from the equatorial region onto the outer cortex throughout life.

The patterns of nucleic acid and protein synthesis correlate well with the cytological characteristics of the lens. DNA synthesis is found only in the low lens epithelium and is most heavily concentrated in the germinative region (Fig. 13-6A). RNA synthesis is heavy in the lens epithelium, the equatorial region, and the cortex of the body, but it is absent in the core of the lens (Fig. 13-6B). Protein synthesis occurs throughout the lens, although in reduced amounts in the core (Fig. 13-6C), but experiments with actinomycin D have shown that protein synthesis in the core, in contrast to that in the epithelium, is due to a long-lived mRNA (Fig. 13-6D). The long-lived mRNA is an adaptation which allows some continued synthesis of crystallin proteins in cells whose nucleic-acid-synthesizing mechanisms have been turned off.

The crystallins constitute a family of lens proteins (α, β, γ, and δ types) which appear to be structural rather than enzymatic in nature. Different combinations of crystallins are found in different animal groups; for example, α, β, and δ crystallins are found in the avian lens, whereas the α, β, and γ varieties are found in the mammalian lens. Different types of crystallins appear at different times during lens development. In the chick, δ-crystallin can first be detected in the elongated cells of the invaginating lens placode at about 50 hours, and the β-crystallins a few hours later (Zwaan and Ikeda, 1968). The last to appear is α-crystallin, which can be found in the early lens vesicle at about 80 hours. Less than a day after lens induction, cells of the lens placode begin to accumulate δ-crystallin mRNA (Fig. 13-7). This mRNA accumulates rapidly, and later in development it is found in both the epithelial and fiber cells of the lens. The factors that lead to its accumulation and stabilization are poorly understood. Much also remains to be learned about the regulation of translation of δ-crystallin, but evidence for translational control is shown by the fact that younger lens epithelial cells produce δ-crystallin at a rate three times faster than that of later cells, which contain the same number of molecules of the mRNA. Although interpretations regarding the relationship of the crystallins to cyto-differentiation in the lens have varied rather widely over the years, there is increasing evidence that these proteins are products of differentiation more than causative factors influencing subsequent differentiation.

Later Development of the Lens

The influence of the retina on the lens does not cease with the initial inductive event. A continuing retinal influence during later embryonic life has been demonstrated by the Coulombres (1963), who surgically reversed the lens in chick embryos so that the low outer-surface epithelium faced the retina and the elongated cells of the deep layer faced the outside. Very rapidly, the low epithelial cells which were facing the retina began to elongate and formed an

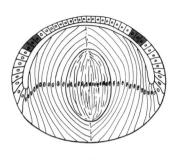

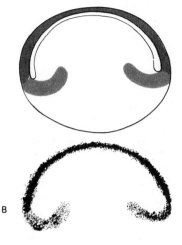

A

B

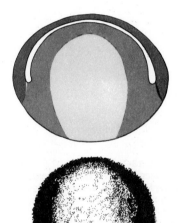

C

D

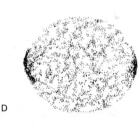

FIGURE 13-6
Synthetic activities in the lens of the 12-day chick embryo. Embryos were
exposed to specific isotopic precursors of proteins and nucleic acids, and
autoradiographs were then prepared from tissue sections. In these figures
drawings of lenses, with intensities of synthetic activities indicated by different levels of gray, are placed over tracings of autoradiographs, showing the
densities of silver grains. (A) Pattern of [^{14}C]thymidine incorporation showing
DNA synthesis concentrated in the germinative region. (B) Pattern of incorporation of [^{14}C]uridine, indicating that most RNA synthesis occurs in the
lens epithelial cells and in the outermost fiber cells of the lens cortex. (C)
[^{14}C]leucine incorporation, showing that protein synthesis is greatest in the
lens epithelium and is progressively less toward the lens nucleus. (D) 8
hours after the administration of actinomycin D (an inhibitor of RNA synthesis) and later exposure to [^{14}C]leucine, cells of the lens epithelium no longer
take up the isotope, whereas those of the lens body continue to do so. This
indicates the presence of long-lived mRNA in the lens fiber cells. (*Adapted
from Reeder and Bell, 1965, Science **150**:71.*)

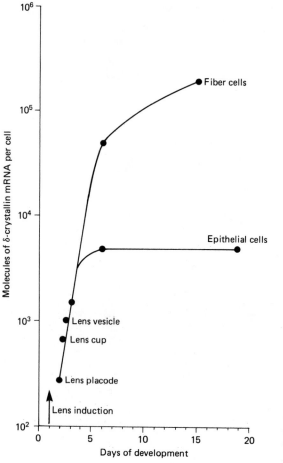

FIGURE 13-7
Quantification of amount of δ-crystallin mRNA in the lens of the developing avian embryo. (*After J. Piatigorsky, 1981.*)

additional set of lens fibers (Fig. 13-8). On the corneal side of the rotated lens, a new lens epithelium formed. These structural adaptations are evidence of a mechanism that ensures a continuous alignment of the lens with the rest of the visual system during development.

Iris and Ciliary Apparatus

As the lens increases in size, it settles back into the optic cup and the margins begin to overlap its edges. We can now recognize the thin overlapping part of the optic cup as the epithelial portion of the iris, and the reduced opening in front of the lens as the pupil. The iris contains muscles (*the dilator* and *sphincter pupillae*) which, as their names imply, regulate the size of the pupil. Embryologically, these muscles are unusual because they arise from neurectoderm instead of mesoderm.

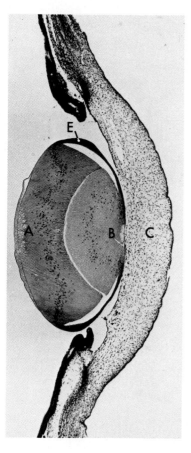

FIGURE 13-8
Axial section through the lens of an 11-day chick embryo. At 5 days of incubation the lens was surgically reversed so that the interior lens epithelium faced the vit-
reous body and retina. The formerly low epithelial cells
elongated to form new lens fibers (A). Owing to the reversal of polarity of the equatorial zone of the lens, new
epithelial cells (E) are added on the corneal face of the
lens, covering the original mass of lens fibers (B). Abbreviation: *C*., cornea. (*Courtesy of A. J. Coulombre,
1965, from Organogenesis, De Haan and Ursprung,
Holt, Rinehart and Winston, New York.*)

In fetuses of the fifth month the ciliary region is readily identifiable by the
marked folding which has involved this portion of the original optic cup (Fig.
13-10). Outside the epithelial layer is the loosely aggregated mesenchyme
which will be organized into the muscular portion of the ciliary body. This
ciliary muscle, by altering the tension on the suspensory ligament of the lens,
controls the lens curvature and thereby helps the eye to change focus so that
objects at different distances can be made to cast sharp images on the retina.
Normal development of the ciliary body appears to depend on the correct
amount of intraocular pressure. If some of the fluid within the developing
eyeball is allowed to escape, a defective ciliary body results (Coulombre and
Coulombre, 1957).

Choroid Coat and Sclera

Outside the optic cup, mesenchymal cells, largely of neural crest origin, early
become massed in a concentrated zone. Reacting to an inductive influence from

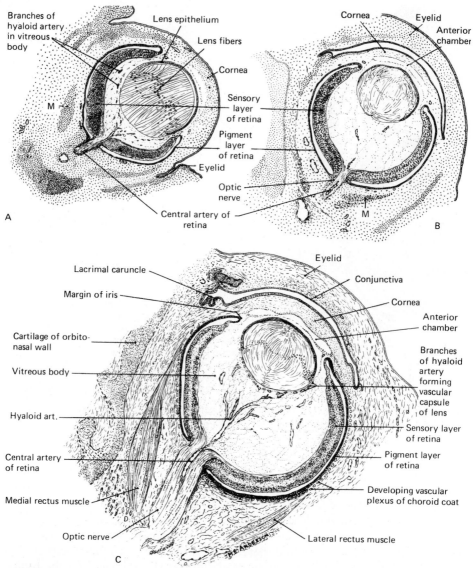

FIGURE 13-9
Three stages in the development of the eye as seen in coronal sections of the head of young human embryos. (A) From an embryo of 17 mm; about 7 weeks. (Projection drawing, ×50, from University of Michigan Collection, EH 14.) (B) From an embryo of 33 mm; about middle of ninth week. (Projection drawing, ×35, from University of Michigan Collection, EH 217.) (C) From an embryo of 48 mm; about middle of tenth week. (*Adapted from Ida Mann, ×25.*) Abbreviation: *M.,* primordial mesenchymal concentration for eye muscle.

the optic cup, this mesenchymal coat becomes differentiated into an inner, highly vascular tunic known as the *choroid coat* (Fig. 13-9C) and an outer tunic composed of a densely woven fibrous connective tissue known as the *sclera*.

The tough sclera molds the eyeball and gives firm attachment to the muscles which move the eye in its socket.

Many inframammalian vertebrates possess a ring of cartilaginous or bony elements (*scleral ossicles*) which surround the outer margin of the cornea. Fourteen ossicles are present in the eye of the chick. Each ossicle arises as the outgrowth of an epithelial papilla from the eye into the surrounding *ectomesenchyme* (mesenchyme derived from neural crest). By about the twelfth day in the chick embryo, ossification begins, and within 2 days it has expanded so that it forms a ring of overlapping bony elements. Deletion experiments suggest that the papillae may induce the ectomesenchyme to form skeletal tissue.

Cornea

Continuous with the sclera in front and forming the part of the eye overlying the lens and the iris is the *cornea* (Fig. 13-10). In postnatal life the cornea is a multilayered structure which, like the lens, must be transparent and without optical aberrations in order for undistorted light rays to reach the retina. The

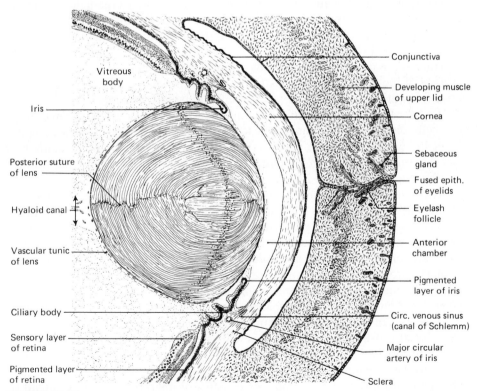

FIGURE 13-10
Anterior part of the eye from a human fetus of about 19 weeks; 174 mm. Vertical section (×20) to show fused eyelids, developing ciliary region, and lens.

mature cornea consists of an outer epithelium underlain by a basement membrane (*Bowman's membrane*) and an inner endothelium, that also is underlain by a basement membrane (*Descemet's membrane*). Between the outer and inner layers is a thick stroma which consists of fibroblasts and layers of collagen fibers arranged perpendicularly to one another (Fig. 13-11). In recent years the cornea has received considerable attention from developmental biologists because it is a very convenient system for studying relationships between cells and the extracellular matrix during development. For a detailed treatment of corneal development, the reader is referred to reviews by Hay (1980) and Hay and Revel (1969).

An inductive influence emanating from the lens vesicle and the optic cup stimulates the transformation of ordinary surface ectoderm to corneal epithelium (Lewis, 1904). This represents one of the terminal events of a long series of inductions involved in the formation of the eye. Some of the major inductive events are summarized in Fig. 13-4.

Shortly after the first inductive event, the future cornea is two cells thick and differs little from the general cutaneous ectoderm, including the presence of an outer peridermal layer on the ectoderm. Soon the basal layer of epithelial cells increases in height because of the elaboration of secretory organelles within the cells. In these cells the Golgi apparatus shifts toward the basal surface, where it is involved in the secretion of a collagenous extracellular matrix known as the *primary stroma* (Fig. 13-11).

The next major step in corneal development is the migration, over the inner surface of the acellular primary stroma, of the *corneal endothelial cells*, which arise from the mesodermal mesenchyme associated with blood vessels around the lip of the optic cup. When the endothelial cells have ceased their migration, the cells change from a squamous to a cuboidal shape and form a complete inner lining of the cornea, which at this time consists of the outer epithelium and the inner endothelium bounding on either side of the still acellular primary stroma (Fig. 13-11, stage 22).

After they have formed a complete layer, the corneal endothelial cells synthesize large amounts of hyaluronic acid and secrete it into the primary stroma. The water-binding properties of hyaluronic acid cause the stroma to swell greatly (Fig. 13-12). The swollen matrix seems to serve as a highly favorable substrate for cell migration, and fibroblasts of neural crest origin then invade the primary stroma and begin proliferating within it (Fig. 13-11, stage 30). The migratory phase of fibroblastic seeding of the primary stroma comes to an end when large amounts of hyaluronidase, probably synthesized by the fibroblasts themselves, are secreted into the stroma and break down much of the hyaluronic acid that is present. This signals a new phase, characterized by the settling in of the fibroblasts in their new location and their addition of coarse collagen fibers to the substance of the stroma (Fig. 13-11, stage 40). (The correlation of high concentrations of hyaluronic acid with cell migration and its removal by hyaluronidase with subsequent stabilization of the cells are not confined to the cornea alone and may have widespread significance in developing systems.)

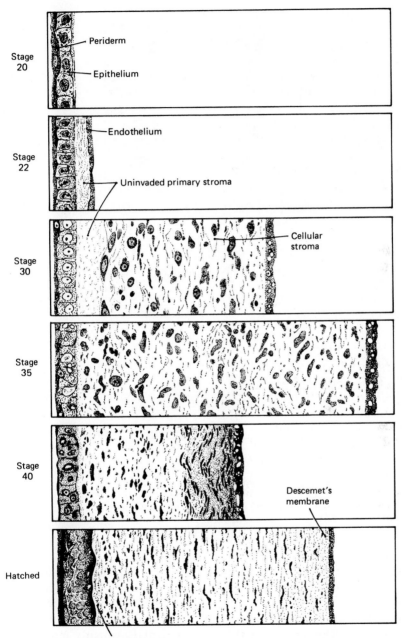

Stage
20

Periderm

Epithelium

Stage
22

Endothelium

Uninvaded primary stroma

Stage
30

Cellular
stroma

Stage
35

Stage
40

Descemet's
membrane

Hatched

Bowman's membrane

FIGURE 13-11
Successive stages in the development of the cornea of the chick embryo. The
stages indicated are those of Hamburger and Hamilton (1951). (*After Hay and
Revel, 1969, Fine Structure of the Developing Avian Cornea*, Monogr. in Devel.
Biol., *vol. 1, S. Karger, Basel.*)

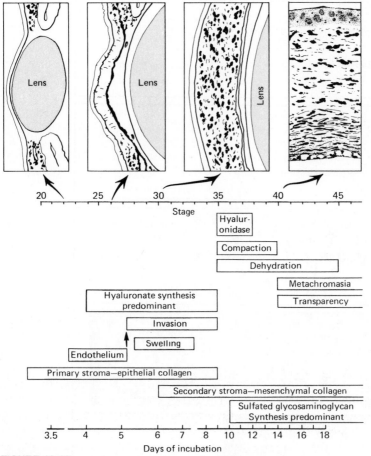

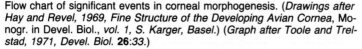

FIGURE 13-12
Flow chart of significant events in corneal morphogenesis. (*Drawings after Hay and Revel, 1969, Fine Structure of the Developing Avian Cornea*, Monogr. in Devel. Biol., *vol. 1, S. Karger, Basel.*) (*Graph after Toole and Trelstad, 1971, Devel. Biol.* **26**:*33.*)

After being populated by fibroblasts, the primary stroma is called the *secondary stroma*. Only a thin layer of acellular primary stroma remains beneath the corneal epithelium. Both the outer epithelium and the inner endothelium of the cornea continue to secrete extracellular material beneath their basal surfaces. This results in the formation of Bowman's membrane beneath the epithelium (Fig. 13-11) and Descemet's membrane under the corneal endothelium.

Once its major components are established, the intially opaque cornea must become transparent and establish architectural properties appropriate for the transmission of a path of undistorted light into the eyeball. In the chick embryo the transparency of the cornea increases from 40 percent transmission in the 14-day embryo to 100 percent in the 19-day embryo. This is accomplished by the dehydration of the corneal stroma, a process which is stimulated by thyroxine secreted into the blood by the maturing thyroid gland. The role of thyroxine has been demonstrated by the premature dehydration of the cornea when thyroid hormone was administered early and the retardation of dehydration by thyroid inhibitors. The clearing of the cornea results from the actions of thyroxine on the corneal endothelium, which pumps out sodium from the corneal stroma into the anterior chamber of the eye. Water molecules follow the sodium ions, thus resulting in the dehydration of the stroma.

The curvature of the cornea is on a smaller radius than that of the remainder of the eyeball and thus appears to bulge out from the eye as a whole. The perfection of the corneal curvature is of great functional importance, for the cornea is what we might call the front lens of the eye and acts in conjunction with the crystalline lens in bringing light rays into focus on the retina. Individuals who develop irregularities in the curvature of the cornea have a condition called *astigmatism*, which causes distortions of the visual image.

Retina

After the formation of the optic cup, the inner and outer walls follow different pathways of differentiation. The outer wall becomes thin and highly pigmented, forming the *pigment layer of the retina* (Fig. 13-9). Cells of the inner wall of the optic cup continue to proliferate, causing this layer to increase in both thickness and circumference. These cells differentiate into neural elements, and the inner wall of the optic cup is then known as the *neural (sensory) layer of the retina*. Cells in the center of the retina mature first, leaving a zone of proliferating cells around the margins of the retina. As long as the retina grows, daughter cells from the marginal growth zone contribute to its substance. During the very early stages of formation of the retina, its polarity becomes fixed in a manner reminiscent of the determination of the limb axes. In the retina the nasotemporal (anteroposterior) axis is fixed first. This is followed by fixation of the dorsoventral axis, and finally radial polarity is established.

The mature neural retina is an extremely complex tissue composed of three well-defined layers of neural cells. From the inside to the outside of the retina, these layers are called the *layer of ganglion cells*, the *layer of bipolar cells*, and the *layer of rods and cones*, the light-sensitive elements (receptors) of the eye. In the postnatal eye, light striking the retina must first pass through the ganglion cell layer and then the bipolar cell layer before it strikes the rod and cone cells. These cells convert the light signal into neural signals, which are transmitted via elaborate networks of synapses through the bipolar cells to the ganglion cells. Long processes of the ganglion cells extend through the optic nerve to the primary visual centers of the brain.

In the embryo, cells of the ganglion layer (i.e., the layer closest to the center of the eyeball) begin to differentiate first. Differentiation of the remaining layers follows, with the rods and cones of the outermost layer of the neural retina taking shape last. The differentiation of cell layers within the retina is accompanied by waves of cell death, but the function of this phenomenon in the retina is poorly understood. Although the development of synaptic connections within the retina is currently receiving considerable attention, the details are beyond the scope of this book. The neurons constituting the ganglion cell layer send out axons which grow out into the ventral wall of the optic stalk and thence toward the optic centers of the brain. In an extremely precise manner they make connections with the next order of neurons in the visual pathway.

Eyelids, Conjunctiva, and Associated Glands

Human eyelids start to develop during the seventh week as folds of skin growing back over the cornea (Fig. 13-9A). Once they have begun to form, the eyelids close over the eye quite rapidly, usually meeting and fusing with each other by the end of the ninth week. This fusion involves only the epithelial layers of the lids (Fig. 13-10), and the eyelashes and the glands that lie along the margins of the lids start to differentiate from this common epithelial lamina before the lids reopen. Signs of the loosening of the epithelial union can be seen in the sixth month, but it is ordinarily well into the seventh month before the eyelids actually reopen.

In the mouse temporary epidermal fusions occur between digits and between the ears and underlying head epidermis, in addition to the fusion of the eyelids. Harris and McLeod (1982) have outlined the differences between these temporary fusions and permanent fusions, such as that in the palate (see Chap. 14). The main features are that in temporary fusions an area of epidermis remains between the two joined tissues and that cell death is not involved. Also, in temporary fusions, specializations of both the epidermal and peridermal cells are found in the area of fusion.

The space between the eyelids and the front of the eyeball is commonly referred to as the *conjunctival sac*. The most massive glands opening into the conjunctival sac are the *lacrimal glands*. They develop from multiple epithelial buds which make their first appearance during the ninth week. The lacrimal glands produce a thin, watery secretion ("tears") which under normal conditions keeps the corneal surface cleaned and lubricated. In normal circumstances the fluid produced by the lacrimal glands, after bathing the conjunctival surfaces, passes into the nasal chamber by way of the *nasolacrimal duct*.

Changes in the Position of the Eyes

During development the eyes undergo a striking change in their relative position. In embryos of the sixth week they are far around on either side of the head like the eyes of a fish (Fig. 14-1C). In such a position there can be no overlapping of their visual fields, thus precluding the binocular type of vision so important to humans in estimating distances. As the facial structures grow, the eyes are carried forward in the head, and as a result, their optical axes begin to converge. By the eighth week the eyes are beginning to look quite definitely forward (Fig. 14-1F), and by the tenth week the angle is approximately 70°, only about 10° wider than it is in the adult.

THE EAR

For convenience, the adult mammalian ear may be divided into three regions: external, middle, and internal. The external ear is essentially a sound-collecting funnel consisting of the *pinna* and the *external auditory canal*. The middle ear is a sound-transmitting mechanism involving a chain of three *auditory ossicles*, which pick up the vibrations received by the eardrum and transmit them across the middle ear, or tympanic cavity, to the receptive mechanism of the internal ear. The internal ear is composed of an elaborate system of fluid-filled, epithelially lined chambers and canals constituting the so-called *membranous labyrinth*. The membranous labyrinth lies within the temporal bone in a similarly shaped but larger series of cavities constituting the *bony labyrinth*. The narrow space between the walls of the bony labyrinth and the membranous labyrinth is known as the *perilymphatic space* and is filled with *perilymphatic fluid*. The sound-receiving portion of the membranous labyrinth is the *cochlea*, a curiously shaped structure spirally coiled in a manner suggestive of a snail shell. Closely associated with the cochlea is the vestibular complex, which is concerned with equilibration. The vestibular portion of the membranous labyrinth is composed of the *sacculus*, the *utriculus*, and the three *semicircular ducts*, or *canals*. It is phylogenetically the most primitive part of the ear; in fact, it is the only part of the ear that has been differentiated in the fishes.

Formation of the Auditory Vesicle

The primordium of the membranous labyrinth, or inner ear, is the first part of the ear mechanism to make its appearance. Formation of the auditory vesicle is stimulated by an inductive action by the hindbrain upon the overlying ectoderm. In human embryos morphological changes are first noticeable early in the third week (Fig. 8-2), when the superficial ectoderm on either side of the still-open neural plate becomes slightly thickened. This thickening is the start of the *auditory placode*, which by the middle of the third week (Fig. 8-3B) becomes quite clearly marked. By the end of the third week the auditory placode has taken shape as a sharply circumscribed thickening in the ectoderm on either side of the developing myelencephalon (Fig. 13-13A). During the fourth week the placode is invaginated to form the *auditory pit* (Fig. 13-13B). The pit becomes deepened, and finally its opening at the surface is closed. After it has been closed off from the surface, the former auditory pit constitutes a sac called the *auditory vesicle*, or *otocyst*. Like the limb bud and retina, the major axes of the auditory vesicle are sequentially determined, with the anteroposterior axis fixed first and the dorsoventral axis fixed next. Grafting experiments

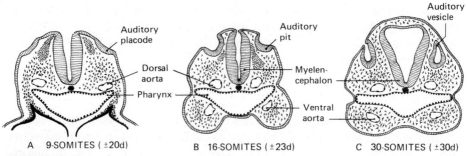

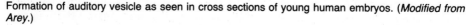

FIGURE 13-13
Formation of auditory vesicle as seen in cross sections of young human embryos. (*Modified from Arey.*)

have suggested that in all three structures (eye, ear, and limb) axial determination is accomplished by the recognition and interpretation of certain general environmental cues that are present in the lateral walls of the embryo. At an early stage, while its walls are still homogeneous, the otocyst is regionally determined. Pieces removed from otocysts of 11- and 12-day mouse embryos and grown in vitro differentiate into predictable structures, based on the locations from which they were removed (Ruben and Van De Water, 1983).

Many of the soft ectodermally derived tissues of the head induce the mesenchyme surrounding them to form protective coverings of skeletal tissue. For example, the early brain induces the formation of the flat bones of the cranium; the eyeball induces the formation of the scleral cartilages; and the olfactory placodes appear to stimulate the formation of hard tissues in the nasal region. Similarly, the auditory vesicle induces the mesenchyme around it to form the cartilaginous ear capsule (Balinsky, 1925). In contrast to the former examples, in which the induced mesenchyme is of neural crest origin, the mesenchyme forming the ear capsule is mainly of mesodermal origin. The inductions involved in formation of the otic capsule represent only a small portion of the tissue interactions and intermingling that are required to build up the inner ear.

Differentiation of the Auditory Vesicle to Form the Inner Ear

As the auditory vesicle enlarges, it changes from its originally spheroidal shape and becomes elongated dorsoventrally. In keeping with its phylogenetically older status, the vestibular part of the inner ear takes shape before the auditory portion (Anniko, 1983). About where the epithelium of the auditory vesicle has been separated from the superficial ectoderm, there develops a tubular extension of the vesicle, known as the *endolymphatic duct* (Fig. 13-15A). As the auditory vesicle expands laterally, the endolymphatic duct is left occupying a progressively more medial position in relation to the rest of the vesicle. Almost from the outset of its differentiation the more expanded dorsal portion of the auditory vesicle with which the endolymphatic duct is connected can be identified as the primordium of the vestibular part of the membranous labyrinth, and the more slender ventral expression is recognizable as the primordium of the cochlea (Fig. 13-14).

By the close of the sixth week of development conspicuous flanges appear on the vestibular portion of the auditory vesicle, foreshadowing the differentiation of the *semicircular ducts*. As the flanges push out from the main vesicle, their central portions become thin and finally undergo resorption so that the original semilunate flange becomes converted into a looplike duct (Fig. 13-14C to 13-14E). Three such ducts are formed, each occupying a plane in space approximately at right angles to the other two. While the semicircular ducts are taking shape, the vestibular portion of the auditory vesicle is being subdivided by a progressively deepening constriction into a more dorsal utricular portion and a more ventral saccular portion (Fig. 13-14E to 13-14G). When this division has occurred, the semicircular ducts open off the utriculus. Near one of their two points of communication with the utriculus, each semicircular canal forms a local enlargement known as an *ampulla*. Within the ampulla there develops a specialized area called a *crista*, containing neuroepithelial cells with hairlike processes projecting into the lumen of the ampulla. These specialized receptors are innervated by branches of the vestibular division of the eighth cranial nerve. Changes in the position of the head are accompanied by a lag in the movement of fluid within the semicircular ducts, resulting in mechanical stimulation of the neuroepithelial cells of the crista. The nerve impulses thus initiated pass over the appropriate central pathways and make us aware of positional changes. In light of this function, the significance of the arrangement of the three semicircular ducts in planes at right angles to each other is self-evident.

Specialized areas called *maculae* are developed in the sacculus and utriculus. The maculae contain neuroepithelial cells similar in general character to those in the cristae of the semicircular ducts and like them are supplied by branches of the vestibular division of the eighth cranial nerve. Impulses initiated in the maculae make us aware of static position, in contrast to the sense of positional change mediated through the mechanism of the semicircular canals.

The *cochlear portion of the membranous labyrinth* is the sound-perceiving part of the ear mechanism. The cochlea (Fig. 13-14F and G) is a tiny snail-shaped structure which consists of three parallel ducts, a central *cochlear duct* and dorsal and ventral *perilymphatic ducts*. Running the entire length of the cochlear duct is the organ of Corti, a band of tissue containing very specialized auditory receptors which are connected to terminal fibers of the auditory nerve in the *spiral ganglion* of the cochlea (Fig. 13-14G). Sound is perceived when sound waves in the air deform the *tympanic membrane* (eardrum; Fig. 13-15). The motions of the tympanic membrane are mechanically transmitted by the auditory ossicles of the middle ear to another membrane covering the *oval window* (Fig. 13-15), which in turn sets up waves

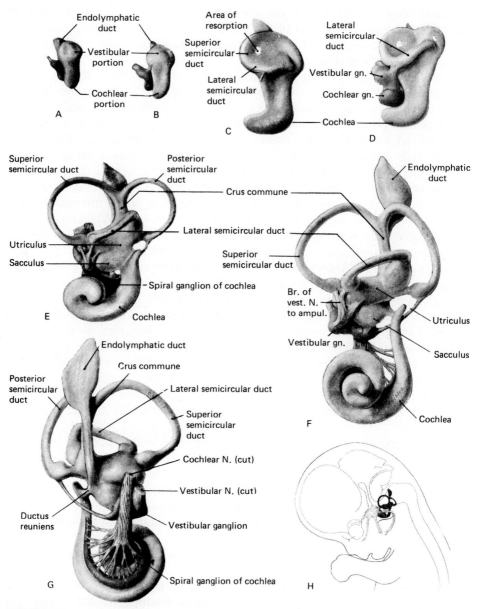

FIGURE 13-14
Development of the membranous labyrinth in human embryos. (*After Streeter, 1906,
Am. J. Anat., vol. 6.*) (A) 6 mm, lateral view; (B) 9 mm, lateral view; (C) 11 mm, lateral view;
(D) 13 mm, lateral view; (F) 30 mm, lateral view; (G) 30 mm, medial aspect; (H) outline of
head of 30-mm embryo to show position and relations of developing inner ear.

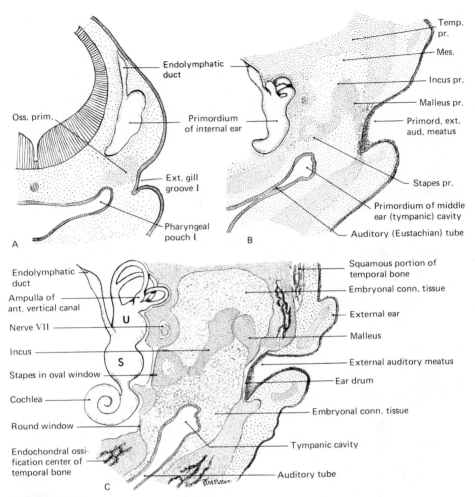

FIGURE 13-15
Schematic diagrams showing three stages in the development of the middle-ear chamber and the auditory ossicles. Abbreviations: *Mes.*, mesenchyme; *Oss. prim.*, mesenchymal concentration which is the first indication of the primordia of the auditory ossicles; *Temp. pr.*, mesenchymal concentration where primordium of temporal bone is taking shape.

in the fluid within the cochlea. The waves are then detected by the appropriate auditory receptors within the organ of Corti, which convert the mechanical stimulus into a neural signal.

Originating from the most ventral part of the auditory vesicle, the cochlea elongates rapidly during the sixth week and shows a sharp forward bend of its distal end (Fig. 13-14C and D). Elongation continues at an accelerated rate during the seventh and eighth weeks, and the initial bend rapidly develops into a spiral of $2\frac{1}{2}$ turns (Fig. 13-14E and F). The cochlear division of the eighth nerve follows the cochlea in its growth changes, and its fibers fan out to be distributed along the entire length of the cochlear duct as the spiral ganglion of the cochlea (Fig. 13-14G).

Middle Ear

While the receiving mechanism of the ear is forming in the manner just described, the transmitting apparatus of the middle ear is also taking shape. Recall that in their initial relations the first pharyngeal pouches extend laterad so that their endodermal lining makes contact with the ectoderm at the bottom of the first gill furrow to form the gill plate. The distal portion of the pouch remains somewhat expanded to form the primordium of the middle-ear chamber, or *tympanic cavity*, but the proximal portion soon becomes narrowed to form the *auditory (eustachian) tube* (Fig. 13-15A and B). The direct contact between the endoderm of the pharyngeal pouch and the ectoderm of the floor of the gill furrow does not last long. The blind outer end of the pouch, which constitutes the primordium of the tympanic cavity, pulls away from the surface, and a conspicuous concentration of mesenchyme appears adjacent to it (Fig. 13-15A). As development progresses, the mesenchymal cells of this primordial mass become organized into the cartilaginous precursors of the *auditory ossicles* lying between the developing inner ear and the retained portion of the gill furrow, which may now be said to constitute the primordium of the external auditory meatus. At this stage the early ossicles lie above the primordial tympanic cavity, embedded in a very loose embryonic connective tissue (Fig. 13-15C).

Phylogenetically, the auditory ossicles are derived from bones involved in the suspension and articulation of the jaw in the lower vertebrates. The *malleus* and *incus* are derivatives of the first branchial arch and are the homologues of the articular (malleus) and quadrate (incus) bones, which are essential components of the articulation between the upper and lower jaws of the lower vertebrates through the reptiles. The articular bone represents the ossified basal end of Meckel's cartilage (see Chap. 16), and it is the articular surface of the primitive lower jaw. The quadrate bone, a dermal bone, is the articular surface for the upper jaw. The third element (the *stapes* in mammals) is derived from the most dorsal portion of the hyoid (II) arch. In crossopterygian fishes this bone is called the *hyomandibula*, and it serves to anchor the jaw apparatus to the neurocranium in the vicinity of the inner ear. In amphibians the hyomandibula is converted into a bone (*columella*) of the middle ear. The columella alone provides the mechanical connection between the tympanic membrane, located on the surface of the head, and the inner ear. In mammals, many of the dermal bones originally associated with the jaw in phylogeny have been eliminated or used for other purposes. In the latter category are the articular and quadrate bones, which have been incorporated into the middle ear as the malleus and incus. The muscles and nerves which are associated with the middle-ear ossicles are appropriately derived from the first and second arches in accordance with their phylogenetic and ontogenetic origins (Fig. 16-3).

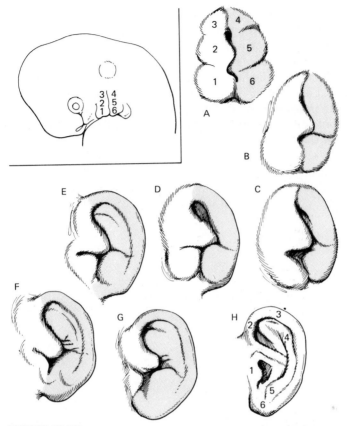

FIGURE 13-16
Stages in development of the external ear. (*After Streeter, 1922, Carnegie Contr. Embryol.* **14**:*111.*) The parts derived from the mandibular side of the cleft are unshaded; the parts from the hyoid side are shaded.

During the latter part of intrauterine life the connective tissue around the auditory ossicles begins to undergo rapid resorption, with a resultant expansion of the tympanic cavity. Eventually the ossicles come to lie suspended within the enlarged tympanic cavity, with only a thin layer of epithelium overlying the periosteal investment. At the time of birth, however, there is still a residue of unresorbed embryonic connective tissue partially filling the tympanic space and more or less damping the free movement of the ossicles. Full mobility of the ossicles is acquired within a few months after birth, when the remaining loose connective tissue is resorbed. When this has occurred, movement imparted by sound waves to the eardrum is freely transmitted by the ossicles to the membrane of the oval window to which the stapes is attached.

External Ear

The external ear is formed by the growth of the mesenchymal tissue flanking the first (hyomandibular) gill furrow of the young embryo. During the second month several nodular enlargements appear, some of them arising from mandibular-arch tissue rostral to the first gill furrow and others arising from the hyoid arch along the caudal border of the furrow. The coalescence of these tubercles and their further development mold the *pinna* of the ear (Fig. 13-16). In view of the number of separate growth centers involved, it is not surprising that the configuration of the fully formed external ear exhibits so wide a range of variations.

CHAPTER 14

THE FACE AND ORAL REGION

Development of the face is intimately connected with the development of the entire head and neck region; it involves a cascade of closely integrated events that start in the early embryo and continue uninterrupted until the phase of postnatal growth has ceased. Much of the substance of the face is derived from the neural crest, which in the head and neck is capable of forming derivatives, such as bone and cartilage, which the neural crest of the trunk is unable to do (Noden, 1984). Therefore, in understanding facial development, it is necessary to go back at least as far as the origin of the cranial neural crest and its imprinting with some type of pattern specifying the face. The next phase of facial development consists of the migration, and concomitant proliferation, of neural crest cells. As they migrate, the neural crest cells become closely associated with the overlying ectoderm and are involved in a series of inductive interactions which are required for building up the skeleton (Hall, 1987). As the neural crest is migrating, a second migration of myogenic cells from the somitomeres provides the branchial arches with a mesodermal core that will ultimately become transformed into the branchiomeric musculature. These cellular migrations result in the establishment of the system of branchial arches (see Chap. 16), which initially appear as simple discrete masses of mesoderm covered by ectoderm. Most of the tissues of the midface and lower face originate from the first branchial arch. There, the neural crest mesenchyme differentiates into connective tissue, cartilage, bone, and tooth pulp.

Early in development, the expanding forebrain and optic cups displace the original head mesoderm caudally, allowing the migrating neural crest to fill in the interstices between the brain and overlying ectoderm (Ross and Johnston, 1972). The boundary between the neural crest of the face and the mesoderm of the rest of the head is a discrete one (Fig. 6-17), with cells of the neural crest surrounding the stomodeum, pharynx, and diencephalon while it constitutes the bulk of the branchial-arch mesenchyme. Mesodermally derived mesenchyme surrounds the mesencephalon, the myelencephalon and the otic capsule. Certain bones of the skull (the temporal bone and others) contain discrete areas of both mesodermal and neural crest ectoderm.

In keeping with its recent evolutionary origins, the mammalian face develops relatively late in embryonic development. At 4 weeks the human face consists of only a few primordial tissue masses partially surrounding the future oral region (Fig. 14-1A). The oral cavity is represented by an ectodermal depression, the *stomodeum*, which abuts upon the blindly ending anterior end of the foregut. A thin *oral* (*stomodeal*) *plate* consisting of apposed layers of ectoderm and endoderm maintains a temporary partition between stomodeum and foregut. Rostral to the stomodeum, the area of the future face is dominated by the frontal prominence of the greatly overhanging forebrain. Laterally the maxillary process of the first branchial arch is visible, and caudal to the stomodeum the mandibular arch and other emerging branchial arches are the dominant structures. Between the forebrain and the maxillary processes, the thickened nasal placodes are becoming visible.

Later development of the face is characterized by growth of the primordial tissue masses surrounding the stomodeum and their ultimate rearrangement into recognizable facial structures. Even as late as the time when the oral plate ruptures and establishes communication between the cephalic end of the gut and the outside world, the stomodeal depression is very shallow (Fig. 8-12). The deep oral cavity of the adult is formed by the forward growth of structures about the margins of the stomodeum. Some idea of the extent of this forward growth can be gained by considering the fact that the tonsillar region of the adult is at about the level occupied by the oral plate before it has ruptured and disappeared. The growth of the structures bordering the stomodeum does not just give rise to superficial parts of the face and jaws but actually builds out the walls of the oral cavity itself. Formation of the midface region involves not only the relative displacement of the growing forebrain away from the mouth but also the filling in of the space between the oral cavity and the forebrain. This is accomplished by the building up of a rim of mesenchymal cells around the pair of thickened ectodermal nasal placodes, the medial migration of the eyes from their original positions at the side of the head, and the ingrowth of the maxillary process into the area. Throughout much of the fetal period and the period of postnatal growth, one of the most characteristic aspects of facial development is the continued increase in relative prominence of the midface region.

THE FACE AND JAWS

The most conspicuous landmarks of a 4-week human embryo (Figs. 8-12 and 14-1A) are the stomodeal depression and the mandibular arch, which constitutes its caudal boundary. Within the next week most of the structures which

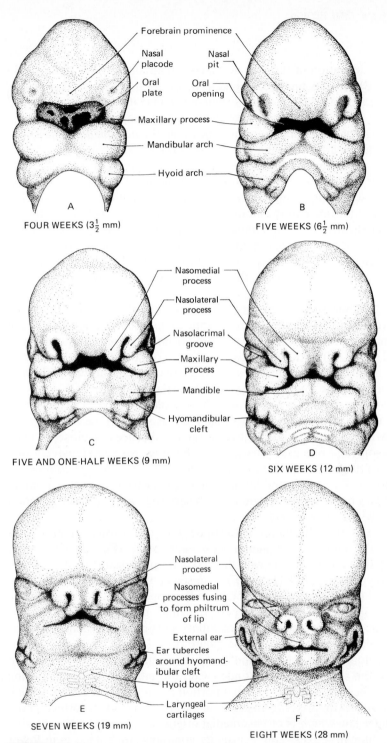

Forebrain prominence

Nasal placode

Nasal pit

Oral plate

Oral opening

Maxillary process

Mandibular arch

Hyoid arch

A

FOUR WEEKS (3½ mm)

B

FIVE WEEKS (6½ mm)

Nasomedial process

Nasolateral process

Nasolacrimal groove

Maxillary process

Mandible

Hyomandibular cleft

C

FIVE AND ONE-HALF WEEKS (9 mm)

D

SIX WEEKS (12 mm)

Nasolateral process

Nasomedial processes fusing to form philtrum of lip

External ear

Ear tubercles around hyomandibular cleft

Hyoid bone

Laryngeal cartilages

E

SEVEN WEEKS (19 mm)

F

EIGHT WEEKS (28 mm)

FIGURE 14-1
Drawings showing, in frontal aspect, some of the important steps in the formation of the face. (*After William Patten, from Morris, Human Anatomy, McGraw-Hill, New York.*)

take part in the formation of the face and jaws are already clearly distinguishable (Fig. 14-1B). In the midline, cephalic to the oral cavity, is a rounded overhanging area known as the *frontal prominence*. On either side of the frontal prominence are horseshoe-shaped elevations surrounding the olfactory pits. The median limbs of these elevations are known as the *nasomedial processes*, and the lateral limbs are called the *nasolateral processes*.

Growing toward the midline from the cephalolateral angles of the oral cavity are the *maxillary processes*. In lateral views of the head (Figs. 8-6, 8-7, and 14-2A) it will be seen that the maxillary process and the mandibular arch merge with each other at the angle of the mouth. Thus the structures which border the oral cavity cephalically are (1) the unpaired frontal prominence in the midline, (2) the paired nasomedial processes on either side of the frontal prominence, and (3) the paired maxillary processes at the extreme lateral angles (Fig. 14-1B). From these primitive tissue masses, the upper lip, the upper jaw, and the nose are derived.

The caudal boundary of the oral cavity is less complex, consisting of the paired primordia of the mandibular arch. Appearing first on either side of the midline are marked local thickenings resulting from the rapid proliferation of mesenchymal tissue. Until these thickenings have extended from either side to merge in the midline, there remains a conspicuous midline notch; with their merging, the arch of the lower jaw is completed (Fig. 14-1B to F).

During the sixth week (Fig. 14-1C and D) marked progress is made in the development of the upper jaw. The maxillary processes become more prominent and grow toward the midline, crowding the nasal processes closer to each other. The nasomedial processes, meanwhile, have grown so extensively that the lower part of the frontal (forebrain) prominence between them is completely overshadowed (cf. Fig. 14-1B and D), and they appear to be almost in contact with the maxillary processes on either side. The groundwork for the completion of the upper jaw is now well laid down.

In the final steps in the formation of the upper jaw, the nasomedial processes move toward the midline and merge with each other (Fig. 14-1D). Soon afterward they fuse on either side with the maxillary processes to complete the arch of the upper jaw (Fig. 14-1E and F). The segment of the upper jaw which is of nasomedial origin gives rise externally to the upper lip in the region of the *philtrum* (Fig. 14-1E and F). A deeper triangular projection of the fused nasomedial processes becomes the premaxillary portion of the dental arch as well as the median (primary) part of the palate (Fig. 14-3).

Toward the close of the second month and the beginning of the third, when the molding of the soft parts is well under way, formation of the deeper-lying bony structures begins. The more medial portion of the maxillary bone, which carries the incisor teeth, arises from separate ossification centers formed in the part of the upper jaw which is of nasomedial origin. This origin of the incisive portion of the human maxilla emphasizes its homology with what is in lower forms a separate bone known as the *premaxillary*, or *intermaxillary*. In the skulls of human infants the sutures separating the incisive portion from the rest

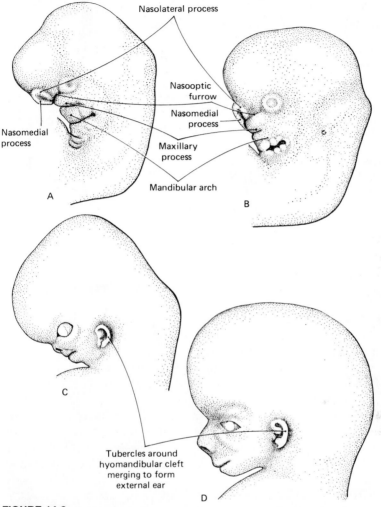

FIGURE 14-2
Lateral views of development of the face and external ears. (*After William Patten.*) The embryos represented are the same as those drawn in face view in Fig. 14-1. (A) 5½ weeks; (B) 6 weeks; (C) 7 weeks; (D) 8 weeks.

of the maxilla are likely to be still evident, and occasionally traces of them may be made out in the adult skull. The rest of the maxillary bone, carrying all the upper teeth behind the incisors, is developed in the part of the upper jaw that arises from the maxillary process.

THE PALATE

Late in the second month, when the upper jaws have been established, the palatal shelves begin to make their appearance. These paired structures

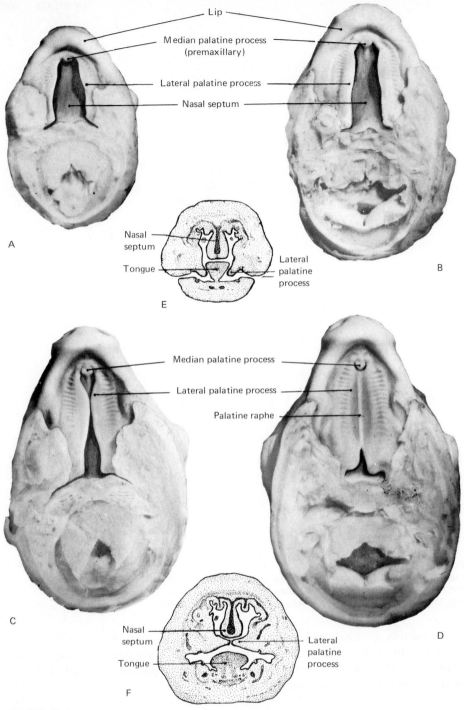

FIGURE 14-3
Photographs (×5) of dissections of pig embryos made to expose roof of mouth and show development of palate. (A) 20.5 mm; (C) 26.5 mm. The diagrams of transverse sections are set in to show (E) the relations before the retraction of the tongue from between the palatine processes; (F) after the retraction.

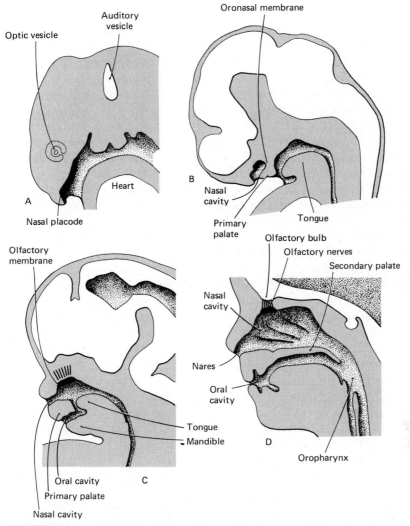

FIGURE 14-4
Parasagittal sections of human heads showing the development of the nasal
chambers. (A) 5 weeks; (B) 6 weeks; (C) 7 weeks; (D) 12 weeks.

subdivide the most rostral portion of the original stomodeal chamber. Since the
nasal pits break through above the level of the shelves, the formation of the
palate in effect elongates the nasal chambers backward so that they open
eventually into the region where the oral cavity becomes continuous with the
pharynx (Fig. 14-4).

Both the nasomedial processes and the maxillary processes contribute to the
palate. From the intermaxillary (nasomedial) region a small, triangular median
portion of the palate (*primary palate*) is formed (Fig. 14-3A and B). The main

part of the palate (*secondary palate*) is derived from the maxillary processes. Shelflike outgrowths arise on either side and toward the midline (Fig. 14-3A and B). When these palatal shelves first start to develop, the tongue lies between them, and they are directed obliquely downward so that the margins lie along the floor of the mouth on either side of the root of the tongue (Fig. 14-3E). As development progresses, the tongue moves down and the margins of the palatal shelves swing upward and toward the midline (Fig. 14-3F). Further growth brings them into contact with each other, and their fusion soon completes the main part of the palate (Fig. 14-3D). In the extreme rostral region the small, triangular premaxillary (median palatine) process lies between the lateral palatine shelves, and they fuse with it instead of with each other. As the palate is being formed, the nasal septum grows toward it and is fused to its cephalic face. Thus the separation of right and left nasal chambers from each other is accomplished at the same time as the separation of the deeper portions of the nasal chambers from the oral cavity.

Sometimes the palatal shelves fail to fuse with one another, resulting in a congenital defect known as *cleft palate*. Cleft palate is often associated with a *cleft lip*, which results from the defective fusion of the nasomedial and maxillary processes. Because of the high incidence of these defects in humans, a great amount of research has been directed toward understanding the basis for the elevation and fusion of the palatal shelves. Nevertheless, we are far from a complete understanding of either normal or defective development of the face and palate.

The developmental potential of the palatal shelves is greater than one might expect. In the overlying ectoderm, for example, although the cells are of the same type early in development, they undergo marked changes as development progresses. For example, as the two palatal shelves fuse, the ectodermal cells along the point of fusion die. On the nasal side of the palate the ectodermal cells specialize into a ciliated, mucus-secreting columnar epithelium, whereas on the oral side the ectodermal cells differentiate into a stratified squamous epithelium that is capable of withstanding the mechanical pressures associated with eating and chewing. Recombination experiments (Ferguson and Honig, 1984) suggest that, as in the skin, the underlying mesenchyme, through an inductive interaction, determines the developmental fate of the overlying ectoderm. The death of ectodermal cells along the line of fusion of the palatal shelves is important for maintaining a permanent fusion. By contrast, localized death of epithelial cells is not seen along lines of temporary fusion, e.g., the eyelids.

Many hypotheses (rev. by Ferguson, 1978) have been proposed to account for elevation and fusion of the palatal shelves. Over the years attention has turned from gross mechanisms, such as swallowing movements that remove the tongue from between the palatal shelves, to mechanisms based on properties of the cells and their surrounding extracellular matrix. It is becoming apparent that morphogenesis and growth not only of the palate but of the other facial processes is a very complex process involving the interaction of many factors at a number of different levels of organization (rev. by Zimmerman, 1984).

THE NASAL CHAMBERS

The first indication of the nose in human embryos is the formation of a pair of thickened ectodermal *nasal placodes* on the frontal aspect of the head (Figs. 14-1A and 14-4A). Almost as soon as they are formed, the nasal placodes sink below the general surface level so that the thickened epithelium constitutes the floor of the *nasal pits* (Fig. 14-4B). The mesenchymal tissue surrounding the nasal pits proliferates rapidly so that the pits are deepened both by their own progressive invagination and by the forward growth of the surrounding tissue. The bordering elevations become horseshoe-shaped, with their own ends toward the mouth. The two limbs of the nasal elevations are named the *nasomedial* and the *nasolateral processes* (Fig. 14-1C). At first the nasal pits are far apart, but as development progresses, the two nasal pits and their associated processes converge toward the midline (Fig. 14-1B to 14-1F). The nasomedial processes on either side eventually merge with each other to form the medial portion of the upper lip and the septum of the nose. The nasolateral processes become the alae (wings) of the nose.

While these external changes are taking place, the nasal pits become progressively deeper and extend backward and downward toward the oral cavity (Fig. 14-4B). During the seventh week the tissue separating the nasal pits from the oral cavity becomes thinned to merely a double layer of epithelium, the *oronasal membrane*. When this breaks through, the nasal pits open freely into the oral cavity just caudal to the arch of the upper jaw (Fig. 14-4C). The formation of the palate by the fusion of the palatal shelves greatly lengthens the original nasal chamber (Fig. 14-4D). The olfactory area differentiates within the roof of each nasal chamber. The olfactory receptor cells, which are actually very primitive bipolar neurons, differentiate within the epithelium itself among tall columnar cells called *sustentacular cells*. The apical process of a receptor cell forms a knob containing highly modified cilia, which are assumed to be the chemical receptor sites (Frish, 1967). The basal process elongates and makes connections with other neurons in the olfactory bulb, through which olfactory stimuli are transmitted in the form of neural signals to the appropriate centers in the brain.

THE TONGUE

While the palate is forming the roof of the mouth, the tongue takes shape in the floor. From a developmental standpoint the tongue may be conveniently described as a sac of mucous membrane which becomes filled with a mass of growing muscle. The reason for making this crude comparison is that the lingual epithelium and the lingual muscles have different origins and undergo such striking changes in relative position that it is desirable to consider them separately. The primordial areas involved in forming the covering of the tongue appear early in the second month of development. In 5-week embryos paired lateral thickenings, which arise as a result of the rapid proliferation of the mesenchyme beneath the overlying epithelium, are known as the *lateral lingual*

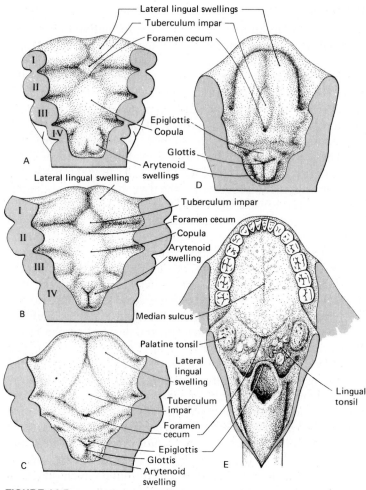

FIGURE 14-5
Stages in the development of the human tongue. The upper part of the
head has been cut away to permit viewing of the oropharyngeal region
from above. The cut visceral arches are indicated by their numbers (I–IV),
respectively. (A) Fourth week; (B) late in the fifth week; (C) early in the
sixth week; (D) middle of the seventh week; (E) adult.

swellings (Fig. 14-5A and B). Between them is a small median elevation known
as the *tuberculum impar.* Behind the tuberculum impar is another median
elevation appropriately called the *copula* (i.e., yoke), for it unites the second
and third arches in a midventral prominence. The copula extends cephalocaudally from the tuberculum impar to the primordial swelling which marks the
beginning of the epiglottis (Fig. 14-5A and B).

With all the shiftings of components that occur as the tongue develops, one
landmark, a small pit known as the *foramen cecum*, serves to delineate the

border between the parts of the tongue formed from the first and second branchial arches. Embryologically, the foramen cecum is a vestige of the invagination from the floor of the pharynx which gives rise to the thyroid primordium.

The innervation of the tongue reflects the multiple origins of its components. The musculature of the tongue is innervated by the hypoglossal (XII) nerve, in keeping with the presumed origin of the tongue muscle from the *occipital (postotic) myotomes*. General sensation of the tongue correlates well with the branchial arch from which the mucosa was derived. Thus, the body of the tongue is innervated by the trigeminal (V) nerve. The root of the fully developed tongue is innervated by the sensory components of the glossopharyngeal (IX) and vagus (X) nerves. Overgrowth of the mucosa derived from the base of the second branchial arch by that of the third arch eliminates the second-arch mucosa and its nerve supply from the root of the tongue.

The function of taste is accomplished by the taste buds, which form on the lingual papilla. The taste buds first appear during the seventh week of gestation as the result of an interaction between the fibers of special visceral afferent nerves (VII and IX) and the overlying epithelium. It may seem surprising that the taste buds covering the main body of the tongue, which is of branchial arch I origin, are innervated by fibers of the seventh nerve, which supplies second-arch derivatives. This is accomplished by a special nerve branch called the *chorda tympani* which leads taste bud fibers from the seventh nerve into a major branch (the mandibular) of the fifth nerve, which is the nerve of the first arch. In many mammals taste buds are not confined to the tongue but are present in significant numbers on the mucosa of the soft palate, the nasoincisor duct, and the epiglottis (Mistretta, 1972). There is good evidence that the fetus is able to taste, and it has been hypothesized that the function of taste may be used by the fetus to monitor its intraamniotic environment (Bradley and Mistretta, 1975).

SALIVARY GLANDS

The human salivary glands originate during the sixth and seventh weeks as ridgelike thickenings of the oral epithelium (Fig. 14-6A). Because of the relatively extensive epithelial shifts in the oral cavity, the germ-layer origins of the individual salivary glands are not definitely known, but in most current opinion the parotid gland is thought to be derived from ectoderm and the submandibular and sublingual glands are thought to be endodermal structures.

The growth and morphogenesis of salivary glands are based on continued interactions between the salivary epithelium and the associated mesenchyme. Surrounding the epithelial lobule in a developing salivary gland is a basal lamina containing laminin and glycosaminoglycans of different ages and compositions. In clefts and around the stalk it is also associated with types I and IV collagen and a characteristic basement membrane-1 (BM-1) proteoglycan. (Remember from Chap. 1 that most tissues produce sets of tissue-specific proteoglycans

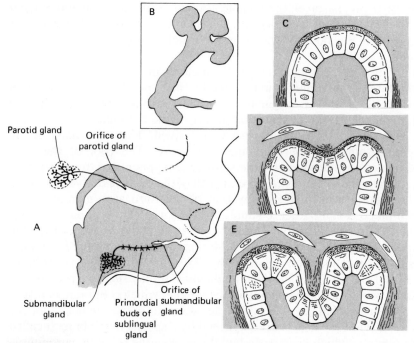

FIGURE 14-6
Development of the salivary glands. (A) Schematic diagram of developing salivary glands in an 11-week human embryo. (B) Early salivary gland developing in culture. (C–E) Model depicting the relationship between cleft formation in the developing gland and the distribution of extracellular materials. (*After Bernfield, 1973, Am. Zool.* **13**:*108*.) (C) Primary lobule showing the accumulation of newly synthesized glycosaminoglycans (stipple) within the basal lamina at the distal end of the lobule. Alongside the lobule collagen fibers (*wavy lines*) accumulate outside the basal lamina. (D) Early cleft formation within the developing lobule. Cleft formation is associated with contraction of bands of microfilaments within the cells at the apex of the lobule and the beginning of deposition of collagen fibers outside the basal lamina at the cleft. (E) Deepening of the cleft continues with an accentuation of the processes noted in D and a reduction of new glycosaminoglycan synthesis within the cleft.

which have a general structure like that of the molecule illustrated in Fig. 1-12.) At growing points of lobules, these latter components are not present. Bernfield et al. (1984) postulated that branching and growth of the salivary glands are controlled to a large extent by the surrounding mesenchyme. Branching of a primary lobule is associated with the preservation of the full complement of basal lamina components in the region of the future cleft and the removal of the collagens and proteoglycans from the areas where outgrowth will occur. Branching is actually accomplished by the contraction of well-ordered microfilaments within the epithelial cells at the point of branching (Fig. 14-6D). The alterations of the basal lamina in this process are accomplished largely by synthetic and degradative activities of the surrounding mesenchyme. Continued mitotic activity and the production of newly synthesized glycosaminogly-cans at the tips of the secondary lobules ensure continued growth of the primordium of the gland.

DEVELOPMENT OF THE TEETH

In primitive vertebrates the teeth are smaller and more numerous and are distributed over much wider areas than is the case in mammals. In their simplest form they are plates with conical protruding tips consisting of a core of calcified material called *dentine* and an apical cap of much harder material called *enamel.* They are true dermal organs, as their dentine is formed by the connective-tissue layer of the skin, and their enamel by the epithelial layer. In the development of our own more highly specialized teeth it is interesting to see that the same dual origin (epithelium and underlying mesenchyme) is retained. Even though human teeth start to form completely inside the gums instead of on a dermal surface, their enamel comes from specialized areas of epithelium which have grown down into the locations where the teeth are formed. Likewise their dentine comes from specialized mesenchymal cells, but of neural crest origin. When it is recalled that the epithelium that lines the tooth-forming part of the oral cavity is infolded stomodeal ectoderm, we can see that, highly specialized as they are as to both structure and development, human teeth have retained fundamentally the same origin in ontogeny that they had in phylogeny.

The Initiation of Tooth Development

Tooth development begins with the migration of neural crest cells into the regions of the future upper and lower jaws. Certain groups of crest cells, which are specified for tooth formation, act on the overlying oral epithelium, causing the local ingrowth of a band of epithelial cells to form the *dental ledge* (*dental lamina*) (Figs. 14-7 and 14-8). This is the first manifestation of a continuing series of interactions between neural crest mesenchyme and oral ectoderm that ultimately leads to the formation of a tooth (Ruch, 1985). Experimental studies have shown that the dental mesenchyme initiates tooth development. How and when the neural crest cells become determined to take part in tooth formation is not known.

Like so many other structures of dual origin, the further development of teeth depends on reciprocal inductive interactions between epithelial and mesenchymal precursors (Koch, 1967). As a result of the inductive processes, morphologically unspecialized ectodermal and mesenchymal cells of the oral region differentiate into highly organized complexes of secretory cells which themselves have only a transient existence. However, the products of their secretory activity—the teeth—are the most morphologically stable elements of the body, and in many mammals they persist throughout the entire lifetime of the individual.

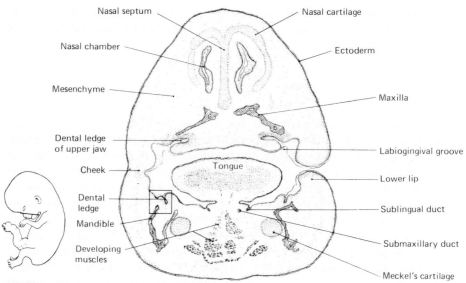

FIGURE 14-7
Projection drawing (×20) of a section through the jaws of a human embryo of the eighth week; C–R 25 mm. (University of Michigan Collection, EH 164.) Outline, lower left, shows actual size of embryo, and the line across the jaws gives the location of the section. The rectangle about the dental ledge indicates the area represented at higher magnification in Fig. 14-8A.

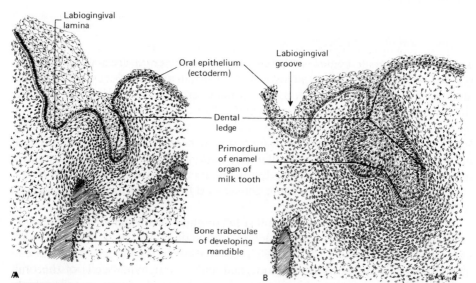

FIGURE 14-8
(A) Projection drawing (×150) of the dental ledge of a human embryo of the eighth week. The drawing was made from the area indicated by the rectangle in Fig. 14-7. (B) Projection drawing (×150) of a comparable area from a somewhat older embryo; C–R 30 mm. (University of Michigan Collection, EH 15.) Note appearance of primordium of enamel organ of the milk tooth as a local bud on the side of the dental ledge.

In the developing tooth, the mesenchymal component determines form (Kollar, 1981). For example, when the mesenchymal component of a molar tooth bud was combined in vitro with the epithelial component of an incisor, a molar tooth developed (Kollar and Baird, 1969). Conversely, after molar ectoderm was combined with incisor mesenchyme, an incisor tooth took form. An intriguing experiment by Silbermann et al. (1977) demonstrated that the differentiative properties of dental mesenchyme are labile, whereas its morphogenetic properties are relatively stable. They combined mesenchyme from a molar tooth primordium of a mouse with ectoderm from the limb bud of a chick and then explanted the combined tissues onto the chorioallantoic membrane in a chicken egg. The explanted tissues grew into a structure with the shape of a tooth, but instead of being made of dentine, the normal differentiated end product of dental mesenchyme, the tooth was made of cartilage. With these demonstrations of the control of dental morphogenesis by the mesenchymal dental papilla, it is interesting to recall that the control of the form of the cranial muscles is also inherent in the neural crest that forms the connective-tissue stroma of the muscles. If in fact cells of the cranial neural crest are able to direct the morphogenesis of both teeth and cranial muscles, it is next necessary to ask how and when the appropriate neural crest cells acquire the ability to direct morphogenesis and how they express or transmit this information to responding tissues.

Formation of the Enamel Organ and Dental Papilla

After the dental ledge is established, local buds arise from it at each point where a tooth is destined to be formed. Since these ectodermal masses give rise to the enamel crown of the tooth, they are called *enamel organs* (Fig. 14-9). As it develops, the enamel organ assumes a shape suggestive of an inverted goblet, with the section of the dental ledge appearing like a distorted stem (Fig. 14-9). The epithelial cells lining the inside of the goblet soon take on a columnar shape. Because they constitute the layer that secretes the enamel cap of the tooth, they are called *ameloblasts* (enamel formers). The outer layer of the enamel organ is made up of closely packed cells which are at first columnar in shape but soon, with the rapid growth of the enamel organ, become flattened. They constitute the so-called outer epithelium of the enamel organ. Between the outer epithelium and the ameloblast layer is a loosely aggregated mass of cells designated as the *stellate reticulum* (Fig. 14-9).

The mass of mesenchymal cells inside the cup of the enamel organ is called the *dental papilla*, and it constitutes the primordium of the pulp of the tooth (Fig. 14-9). The cells of the dental papilla proliferate rapidly and soon form a dense aggregation. A little later, as the enamel organ begins to assume the shape characteristic of the crown of the tooth it is to lay down, the outer cells of the dental papilla take on a columnar form similar to that of the ameloblasts (Figs. 14-10 and 14-13). These cells are now called *odontoblasts* (dentine formers) because they are about to become active in the secretion of dentine. By this time the dental ledge has begun to degenerate (Fig. 14-11).

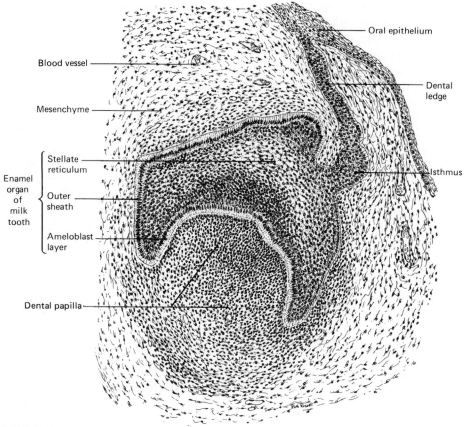

FIGURE 14-9
Projection drawing (×150) of tooth primordium in a human embryo of the eleventh week.

From the functional standpoint, one of the most important components of the tooth primordium is the extracellular matrix that forms a basement-membrane-like structure between the enamel organ and the dental papilla (Thesleff and Hurmerinta, 1981). A number of studies have shown that the inductive interactions in tooth development are mediated by the extracellular matrix that lies between the sheets of ameloblasts and odontoblasts. Beneath the inner epithelium the matrix takes the form of a basal lamina consisting of type II collagen, proteoglycans, fibronectin, and laminin. Fibronectin and types I and II collagen are secreted by the odontoblasts; the ameloblasts contribute type IV collagen, laminin, and proteoglycans.

The Formation of Dentine and Enamel

During the final stages of their differentiation the odontoblasts withdraw from the cell cycle, become polarized, and begin to secrete a material called

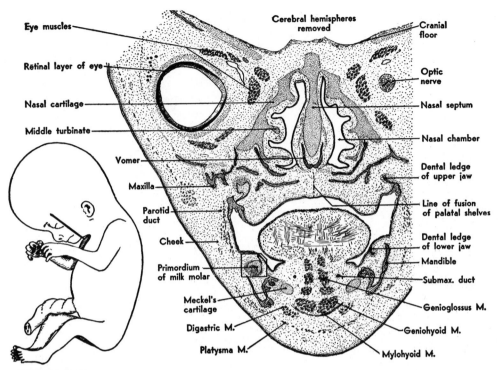

Eye muscles

Retinal layer of eye

Nasal cartilage

Middle turbinate

Vomer

Maxilla

Parotid duct

Cheek

Primordium of milk molar

Meckel's cartilage

Digastric M.

Platysma M.

Cerebral hemispheres removed

Cranial floor

Optic nerve

Nasal septum

Nasal chamber

Dental ledge of upper jaw

Line of fusion of palatal shelves

Dental ledge of lower jaw

Mandible

Submax. duct

Genioglossus M.

Geniohyoid M.

Mylohyoid M.

FIGURE 14-10
Projection drawing (×50) of a parasagittal section of the lower jaw of a human embryo of the fourteenth week, passing through the primordium of a lower central incisor; C–R 104 mm. (University of Michigan Collection, EH 145.) The small sketch, lower left, indicates the relations of the area represented.

predentin from the surface of the cells facing the enamel organ. With the shift toward producing a terminal differentiation product that will become an integral part of the mature tooth, the odontoblasts cease producing large amounts of type III collagen and fibronectin and produce larger amounts of type I collagen and other molecules that form the organic matrix that constitutes roughly one-third of the mass of the dentin of the mature tooth. Upon this organic basis, the inorganic components of the dentine are deposited. The first dentine is deposited against the inner face of the enamel organ, starting at the crown of the tooth (Fig. 14-11). As the odontoblasts continue to secrete additional dentine, the accumulation of its own product forces the layer of odontoblasts back, away from the material previously deposited (Figs. 14-12 and 14-13).

The terminal differentiation of ameloblasts occurs after the differentiation of the odontoblasts, and it probably does so in response to the presence of the predentin secreted by the latter (Ruch, 1985). After withdrawing from the cell cycle, the ameloblasts switch from the production of basal lamina components to the secretion of *amelogenins* and *enamelins*, the proteins that constitute the organic matrix of enamel. Only about 5 percent of mature enamel is organic,

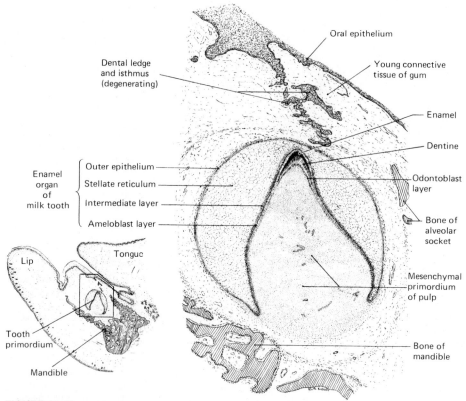

FIGURE 14-11
Projection drawing (×40) of primordium of lower central milk incisor from a human embryo of 19 weeks' presumptive fertilization age; C–R 174 mm. (University of Michigan Collection, EH 143.) (*Orienting sketch, lower left, ×5.*)

and of that amount 90 percent is amelogenin. Recently, the gene that encodes the amelogenin message has been cloned (Snead et al., 1983). Like dentine, enamel is first laid down at the crown of the tooth and its deposition progresses toward the host.

The enamel genes have been highly conserved throughout vertebrate phylogeny, from the hagfish to mammals. In fact, it has been postulated that among the early vertebrates enamel served as part of an electroreceptor apparatus (Northcutt and Gans, 1983). During phylogeny, there has been a shift from enamelin as the predominant component of enamel in the lower aquatic vertebrates to amelogenin in terrestrial forms, starting with the reptiles. Modern birds do not have teeth, nor do they form enamel, but an interesting recombination experiment by Kollar and Fisher (1980) may shed some light on the genetic ancestry of birds. They combined molar dental papillae of mouse embryos with ectoderm from the jaw of chick embryos and allowed the tissues to develop in the anterior chamber of the eye of a mouse host. In time the

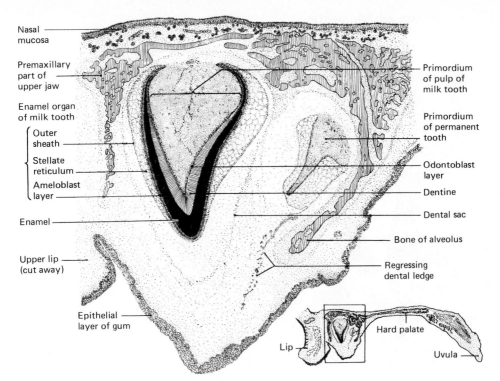

Labels for Figure 14-12 (clockwise from top left):
Nasal mucosa
Premaxillary part of upper jaw
Enamel organ of milk tooth
Outer sheath
Stellate reticulum
Ameloblast layer
Enamel
Upper lip (cut away)
Epithelial layer of gum
Lip
Hard palate
Uvula
Regressing dental ledge
Bone of alveolus
Dental sac
Dentine
Odontoblast layer
Primordium of permanent tooth
Primordium of pulp of milk tooth

FIGURE 14-12
Projection drawing (×8) of upper jaw of a human fetus at term, showing developing central incisor tooth. Orienting sketch in lower right-hand corner is actual size.

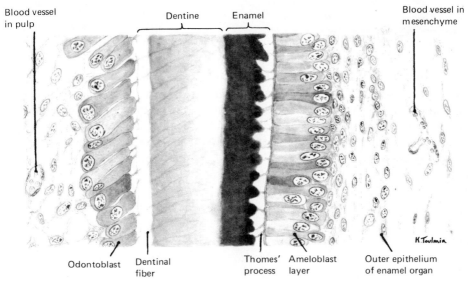

Labels for Figure 14-13:
Blood vessel in pulp
Dentine
Enamel
Blood vessel in mesenchyme
Odontoblast
Dentinal fiber
Thomes' process
Ameloblast layer
Outer epithelium of enamel organ
K Toulmin

FIGURE 14-13
Projection drawing (×425) of small segment of developing incisor from 130-mm pig embryo, showing formation of enamel and dentine. The conditions illustrated here are closely comparable with those seen in human embryos late in the fifth month.

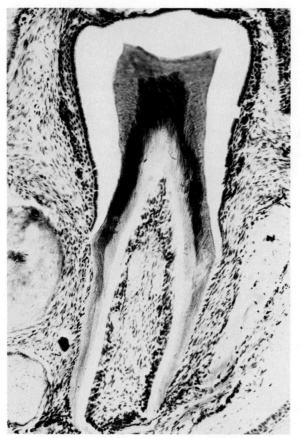

FIGURE 14-14
"Hen's tooth" formed as the result of combining pharyngeal-arch epithelium from a 5-day chick embryo with molar mesenchyme from a 16-day mouse embryo. (*Courtesy of E. Kollar, from Kollar and Fisher, 1980.*)

tissues produced a "hen's tooth" (Fig. 14-14), suggesting that the enamel genes are preserved in the chicken even though the expression of the genes was lost as a casualty of phylogenetic specialization. Although this conclusion could be strengthened by verification of the enamel by recently available probes, it has stimulated a number of other experimenters to look for unexpressed genes in other animals.

CHAPTER 15

THE DIGESTIVE AND RESPIRATORY SYSTEMS AND THE BODY CAVITIES

Development of the digestive system, its associated structures, and the body cavities can be viewed at several levels. At the level of gross structure, the formation of these structures is a logical consequence of the folding over of the body walls, differential growth, and even relations of major vascular channels to the heart. At another level, the major gastrointestinal glands, the respiratory system, and the character of the lining of the gut depend heavily on epithelio-mesenchymal interactions. Finally, the functional requirements of the digestive tract and its associated glands necessitate the biochemical differentiation of numerous enzyme systems.

THE DIGESTIVE SYSTEM

Early in the period of organogenesis the digestive system consists of a relatively simple tube (Fig. 8-10D) that is bounded at the cranial end of the foregut by the degenerating oral plate (Figs. 8-13 and 15-1A) and at the caudal end of the hindgut by the still-intact cloacal membrane. The midgut retains an open ventral connection with the yolk sac through the yolk stalk. At this stage the allantois is a prominent diverticulum from the ventral surface of the hindgut (Figs. 8-10D and 8-11D). Along the cranial half of the gut may be seen the earliest primordia of the glands associated with the digestive tract. So many new structures arise in the region of the pharynx that their embryology will be considered separately in Chap. 16.

Further development of the gut proper involves the process of elongation, herniation of part of the gut into the body stalk, rotation of several local regions of gut, and finally, histogenesis and functional maturation. While these are

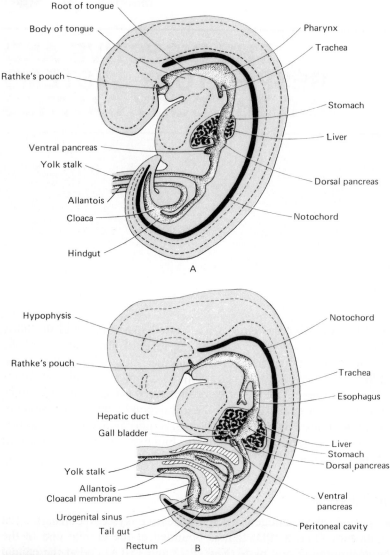

FIGURE 15-1
Sagittal sections through early human embryos, showing early stages in the formation of the gut and associated structures. (A) Early in the fifth week; (B) early in the sixth week.

taking place, the primordial digestive glands and respiratory system are growing out in complex branching patterns as the result of interactions between local gut endoderm and its enveloping mesoderm, with the deep involvement of extracellular matrix material produced by these tissues.

Esophagus

The pharynx becomes abruptly narrowed just caudal to the most posterior pouches. At this point the primitive gut gives rise ventrally to the tracheal outgrowth (Figs. 8-13 and 12-10). The region of narrowing where the trachea becomes confluent with the gut tract may be regarded as the posterior limit of the pharynx. From this point to the dilation that marks the beginning of the stomach, the gut remains of relatively small and uniform diameter and becomes the esophagus (Fig. 15-1). The original endodermal lining of the primitive gut gives rise only to the epithelial lining of the esophagus and its glands. During the seventh and eighth embryonic weeks, the human esophageal epithelium proliferates and nearly occludes the lumen. In a number of species (e.g., the chick) the esophagus becomes completely occluded, only to reopen later as a result of cell degeneration within the epithelium. The connective tissue and muscle coats of the esophagus are derived from mesenchymal cells which gradually become concentrated about the original epithelial tube (Fig. 16-8D).

Stomach

The region of the primitive gut destined to become the stomach is more or less clearly marked by a dilation (Fig. 15-1). Its shape, even at this early stage, is strikingly suggestive of that of the adult stomach. Its position, however, is quite different.

In young embryos, the concave border of the stomach faces ventrally and the convex border faces dorsally. In order to reach its adult relations, two positional shifts occur concomitantly.

1 If one looks at the stomach down the line of the esophagus, it rotates approximately 90°, so that the originally dorsal convex border is now left and the ventral concave border is facing right (Fig. 15-3).

2 The pyloric (caudal) end of the stomach tips somewhat cranially, so that the stomach is aligned diagonally across the body (Fig. 15-2).

As the rotation of the stomach is occurring, the *dorsal mesogastrium* (part of the primary mesentery) to which it is attached swings out with it, forming the pouchlike *omental bursa* (Figs. 15-2C to 15-2E and 15-3A to 15-3D). Both the tail of the pancreas and the spleen are embedded in the dorsal mesogastrium. The stomach also retains an intact portion of the primitive *ventral mesentery* (Fig. 15-11) which encloses the massive liver (Fig. 15-3A to 15-3D).

Histogenesis of the human gastric mucosa begins toward the end of the second month with the appearance of folds (*rugae*) and the first gastric pits. During the next few weeks the formation of gastric pits and the glands associated with them spreads throughout the wall of the stomach. Many of the specific cells involved in secretion begin to differentiate both morphologically and cytochemically early in the fetal period, but neither hydrochloric acid nor pepsin is present in the gastric contents until near term (Johnson, 1985).

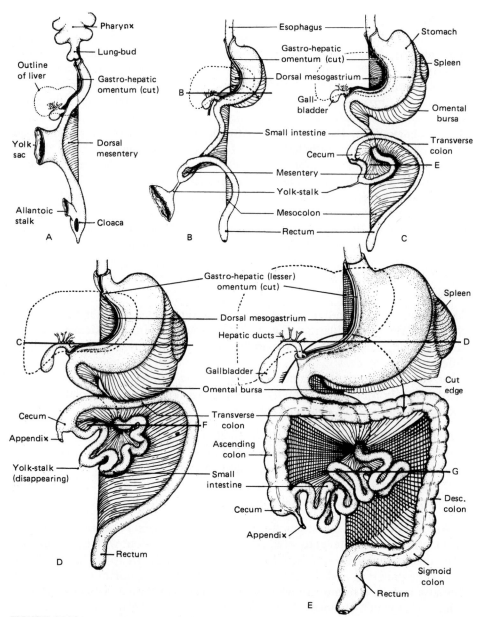

FIGURE 15-2
Frontal-view plans, schematically summarizing major developmental changes in position of gut and in relations of mesenteries. Crosshatched areas in E indicate the part of mesentery of the duodenum and the parts of the mesocolon which become fused to body wall. The heavy lines marked B, C, D, E, F, and G indicate the locations of the cross sections designated by the same letters in Fig. 15-3.

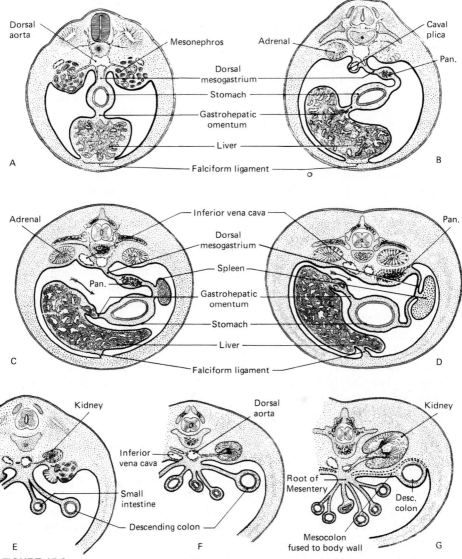

FIGURE 15-3
Cross-sectional plans of developing body to show changes in relations of mesenteries. (A–D) Sections at level of stomach and liver to show formation of omental bursa. (E–G) Sections at level of kidney to show fusion of parts of mesocolon to body wall. This illustration should be studied in comparison to Fig. 15-2, which shows corresponding stages in frontal plan. The arrows in C and D indicate the epiploic foramen.

Intestines

The primitive gut is at first a fairly straight tube extending throughout the length of the body. Near its midpoint it opens ventrally into the yolk sac (Figs. 8-10D and 8-13). The first conspicuous departure from this condition is the formation of a hairpin-shaped loop extending into the belly stalk (Fig. 15-14). The yolk stalk connects with the gut at the bend of the loop and forms an excellent point of orientation in following the series of foldings and kinkings by which the definitive configuration of the intestinal tract is established. The attachment of the yolk stalk is just cephalic to what will be the point of transition from small to the large intestine. Thus all the gut between the yolk stalk and the stomach becomes small intestine, and except for about 2 feet of the terminal part of the small intestine, the gut caudal to the yolk stalk forms the large intestine.

The positional changes which bring about adult relationships are initiated by the throwing of a twist in the primary U-shaped bend of the gut which extends into the belly stalk. Viewed in ventral aspect, the twist in the gut tract is counterclockwise (Fig. 15-2B and C). The immediate result of this twist is to bring a considerable proportion of the original cephalic limb of the gut loop posterior to the segment of the caudal limb which was twisted across it (Fig. 15-2D). This initial twist is the primary factor in establishing the fundamental positional relations of the large and small intestines. We can recognize immediately in the crossing segment of the caudal limb of the gut loop what we know in the adult as the *transverse colon*. We can see also just how it comes about that the adult jejunum and ileum lie in the abdomen below the level of the transverse colon (Fig. 15-2E).

The coiling so characteristic of the small intestine begins soon after the primary twist in the gut loop has occurred (Fig. 15-2D). That portion of the cephalic limb of the primary loop which emerges below the transverse colon is destined to become the jejunum and ileum. This part of the intestine now begins to increase in length exceedingly rapidly and consequently becomes freely coiled on itself. The coiling begins while the twisted primary gut loop still projects out into the extraembryonic coelom of the belly stalk, giving the embryo at this age the appearance of having an umbilical hernia. By about the tenth week of human development the abdomen has enlarged sufficiently to accommodate the entire intestinal tract, and the protruding part of the intestinal loop moves back through the umbilical ring into its definitive position within the peritoneal cavity. In this retraction the coils of small intestine tend to slip into the abdominal cavity ahead of the protruding part of the colon. In doing so they crowd to the left the lower part of the colon which has remained from the first in the abdominal cavity. This establishes the descending colon in its characteristic position close against the body wall on the left (Figs. 15-2D and E and 15-3F and G). When the upper part of the colon which projected into the belly stalk is finally drawn into the peritoneal cavity, its cecal end swings to the right and downward (Fig. 15-2E).

Colony (1983) has delineated three major phases of the histogenesis of the intestinal epithelium: (1) an early phase of epithelial proliferation and morphogenesis, (2) an intermediate period of cellular differentiation in which the distinctive cell types characteristic of the intestinal epithelium appear, and (3) a later phase of physiological maturation of the different types of epithelial cells.

The intestinal epithelium begins a phase of rapid proliferation early in the second month, and by 6 to 7 weeks the exuberant epithelial growth temporarily occludes the lumen. By the end of the second month the continuity of the intestinal lumen is usually restored. At about this time aggregates of mesodermal cells begin to invade the stratified intestinal epithelium, which is developing small secondary lumina beneath its surface (Fig. 15-4). Coalescence of secondary lumina and continued mesodermal upgrowth result in the formation of the

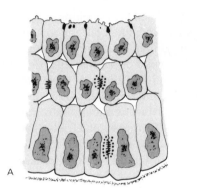
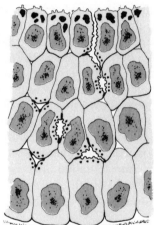
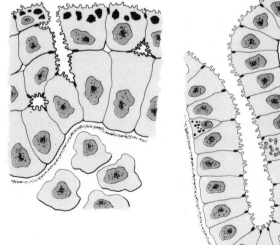

FIGURE 15-4
Steps in the formation of duodenal villi in the fetal rat. (A) 15–16 days; (B) 17 days; (C) 18 days; (D) 19 days. (*After Mathan et al., 1976, Am. J. Anat.* **146**:73.)

minute fingerlike intestinal villi, which greatly increase the absorptive surface of the intestine. At the base of the villi are tubular invaginations called *intestinal crypts*. The epithelial cells in the crypts have a high rate of mitosis, and autoradiographic studies have shown that over 3 to 4 days the daughter cells of crypt cells migrate up the villi as the villus epithelium and are finally shed from the tips of the villi. In this manner the epithelium of the small intestine is continuously removed. Studies of chimeric mouse embryos have shown that the epithelial cells of each crypt are derived from a single progenitor cell (Pander et al., 1985).

The period of differentiation of specific epithelial cell types begins at the time when villi form. By the end of the second trimester of human pregnancy, all cell types found in the intestinal lining have differentiated, but they do not have adult patterns of function. In humans, the last trimester of pregnancy is devoted to the functional maturation of enzyme systems and secretions of the epithelial cells. Recombination studies involving the epithelium and mesoderm of chick and rat intestines have shown that the controls for biochemical differentiation of the epithelium are inherent in the epithelium (Kedinger et al., 1981). The intestinal tracts of many mammals are adapted for the digestion of milk as the time of birth approaches. For example, *lactase*, adapted for the breakdown of *lactose*—a disaccharide called *milk sugar*—is one of the active enzymes formed in the late fetus. Much of the biochemical differentiation to accommodate the adult diet occurs after birth in many mammals (Thomson and Keelan, 1986).

Histochemical studies have shown that many of the intestinal secretions are formed in small amounts within the mucosa during the fetal period, but secretion into the gut occurs later, and only in small amounts until birth. The intestines of mammalian fetuses contain a greenish material called *meconium* which consists of desquamated cells from the gut, hairs, and other materials swallowed along with the amniotic fluid and bile secretion.

Liver

Very early in development the endodermal *hepatic diverticulum* from the floor of the foregut extends into the mesenchyme of the *septum transversum* (Fig. 15-13). This early hepatic outgrowth represents the first morphological evidence of a series of inductive processes which had already begun at an earlier embryonic stage (Fig. 15-5). The original diverticulum has become clearly differentiated into several parts (Figs. 8-13 and 15-6), and a maze of branching and anastomosing cell cords grows out from it.

In addition to the mesenchyme of the septum transversum, any mesenchyme derived from either the splanchnopleural or somatopleural components of the lateral plate mesoderm is capable of supporting continued hepatic outgrowth and differentiation. In contrast, axial mesoderm allows only minimal survival and little progression of development of hepatic endoderm (Le Douarin, 1975). The distal portions of the hepatic cords give rise to the secretory tubules of the

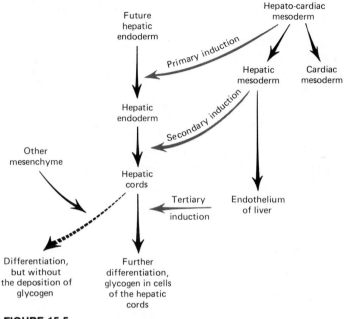

FIGURE 15-5
Tissue interactions in the morphogenesis of the endodermal component of the liver. (*Modified from Deuchar, after Wolff.*)

liver, and their proximal portions form the hepatic ducts. A dilation which is the primordium of the *gallbladder* originates from where the hepatic ducts become confluent. Closer to the gut tract a separate outgrowth of cells forms the ventral primordium of the pancreas.

The branching and anastomosing tubules which are distal continuations of the hepatic ducts constitute the actively secreting portion of the liver. Their position and extent in embryos of various ages are shown in Figs. 8-13, 12-10, 15-7C, and 15-14. The organization of these secreting units in the liver is quite characteristic. The *hepatic tubules* are not packed as closely together in a framework of dense connective tissue as is usually the case in massive glands. Surprisingly little connective tissue is formed between them, and the intertubular spaces become pervaded by a maze of dilated and irregular capillaries known as *sinusoids*.

The development of the liver from the early endodermal buds to its mature form involves not only an increase in mass and structural complexity but also a gradual acquisition of the metabolic pathways which enable it to carry out its manifold functions in postnatal life. A major function of the liver is the synthesis and storage of glycogen, which serves as a carbohydrate reserve for the entire body. The embryonic liver, particularly during the late fetal period, actively stores glycogen, and in mammals this function has been shown to be under the control of adrenocortical steroid hormones in conjunction with an

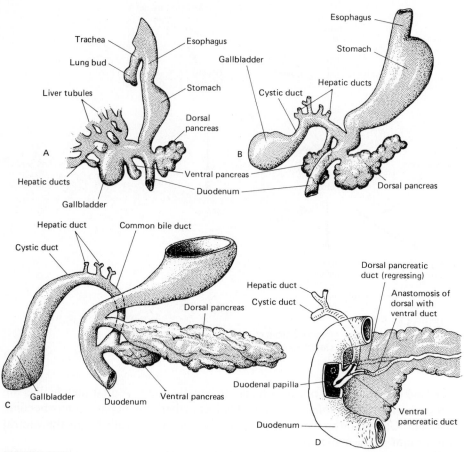

FIGURE 15-6
Development of hepatic and pancreatic primordia. All diagrams are viewed from the ventral aspect. (A) Semischematic diagram based in part on Thyng's reconstructions of a 5.5-mm pig embryo. This is comparable to a human embryo early in the fifth week. (B) Reconstruction from a 9.4-mm pig embryo. This is comparable to a human embryo of 7 weeks. (C) Schematized drawing equivalent to a 20-mm pig embryo or a human embryo of 7 weeks. (D) Manner in which common bile duct and pancreatic duct become confluent in the ampulla of Vater and discharge through the duodenal papilla.

influence by the pituitary (Jost, 1962). The system of enzymes involved in the synthesis of urea from nitrogenous metabolites gradually assumes prominence in the fetal liver and attains its full functional capacity close to the time of birth. As we shall see later, the embryonic liver also serves as a transient site of blood-cell formation.

The great synthetic activity of the postnatal liver is served by an unusual adaptation of the vascular system, the *portal vein*. This is a vein which arises from a confluence of smaller veins and their capillaries along much of the intestinal tract and which breaks up into a capillarylike network within the

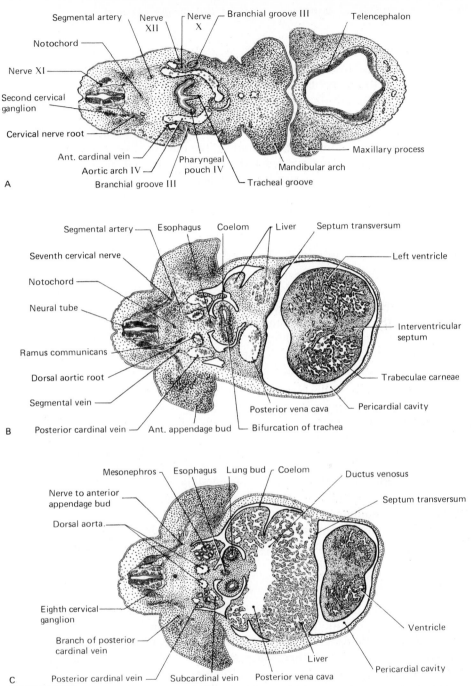

FIGURE 15-7

Three transverse sections from the series of the 9.4-mm pig embryo used in making the reconstruction illustrated in Fig. 18-9. By laying a straightedge across this reconstruction at the level of the marginal line bearing the serial number of a cross section, the relations of that section within the body as a whole are precisely indicated. (A) Section 180 at the level of the laryngotracheal groove; (B) section 293 at the level of the bifurcation of the trachea; (C) section 309 through the cephalic part of the liver and the lung buds.

parenchymal substance of the liver. By this arrangement, the postnatal liver is the first organ to receive blood rich in protein, carbohydrate, and fat metabolites absorbed through the intestinal walls. The mammalian embryo possesses a different anatomical adaptation which fulfills a similar function. In embryonic life food materials are not carried through the portal vein because the digestive tract is nutritionally nonfunctional. Rather, the umbilical vein, arising from the placenta, carries foodstuffs extracted from the maternal blood to the liver, where much of its blood empties into a hepatic capillary network. By this adapatation in its circulation, the embryonic liver is thus afforded first access to metabolites essential for its synthetic functions.

Pancreas

The pancreas makes its appearance in the same region and at about the same time as the liver. It is derived from two separate primordia which later become fused. One primordium arises dorsally, directly from the duodenal endoderm; the other arises ventrally, from the endoderm of the hepatic diverticulum (Fig. 15-6A). As the duodenum rotates, the ventral pancreatic bud is carried into the dorsal mesentery, where it approaches and ultimately fuses with a portion of the more extensive dorsal pancreas (Fig. 15-6B to D).

Like the salivary glands and the other organs which bud off from the gut, the appearance of the pancreas is the result of a coordinated interplay between localized endodermal cells and their surrounding mesenchyme. The earliest stages in pancreatic development are not well understood, but in the very early gut, before obvious differentiation has occurred, a small population of approximately 300 cells (Wessels, 1977) has become specified to form pancreas. The glandular epithelium of the pancreas is formed by the budding and rebudding of cords of cells derived from this population of primordial cells (Fig. 15-8). Although surrounding mesenchyme is required for epithelial outgrowth and branching, in vitro recombination experiments have shown that the developing pancreas, in contrast to the liver, seems content with mesoderm from a variety of sources.

The mature pancreas is a dual organ, consisting of an exocrine and an endocrine portion, the latter being embodied in the million or so small clusters of secretory cells, the *islets of Langerhans*, which are scattered among the acini of the exocrine portion. The acinar cells, the principal secretory cells of the exocrine part of the pancreas, produce a variety of digestive enzymes which are carried into the small intestine via a system of ducts. The islets consist of several varieties of cells, the most prominent of which are the glucagon-secreting α cells and the insulin-secreting β cells. These hormones are secreted directly into the capillaries, which provide a rich blood supply to the islets. Insulin lowers and glucagon raises the level of glucose in the blood.

In the developing pancreas there is a close correlation between morphogenesis and the synthesis of its specific secretory proteins. Several phases of maturation have been described (Fig. 15-8). The first phase precedes outgrowth

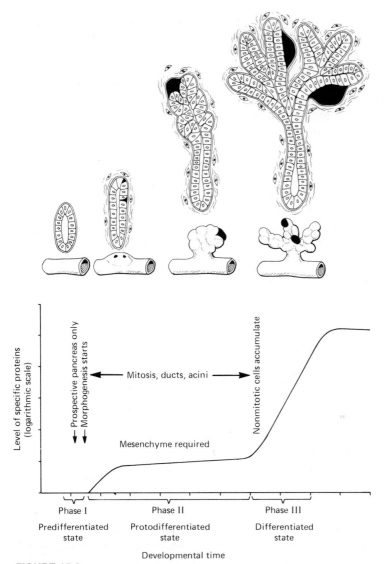

FIGURE 15-8
Stages of structural and functional differentiation of the pancreas (*After Rutter*). The graph and drawings stress the exocrine component of the pancreas. The dark areas in the drawings represent primitive islets. (*Drawings modified from Rutter, 1972*, in Handbook of Physiology. Endocrinology I, *p. 32.*)

and consists principally of the establishment of the primordial population of pancreatic cells. A transition to the second level of maturation occurs with the appearance of the original pancreatic diverticulum. This phase is characterized by the synthesis of low levels of many of the hydrolytic enzymes of the exocrine cells as well as low levels of insulin and relatively high levels of glucagon. During this phase mesenchymal cells form a close association with the budding epithelial cells. The third phase of differentiation involves the production of an elaborate protein-synthesizing and secreting mechanism by the acinar cells and a marked increase in the synthesis of digestive enzymes. Meanwhile, the islets of Langerhans are forming by budding off of the developing acini. Both the α and the β cells elaborate large numbers of secretory granules containing glucagon and insulin, and some quantity of the newly synthesized hormones enters into the fetal circulation.

In the fetus, the duct systems of the exocrine part of the pancreas drain into a main duct running most of the length of the dorsal pancreas. In humans, the dorsal pancreatic duct becomes anastomosed with the main duct of the original ventral component of the pancreas, and through it (the *duct of Wirsung*) the pancreatic enzymes are emptied into the duodenum. The terminal part of the dorsal pancreatic duct regresses (Fig. 15-6D). In some species (e.g., dog and horse) there are two ducts, the ventral pancreatic duct and the nonregressed dorsal pancreatic duct (*duct of Santorini*), which also empties into the duodenum.

THE RESPIRATORY SYSTEM

Trachea

The first indication of the differentiation of the respiratory system is the formation of the *laryngotracheal groove* at the posterior limit of the pharyngeal region (Figs. 8-14D and 16-2). Once established as a separate diverticulum, it grows caudad as the *trachea*, ventral to and roughly parallel with the esophagus (Figs. 8-14E and 12-10). The anatomical relations of the trachea in the embryo, even in early stages, are quite similar to those in the adult. We can recognize its communication with the posterior part of the pharynx as the future *glottis* (Fig. 14-5C and D).

Only the epithelial lining of the adult trachea is derived from foregut endoderm. The cartilage, connective tissue, and muscle of its wall are formed by mesenchymal cells which become massed about the growing endodermal tube (Figs. 16-8D and 18-10C).

Bronchi and Lungs

As the tracheal outgrowth lengthens, it bifurcates at its caudal end to form the two lung buds (Fig. 15-9A and B). These in turn continue to grow and branch, giving rise to the bronchial trees of the lungs (Fig. 15-9C to 15-9F). The characteristic budding pattern of the endodermal bronchial tree is the result of a continuous inductive effect by the surrounding mesoderm. In the absence of its mesodermal investment, budding of the bronchial tree does not occur (Rudnick, 1933). More recent work (Wessels, 1970) has shown that there are

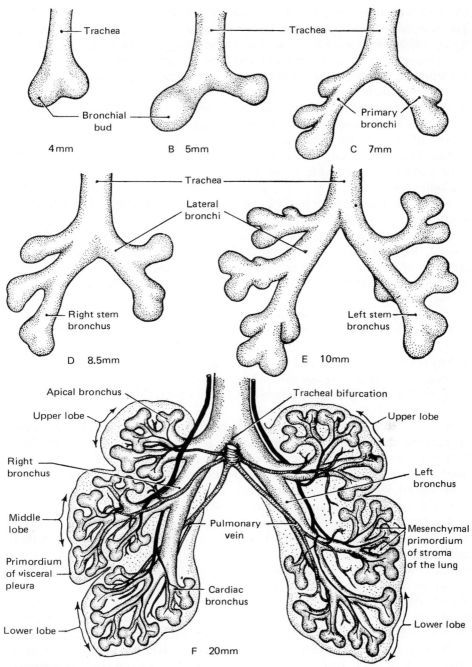

FIGURE 15-9
Diagrams showing development of major bronchi of human lungs. (*Ventral view, adapted from various sources, chiefly Corning and Arey.*)

two forms of mesoderm involved in the formation of the respiratory system. Tracheal mesoderm, possessing highly ordered sheaths of cells and collagen fibers, supports outgrowth of the trachea but inhibits branching, whereas the more loosely organized bronchial mesoderm promotes budding. In exchange experiments, bronchial mesoderm induces budding from the trachea and, conversely, tracheal mesoderm inhibits bronchial budding. In some circumstances salivary gland mesenchyme combined with bronchial endoderm will permit budding (Lawson, 1983). In this case, the branching pattern is characteristic of the salivary gland but the epithelium differentiates according to its origin. The same study showed that in order to promote branching, the mesenchyme must be able to stimulate a high rate of proliferation of the epithelial cells.

The terminal portions of the branches where cell proliferation is exceedingly active tend to remain somewhat bulbous (Fig. 15-10B and C). Later in development these terminal portions of the bronchial buds become still more dilated, their epithelium thins markedly, and they give rise to the characteristic air sacs of the lungs. The pleural covering of the lungs is derived from splanchnic mesoderm pushed ahead of the lung buds in their growth (Fig. 15-10C and D).

The lungs do not at first occupy the position characteristic of adult anatomy. In very young embryos they lie dorsal to the heart (Figs. 8-13 and 8-19). A little later, when they have extended caudad, they are situated dorsal to the heart and liver (Figs. 15-13 and 15-14). The changes by which they eventually come to occupy their definitive position in the thorax will be discussed in connection with the partitioning of the primitive coelom to form the body cavities of the adult.

As is also true of the embryonic heart and circulatory system, the lungs must prepare themselves well before birth for the important physiological role which they must assume with the newborn infant's first breath. Within the first few minutes after birth they must become transformed from organs resembling waterlogged sponges to air-filled sacs which are capable of retaining an adequate air space without collapsing at each exhalation.

The fetal human lung is filled with a fluid resembling blood plasma instead of amniotic fluid, as was often formerly supposed. Secreted by the lungs, this fluid undergoes some distinct changes, particularly in its lipid composition, during the latter weeks of pregnancy. The lipids, primarily lecithin and sphingomyelin, constitute what is known as *pulmonary surfactant*. Surfactant acts to reduce the surface tension of the fluid layer lining the interior of the air sacs (alveoli) within the lungs. This enables the air sacs to remain expanded with minimal effort in breathing. Although surfactant in the human can be detected as early as 26 weeks, it is not accumulated in sufficient quantities in the lungs until shortly before the normal time of birth. It is for this reason that prematurely born infants have a tendency to develop respiratory difficulties.

Another adaptation of the fetal respiratory tract for its function at birth is the relatively great size of the trachea and other respiratory passages in relation to

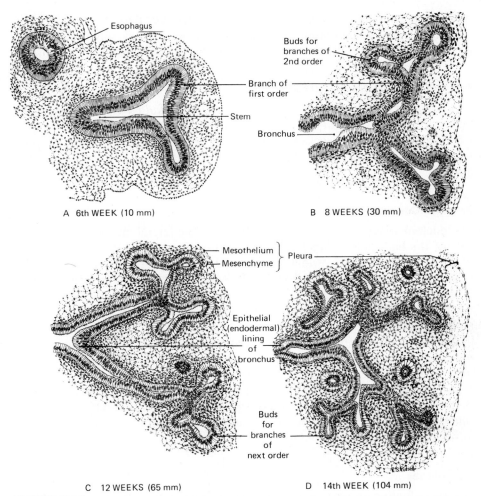

Esophagus

Buds for
branches of
2nd order

Branch of
first order

Stem

Bronchus

A 6th WEEK (10 mm)

B 8 WEEKS (30 mm)

Mesothelium
Mesenchyme

Pleura

Epithelial
(endodermal)
lining
of
bronchus

Buds
for
branches
of
next order

C 12 WEEKS (65 mm)

D 14th WEEK (104 mm)

FIGURE 15-10
Projection drawings (×100) showing four early stages in the histogenesis of the human lung. (A)
Embryo of sixth week (EH 56; C–R 10 mm). (B) Embryo of 8 weeks (EH 15; C–R 30 mm). (C)
Embryo of 12 weeks (EH 173, B; C–R 65 mm). (D) Embryo of fourteenth week (EH 145, C; C–R
104 mm). All embryos from University of Michigan Collection.

the total size of the fetus. Although a normal newborn infant is approximately
one twenty-fifth the size of an adult, the respiratory passages are considerably
larger, being between one-fourth and one-third the adult diameter. Such
precocious growth is a physical necessity because if the respiratory passages
were any narrower, the physical resistance to breathing would be too great to
be overcome (Avery et al., 1973). By the time of the baby's first breath, the
fluid that was present in the lungs has been eliminated by a combination of
physical expulsion during delivery and resorption into the lymphatic and blood
vessels of the lungs. If the newborn infant is able to overcome the great

physical resistance involved in taking the first breath or two, the blood flow in the lungs is quickly increased to postnatal levels with the closing of the ductus arteriosus (see Chap. 18) and the normal breathing pattern is quickly stabilized.

THE BODY CAVITIES AND MESENTERIES

The body cavities of adult mammals are the *pericardial cavity*, containing the heart; the paired *pleural cavities*, containing the lungs; and the *peritoneal cavity*, containing the viscera lying caudal to the diaphragm. All three of these regional divisions of the body cavity are derived from the coelom of the embryo.

Primitive Coelom

The coelom arises by means of the splitting of the lateral mesoderm on either side of the body into splanchnic and somatic layers (Figs. A-47A and B and 15-11A). It is therefore primarily a paired cavity bounded on one side by splanchnic mesoderm and on the other by somatic mesoderm. In forms such as birds and mammals, which have highly developed extraembryonic membranes, the coelom extends between the mesodermal layers of the extraembryonic membranes beyond the confines of the developing body. The splitting of the mesoderm in mammals occurs first extraembryonically and progresses thence toward the embryo (Fig. 6-23). When the body of the embryo is folded off from the extraembryonic membranes, the extra- and intraembryonic portions of the coelom are thereby separated from each other, with the last place of confluence to be closed off being in the region of the belly stalk (Figs. 15-11C and 15-12). The intraembryonic portion of the primitive coelom, thus delimited, gives rise to the body cavities of the adult.

Mesenteries

The same folding process that separates the embryo from the extraembryonic membranes completes the floor of the gut (Figs. 8-10 and 15-11). Coincidentally the splanchnic mesoderm of either side is swept toward the midline, enveloping the now tubular digestive tract. The two layers of splanchnic mesoderm which thus become apposed to the gut and support it in the body cavity are known as the *primary*, or *common*, *mesentery*. The part of the mesentery ventral to the gut, attaching it to the ventral body wall, is the *ventral mesentery* (Fig. 15-11D). The primary mesentery, while intact, keeps the original right and left halves of the coelom separate. But the part of the mesentery ventral to the gut breaks through very early, bringing the right and left coelom into confluence and establishing the unpaired condition of the body cavity characteristic of the adult (Fig. 15-11F). Farther cranially, the mesocardium, which supports the primitive heart, follows a roughly similar portion of development (Fig. A-21).

In the hepatic region the ventral mesentery does not disappear. The liver arises, as we have seen, from an outgrowth of the gut and in its development

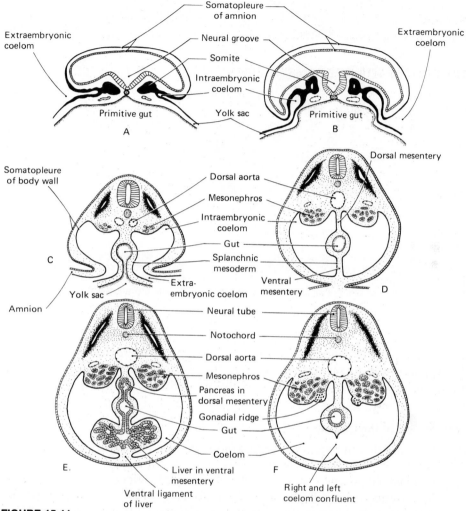

FIGURE 15-11
Diagrams illustrating early stages in the development of the coelom and mesenteries.

pushes into the ventral mesentery (Fig. 15-11E). The portion of the ventral mesentery between the liver and the stomach persists as the *ventral mesogastrium*, and the portion between the liver and the ventral body wall, although reduced, remains in part as the *falciform ligament* of the liver (Fig. 15-14).

Whereas the ventral mesentery, except in the region of the liver, eventually disappears, almost the entire original dorsal mesentery persists. It serves at once as a membrane supporting the gut in the body cavity and a path over which nerves and vessels reach the gut from main trunks situated in the dorsal body wall. Its different regions are named according to the part of the digestive

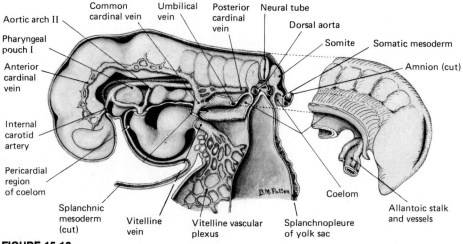

FIGURE 15-12
Schematic plan of lateral dissection of young mammalian embryo to show the relations of the pericardial region of the coelom to the primary paired coelomic chambers caudal to the level of the heart. The proportions of the illustration were based in part on Heuser's study of human embryos about 3 weeks old, but all the essential relationships shown are equally applicable to pig embryos of 3 to 4 mm.

tube with which they are associated, for example, the *mesogastrium*, referring to the part of the dorsal mesentery supporting the stomach, and the *mesocolon*, referring to the part of the dorsal mesentery supporting the colon (Fig. 15-14).

Partitioning of the Coelom

The structure that initiates the division of the coelom into separate chambers is the *septum transversum*. The septum transversum appears very early in development (Figs. 8-13 and 15-13B) and is a conspicuous structure (Fig. 15-14). Extending dorsally from the ventral body wall, it forms a sort of semicircular shelf. Fused to the caudal face of the shelf is the liver, and on its cephalic face rests the ventricular part of the heart. The septum transversum is the beginning of the diaphragm; however, it must be kept in mind that the diaphragm is a composite structure of which only the ventral portion is derived from the septum transversum.

As it expands from the ventral body wall, the septum transversum acts as a partial partition between the pericardial and peritoneal portions of the coelom. When it has grown to an extent where it makes contact with the floor of the foregut, the septum transversum has almost cut the coelomic cavity into two parts. The separation is not complete, however, for on either side of the gut dorsal to the septum transversum small portions of the coelomic cavity remain (Fig. 15-13C). These remnants of the coelom are actually short channels called the *pleural canals*. They connect the thoracic portion of the coelom, which is

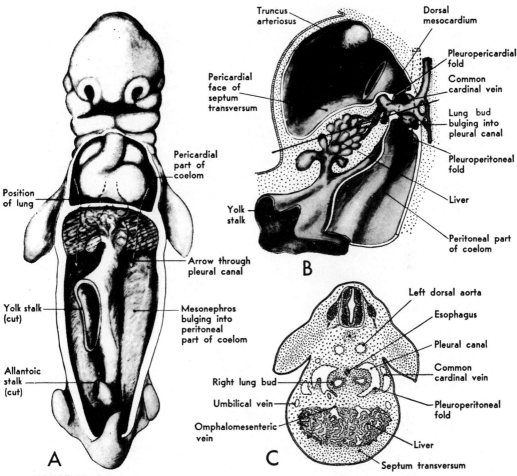

FIGURE 15-13
Relations of various parts of coelom during fifth week. (A) Semischematic frontal plan. Arrows indicate location of pleural canals, on either side, dorsal to liver. (B) Lateral dissection to show left pleural canal opened with lung bud bulging into it medially. (*Modified from Kollmann.*) (C) Section through body of an 8-mm embryo, schematized to show relations at level of pleural canal. Level of section is indicated by heavy line across B.

the single pericardial cavity, with the peritoneal part of the coelom, which is also a single cavity (Fig. 15-13A and B). At first the pleural canals are relatively small channels, but as the developing lung buds grow into them, they increase greatly in size and become the third major component of the coelom, the pleural cavities.

The pleural canals are in effect partially bounded by two paired folds of tissue, the pleuropericardial and pleuroperitoneal folds. The *pleuropericardial folds* are ridges of tissue arising from the dorsolateral body walls where the common cardinal veins bulge into the coelom as they swing toward the midline

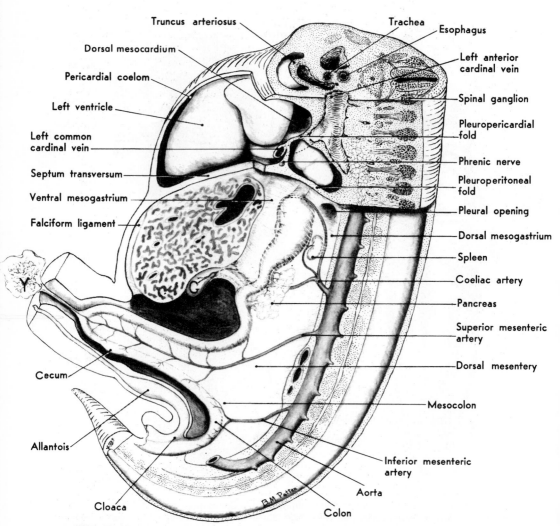

FIGURE 15-14
Semidiagrammatic drawing showing the arrangement of the viscera, body cavities, and mesenteries in a human embryo early in the seventh week of development. In all essentials the conditions represented here will be found in pig embryos of 15 mm. In the region of the developing lungs the body is cut parasagittally, well to the left of the midline, in order to show the relations of pleuropericardial and pleuroperitoneal folds. Below the developing diaphragm, dissection has been carried to the midline. Abbreviations: *G*, gallbladder; *Y*, yolk sac.

to enter the sinus venosus of the heart (Figs. 15-13B, 15-14, and 15-16B and C). In time these folds effectively partition the anterior ends of the pleural canals from the pericardial cavity. Posteriorly, the pleural canals are delimited by the *pleuroperitoneal folds* (Fig. 15-13B), which project from the body wall to demarcate the posterior pleural canals from the peritoneal coelom.

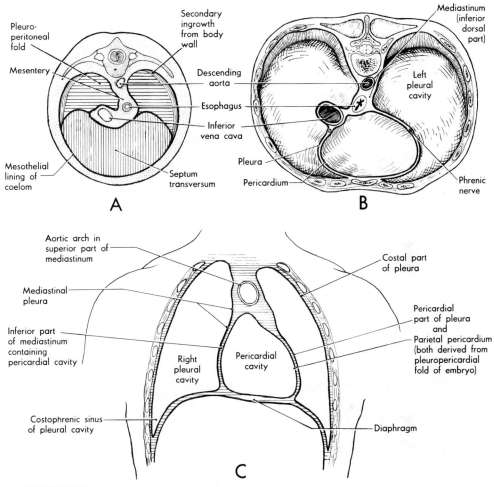

FIGURE 15-15
Diaphragm in embryo and in adult. (A) Diagram indicating embryological origin of various regions of diaphragm. (*Modified from Broman.*) (B) Thoracic face of adult diaphragm. (*Modified from Rauber-Kopsch.*) (C) Relations of diaphragm as seen in a frontal section of adult body. (*Modified from Rauber-Kopsch.*)

The formation of the definitive pleural cavities and their isolation from the pericardial and peritoneal cavities involve the expansion of the lungs into the pleural cavities as well as the sealing off of both ends of the pleural canals by the pleuropericardial and pleuroperitoneal folds. The early lung buds expand caudally and laterally in the mesenchyme ventral to the esophagus. This expansion results in a bulging of the lung buds into the pleural canals (Fig. 15-13C). As the developing lungs continue to grow, they expand dorsally, laterally, and ventrally, thus displacing the mesenchyme of the body wall (Fig. 15-16B to 15-16D). This expansion of the lungs greatly increases the size

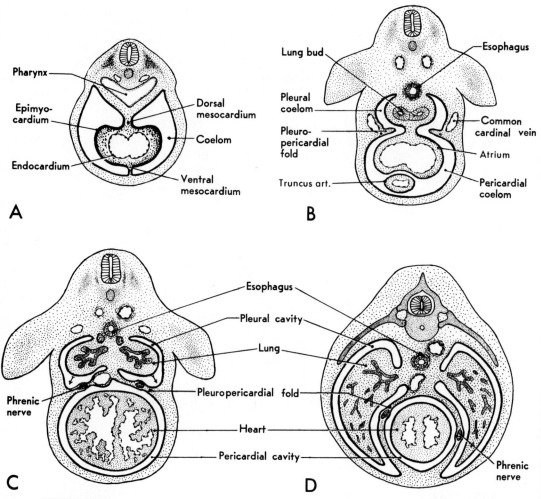

FIGURE 15-16
Schematic diagrams showing the manner in which the pleural and pericardial regions of the coelom become separated.

of the original pleural canals, so that they now partially surround the pericardial cavity and also extend somewhat dorsal to the esophagus (Fig. 15-16D).

As the lungs are expanding, the anterior portion of the pleural canals is becoming separated from the pericardial cavity by the pleuropericardial folds. It will be recalled that the pleuropericardial folds are actually ridges of mesenchymal tissue which accompany the common cardinal veins as they converge toward the heart (Fig. 15-16B and C). As the lungs expand into the lateral body walls and the common cardinal veins shift toward the midline with the descent of the heart into its final position, the pleuropericardial folds converge and fuse with one another, thus effecting a dorsal separation between the lungs and the heart.

Meanwhile, the pleuroperitoneal folds are converging from the body walls toward the septum transversum to complete the separation of the posterior ends of the pleural canals from the peritoneal cavity. The dorsal mesentery is caught between the septum transversum and the pleuroperitoneal folds. Fusion along the lines of contact of these structures essentially completes the diaphragm, especially dorsolaterally (Fig. 15-15A). Later in development the edges of the diaphragm, especially dorsolaterally, are invaded by body-wall mesenchyme, which contributes the marginal part of the diaphragmatic musculature (Fig. 15-15A).

CHAPTER 16

DUCTLESS GLANDS, PHARYNGEAL DERIVATIVES, AND THE LYMPHOID ORGANS

Some functionally integrated systems in the body defy neat anatomical packaging because the individual organs are widely scattered in location and quite diverse with respect to their mode of origin. Because of their functional ties, it is preferable to present them as groups. Two such systems, the endocrine system and the lymphoid system, will be presented in this chapter. Because of the intimate association between the origin of many endocrine and lymphoid organs and the development of the pharynx, the embryology of the entire pharyngeal region has been left to this chapter.

DEVELOPMENT OF THE HYPOPHYSIS

The hypophysis is formed from two separate primordial ectodermal parts which secondarily unite. One of these primordia, known as *Rathke's pocket*, arises from the stomodeal ectoderm and extends in the midline toward the diencephalic floor (Fig. 16-1A). The other primordial part of the hypophysis is the *infundibular process*, which forms from neural ectoderm in the floor of the diencephalon. In time, without losing its basic relationship to the diencephalon, the infundibular process comes to constitute the pars neuralis, or *neural lobe*, of the adult hypophysis (Fig. 16-1D and E).

In later development, Rathke's pocket elongates, and its blind end partially enfolds the infundibular process (Fig. 16-1B). At the same time, the original stalk which connected it to the stomodeum becomes narrowed (Fig. 16-1B) and ultimately regresses (Fig. 16-1C and D). The stomodeal portion of the hypophysis becomes molded into a double-layered cup about the neural lobe. The outer wall of the cup thickens and takes on a glandular appearance as it

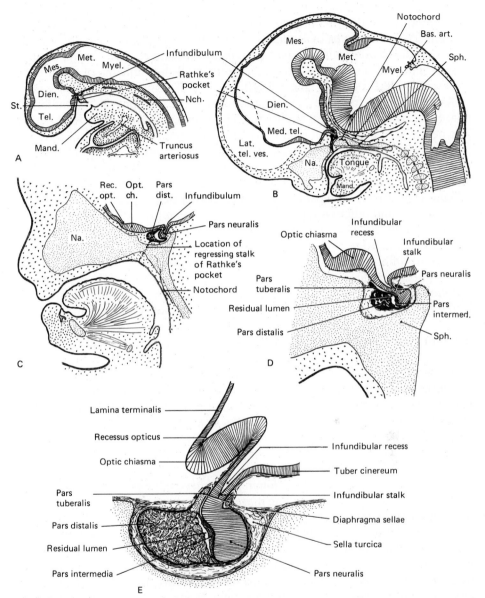

FIGURE 16-1
Diagrams showing the changing relations of the neural and stomodeal portions of the hypophysis during its development. (A) Semischematic sagittal section of head of 4-week (4 to 5 mm) human embryo. (B) Projection diagram of sagittal section of head of human embryo of about 6½ weeks. (C–R 15 mm, University of Michigan Collection, EH 4.) (C) Projection diagrams of sagittal section of head of human embryo of eighth week. (C–R 25 mm, University of Michigan Collection EH 33.) (D) Sagittal section of hypophyseal region of embryo of eleventh week. (C–R 60 mm, University of Michigan Collection, EH 23.) (E) Schematic plan of adult hypophysis.

differentiates into the *pars distalis* (*anterior lobe*) of the hypophysis (Fig. 16-1D and E). The inner layer of Rathke's pocket becomes closely adherent to the pars neuralis to constitute the *pars intermedia*. A thin *residual lumen* separating the pars distalis and the pars intermedia is all that remains of the original lumen of Rathke's pocket (Fig. 16-1E).

As pregnancy progresses, the components of the hypophysis first undergo a phase of cytodifferentiation and then begin to function late in the fetal period. The neural lobe becomes organized as a neurosecretory tissue and releases two polypeptide hormones: *antidiuretic hormone*, which aids in water retention by the kidneys, and *oxytocin*, which stimulates the contraction of uterine smooth muscle during parturition and also stimulates the release of milk in lactating females (Fig. 10-27). The pars intermedia is poorly developed in humans, but in both humans and other vertebrates it secretes *melanocyte-stimulating hormone* (MSH), which increases the intensity of skin pigmentation, and β-*endorphin*, an opiatelike peptide.

Several different types of secretory cells differentiate in the anterior pituitary. These produce a variety of "tropic" hormones which promote the secretory activity of other endocrine glands throughout the body. Some, such as the gonadotropic hormones (LH and FSH), are not prominent in the fetus, but both *thyroid-stimulating hormone* (TSH) and *adrenocorticotropic hormone* (ACTH) have important functions in the fetus. During the second half of pregnancy TSH secreted by the pituitary takes control over hormonal secretions by the thyroid gland. Similarly, a functional pituitary-adrenal axis is set up late in pregnancy, and among other functions it is important for the onset of parturition. Removal of either the fetal pituitary or the adrenal glands results in prolongation of gestation in sheep (Challis et al., 1977).

BRANCHIAL REGION

In 1-month human embryos (4- to 6-mm pig embryos) the pharyngeal part of the foregut and the tissues surrounding it are organized in a manner reminiscent of the branchial region of the lower vertebrates, with a paired series of branchial arches alternating with branchial clefts and pharyngeal pouches along either side of the pharynx (Figs. 8-14B and 16-2). This stage of development of the branchial region is a recapitulation of conditions which had an obvious functional significance in water-living ancestral forms, for the branchial clefts and the pharyngeal pouches of the mammalian embryo are homologous with the outer and inner portions of the gill slits in lower forms. As so often happens in the embryo, the repetition of phylogenetic history is slurred over. Although in the mammalian embryo the tissue closing the gill clefts is reduced to a thin membrane consisting of a layer of endoderm and ectoderm with little or no intervening mesoderm (Fig. 12-16B), this membrane rarely disappears altogether. Although aspects of the development of certain branchial structures have been dealt with in other chapters, this section will summarize the development of the region as a whole. Figure 16-3 presents in diagrammatic

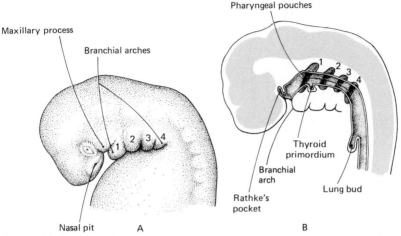

FIGURE 16-2
Topography of the branchial region in a human embryo in the fifth week. (*In part after Langman, Medical Embryology, 1963, Williams & Wilkins, Baltimore.*) (A) External configuration to show the branchial arches. (B) Head opened medially to show pharyngeal pouches.

form the main adult derivatives of key branchial structures, a number of which arise from neural crest cells (see Table 6-2, p. 239).

BRANCHIAL ARCHES

There are four recognizable branchial arches, each of which contains a cranial nerve, a blood vessel (aortic arch; Fig. 16-4), a skeletal primordium of neural crest origin, and premuscle mesenchyme (Fig. 16-5).

The first arch (mandibular), which gives rise to both the mandibular and the maxillary portions of the face, is innervated by the trigeminal (V) nerve. Its cartilaginous rod (Meckel's cartilage) persists in the lower jaw as Meckel's cartilage, later to be surrounded by the intramembranously formed bone that forms the mandible. The dorsal part of the rod breaks up to form one of the middle-ear ossicles (malleus) (see Chap. 13). The premuscular mesenchyme (mesodermal) of the first arch forms the muscles of mastication and others (see Fig. 16-3), all innervated by the fifth cranial nerve.

The second arch (hyoid), which is innervated by the seventh cranial nerve, forms a string of skeletal structures, starting with the stapes of the middle ear and extending down to the body of the hyoid bone (Fig. 16-5D). Much of the mesoderm migrates to the face to form the muscles of facial expression, but other second-arch muscles are associated with second-arch skeletal derivatives. All the muscles of the second arch are innervated by the seventh cranial nerve.

The skeletal and muscular derivatives of the third arch are related to the hyoid bone and upper pharynx. The one muscle derived from this arch is

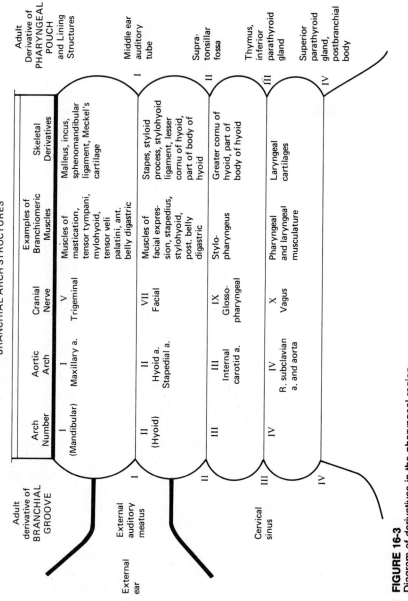

FIGURE 16-3
Diagram of derivatives in the pharyngeal region.

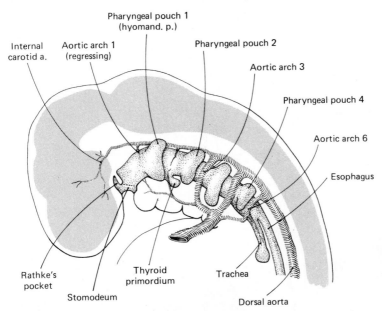

Pharyngeal pouch 1
(hyomand. p.)

Pharyngeal pouch 2

Internal
carotid a.

Aortic arch 1
(regressing)

Aortic arch 3

Pharyngeal pouch 4

Aortic arch 6

Esophagus

Rathke's
pocket

Thyroid
primordium

Trachea

Stomodeum

Dorsal aorta

FIGURE 16-4
Schematic diagram of the branchial region with the overlying ectoderm and
mesenchyme removed to show the pharyngeal pouches and the relations of
the aortic arches to them.

innervated by the ninth cranial nerve. The fourth branchial arch, which is
innervated by the vagus (X) nerve, supplies some cartilaginous and muscular
structures in the larynx and lower pharynx, but the vagus nerve grows into the
thoracic and abdominal cavities as well.

BRANCHIAL CLEFTS

Only the first branchial cleft contributes to a recognizable structure, persisting
as the external auditory meatus. Clefts II to IV temporarily become overshadowed by the hyoid arch (the homologue of the operculum, or gill cover, of
fishes) and are collectively called the *cervical sinus* (Figs. 8-7 and 16-6A). Later
in embryonic development, the external contours of the neck smooth out and
the cervical sinus disappears without a trace.

PHARYNX

The main pharyngeal chamber of the embryo is converted directly into the
pharynx of the adult. In this process the configuration of its lumen is simplified
and relatively reduced in extent. An important factor in these changes is the
separation of various pouches from the main part of the pharynx. The cell
masses thus originating migrate into the surrounding tissues and there undergo
divergent differentiation.

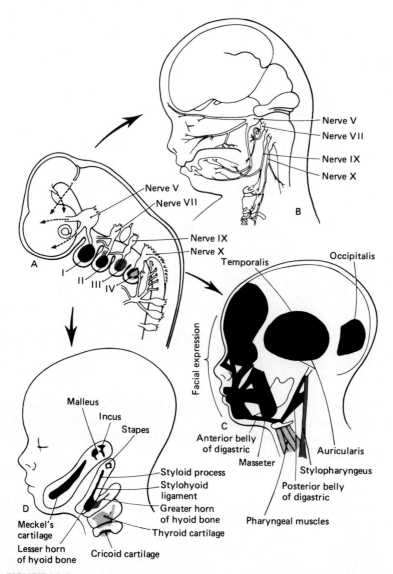

FIGURE 16-5
Schematic diagrams showing major derivatives of structures that constitute the
branchial arches. (A) 5-week embryo; (B–D) 4- to 5-month fetuses. The gray tones
of structures in C and D correspond to those of the branchial arches depicted in
A. (See colored insert.)

The first pair of pharyngeal pouches, extending between the mandibular and
hyoid arches, comes into close relation to the distal ends with the auditory
vesicles. They give rise on either side to the *tympanic cavity* of the middle ear
and the *auditory (eustachian) tube* (Fig. 13-15).

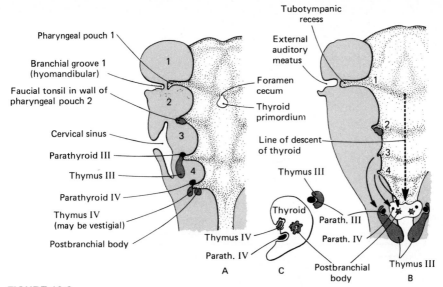

FIGURE 16-6
Diagrams showing the origin of the pharyngeal derivatives. (*Adapted from several sources.*) (A) The primary relations of the several primordia to the pharyngeal pouches. (B) Course of migration of some of the primordia from their place of origin. (C) Definitive relations of parathyroids, postbranchial body, thymus, and thyroid as they appear in a transverse section of the right lobe of the thyroid taken above the level of the isthmus. Abbreviation: *Parath.*, parathyroid.

The second pair of pouches becomes progressively shallower and less conspicuous. Late in fetal life the *faucial (palatine) tonsils* are formed by the aggregation of lymphoid tissue in their walls, and vestiges of the pouches themselves persist as the *supratonsillar fossae*.

From the floor of the pharynx, at about the level of the constriction between the first pair and the second pair of pharyngeal pouches, a midline diverticulum gives rise to the thyroid gland (Figs. 16-4 and 16-6). The endodermal cells making up the walls of this evagination push into the underlying mesenchyme, break away from the parent pharyngeal epithelium, and migrate down into the neck (Figs. 16-7, 16-8D, and 18-27). As with most developing glands, differentiation of the epithelial follicles of the thyroid is dependent on a specific inductive interaction with the surrounding mesenchyme. Only after arriving in its definitive location, relatively late in development, does the thyroid primordium undergo its final characteristic histogenetic changes.

The third and fourth pairs of pharyngeal pouches give rise to outgrowths which are involved in the formation of the parathyroid glands, the thymus, and the postbranchial bodies. In most mammals there are two pairs of parathyroid glands, usually spoken of as parathyroids III and parathyroids IV because they arise from the third and fourth pharyngeal pouches (Figs. 16-6A and 16-7). As was the case with the thyroid, the parathyroid primordia soon break away from their points of origin and migrate into the neck. Here, they are positionally more or less closely associated with the thyroid. Parathyroids IV are particularly likely to become adherent to the thyroid capsule or even to become partially embedded in the substance of the gland (Fig. 16-6B), whereas parathyroids III migrate to a position caudal to parathyroids IV.

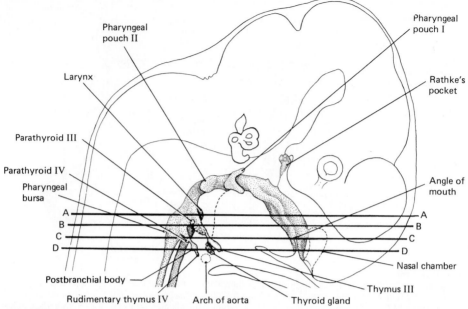

FIGURE 16-7
Pharynx of 15-mm pig embryo schematically represented in relation to the outlines of other cephalic structures. (*Adapted from several sources.*) The heavy horizontal lines indicate the levels of the correspondingly lettered sections in Fig. 15-8.

The *thymus* in the mammalian group is derived from outgrowths from the more ventral portions of the third and fourth pharyngeal pouches (Figs. 16-6A and 16-7). In different species there is a considerable difference in the relative conspicuousness of the two pairs of primordia. In most of the higher mammals the primordia arising from the third pouches are much more important as thymic contributors. This is the situation for the pig as well as for humans. The characteristic histogenetic changes in the thymus occur relatively late in development. The original epithelial character of the thymus becomes obscured because of an extensive ingrowth of mesenchyme, some of which is of neural crest origin. Soon thereafter lymphoid cells invade the organ and proliferate extensively, thus imparting to the thymus the histological characteristics of a true lymphoid organ. Even before its invasion by lymphoid cells, the thymic primordium is penetrated by autonomic nerve fibers, which may play an important role in linking neural and immune function. As the thymus matures, the original endodermal cells become converted into a specialized reticular connective tissue that embraces the thymic lymphocytes.

The *postbranchial (ultimobranchial) bodies*, structures that are probably of neural crest origin, produce a polypeptide hormone, *calcitonin*, which acts to reduce the concentration of calcium in the blood. The activity of this hormone serves to counter the long-recognized function of *parathyroid hormone*, which causes an increase in blood levels of calcium. In mammalian development, the calcitonin-producing cells of the postbranchial bodies are incorporated into the thyroid gland, whereas in birds and the other lower vertebrates the postbranchial bodies persist as distinct organs.

ADRENAL GLANDS

Certain cells that migrate ventrally from the neural crest when the sympathetic ganglia are formed do not become nerve cells but rather gland cells which are active in the production of the specific hormones epinephrine and norepinephrine. They exhibit a characteristic reaction with chromic acid salts, which has led to their designation as *chromaffin cells*. Clusters of chromaffin cells become located close to each sympathetic ganglion, and other masses of chromaffin tissue from the same source appear in various places beneath the mesoderm lining the coelom. The largest mass of extrasympathetic chromaffin tissue appears just cephalic to the kidney and becomes converted into the *medulla of the adrenal* (Fig. 12-20).

The cortical portions of the adrenal glands appear very early in development as local concentrations of cells in the splanchnic mesoderm medial to the gonadal ridge (Fig. 16-9A). These cells push into the underlying mesenchyme and show a tendency to become arranged in cords. Later in development the migrating cells that give rise to the medulla of the adrenal invade the cortical primordium and become encapsulated within it (Fig. 16-9B). In view of its origin very near the gonadal tissue, it is noteworthy that the adrenal cortex and the gonads produce certain steroid hormones that have very similar actions.

The cells first forming the adrenal cortex proliferate extensively to form a large fetal cortex, the function of which is poorly understood. A few days after the fetal cortex first forms, a second wave of cells surrounds it. These cells, which constitute the germinal layer of the definitive cortex, also proliferate and ultimately form the three steroid hormone–producing cortical layers characteristic of the mature adrenal gland. The fetal cortex, by contrast, remains quite prominent until birth, after which it rapidly involutes and ultimately disappears (Fig. 16-9).

THE LYMPHOID SYSTEM

A major functional characteristic of the vertebrate body is the ability to defend itself against foreign cells and organisms. In response to exposure to disease-producing bacteria or viruses, higher vertebrates produce *antibodies*, protein molecules which react very specifically against microorganisms or their toxic products and thus prevent recurrence of disease. Another form of defense is operative when a transplanted kidney in a human becomes infiltrated by lymphocytes and is rejected by the host. These defense reactions are examples of the functioning of the *lymphoid (immune) system*. The lymphoid system is not embodied in a single anatomical structure. Like the endocrine system, it is structurally diffuse but functions as an integrated unit. Very little was known about the ontogeny of the lymphoid system before 1960. Since that time, our understanding has advanced rapidly, and although a number of important questions remain unanswered, it is now possible to present a general scheme of the development of the lymphoid system.

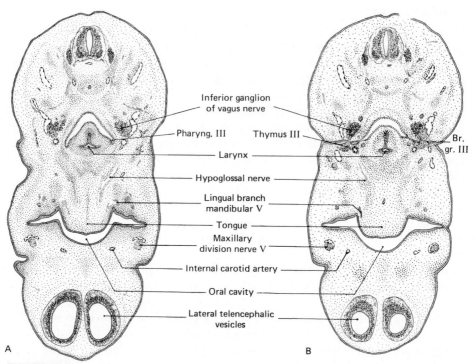

Inferior ganglion
of vagus nerve

Pharyng. III   Thymus III

Br.
gr. III

Larynx

Hypoglossal nerve

Lingual branch
mandibular V

Tongue

Maxillary
division nerve V

Internal carotid artery

Oral cavity

Lateral telencephalic
vesicles

A   B

FIGURE 16-8
Drawings (×11) of transverse section through the pharyngeal region of a 15-mm pig embryo. (A) Upper laryngeal level; (B) level of third pharyngeal pouch; (C) level of fourth pharyngeal pouch; (D) through neck, caudal to level of pharynx and larynx. The level of each of the sections represented in this figure is indicated by the correspondingly lettered lines in Fig. 16-7. Abbreviations: *Br. gr. III*, third branchial groove; *N. X*, vagus nerve; *N. XII*, hypoglossal nerve; *Pharyng. III*, *Pharyng. IV*, third and fourth pharyngeal pouches; *Postbranch.*, postbranchial body; *Premusc.*, premuscular concentration of mesenchyme.

Many of the lymphoid structures, in particular the lymph nodes, are an integral part of the lymphatic system, a system of delicate thin-walled vessels that are distributed throughout the body. Lymphatic vessels start as blindly ending tubules which, like the veins, coalesce into progressively larger vessels. At strategic points throughout the body (e.g., the groin and the armpit) the lymphatic vessels empty into lymph nodes, which can act as mechanical filters to remove certain foreign invaders. At the same time, antibody-producing cells can react to the foreign material by dividing and forming specific antibodies against it. Past the regional lymph nodes, the lymph, consisting of an ultrafiltrate of blood plasma plus some white blood cells, flows through still larger vessels and ultimately empties into large veins near the heart.

Wherever it is found, lymphoid tissue presents a roughly similar microscopic appearance. It contains a stroma consisting of stellate mesenchymelike reticular cells which produce a loose meshwork of very fine reticular fibers. Embedded in this meshwork are lymphocytes and other cells involved in

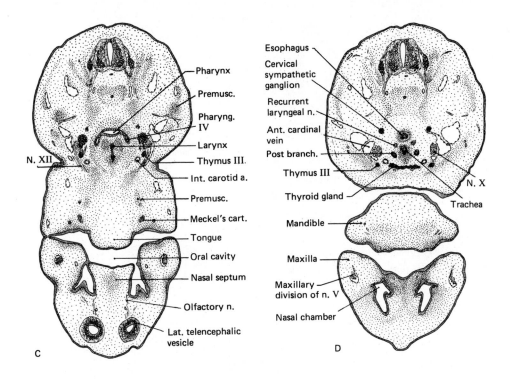

immunological defense reactions. Autonomic nerve fibers in the thymus, spleen, and other lymphoid organs may provide a functional anatomical link between immune function and the activities of the nervous system (Marx, 1985). Detailed anatomical features of the lymphoid system will not be considered here.

Several organs, such as the thymus gland, lymph nodes, and tonsils, are primarily lymphoid in nature, but the remainder of the lymphoid tissue in the body is closely associated with tissues fulfilling other functions. Thus in bone marrow it is intermingled with hematopoietic tissue, whereas in the spleen it is associated with tissue primarily concerned with the destruction of old blood cells. The epithelial linings of the gastrointestinal tract, and to a much lesser extent those of the respiratory and female genital tracts, are underlain by varying amounts of lymphoid cells which produce a special type of antibody and effectively serve as a front line of immunological defense for these internal body surfaces.

Development of the Lymphoid System

The ontogeny of the immune system has been studied intensively in both birds and mammals, and a general scheme of lymphoid development based on findings in both of these groups is presented in Fig. 16-10.

In terms of function, the immune system can be divided into two parts, one of which is concerned with *humoral immunity*, that is, the production of antibodies, and the other with *cell-mediated immunological responses*. Cell-

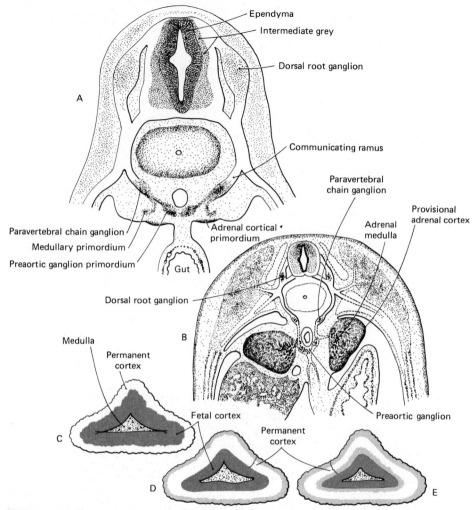

FIGURE 16-9
Stages in the formation of the sympathetic ganglia and the adrenal gland. (A) Cortical and medullary components still separate; (B) medullary cells migrating into the provisional adrenal cortex; (C) fifth month; (D) eighth month; (E) newborn infant (all zones of mature cortex present). (*C–E after Gray and Skandalakis, 1972, Embryology for Surgeons, Saunders, Philadelphia.*)

mediated responses are those in which immunological reactivity is bound to the cells themselves and not with antibody secreted by these cells, as is the case in humoral immunity.

The origin of lymphoid cells has not been definitively established (Owen and Jenkinson, 1981), but it is thought that they arise from the mesoderm of the liver or bone marrow. The *lymphoid stem cells* then migrate into what are called the *central lymphoid organs* (Fig. 16-10), where they are subjected to an inductionlike influence which determines the subsequent functional properties of the cells. There appear to be two central lymphoid organs, the thymus and the *bursa of Fabricius*.

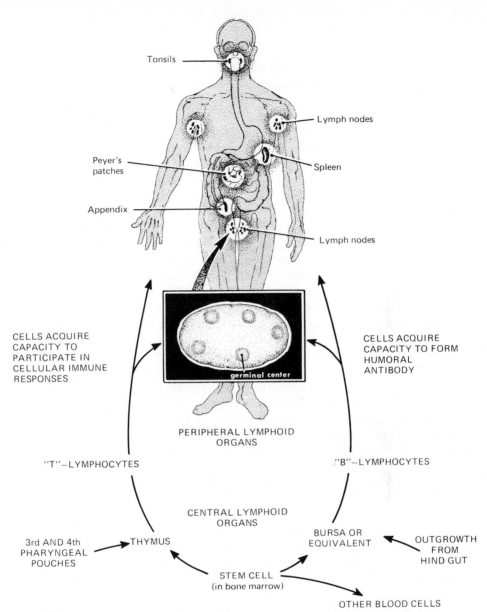

FIGURE 16-10
Simplified summary of the functional development of the immune system. Hemopoietic stem cells pass from the bone marrow into either the thymus or the bursa. During passage through these central lymphoid organs they acquire the capacity to participate in either cell-mediated or humoral immune responses. The "T" (thymic) or "B" (bursal) lymphocytes then migrate to the various peripheral lymphoid organs (top of figure), where they await a specific antigenic stimulus. Further interactions between T and B lymphocytes are necessary for the development of full immunological function.

Not long after its formation from the endoderm of the third and fourth pharyngeal pouches, the epithelial thymus is invaded by lymphoid stem cells. While in the thymus these cells become conditioned to take part in cell-mediated immune responses rather than to produce antibodies (Rothenberg and Lugo, 1985). Many of the lymphocytes within the thymus become programmed to die within the thymus. Those which leave the thymus are called T lymphocytes. There are several populations of T cells, each with differentfunctions. Some are called *killer cells*, and they are responsible for the cell-mediated immune defenses of the body. One of their main functions seems to be to destroy cancer cells that spontaneously form in the body (and are antigenically different from other cells in the body). Another major type of T cell is the *helper T cell*, which interacts closely with B cells, enabling the latter to form humoral antibodies.

In birds, lymphoid stem cells not entering the thymus migrate to the *bursa of Fabricius*, the other central lymphoid organ. There they are transformed into the line of B lymphocytes that will produce humoral antibody. The bursa arises as an endodermal bud from the hindgut of the 13-day chick embryo, and its manner of histogenesis is somewhat similar to that of the thymus. Mammals do not possess a bursa of Fabricius. After years of searching for its functional equivalent, immunologists currently feel that in the mammal lymphocytes become conditioned for B-cell function (antibody production) in general sites of blood-cell formation, specifically the liver in the embryo and the bone marrow after birth.

From the central lymphoid organs the lymphoid cells, recognizable as B and T lymphocytes, migrate to *peripheral lymphoid tissues*, for example, the lymph nodes, spleen, tonsils, and appendix (Fig. 16-10). B cells of the humoral part of the lymphoid system settle down into aggregations called *germinal centers*. When stimulated by foreign antigens, cells of the germinal centers undergo additional changes (proliferation and transformation into plasma cells) and then begin to produce antibodies. T lymphocytes emanating from the thymus go to the same peripheral lymphoid tissues as do the B cells, but instead of settling down into discrete germinal centers, they form dense masses of small cells around the germinal centers or in other areas. These represent the cells that can be mobilized to face the challenge of foreign cells, for example, those of a transplanted kidney, which differ immunologically from the host.

In early embryos, the protective function of the immune system is minimal and foreign cells or large molecules are not met with an immunological rejection. Later in fetal development and shortly after birth, the embryo begins to distinguish "self" from "nonself" (Burnet, 1969), and at that time the immune defense system begins to become functional. The T lymphocytes of humans are capable of generating cell-mediated immune responses at birth, but the antibody-secreting function of the B lymphocytes does not mature until varying periods after birth, depending on the specific class of antibody. Part of the deficiency in antibody production in newborn infants is made up by the presence of maternal antibody (IgG), which passes from the mother through the placenta in utero. In some mammals, e.g., rodents and ruminants, maternal antibodies are passively transferred postnatally to the young animal via the mother's milk. The intestinal epithelium is specially adapted for the passage of maternal antibodies. The maternal antibody is gradually degraded over the first 2 to 3 months of life, after which it is replaced by antibody synthesized by the infant.

CHAPTER 17

THE DEVELOPMENT OF THE UROGENITAL SYSTEM

In both their adult anatomy and their embryological development, the urinary and reproductive systems are so closely linked that they must inevitably be considered together. Both the urinary and genital tracts share a common origin from the intermediate mesoderm of the early embryo. The urinary system begins to take shape first, following a series of stages that reveal its phylogenetic history. Only later, after a certain degree of function already exists within the developing urinary system, does the genital system appear. Genital development is initiated in the gonad, which differentiates into an ovary or a testis. The formation of genital ducts and glands occurs later, with their male and female configuration coming about as a result of specific secretions, or lack thereof, from the gonads.

THE URINARY SYSTEM

General Relationship of Pronephros, Mesonephros, and Metanephros

Three major types of urinary organs occur in adult vertebrates. Although they are often treated as distinct structures in elementary textbooks, one must recognize when reviewing both the phylogenetic and ontogenetic history of the vertebrate kidney that the major types of kidneys actually represent dominant forms, with intergrades between them. Some authors feel that the amount of intergradation is so great that the vertebrate kidney should be looked upon as a true continuum of varying morphological and functional units. According to this concept, the aggregate of all forms of renal units is called the *holonephros*.

For a more specialized treatment of this concept, the reader is referred to the works of Torrey (1965, 1971).

The most primitive of the major types of kidney is the *pronephros*, which exists as a functional excretory organ only in some of the lowest fishes. It is located far cephalically in the body. In all the higher fishes and in the Amphibia, the pronephros has degenerated, and its functional role has been assumed by the *mesonephros*, a new organ located farther caudally in the body. In reptiles, birds, and mammals a third urinary organ develops caudal to the mesonephros. This is the *metanephros*, or permanent kidney. All three of these organs are paired structures located retroperitoneally in the dorsolateral body wall.

In a recapitulation of phylogeny, three major types of kidneys—the pronephros, mesonephros, and metanephros—are formed in succession during the development of avian and mammalian embryos. The pronephros is the most anterior of the three and the first to be formed. It is wholly vestigial, appearing only as a slurred-over recapitulation of structural conditions which exist in the adults of the most primitive of the vertebrate stock. The mesonephros makes its appearance in the embryo somewhat later than the pronephros does and is formed caudal to it. The mesonephros is the principal organ of excretion during early embryonic life, but like the pronephros, it also disappears in the adult, except for parts of its duct system, which become associated with the male reproductive organs. The metanephros is the most caudally located of the excretory organs and the last to appear. It becomes functional late in embryonic life when the mesonephros is regressing, and persists permanently as the functional kidney of the adult.

As a general introduction, the main stages in the ontogenetic history of the nephric organs are schematically illustrated in Fig. 17-1. Kidney formation begins with the appearance of the pronephros, several pairs of tubules located in the cephalic portion of the intermediate mesoderm. These small tubules, forming in a cephalocaudal sequence, empty laterally on each side into a common duct called the *primary nephric (pronephric) duct* (Fig. 17-1B). The primary nephric ducts elongate toward the cloaca, and with their progression caudad, new pairs of tubules form and join the ducts.

After only a few segments, the character of the newly forming tubules changes, providing a gradual transition to the next major type of tubule, the mesonephric tubule (Fig. 17-1C). As the mesonephric tubules continue to form, the pronephric tubules, which are rudimentary in all higher vertebrates, degenerate. With the mesonephric tubules now constituting the dominant variety of nephric tubule, the primary nephric ducts are called the *mesonephric (Wolffian) ducts* because of their new associations.

While the posterior parts of the mesonephros are still becoming established in their final form, a small outgrowth appears at the cloacal end of the mesonephric duct (Fig. 17-1D). These ducts, called the *metanephric diverticula* or *ureteric buds*, grow into the most posterior part of the intermediate mesoderm and stimulate the formation of the third major type of tubule, the *metanephric tubule*. Instead of being strung out along the dorsal wall of the

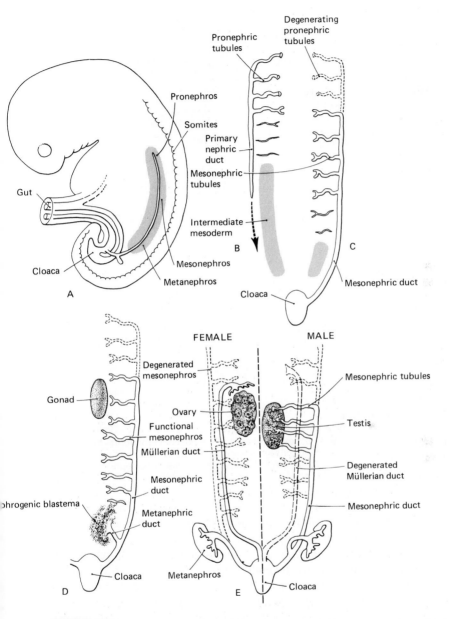

FIGURE 17-1
Schematic diagrams to show the relations of pronephros, mesonephros, and metanephros at various stages of development. (A) Diagram showing the subdivision of the intermediate meso-derm into pronephric, mesonephric, and metanephric segments. (B) Early stage, showing the primary nephric duct extending toward the cloaca. (C) Establishment of the mesonephros. (D) Early appearance of the metanephros. (E) Urogenital system after sexual differentiation. The Müllerian ducts are structures which arise later in development and are not part of the urinary system proper.

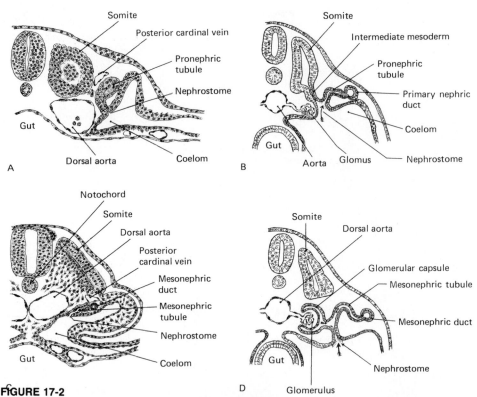

FIGURE 17-2
Drawings of nephric tubules. (A) Transverse section through twelfth somite of 16-somite chick to show pronephric tubule. (*After Lillie.*) (B) Schematic diagram of functional pronephric tubule. (*After Wiedersheim.*) (C) Transverse section through seventeenth somite of 30-somite chick to show primitive mesonephric tubule. (D) Schematic diagram of functional mesonephric tubule of primitive type. (*After Wiedersheim.*)

body cavity like the pronephric and mesonephric tubules, the metanephric tubules become aggregated into a compact mass which ultimately becomes the definitive kidney. With the establishment of the metanephroi, or permanent kidneys, the mesonephroi begin to degenerate. The only parts of the mesonephric system to persist, except in vestigial form, are some of the ducts and tubules, which in the male are appropriated by the testes as a duct system.

Despite the variations in morphology, the different varieties of nephric tubules all function in a roughly similar manner. Fluid enters the proximal end of the tubule either as a filtrate from the blood, in the case of tubules supplied with glomeruli, or as a coelomic fluid, in the case of pronephric and primitive mesonephric tubules which have openings (*nephrostomes*) into the body cavity (Fig. 17-2). All functional tubules are closely associated with networks of capillaries, and as the fluid flows down the tubules, it is processed by the actions of the epithelial cells lining the tubules and the associated capillaries.

Ultimately the fluid is transformed into urine and flows down the appropriate duct systems to the exterior. The composition of the urine varies in accordance with both functional needs and the arrangement of the tubules that produce the urine.

Pronephros

The pronephros is well developed and functional in the embryos and larvae of fishes and amphibians, but in the embryos of birds and mammals it is represented only by several pairs of rudimentary tubules or cords of cells. In amphibians, the pronephros is made up of only three to five pairs of tubules, arising at the levels of the second through fourth somites.

A typical pronephric tubule arises from a *nephrotome*, a hollow mass of intermediate mesodermal cells which come off of the somite like a stalk. A ventral opening of the nephrotome, called the *nephrostome*, is continuous with the coelom, which separates the visceral and parietal layers of the lateral plate mesoderm (Fig. 17-2B). Close to the nephrostome is a vascular ridge called the *glomus*, through which waste materials from the blood are excreted (Fig. 17-2B). These are swept into the nephrostome for processing in the pronephric tubules. From the nephrostome the pronephric tubule extends laterally. The lateral (distal) ends of several pronephric tubules come together to form the primordium of the primary nephric duct. In its early form it is a solid cord of cells, but later it hollows out into a duct.

The primary nephric duct extends backward toward the cloaca, following environmental cues as it pushes caudally (Fig.17-3). If the normal pathway of extension is disturbed by partially transecting the embryo or even rotating the

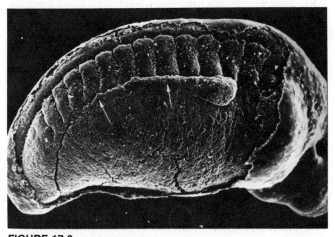

FIGURE 17-3
Scanning electron micrograph of an *Ambystoma* embryo, showing the pronephric duct (arrows) just beneath the somites. (*Courtesy of T. Poole, from Poole and Steinberg, 1983.*)

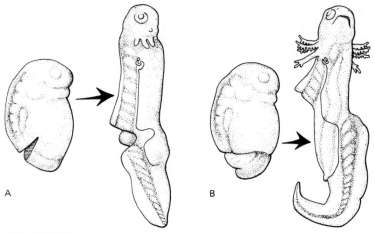

FIGURE 17-4
Experiments involving interference with the pathway of migration of the pronephric duct (black lines) in amphibian larvae. (A) After partial transection of
the embryo, the duct courses around the defect and then returns to its customary route of extension. (B) After rotation of the posterior part of the embryo, the duct assumes its normal course through the tissues of the rotated
segment. (*After Holtfreter*, 1943, *Rev. Canad. Biol.* **3**:*220.*)

posterior end (Fig. 17-4), the primary nephric duct often makes its way back to
the correct path and continues its course toward the cloaca (Holtfreter, 1943).
The pronephric duct has different modes of extension in different species. For
example, in salamanders it begins as an ovoid solid mass of cells, which then
undergoes extensive remodeling by means of active cell migration and cell
rearrangements to form an elongated, narrow duct (Fig. 17-5). There is
evidence that caudad extension of the duct may be guided by differential
adhesion of the tip to the mesodermal substrate, which itself is undergoing a
series of craniocaudal developmental changes. In contrast, the pronephric duct
of *Xenopus* appears to be sculpted along its entire length by segregation of local
mesodermal cells rather than by a caudal extension mechanism (Poole and
Steinberg, 1984). More recent work has shown that grafted cranial neural crest
cells are able to follow the same set of directional cues on the lateral mesoderm
substrate as does the pronephric duct (Zackson and Steinberg, 1986).

The pronephros in the chick is represented by rudimentary tubules which
first arise from the intermediate mesoderm as solid buds of cells at about 36
hours of incubation (Fig. 17-2A). The pairs of pronephric tubules appear at the
levels of the fifth to sixteenth somites. The distal ends of the tubules give off
extensions which form a continuous cord of cells, the primary nephric duct
(Fig. 17-2C). The duct in the chick elongates by means of a combination of cell
locomotion and cell proliferation at its tip.

In the pronephric tubules of the chick embryo there are vestiges of a
nephrostome opening into the coelom (Fig. 17-2A), but the tubules never

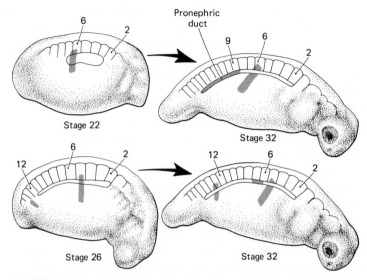

FIGURE 17-5
Vital staining study illustrating the mechanism of extension of the pronephric duct in the amphibian (*Ambystoma*) embryo. Upper row. When a
dye mark is placed below somite 6 in the stage 22 embryo and the duct
is allowed to elongate, the dyed portion of the duct has moved caudal to
the mark. Lower row. If a dye mark (gray) is placed caudal to the tip of
the pronephric duct in a stage 26 embryo, the extending pronephric duct
at stage 32 does not include any dyed cells. This shows that the pronephric duct is derived from cells of the duct itself rather than by assimi-
lating mesenchymal cells as it moves into new territories. (*Adapted from
Poole and Steinberg, 1983.*)

become completely patent and never acquire the vascular relations characteristic
of the functional pronephros in primitive vertebrates. Shortly after their initial
appearance the pronephric tubules begin to undergo regressive changes, and by
the end of the fourth day of incubation a few isolated epithelial vesicles are all that
remain to chronicle the transitory appearance of the avian pronephros.

Mesonephros

The mesonephric tubules develop from the intermediate mesoderm caudal to
the pronephros (Friebova, 1975). A cephalocaudal gradient of development in
the mesonephric tubules can be explained by the relationship between the
primary nephric duct and the mesonephrogenic mesoderm. Waddington (1938)
found that if the caudal extension of the primary nephric duct is interrupted,
mesonephric tubules do not form or develop only poorly caudal to the level of
interruption. It is likely therefore that as the primary nephric duct pushes back
toward the cloaca, it stimulates or induces the intermediate mesoderm along
the way to form mesonephric tubules.

A mesonephric tubule differs from a pronephric one chiefly in its relation to the blood vessels associated with it. It develops a cuplike outgrowth into which a knot of capillaries is pushed. The cup-shaped outgrowth from the tubule is called the *glomerular (Bowman's) capsule*, and the tuft of capillaries is termed a *glomerulus* (Fig. 17-2D). The diminuitive form *glomerulus* is used to distinguish such small, localized capillary tufts from the continuous vascular ridge of the pronephros, which is called a *glomus*.

Chick embryos show some cephalically located mesonephric tubules suggestive of the more primitive type diagramed in Fig. 17-2D, with a ciliated nephrostome which draws in fluid emanating from the coelom. This fluid mingles in the tubule with the fluid from the glomerulus. In the embryos of higher vertebrates only a few of the more cephalic mesonephric tubules ever show nephrostomes, and these are rudimentary (Fig. 17-2C). The great bulk of the functioning mesonephros is made up of tubules which form no nephrostomes (Fig. 17-6) but depend entirely on their glomerular apparatus for their fluid intake.

The early stages in the formation of the definitive type of mesonephric tubules without nephrostomes are much the same in mammalian material as those seen in 2- to 4-day chicks. Once having established their connection with the primary nephric duct (Fig. 17-6B), the primordial tubules elongate and take on an S-shaped configuration (Fig. 17-6C and D). Continued increase in length brings about a series of secondary bendings (Fig. 17-6E) which increase their surface area, thereby enhancing their capacity for interchanging materials with the blood in the adjacent capillaries.

The mesonephric tubules function much like the nephrons of the adult kidney. A filtrate of blood from the glomerulus enters the capsule and flows into the tubule, where selective resorption of ions and other substances occurs. Substances that are resorbed enter the capillary plexus that is closely applied to the mesonephric tubules (Fig. 17-6E) and then drain into the subcardinal veins, which return the renal blood into the general circulation. A major difference between the mesonephric kidney and the permanent kidney of higher vertebrates is the relative inability of the mesonephros to concentrate urine. This is related to the elongated structure of the mesonephros and the absence of a well-developed renal medulla, a structural adaptation of land animals to preserve water by concentrating it through an elaborate countercurrent exchange mechanism. Such a fluid-conserving mechanism is not needed by the embryo, which lives in a bath of amniotic fluid, just as preservation of body water is not a problem for the mesonephric kidney of freshwater fishes and aquatic amphibians. Many details of the function of the mesonephros in the embryo are poorly understood.

Although it is relatively more conspicuous early in development, the mesonephros does not attain its greatest actual bulk until later. When the metanephros becomes well developed, the mesonephros undergoes rapid involution and ceases to be of importance in its original capacity, but in the male its ducts and some of its tubules still persist and give rise to structures of vital functional importance.

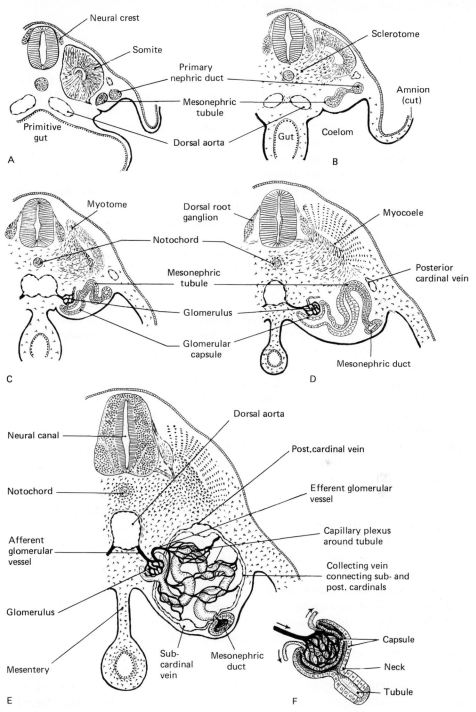

FIGURE 17-6

Development of mesonephric tubules and their vascular relations. (A) Tubule primordium still independent of duct. (B) Union of tubule with primary nephric duct. (C) Early stage in development of glomerulus and capsule. (D) Further development of capsule and lengthening of tubule. (E) Relations of blood vessels to well-developed mesonephric tubule. (F) Glomerulus and capsule, enlarged.

Metanephros

The development of the metanephros begins with the appearance of a tiny budlike outgrowth from the mesonephric duct just cephalic to the point where the duct opens into the cloaca (Fig. 17-7A). The outgrowth, called the *metanephric diverticulum*, pushes into the posterior portion of the intermediate mesoderm, which condenses around the diverticulum to form the *metanephrogenic blastema*. Thus, from the very beginning the permanent kidney has a dual origin: the metanephric diverticulum, which gives rise to the ureter, the renal pelvis, and the collecting duct system; and the intermediate mesoderm, from which the tubular units of the kidney arise.

The essential feature of metanephric development is the elongation and dichotomous branching (up to 14 or 15 times in the human) of the metanephric diverticulum and the formation of tubular excretory units around the tips of the branches. This is accomplished through a series of reciprocal inductive interactions between the two components of the metanephros (Saxén et al., 1986). The terminal portions of the metanephric duct induce the formation of metanephric tubules in the metanephrogenic mesoderm which surrounds it (Grobstein, 1955). In the absence of the metanephric duct, tubules do not appear. Conversely, the metanephrogenic mesoderm, acting in turn on the metanephric duct, induces the characteristic branching pattern of the duct system (Erickson, 1968). These reciprocal inductive interactions continue throughout the tubule-forming stage of kidney development. The morphological characteristics of the tubules formed from the metanephrogenic mesoderm are not rigidly fixed, however, for if this mesoderm is brought into contact with mesonephric ducts, tubules of the mesonephric variety are formed. This lability of developmental fate has been used by Torrey (1971, p. 340) as evidence in favor of the holonephric concept of the vertebrate kidney.

The developing metanephros has been an important model for experimental embryologists trying to understand the nature of inductive interactions. Analysis of this inductive process has been greatly facilitated through an experimental model by which metanephric induction can be demonstrated in vitro. Grobstein (1955) isolated tissue of the metanephric diverticulum and mesenchyme of the metanephrogenic blastema and placed them in culture, with a porous filter separating the two tissues. Despite the interposition of the filter, metanephric tubules formed in the mesenchyme, suggesting that the inductive reaction is mediated by diffusion of a chemical. More recent work with different varieties of filters has shown that cellular processes penetrate the small (<1.0 μm) pores of the filter so that the cells on either side of the filter are actually in close contact with one another (Lehtonen, 1976). Nevertheless, the experimental evidence suggests that the inductive effect is mediated by the short-range transmission of active compounds from one component of the system to the other (Saxén and Lehtonen, 1978). The development in vivo of Danforth's short tail mutant mice supports this conclusion. Although the ureteric bud grows to within one cell's diameter of

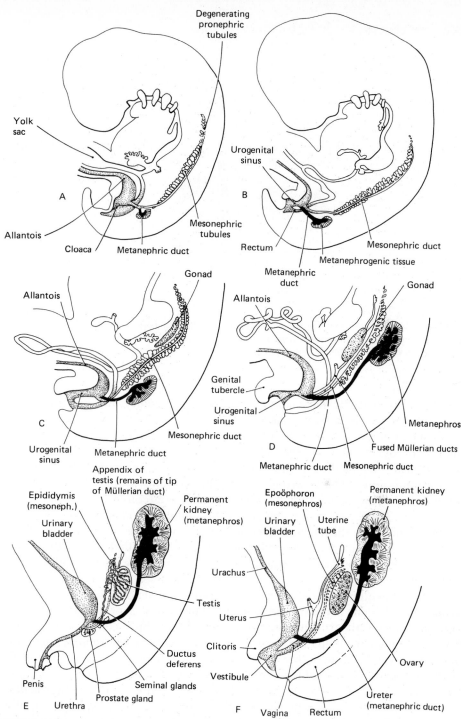

FIGURE 17-7

Diagrams showing relative sizes and positions of nephric organs of human embryo at various stages of development. (A) Early in fifth week. (*Adapted from several sources covering 5- to 6-mm embryos.*) (B) Early in sixth week. (*Modified from Shikinami's 8-mm embryo.*) (C) Seventh week. (*Modified from Shikinami's 14.6-mm embryo.*) (D) Eighth week. (*Adapted from Shikinami's 23-mm embryo and the Kelly and Burnam 25-mm stage.*) (E) Male at about 3 months—schematized. (F) Female at about 3 months—schematized.

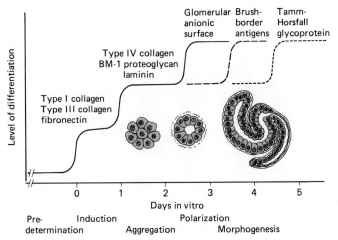

FIGURE 17-8
Scheme illustrating the multiphasic determination and differentiation
of mouse metanephric mesoderm in vitro (*Adapted from Saxén et
al., 1981, in Biology of Human Growth, Raven Press.*)

the metanephrogenic mesenchyme, induction is faulty and kidneys fail to
develop (Gluecksohn-Schoenheimer, 1943).

In vitro studies have shown several phases in the differentiation of
metanephrogenic mesenchyme (Fig. 17-8). Although it has the character of
any typical mesenchyme, the metanephrogenic mesenchyme is already
predetermined and has a strong bias toward forming kidney tissue. For this
reason heteroinductors such as spinal cord are effective in eliciting tubule
formation. The extracellular matrix of the uninduced mesenchyme contains
types I and III collagen and fibronectin. In response to induction by the
uterine bud the mesenchyme becomes converted to epithelial structures (Fig.
17-9), and in the process profound changes occur in the extracellular matrix
(Ekblom, 1984). Types I and III collagen are removed from the mesenchyme
and are replaced by matrix components characteristic of epithelial structures
(type IV collagen, laminin, and basement membrane-1 proteoglycan) as the
mesenchyme begins the conversion to epithelial tubules.

While the early inductive changes are taking place the *metanephric duct*
(*ureteric bud*) elongates rapidly (Fig. 17-7). Meanwhile the cells of the
metanephrogenic mesenchyme become concentrated around the distal end of
the metanephric duct and lose their original relations with the intermediate
mesoderm (Fig. 17-10B). The pelvic end of the diverticulum expands within its
investing mass of mesoderm and takes on a shape suggestive of the pelvic
cavity of the adult kidney (Fig. 17-12). From this early pelvic dilation arise the
numerous outgrowths of the future system of collecting tubules (Fig. 17-12E).
These push radially into the surrounding mass of nephrogenic mesoderm, and
the histogenesis of the renal tubules begins.

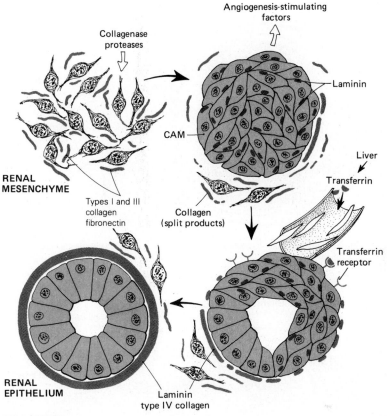

FIGURE 17-9
Major sequential events in the conversion of metanephrogenic mesenchyme to epithelial kidney tubules. (*Adapted from Ekblom, 1984.*)

Histogenesis of the Metanephros

Individual uriniferous tubules arise near the distal ends of the terminal branches of the collecting duct system. Cells of the metanephric blastema become arranged into small vesicular masses which lie close to the blind end of a collecting duct (Fig. 17-13A). Each of these masses becomes an elongated and highly convoluted tubule. One end of the tubule attaches to the collecting duct, with the site of junction being the boundary between the distal convoluted tubule and the arched collecting duct in the differentiated kidney. The other end of the developing tubule becomes associated with a blood vessel. Mouse-quail recombination experiments (Sariola, 1985) have shown that the glomerular endothelium arises from outside the kidney rather than from the metanephric mesenchyme. Induced metanephric mesenchyme is able to stimulate the ingrowth of capillaries, whereas uninduced mesenchyme cannot. The blood vessels, which are ultimately converted to the renal artery, form a small

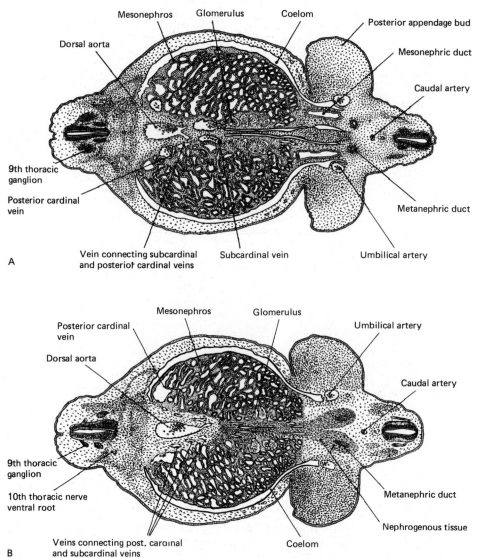

FIGURE 17-10
Two transverse sections from the series of the 9.4-mm pig embryo used in making the reconstructions illustrated in Fig. 18-9. (*Projection drawings, ×17.*) By laying a straightedge across
either of these reconstructions at the level of the marginal line bearing the serial number of a
cross section, the relations of that section within the body as a whole are precisely indicated. (A)
Section 518 passing through meso- and metanephric ducts; (B) section 529 passing through the
main concentration of metanephrogenic tissue.

glomerulus (Fig. 17-13D), and the tubule expands around it to form the
glomerular capsule. Differentiation of the tubule progresses from the glomerular capsule to the proximal to the distal convoluted tubule. One portion of the

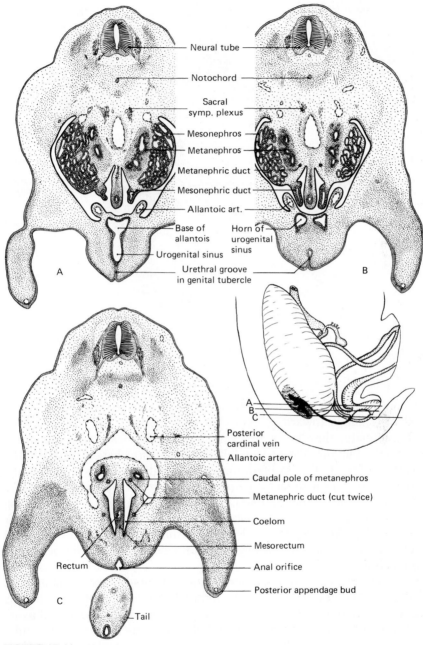

Neural tube

Notochord

Sacral
symp. plexus

Mesonephros

Metanephros

Metanephric duct

Mesonephric duct

Allantoic art.

Base of
allantois

Horn of
urogenital
sinus

Urogenital sinus

Urethral groove
in genital tubercle

A

B

Posterior
cardinal vein

Allantoic artery

Caudal pole of metanephros

Metanephric duct (cut twice)

Coelom

Mesorectum

Anal orifice

Posterior appendage bud

Rectum

C

Tail

FIGURE 17-11
Projection drawings (×15) of transverse sections through the pelvic region of a 15-mm
pig embryo. The level of each section is indicated on the inset.

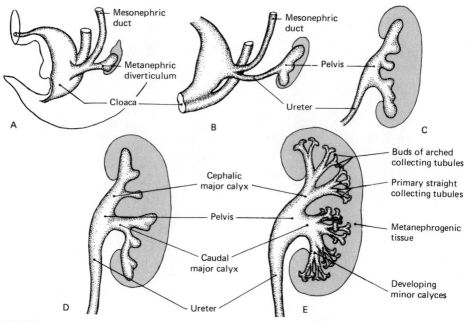

FIGURE 17-12
Diagrams showing a series of stages in the growth and differentiation of the metanephric diverticulum.

tubule becomes greatly elongated into a hairpin shape to form the *loop of Henle*, which extends toward or into the renal medulla. While the kidney tubules differentiate, they begin to express molecular features (e.g., brush border antigens and the Tamm-Horsfall glycoprotein) characteristic of the mature kidney (Fig. 17-8). Proliferation of the epithelial cells of the renal tubules depends on a growth factor (*transferrin*, an iron-binding protein) from the liver (Thesleff and Ekblom, 1984). As the mesenchymal cells are being transformed into tubular epithelial cells, transferrin receptors appear on their surfaces (Fig. 17-9). Marking the final stages of the mesenchymal to epithelial conversion, a basement membrane begins to form along the outer border of the tubular epithelium.

As the kidney grows, additional generations of tubules are formed in its peripheral zone. The development of the internal architecture of the kidney is structurally very complex. The collecting duct system expands outward as its pattern of branching continues, and new tubules continue to form at the ends. Histogenesis of the metanephros is structurally very complex. A detailed, well-illustrated summary of renal histogenesis can be found in Hamilton et al. (1972, p. 384). Upon completion of its development, the metanephros has about 15 generations of nephrons, and Osathanondh and Potter (1966) reported that the kidneys at birth contain about 822,300 nephrons.

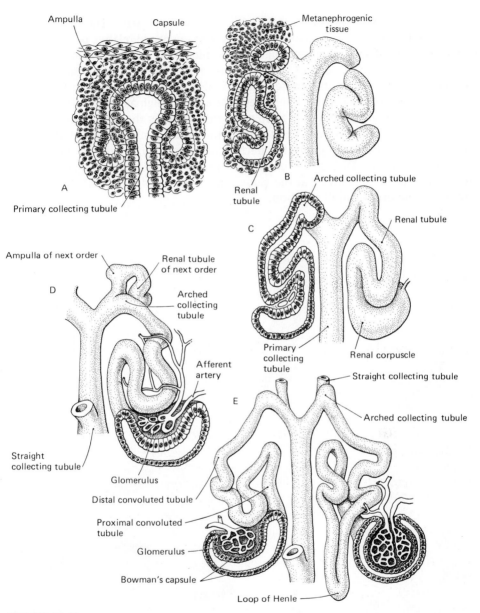

FIGURE 17-13
Diagrams showing the development of the metanephric tubules of mammalian embryos. (*After Huber, from Kelly and Burnam, Diseases of Kidneys, Ureters, and Bladders. Courtesy of Appleton-Century-Crofts, New York.*)

Later Positional Changes in the Kidney

When the metanephric kidneys are first established they are located far caudally in the growing body (Fig. 17-7A and B), but as development progresses they come to lie relatively much farther cephalad (Fig. 17-7C to 17-7F). Their own actual movement headward is not quite so great as their change in relative position would indicate. Part of their apparent migration is due to the marked expansion of the portion of the body caudad to them. The metanephros in human embryos arises opposite the twenty-eighth somite (fourth lumbar segment). At term it has moved up to the level of the first lumbar vertebra or even as high as the twelfth thoracic.

More striking than the change in segmental level is the shift of the kidneys out of the pelvic part of the coelom. In young embryos the kidneys lie retroperitoneally, bulging into the narrow pelvic cavity, caudal to the bifurcation of the aorta. During the early fetal period, they slide craniad over the umbilical arteries, rotating at the same time (Fig. 17-14), and ultimately reach their characteristic adult position.

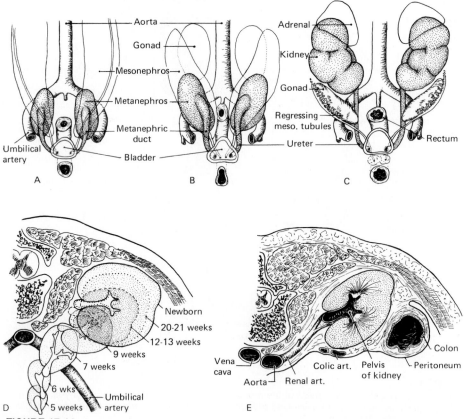

FIGURE 17-14
Diagrams showing changes in position of kidney during development. (*Redrawn from Kelly and Burnam, Diseases of Kidneys, Ureters, and Bladder. Courtesy Appleton-Century-Crofts, New York.*) (A–C) Frontal views showing ascent of kidneys out of pelvis. Note their rotation, occurring chiefly as they rise above the common iliac arteries. (D) Schematic composition diagram showing rotation of kidney as seen in a cross section of the body. (E) Location of kidney as seen in a cross section of the adult body at lumbar level.

Formation of the Bladder and Early Changes in the Cloacal Region

In dealing with the development of the extraembryonic membranes, we have already taken up the formation of the allantois as an evagination from the primitive gut (Fig. 8-10C and D). Caudal to the point of origin of the allantois the gut becomes enlarged to form the *cloaca* (Fig. 17-15). When the cloacal dilation is first formed, the hindgut still ends blindly. Under the root of the tail, the ectoderm sinks in toward the gut to form the *proctodeum*. The thin plate of tissue formed by the apposition of the proctodeal ectoderm and the endoderm of the hindgut is known

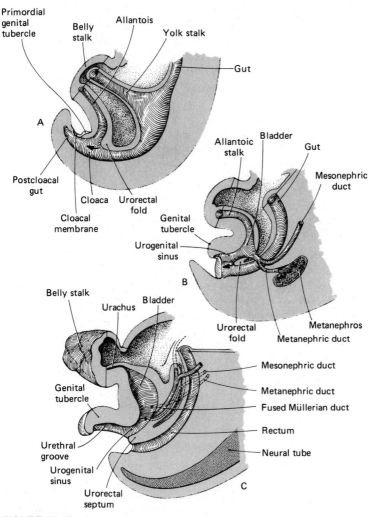

FIGURE 17-15
Stages in the subdivision of the human cloaca by the urorectal fold.

as the *cloacal membrane*, which eventually ruptures, establishing a caudal outlet for the gut in much the same manner that rupture of the oral plate has previously established communication between the stomodeum and the foregut.

Before this occurs, important changes take place internally. The *urorectal fold*, a crescentic fold which cuts into the cephalic part of the cloaca where the allantois and the gut meet (Fig. 17-15), grows caudally toward the cloacal membrane. This results in the partitioning of the cloaca into two parts, the *rectum* and a ventral *urogenital sinus* (Fig. 17-15). As the urorectal septum approaches the protodeal end of the cloaca, the cloacal membrane ruptures. With this, the separation between the digestive system and the urogenital system is complete. Meanwhile the proximal part of the allantois has become greatly dilated and can now quite properly be called the *urinary bladder*.

In the growth of the bladder the caudal portion of the mesonephric duct is absorbed into the bladder wall. This absorption progresses until the part of the mesonephric duct caudal to the point of origin of the metanephric diverticulum has disappeared. The end result of this process is that the mesonephric and metanephric ducts open independently into the urogenital sinus. The metanephric duct, possibly because of traction exerted by the kidney in its migration cephalad, acquires its definitive opening somewhat laterally and cephalically to that of the mesonephric duct. It then discharges into the part of the urogenital sinus that was incorporated into the bladder. The mesonephric ducts open into the part of the urogenital sinus that remains narrower and gives rise to the urethra (Fig. 17-16).

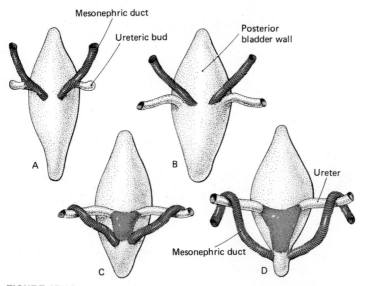

FIGURE 17-16
Dorsal view of the developing urinary bladder, showing the changing relationships of the mesonephric duct and uterus. In (D), note the incorporation of portions of the mesonephric duct (gray) into the wall (trigone) of the bladder. (*Adapted from Sadler, 1985, Langman's Medical Embryology, Williams & Wilkins.*)

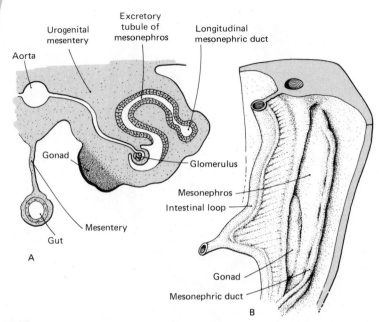

FIGURE 17-17
Diagrams showing the organization of urogenital structures in a 5-week human embryo. (A)
Cross section through the thoracic region. (B) Dissection of an embryo. (*After Langman, 1975,
Medical Embryology, 3rd ed., Williams & Wilkins, Baltimore.*)

THE DEVELOPMENT OF THE REPRODUCTIVE ORGANS

Factors Involved in Sexual Differentiation

Differentiation of the reproductive system is a complex process involving a
number of different mechanisms operating at several stages of development
(Naftolin et al., 1981; Segal, 1985). Two important generalizations regarding
mammalian sexual differentiation should be kept in mind. One is that several of the
major genital structures (e.g., gonads, sexual ducts, external genitalia) first pass
through a morphologically *indifferent stage* in which they cannot be identified as
being either male or female (Figs. 17-17 and 17-18). Later, a course of development
characteristic of one or the other sex occurs (Table 17-1). The other major
generalization is the inherent tendency of genital structures to develop into the
female type in the absence of specific masculinizing influences (Fig. 17-19).

The first critical stage of sexual differentiation occurs at the moment of
fertilization, when the genetic sex of the zygote is determined by the nature
of the sex chromosome contributed by the sperm.[1] Although an XY zygote is

[1]The genetic basis for sex determination is not the same among all vertebrates. In most mammals,
maleness is specified by the XY combination, but in birds, reptiles and some amphibia the female
instead of the male is the heterogametic sex, with the chromosomal designation of ZW for females and
ZZ for males. Numerous other combinations have been identified in other animal groups.

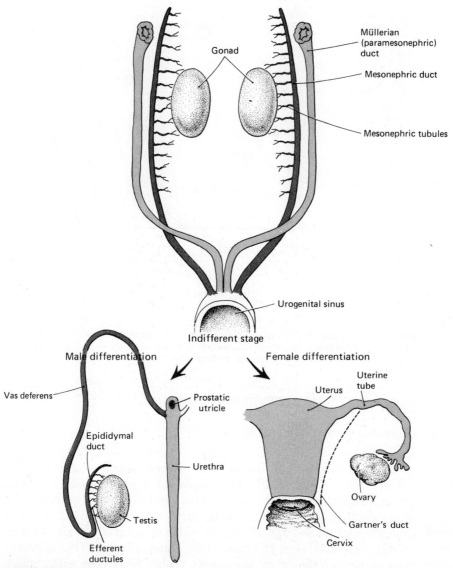

FIGURE 17-18
Schematic diagram showing the differentiation of the male and female reproductive organs from an indifferent stage. The prostatic utricle in the male represents the caudal remains of the Müllerian duct system. Gartner's duct in the female is the remains of the mesonephric duct.

destined to become a male, no distinctive differences between the early development of male and female embryos have been noted. The principal function of the Y chromosome is to direct the differentiation of the indifferent gonad into a testis. This is accomplished after migration of the primordial germ cells into the early gonad (Fig. 17-20).

TABLE 17-1
MAJOR HOMOLOGIES IN THE UROGENITAL SYSTEM

Male derivative	Indifferent structure	Female derivative
Testis	Gonad	Ovary
Spermatozoa	Primordial germ cells	Ova
Seminiferous tubules (Sertoli cells)	Sex cords	Follicular cells
Efferent ductules (paradidymis)	Mesonephric tubules	Epoophoron
Epididymal duct, ductus deferens	Mesonephric (Wolffian) duct	Degenerates (canals of Gartner)
Degenerates (appendix of testes)	Paramesonephric (Müllerian) duct	Uterine tubes, uterus, part of vagina (?)
Bladder, prostatic urethra	Early urogenital sinus (upper)	Bladder, urethra
Lower urethra	Definitive urogenital sinus (lower)	Vestibule
Penis	Genital tubercle	Clitoris
Floor of penile urethra	Genital folds	Labia minora
Scrotum	Genital swellings	Labia majora

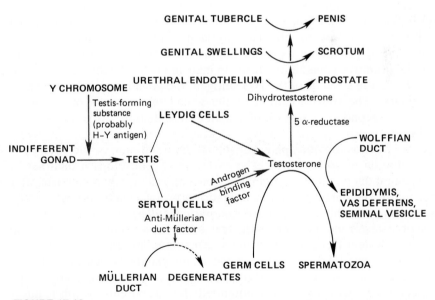

FIGURE 17-19
Summary of the major events leading to the formation of the male phenotype from the indifferent gonad in mammals. (*Adapted from S. Gilbert, 1985, Developmental Biology, Sinauer.*)

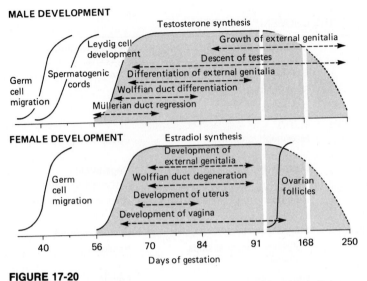

FIGURE 17-20
Summary of the events of sexual differentiation in male and female human embryo (*Adapted from Wilson et al., 1981, Science, 211:1279.*)

The sex-determining function of the Y chromosome is intimately bound with the activity of the H-Y antigen (Ohno, 1985). For many years this antigen had been recognized as a minor Y-linked histocompatibility antigen that was known to result in the rejection of male skin grafted onto females of the same inbred strain of mice. This role of the H-Y antigen seems in a sense to be an immunological by-product of its major function, which is to cause the organization of the primitive gonad in the testis in male animals (Wachtel et al., 1975; Ohno, 1978). In the absence of the H-Y antigen, the gonad later is transformed into the ovary. The transformation of the gonad into a testis or an ovary completes the second critical stage of sexual differentiation: the establishment of gonadal sex.

The next and most obvious phase in sexual differentiation of the embryo is the differentiation of somatic sex. The early embryo develops a dual set of potential genital ducts, one the original mesonephric duct, which persists after degeneration of the mesonephros as an excretory organ, and another, newly formed pair of ducts called the *paramesonephric (Müllerian) ducts*. Under the influence of testosterone secreted by the testes, the mesonephric ducts develop into the duct system through which the spermatozoa are conveyed from the testes to the urethra. Differentiation of the major glands associated with the ducts (prostate and seminal vesicle) also depends on *testosterone* or its derivative, *dihydrotestosterone*. The potentially female paramesonephric ducts regress under the influence of another product of the embryonic testes, the *Müllerian inhibitory factor*, a glycoprotein secreted by the Sertoli cells (Josso and Picard, 1986). In genetically female embryos, neither testosterone nor

Müllerian inhibitory factor is secreted by the gonads. In the absence of testosterone the mesonephric ducts regress, and the lack of Müllerian inhibitory factor permits the paramesonephric ducts to develop into the oviducts, the uterus, and part of the vagina (Jost, 1972). The external genitalia also first take form in a morphologically indifferent condition and then develop either in the male direction under the influence of testosterone or in the female direction if the influence of testosterone is lacking.

A good example of the natural tendency of the body to develop along female lines in the absence of other modifying influences is seen in *Turner's syndrome*, a rare human condition characterized by the deletion of one of the sex chromosomes (XO). Although individuals with Turner's syndrome are sterile and gonadally undifferentiated, the internal and external genitalia are easily recognizable as female in type. Testosterone also acts on the developing brain and affects behavior in a sexually dimorphic manner (De Vries et al., 1984).

The last stages of sexual differentiation occur after birth. The newborn baby is assigned a sex on the basis of its sexual phenotype, and in normal circumstances it develops psychologically as a member of that sex. The final major events in sexual differentiation occur at puberty, when the overall body configuration is transformed from what is essentially an indifferent form to a mature male or female type with the appearance of the secondary sexual characteristics. For both sexes the development of secondary sexual characteristics requires specific input from gonadal hormones. In the male this follows the pattern set early in embryonic life, but in the female this is the first stage at which specific ovarian hormonal influences come into play in the process of sexual differentiation.

It has long been recognized that sometimes the genital structures of an embryo differentiate in a way counter to that which would be predicted by the genetic sex. A classic example of this is the freemartin in cattle (Lillie, 1917). If a heterosexual pair of twin cattle develops in utero with fusion of the extraembryonic blood vessels and intermingling of the blood, the male develops normally. In contrast, there is a large-scale reversal of many sexual structures in the female, resulting in their resembling those of the male. Such a modified female is sterile and is called a *freemartin*. Lillie attributed the sex reversal in freemartins to an overcoming of normal female sexual development by hormones produced by the male, and his theory provided the major impetus to subsequent studies of factors controlling sexual differentiation.

Another example of a disparity between genetic and somatic sex is a condition called *testicular feminization*. Here the genetic sex of the individual is male and the gonads (internal testes) produce large amounts of testosterone, but the external somatic sex develops into that of a typical female. This condition is occasionally seen in humans and is present in a strain of mice. It is due to the lack of development of specific cellular receptors for testosterone, so that despite a plentiful supply of circulating testosterone, the hormone cannot be utilized by the testosterone-sensitive tissues which would normally become male genital structures. The external genitalia then develop into the typical female phenotype by default. The testes, however, still produce Müllerian inhibitory factor. As a result, the uterine tubes and uterus fail to form.

Early Differentiation of the Gonads

From their earliest appearance the gonads are intimately associated with the nephric system. While the mesonephros is still the dominant excretory organ, the gonads arise as ridgelike thickenings (*gonadal ridges*) on its ventromedial face (Fig. 17-17). Histologically, the early gonad consists essentially of a mesenchymal core covered by a mesothelium, which is called the *germinal epithelium* because it was at one time presumed to give rise to the germ cells.

Differentiation of the indifferent gonads into ovaries or testes occurs after the arrival of the primordial germ cells (Jost, 1972). The mesenchyme of the indifferent gonad shows no obvious organization (Fig. 17-21). Testicular cells, under the influence of the H-Y antigen, aggregate very early into primitive seminiferous tubules containing both germ cells and Sertoli cells (Ohno, 1978). The testes soon become hormonally functional, and by their secretions they influence the development of the genital ducts and the external genitalia to differentiate in

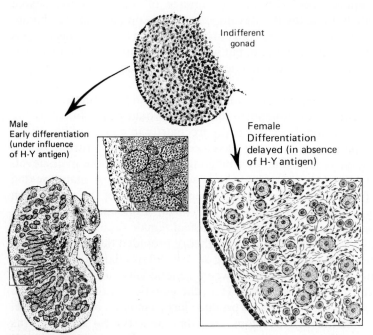

FIGURE 17-21
Contemporary view of gonadal differentiation. According to Jost (1972), the internal structure of the indifferent gonad is not so highly organized as was once believed. The testis (lower left) is that of a 14-week human embryo. The box in the center shows details of the structure of the early seminiferous tubules. The section of ovary is from a newborn infant.

the male direction. Not until late in pregnancy, after meiosis has begun, do the ovaries depart from their indifferent configuration as primordial follicles begin to form. By the time this has occurred, differentiation of the internal and external genitalia into the female type has already largely been accomplished.

Early in development a *rete* system appears in the gonads of both sexes. The origin of the rete system is not completely clear. An extragonadal component may represent persisting portions of mesonephric tubules, and the intragonadal component may be an extension of the extragonadal rete or may arise in situ. In the testis the *rete testis* persists as a network of channels which connect the seminiferous tubules to the efferent ductules, whereas in the ovary, the *rete ovarii* gradually loses prominence.

Some evidence suggests that the embryonic rete may secrete a diffusible factor which acts as a trigger for meiosis (Byskov, 1978). According to this hypothesis the rete in both the ovaries and testes of the early gonad secretes a *meiosis-inducing factor*, but the earlier isolation of the male germ cells within the seminiferous tubules prevents them from being exposed to the effects of this factor (and therefore from entering meiosis), whereas the exposed female germ cells begin the first meiotic prophase in the embryo. Experiments involving the culture of fetal mouse ovaries and testes (Byskov and Saxén, 1976) have provided some evidence in favor of a meiosis-preventing effect by the cells of the seminiferous epithelium. By analogy, the first meiotic block in the female germ cells has been attributed to a similar blocking effect by the follicular epithelium which surrounds the ova later in the fetal period.

THE SEXUAL DUCT SYSTEM OF THE MALE

The ducts that convey the spermatozoa away from the testis are, with the exception of the urethra, appropriated from the mesonephros—a developmental opportunism facilitated by the proximity of the growing testes to the degenerating mesonephros (Fig. 17-22). The mesonephric structures which are taken over by the testes are shown schematically in Fig. 17-23.

A few mesonephric tubules near the testis are retained and converted into the *efferent ductules* (Fig. 2-5), which carry spermatozoa from the testis into the *epididymal duct* (Fig. 17-23), which is the cephalic portion of the mesonephric duct. Remnants of partially degenerated mesonephric tubules may persist as the *appendix of the epididymis* and the *paradidymis* (Fig. 17-23). Distal to the epididymis the mesonephric duct becomes surrounded with a thick layer of smooth muscle and is converted to the *ductus deferens*. The conversion of the mesonephric duct and tubules is accomplished by the action of testosterone secreted by the embryonic testes (Fig. 17-24). Just before the ductus deferens enters the urethral part of the urogenital sinus, it develops local sacculations which, under the influence of testosterone, go on to form the *seminal vesicles* (Fig. 17-23).

In the embryo, the tissues around the urogenital sinus convert testosterone to dihydrotestosterone through an enzyme (5-α-reductase) which is

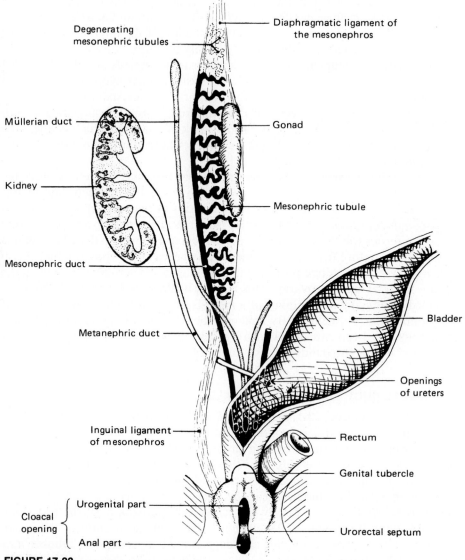

FIGURE 17-22
Schematic diagram showing plan of urogenital system at an early stage when it is still sexually undifferentiated. (*Modified from Hertwig.*)

produced locally (Bardin and Catterall, 1981). Under this influence the dihydrotestosterone (Fig. 17-24), the prostate, and the bulbourethral glands develop from the endodermal urethral epithelium (Fig. 17-23). Like other glands, these arise as a result of mesenchymal-epithelial interactions. Not so usual, however, is the finding that urogenital sinus mesenchyme can induce

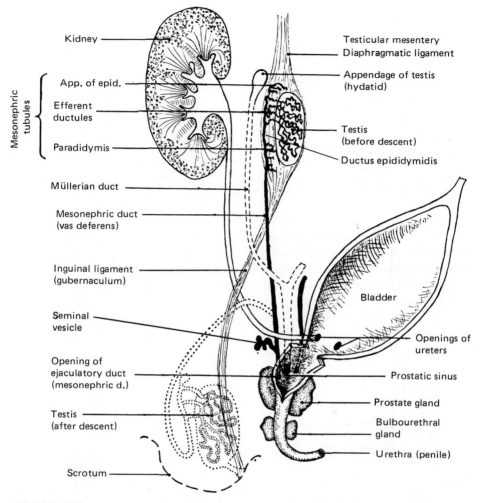

FIGURE 17-23
Diagram of the male sexual-duct system in mammalian embryos. (*Modified from Hertwig.*) The dotted lines indicate the position of the testis and its ducts after descent into the scrotum.

adult urinary bladder epithelium to form prostatelike glandular structures (Cunha et al., 1983).

Although the Müllerian ducts regress in the male, their final distal ends persist as a minute diverticulum (*prostatic utricle*, *vagina masculina*) embedded in the prostate.

THE SEXUAL DUCT SYSTEM OF THE FEMALE

The Müllerian (paramesonephric) ducts first appear close beside and parallel to the mesonephric ducts (Figs. 17-25 and 17-27A and B). They are the primordial

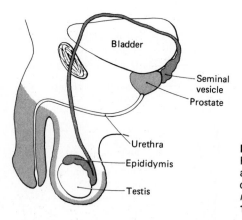

FIGURE 17-24
Regions of the male reproductive system that are sensitive to testosterone (dark gray) and dihydrotestosterone (light gray). (*After Imperato—McGinley et al., 1974, Science,* **186**:*1214.*)

structures from which the uterine tubes (oviducts) and uterus arise in the female (Fig. 17-26). The Müllerian ducts come together caudally and approach the urogenital sinus at a point where the wall of the sinus has thickened to form a Müllerian tubercle. Flanking the fused Müllerian ducts are the unfused ends of the mesonephric ducts, which enter the urogenital sinus and the Müllerian tubercle. In females, the mesonephric ducts degenerate and are represented occasionally by the *canals of Gartner* in the broad ligament by the uterus and vagina.

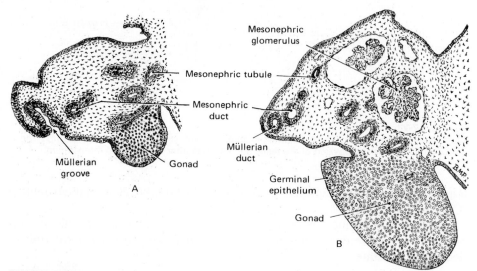

FIGURE 17-25
Projection drawings (×125) at the upper mesonephric level of a 6½-week embryo to show the formation of the Müllerian duct. (University of Michigan Collection, EH 707, CR 15 mm.) (A) Open groove in the coelomic mesothelium of the mesonephros, near its cephalic pole. (B) Slightly caudal to the open groove, the Müllerian duct has become a completed tube, lying in the margin of the mesonephros just lateral to the mesonephric duct.

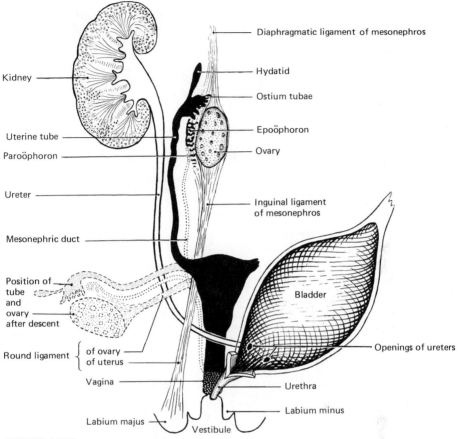

FIGURE 17-26
Schematic diagram showing plan of developing female reproductive system. (*Modified from Hertwig.*)
The dotted lines indicate the position of the ovary and uterine tube after their descent into the pelvis.

Vagina

The origin of the vagina has not been satisfactorily determined. O'Rahilly (1977) reviewed the varying opinions in the literature. One viewpoint is that the vagina originates almost entirely from a vaginal plate arising from the dorsal wall of the urogenital sinus and connecting with the fused ends of the Müllerian ducts (Fig. 17-27). Other authors, however, feel that the terminal portions of the Müllerian ducts make some contribution to the vagina. Still other authors feel that the most distal segments of the mesonephric ducts link the Müllerian ducts to the urogenital sinus (Bok and Drews, 1983).

Uterus

The extent of fusion of the Müllerian ducts varies from one group of mammals to another. Marsupials have paired uteri formed by enlargement of the

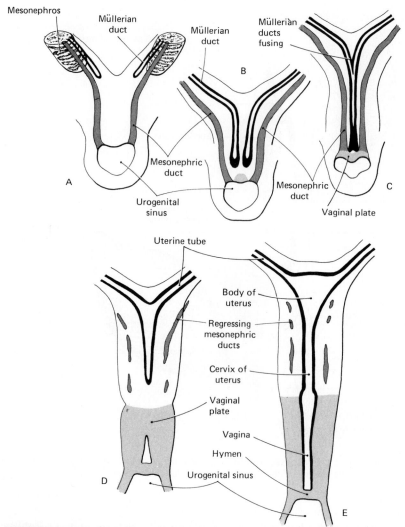

FIGURE 17-27
Fusion of Müllerian ducts in human embryos and formation of the uterus and vagina. (A) Early eighth week; (B) mid-eighth week; (C) ninth week; (D) end of the third month; (E) midfetal period (*A–C after Koff.*)

Müllerian ducts cephalic to their entrance into the vagina (Fig. 17-28). In all the higher mammals, fusion of the Müllerian ducts involves the caudal end of the uterus so that it opens into the vagina in the form of an unpaired neck or cervix. Toward the ovary from the cervix there is great variation in the degree of fusion encountered in the different groups (Fig. 17-28B to 17-28D). In the sow the fusion is carried only a short way beyond the cervix to form a typical *bicornate uterus* (Fig. 17-28). In the human female fusion of the paired Müllerian ducts is

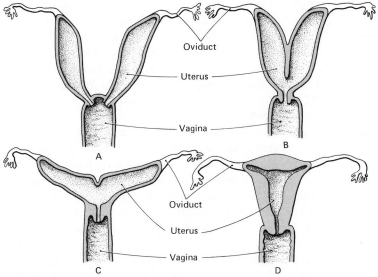

FIGURE 17-28
Four types of uteri occurring in different groups of mammals. (A) Duplex, the type found in marsupials; (B) bipartite, the type found in certain rodents; (C) bicornate, the type found in most ungulates and carnivores; (D) simplex, the type characteristic of the primates. (*After Wiedersheim.*)

complete in the uterine region. As a result the uterus is pear-shaped with a single lumen (Fig. 17-28D). In distinction to a bipartite, or bicornate type, this is called a *simplex uterus*.

Uterine Tubes

The part of the Müllerian duct between the uterus and the ovary remains slender and forms the uterine tube. Near its cephalic end a funnel-shaped opening, the *ostium*, develops. In different forms the detailed configuration of the ostium and its relation to the ovary are quite variable, ranging from a pouchlike dilation that almost completely invests the ovary (sow) to an elaborately fringed, funnel-shaped ostium which opens in the general direction of the ovary (human). Regardless of the configuration, the various forms of ostia are quite efficient in capturing ovulated ova, because abdominal pregnancies, resulting from fertilization of an egg outside the reproductive tract, are uncommon.

DESCENT OF THE GONADS

Descent of the Testes

Neither the testes nor the ovaries remain located in the body at their place of origin. The excursion of the testes is particularly extensive. When the mesonephros begins to grow rapidly, it bulges out into the coelom, pushing ahead of itself a covering of peritoneum. At either end of the mesonephros the peritoneum is in this process thrown into folds. One of them extends cephalad to the diaphragm and is known as the *diaphragmatic ligament of the meso-nephros* (Fig. 17-23). The other, which extends to the extreme caudal end of the coelom, becomes fibrous and is then known as the *inguinal ligament of the mesonephros* (Fig. 17-23). The inguinal ligament is destined to play an important part in the descent of testes.

When the testis develops, it causes a local expansion of the peritoneal covering of the mesonephros to accommodate its increasing mass. As the testis grows, the mesonephros decreases in size and the testis takes to itself more and more of the peritoneal coat of the mesonephros. In this process it becomes closely related to the inguinal ligament of the mesonephros. In effect, the inguinal ligament extends its attachment to include the growing testis as well as the shrinking mesonephros. With this change the ligament is spoken of as the *gubernaculum* (Figs. 17-23 and 17-29).

In the meantime bilateral coelomic evaginations are formed, one in the inguinal region of each side of the pelvis where the caudal end of the gubernaculum is attached. These are the *scrotal pouches*. Perhaps in part because of traction exerted by the gubernaculum but primarily through differential growth, the testes and the mesonephric structures which give rise to the epididymis begin to shift their relative positions progressively farther caudad (Fig. 17-29). Eventually they come to lie in the scrotal pouches.

In its entire descent, the testis moves caudad beneath the peritoneum. It does not therefore enter the lumen of the scrotal pouch directly but slips down under the peritoneal lining and protrudes into the lumen, covered by a layer of peritoneum (Fig. 17-29). In most mammals when the testis has come to rest in the scrotal sac, the canal connecting the sac with the abdominal cavity becomes closed. In some rodents, however, it remains patent and the testes descend into the scrotum only during the breeding season, to be retracted again into the abdominal cavity until the next period of sexual activity.

Formation of the Broad Ligament

In the young female embryo, as in the male, the mesonephroi and the gonads arise retroperitoneally and bulge into the coelom, carrying a fold of peritoneum about themselves (Fig. 17-30). The mesonephroi degenerate more completely in the female than in the male, and their decreasing bulk leaves the peritoneal folds quite thin. At this stage they more or less resemble a pair of mesenteries suspending the Müllerian ducts in their ventral margins and the ovaries on their medial faces (Fig. 17-30). With further degeneration of the mesonephros and its replacement by fibrous tissue, these folds become the part of the broad ligament supporting the uterine tubes.

Farther caudally in the body, where the Müllerian ducts fuse with each other in the midline to form the uterus, the supporting peritoneal folds coalesce

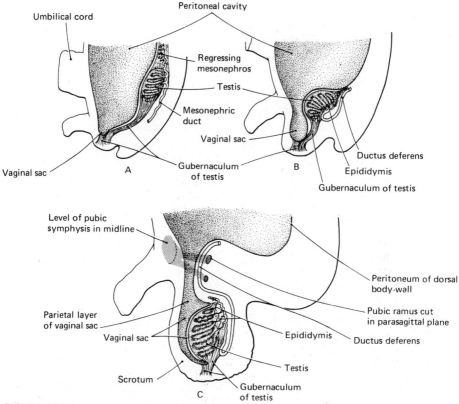

FIGURE 17-29
Schematic diagrams illustrating the relations of the testes and epididymis to the peritoneum during the descent of the testes.

medially to form the part of the broad ligament supporting the uterus (Fig. 17-30C and D).

Descent of the Ovaries

Although the ovaries do not move as far as the testes, their change in position is quite characteristic and definite. As they increase in size, both the gonads and the ducts sag farther into the body cavity. In so doing they pull with them the broad ligament, which, as it is stretched out, allows the ovaries, uterine tubes, and uterus to move caudally and somewhat ventrally (Fig. 17-26). The inguinal ligament of the mesonephros, which in the male forms the gubernaculum, is in the female embedded in the broad ligament. When the ovaries move caudad and laterad, the inguinal ligament is bent into angular form. Cephalic to the bend it becomes the *round ligament of the ovary*, and caudal to it, the *round ligament of the uterus* (Figs. 2-2 and 17-26). It should be noted that the caudal end of the

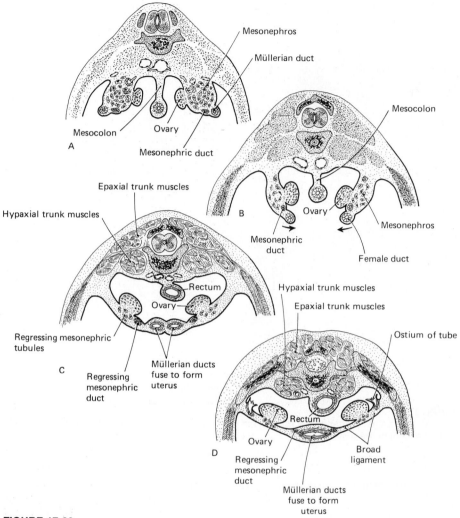

FIGURE 17-30
Schematic cross-sectional diagrams to show some of the main steps in the formation of the broad ligament.

round ligament of the uterus is embedded in the connective tissue of the labium majus in a position homologous with the anchorage of the gubernaculum in the scrotal pouch of the male (cf. Figs. 17-23 and 17-26).

THE EXTERNAL GENITALIA

Indifferent Stages

In very young embryos, a vaguely outlined elevation known as the *genital eminence* can be seen in the midline, just cephalic to the proctodeal depression.

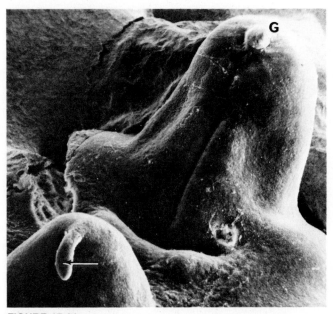

FIGURE 17-31
Scanning electron micrograph of a human embryo (stage 18)
showing the genital tubercle (G) and the regressing tail (arrow).
×38. (*From Fallon and Simandl, 1978, Am. J. Anat.* **152**:*111.*
Courtesy of the authors.)

This is soon differentiated into a central prominence (*genital tubercle*) closely
flanked by a pair of folds (*genital folds*) extending toward the proctodeum (Fig.
17-31). Somewhat farther to either side are rounded elevations known as the
genital swellings (Figs. 17-32A and 17-33A). Between the genital folds is a
depression which attains communication with the urogenital sinus to establish
the urogenital orifice (*ostium urogenitale*). This opening is separated from the
anal opening by the urorectal fold (Fig. 17-15). From this common starting point
the external genitalia of both sexes differentiate.

Male Genitalia

If the individual is to develop into a male, the genital tubercle, under the
influence of dihydrotestosterone (Fig. 17-24), becomes greatly elongated to
form the penis and the genital swellings become enlarged to form the scrotal
pouches (Fig. 17-32B to 17-32D). During the growth of the penis a groove
develops along the entire length of its caudal face and is continuous with the
slitlike opening of the urogenital sinus. This groove later becomes closed over
by a ventral fusion of the genital folds, establishing the penile portion of the
urethra. The portion of the urogenital sinus between the neck of the bladder and
the original opening of the urogenital sinus becomes the prostatic urethra. The

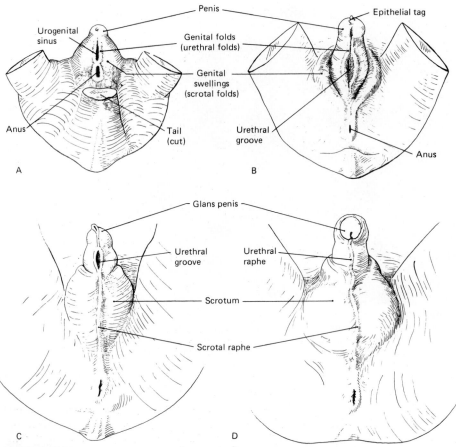

FIGURE 17-32
Stages in the development of the external genitals in the male. (A) At 7 weeks, 17 to 20 mm; (B) in tenth week, 45 to 50 mm; (C) early in twelfth week, 58 to 68 mm; (D) toward close of gestation. (*Adapted from several sources, especially Spaulding, 1921,* in *Carnegie Cont. to Emb., vol. 13.*)

most distal portion of the male urethra arises as a solid cord of ectodermal cells growing in from the glans to meet the penile urethra. The cord then canalizes, completing the urogenital outlet in the male. The line of fusion in the urogenital sinus region and along the caudal surface of the penis is clearly marked by the persistence of a ridgelike thickening known as the *raphe.*(Fig. 17-32C and D).

Female Genitalia

In the female the genital tubercle becomes the clitoris, the genital folds become the labia minora, and the genital swellings become the labia majora (Fig. 17-33). The original opening of the urogenital sinus does not undergo such changes as

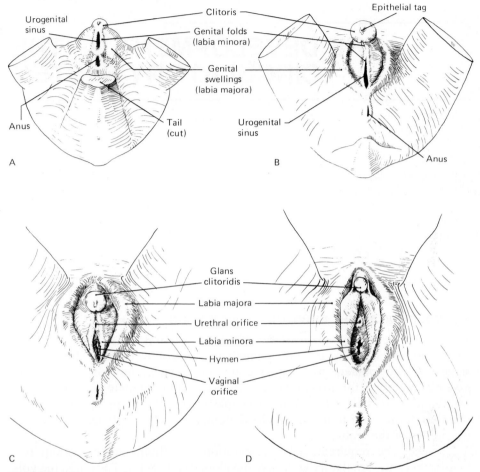

FIGURE 17-33
Stages in the development of the external genitals in the female. (A) At 7 weeks, 17 to 20 mm;
(B) in the tenth week, 45 to 50 mm; (C) at 12 weeks, 75 to 80 mm; (D) toward close of gestation. (*Adapted from a number of sources, especially Spaulding, 1921*, in *Carnegie Cont. to Emb.*,
vol. 13.)

occur in the male but persists nearly in its original position. Its orifice, enlarged
and flanked by the labia, becomes the vestibule into which the vagina and the
urethra open (Fig. 17-7F). The urethra in the female is derived from the
urogenital sinus, being homologous with the prostatic portion of the male
urethra.

CHAPTER 18

THE DEVELOPMENT OF THE CIRCULATORY SYSTEM

INTRODUCTION TO THE EMBRYONIC CIRCULATION

The plan of the embryonic heart and vascular system is bound by three major constraints. First it must meet the immediate needs of the embryo at its various stages of development by supplying it with oxygen, nutrients, and other essential materials for growth while at the same time removing CO_2 and other metabolic wastes. For this, only a simple unidirectional pumping action is required of the heart to move the blood along channels supplying the developing organs of the embryo.

To meet the requirements of respiration, nutrition, and excretion, two extraembryonic circulatory arcs have developed (Fig. 8-17). The arc to the yolk sac, the vitelline arc, supplies foodstuffs to all large-yolked vertebrate embryos; however, in mammals the yolk sac and vitelline circulation persist for what were originally probably subsidiary functions, the origin and transport of primordial germ cells and primitive blood cells. The large allantois of amniote eggs (with its circulatory arc) was originally the chief organ of respiration and deposition of excretory wastes, but as the placenta in mammals evolved into a more efficient organ of exchange, the allantois correspondingly became reduced in prominence. The allantoic circulation, however, has been incorporated into the placenta and continues to serve its original functions.

In addition to satisfying the relatively simple requirements of the embryo, the plan of the embryonic circulation in amniotes must also anticipate the immediate needs of the embryo once it hatches from the egg or is delivered from the mother's uterus (Fig. 18-1). The most immediate and critical adjustment to birth is the need of the newly born individual to breathe independently. Breathing requires that the lungs, which have developed tardily from the

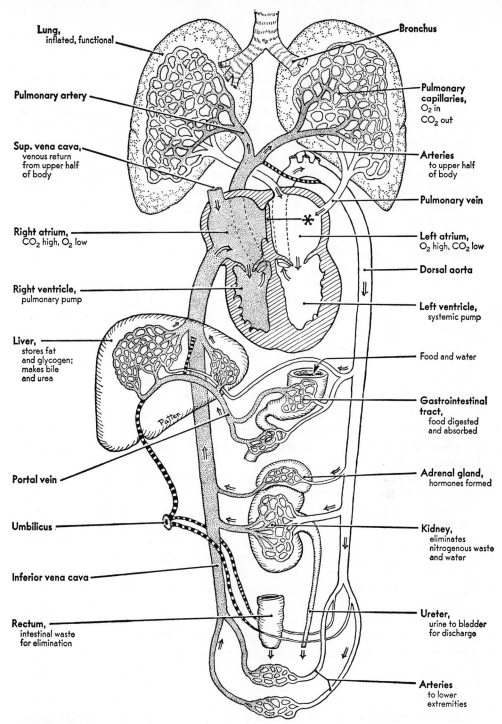

Lung, inflated, functional

Bronchus

Pulmonary artery

Pulmonary capillaries, O_2 in CO_2 out

Sup. vena cava, venous return from upper half of body

Arteries to upper half of body

Pulmonary vein

Right atrium, CO_2 high, O_2 low

Left atrium, O_2 high, CO_2 low

Dorsal aorta

Right ventricle, pulmonary pump

Left ventricle, systemic pump

Liver, stores fat and glycogen; makes bile and urea

Food and water

Patten

Gastrointestinal tract, food digested and absorbed

Portal vein

Adrenal gland, hormones formed

Umbilicus

Kidney, eliminates nitrogenous waste and water

Inferior vena cava

Rectum, intestinal waste for elimination

Ureter, urine to bladder for discharge

Arteries to lower extremities

FIGURE 18-1

Plan of the postnatal circulation. (*After Patten, 1963, in Fishbein, Birth Defects. Courtesy of the National Foundation and the J. B. Lippincott Company, Philadelphia.*) The heavily cross banded structures were important fetal vessels (cf. Fig. 18-29) which after birth ceased to carry blood and gradually became reduced to fibrous cords. The asterisk indicates the valvula foraminis ovalis in the closed position characteristic for postnatal life. (See colored insert.)

morphological standpoint and are untested functionally, begin to function at full capacity immediately after birth. Because of this, the embryonic heart cannot be content with remaining in its original condition, as a simple tube with the blood passing through it in an undivided stream. Early in embryonic life it must be converted into an elaborately valved four-chamber organ, partitioned in the midline and pumping from its right side a pulmonary stream which is returned to the left side and pumped out again as a systemic bloodstream. And the heart cannot cease work while making its internal alterations; there can be no interruption in the current of blood which it pumps to the growing embryo. The systemic, as well as the pulmonary, part of the circulation must be prepared. Because of the delayed development and restricted vascular bed of the embryonic lungs, the left side of the heart receives less blood from the pulmonary veins than the right side of the heart receives from the venae cavae. Yet after birth the left ventricle is destined to do more work than the right ventricle. On the other side of the coin, the right ventricle receives more blood than can be accommodated by the pulmonary circulation. These and other problems of the embryonic heart are solved by the presence of shunts, which act like safety valves, allowing the various chambers of the heart to obtain the exercise they need for their required development but not overloading the pulmonary vasculature beyond its limited carrying capacity.

Besides meeting immediate and future physiological needs, the circulatory plan of the embryo cannot escape the influence of its phylogenetic history. This third constraint causes the pattern of the embryonic circulatory system to take on a form that would not be predicted by its physiological needs alone. This is particularly evident in the pharyngeal region, where in the system of aortic arches there is an unmistakable phylogenetic impress in the arrangement and manner of development of the blood vessels in that region. The blood leaving the ventrally located heart must pass around the gut to reach the dorsally located aorta. In primitive fishes six pairs of aortic arches encircle the pharynx, breaking up the gills into capillaries which carry out the indispensable function of oxygenating the blood. Once an animal has replaced its gills with lungs, it makes little difference functionally whether the blood passes by each gill cleft on its way from heart to aorta. In adult birds and mammals we find this communication simplified to a single main aortic arch. But in the embryos of both birds and mammals a whole series of symmetrical aortic arches appears, for a time encircling the pharynx and passing in close relation to vestigial gill clefts. This can be interpreted only as a recapitulation of ancestral conditions— conditions which, although they have ceased to be of functional importance, appear nevertheless as a developmental phase on the way to a more highly differentiated plan of adult structure. One interpretation of the aortic arches, then, would take into consideration first the fundamental functional necessity of a channel connecting the ventrally located heart with the dorsally located aorta. Second, looking at the striking arrangement of the series of aortic arches in relation to gill arches and clefts, one sees a repetition of the structural plan that existed in water-living ancestral forms. Thus in the end we are led back

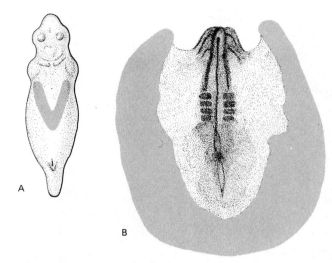

FIGURE 18-2
Sites of primary
erythropoiesis (shaded gray)
in amphibian (A) and bird (B)
embryos.

again to functional significance, for the relationship of the aortic arches to the gill arches writes into the story of individual development an unequivocal record of the evolutionary phase when the gills were a center of primary metabolic importance. Other examples of recapitulation in the circulatory system are the previously mentioned persistence of the vitelline circulatory arc long after the yolk sac has abandoned its original nutritive function and the series of highly developed venous channels in the mesonephros, even when it is destined to degenerate later in development.

EMBRYONIC HEMATOPOIESIS

The first blood cells are produced in extraembryonic sites as small groups of mesodermal cells called *blood islands* (Fig. 18-2). The early blood islands are located next to the endodermal wall of the yolk sac (Fig. 18-3), and in the chick the differentiation of blood islands has been shown to depend on an interaction between the splanchnic mesoderm and the underlying endoderm (Wilt, 1965). Cells in the outer zone of the primordial blood island become flattened as a young vascular endothelium and enclose the more centrally located cells, which become *hematopoietic stem cells* (Fig. 18-3B). Within the endothelial vesicles, fluid accumulates and suspends the developing blood cells.

There is a temporal progression of major sites of hematopoiesis in the embryo (Fig. 18-6). The first hematopoietic activity is seen in the yolk sac. Later in development dominant sites of hematopoiesis are found in the liver and spleen, body mesenchyme, and finally, the bone marrow. Recent experiments on chimeric bird embryos (quail body grafted to a chick yolk sac) have shown that there are two separate populations of hematopoietic precursor cells in the embryo (Dieterlen-Lievre, 1984). The first population (extraembryonic) arises

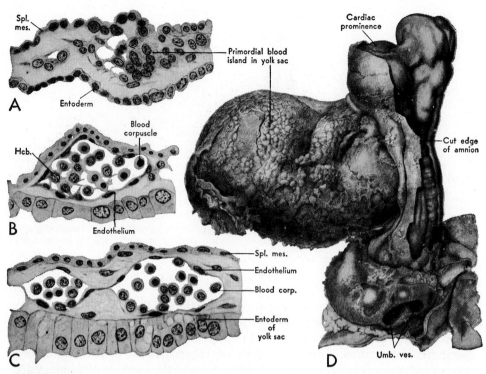

FIGURE 18-3
Development of yolk-sac blood islands in human embryos. A–C are camera lucida drawings, re-produced ×355. (A) Early stage in aggregation of cells between endoderm and splanchnic me-soderm in yolk sac of an embryo early in fourth week (17 somites). (B) Beginning of differentia-tion of endothelium and primitive blood cells, from an embryo of about 4 weeks (4.5 mm). (C) A more advanced area from a 4-week embryo showing endothelium well differentiated and corpus-cles suspended, free, in plasma. (D) The Corner 10-somite embryo showing location of young blood islands on yolk sac. (*Carnegie Cont. to Emb., 1929, vol. 20.*) Abbreviation: *Hcb.*, primitive blood-mother-cell, or hemocytoblast.

in the yolk sac and constitutes the primitive population of erythrocytes. Later, a new population of hematopoietic stem cells, arising from the intraembryonic mesoderm, produces a population of erythrocytes that progressively replaces the yolk sac-derived erythrocytes. The cells in the liver and spleen, and ultimately the bone marrow, are those which give rise to the erythroid series (cells forming red blood cells, or erythrocytes) and the granulocytic series (cells forming the granulated white blood cells, or neutrophils, eosinophils, and basophils). The central lymphoid organs, the thymus and the bursa of Fabri-cius, are populated by cells which are derived from intraembryonic mesoderm and which later differentiate into lymphoid cells (lymphocytes and monocytes). After further developmental processing in the central lymphoid organs, the lymphocytes secondarily seed the peripheral lymphoid tissues (see Chap. 16). In later life other pathways of seeding the central lymphoid organs, e.g., from the bone marrow, may also be set up.

ERYTHROPOIESIS AND HEMOGLOBIN FORMATION

Erythropoiesis is the process by which a mature red blood cell (*erythrocyte*) loaded with hemoglobin molecules differentiates from primitive hemocytoblastic stem cells. Erythropoiesis can be viewed from several aspects and levels of organization. At the tissue and organ level, there are three major phases in erythropoiesis (Fig. 18-6, bottom). For a brief period the only blood-forming activity occurs extraembryonically, in the yolk sac. After the second month the major sites of erythropoiesis shift from the yolk sac to intraembryonic organs. The second major phase is the hepatic period (roughly the third through seventh months of human pregnancy), during which the liver and spleen are the dominant hematopoietic organs. Finally, late in pregnancy the bone marrow takes over as the definitive site for erythropoiesis in higher vertebrates.

The differentiation of individual erythrocytes must also be viewed in the context of their site of origin. The first erythrocytes are derived from the original population of yolk-sac precursors. These cells differentiate relatively synchronously and are released into the bloodstream at an early stage of differentiation. Maturation is completed in the bloodstream, and in mammalian embryos the yolk sac-derived erythrocytes are nucleated.

Differentation of the erythrocytes derived from intraembryonic precursor cells has been extensively studied with respect to both cellular morphology and production of hemoglobin. Both light and electron microscopic preparations show a series of developmental stages which are classical for a cell that is heavily engaged in the production of intracellular protein (Fig. 18-4). From the *hemocytoblast*, a primitive cell that is considered by many to be the stem cell for all families of blood cells, the erythrocyte line first appears as a highly basophilic cell called a *proerythroblast*. These cells have undergone sufficient restriction to be firmly committed to the production of red blood cells, but they have not yet begun to produce hemoglobin in amounts sufficient to be detected by cytochemical analysis. They have large nuclei with prominent nucleoli and largely uncondensed nuclear chromatin. Synthesis of globin mRNA is high. The cytoplasm contains aggregates of mainly free ribosomes, which will be used in intracellular protein synthesis.

Subsequent stages of erythroid differentiation (*basophilic, polychromatophilic*, and *orthochromatic erythroblasts*) show a progressive change in the balance between the accumulation of newly synthesized hemoglobin molecules and the decline of first the RNA-producing machinery and later the protein-synthesizing apparatus. During these stages the cytoplasm continuously stains less basophilically as its eosinophilic staining (characteristic of accumulated protein) increases in intensity. Corresponding to these changes is a decrease in the concentration of ribosomes in the cytoplasm. The nucleus shows signs of its inexorable pathway toward inactivation and elimination by its steadily decreasing size, the progressive condensation of its chromatin, and the elimination of the nucleolus. Finally, the late orthochromatic erythroblast extrudes its pycnotic nucleus. In contrast to yolk-sac erythro-

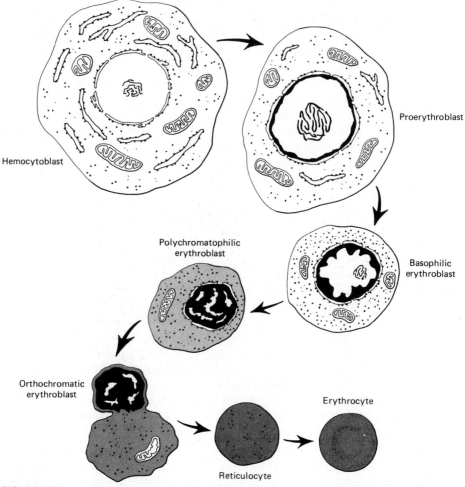

Hemocytoblast

Proerythroblast

Polychromatophilic
erythroblast

Basophilic
erythroblast

Orthochromatic
erythroblast

Erythrocyte

Reticulocyte

FIGURE 18-4
Successive stages in the differentiation of an erythrocyte. Increasing concentrations of hemoglobin are indicated by the intensity of the gray shading in the cytoplasm. (*Adapted from Rifkind,
1974, in Lash and Whittaker, eds., Concepts of Development, Sinauer Assoc.*)

poiesis, the changes normally occur within the hematopoietic tissues, and
only after it has lost its nucleus is the cell, now called a *reticulocyte*, released
into the bloodstream. Reticulocytes still contain small numbers of polysomes, and they continue to produce hemoglobin for the first day or two after
their release into the bloodstream. The mature erythrocyte is a terminally
differentiated cell that lacks both a nucleus and the intracellular apparatus for
macromolecular synthesis. It has often been likened to a bag of hemoglobin,
but such an appellation does not do justice to the extent to which the
structure of the erythrocyte is adapted to fulfilling its function.

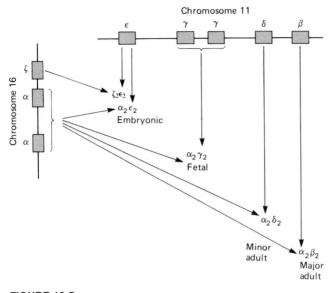

FIGURE 18-5
Sequential gene activation along chromosomes 11 and 16 in hemoglobin synthesis during development. (*After S. Gilbert, 1985, Developmental Biology, Sinauer.*)

There are developmental changes in the hemoglobin molecule itself. The *hemoglobin* molecule (MW 64,500) contains four polypeptide chains (*globin*) complexed to a molecule of heme. The human globin genes are located on chromosomes 11 and 16 (Fig. 18-5), and during embryonic development there is an orderly succession of pairs of globin types (with a pair derived from each chromosome) contributing to the hemoglobin molecule. The earliest hemoglobin molecule (embryonic hemoglobin) contains a pair of ζ-globin molecules derived from chromosome 16 and a pair of ϵ-globin chains from chromosome 11 attached to the heme molecule. A pair of α-globin molecules, specified by genes located on chromosome 16, soon supplants the original pair of ζ-globin chains, and from this point on, regardless of the type of hemoglobin, two of the globin chains are always α molecules.

Embryonic hemoglobin is present for only a couple of months, after which it is supplanted by *fetal hemoglobin*, a form that contains two α chains and two γ-chains (Fig. 18-5) and is dominant throughout the rest of embryonic life (Fig. 18-6). The sequential switching of globin genes on chromosome 11 is carried out in a linear order (ϵ, γ, δ, and β) and it suggests that these genes are linked by a switching mechanism. At about the time of birth, small amounts of δ- and β-globins begin to combine with α chains to form *adult hemoglobin*. Shortly after birth the amount of fetal hemoglobin rapidly declines as the amount of adult hemoglobin increases.

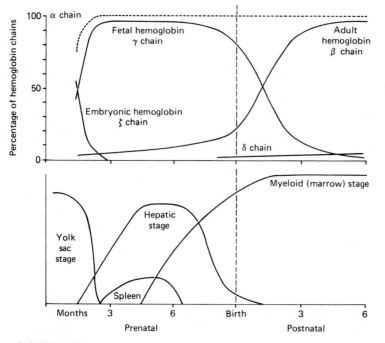

FIGURE 18-6
Periods of erythropoiesis and hemoglobin synthesis in the human. The
lower graph highlights dominant sites of erythropoiesis. The upper graph
shows the percentages of hemoglobin polypeptide chains present in the
blood at a given time. The α chain is placed in a separate category in this
graph. (*Upper graph after Huehns et al., 1964, Cold Spring Harbor Symp.
Quant. Biol. **29**:327. Lower graph after Wintrobe et al., 1974, Clinical He-
matology, 7th ed., Lea & Febiger, Philadelphia.*)

Embryonic and fetal hemoglobin have a higher affinity for oxygen than does
the adult form. This represents a significant adaptation for intrauterine life
because the fetal hemoglobin is able to extract oxygen diffusing across the
placental barrier more efficiently than could adult hemoglobin. During late fetal
life increasing amounts of adult hemoglobin are manufactured. After birth the
proportion of fetal hemoglobin rapidly decreases until by 4 to 6 months it is
present in only minute amounts in the blood.

The production of new erythrocytes is regulated by a humoral substance
called *erythropoietin*. In response to hypoxia, which could result from blood
loss, a relative deficiency of erythrocyte production, or a move to a higher
altitude, the levels of erythropoietin in the blood rise and stimulate the
proliferation of erythroid stem cells. Erythropoiesis within the body of the
embryo is responsive to the effects of erythropoietin, but yolk-sac erythropoiesis is not. In the chick embryo there is evidence in favor of a separate
variety of erythropoietin to which the erythroid stem cells in the yolk sac
respond (Knezevic et al., 1971).

THE ARTERIES

Derivatives of the Aortic Arches

In vertebrate embryos six pairs of aortic arches connect the ventral with the dorsal aorta (Fig. 18-7A). The portions of the primitive paired aortae that bend around the anterior part of the pharynx through the tissues of the mandibular arch (first branchial arch) constitute the first aortic arch (Fig. A-25). The other aortic arches develop later, one aortic arch for each branchial arch. However,

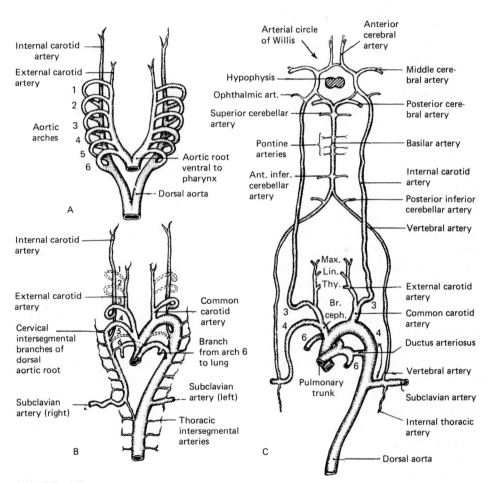

FIGURE 18-7
Diagrams illustrating the major changes which occur in the aortic arches of mammalian embryos. (*Adapted from several sources.*) (A) Ground plan of complete set of aortic arches. (B) Early stage in modification of arches. (C) Adult derivatives of aortic arches. Abbreviations: *Br. ceph.*, brachiocephalic (innominate) artery; *Lin.*, lingual artery; *Max.*, maxillary artery; *Thy.*, thyroid artery. Arrow in (C) indicates change in position of origin of left subclavian artery which occurs in the later stages of development.

in mammalian embryos we never find the entire series of aortic arches well developed at the same time. The two most cephalic arches degenerate as main channels before the most caudal arches have been established. In mammalian embryos the fifth arch never develops beyond a vestigial vessel. As we deal with the developmental reorganization of the aortic arch system, it is important to recognize that remnants of arches commonly persist as arteries that supply specific regions of the head and neck.

The first two arches break down, but the paired ventral and dorsal aortic roots persist as the *external* and *internal carotid arteries* (Fig. 18-7B), which continue to provide the major blood supply to the head. The internal carotid arteries incorporate the third aortic arch on either side. The ventral aortic root between the third and fourth aortic arches persists as the *common carotid artery* (Fig. 18-7B and C). In contrast, the dorsal aortic segments between the third and fourth aortic arches break down to effect a clean separation of blood flow between the head and the body.

The fourth aortic arch has a different fate on opposite sides of the body. On the left, it becomes greatly enlarged and persists as the arch of the adult aorta (Fig. 18-7). On the right, it persists as the root of the subclavian artery, which supplies the right arm.

The sixth aortic arch changes its original relationships somewhat more than the others do. At an early stage of development branches extend from each arch toward the lungs (Fig. 18-7B). After these pulmonary branches have been established, the right side of the sixth aortic arch loses communication with its dorsal aortic root and disappears. On the left, however, that same segment persists as the *ductus arteriosus* (Fig. 18-7C), which acts as an important adaptation to accommodate the fetal circulatory pattern. During the fetal period, the pulmonary vessels to the lungs are not equipped to handle the volume of blood that they will receive postnatally. The ductus arteriosus serves as a shunt to divert the excess blood entering the pulmonary artery directly into the aorta, thus bypassing the poorly developed fetal lungs. The functional importance of this channel will be more fully discussed in connection with the development of the heart and changes in the circulation at birth.

The paired dorsal aorta continues for several body segments caudal to the sixth aortic arch before the right and left dorsal aortae fuse into the single midline vessel (Fig. 18-7A). As the modifications of the aortic arches are taking place, the caudalmost segment of the right dorsal aorta degenerates (Fig. 18-7B), leaving the remainder of the right dorsal aortic root along with the right fourth aortic arch as the proximal portions of the right subclavian artery.

Branches of the Aorta

A series of metamerically arranged vessels coming off the dorsal aorta (the *dorsal intersegmental vessels*) (Fig. 18-7B) give rise to a number of major adult vessels. Some of these are preserved in roughly their original configuration as the intercostal arteries, which run between the ribs (Fig. 18-8C and D). The

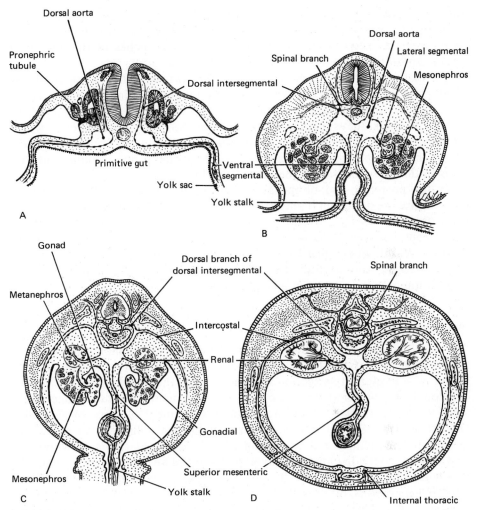

FIGURE 18-8
Cross-sectional plans of the body showing relations of segmental branches of aorta at different stages of development. (See colored insert.)

seventh intersegmental arteries, which arise in the region of the forelimb bud, enlarge greatly to form the *subclavian arteries*, the main arterial trunks to the arms (Fig. 18-7B and C). Cephalic to the subclavian arteries, series of longitudinal branches form the cervical intersegmental arteries connect with one another to form the vertebral arteries (Fig. 18-7C), which send blood to the head. As the vertebral arteries form, the cervical intersegmental roots drop out, leaving the vertebral artery as a branch from the base of the subclavian artery.

Farther caudally, the original *vitelline arteries*, which in younger embryos supplied the yolk sac (Fig. 8-17), undergo a reorganization into a single ventral branch of the aorta (the *superior mesenteric artery*), which constitutes the pivotal point around which rotation of the gut takes place. In a similar manner (Fig. 18-8) the other two major ventral unpaired branches of the aorta (the *coeliac* and *inferior mesenteric arteries*; Fig. 15-14) arise.

The mesonephros is supplied by many small arteries which arise ventrolaterally from the aorta. The early metanephric kidneys are initially fed by small arteries which arise from the aorta along with the mesonephric vessels (Fig. 18-8C). As the kidneys migrate cephalad from their early position deep within the pelvic cavity, they are supplied by more cephalic aortic branches. These vessels become progressively enlarged as the kidneys gain in bulk, and they become the *renal arteries* of the adult (Fig. 18-8D).

At the posterior end of the aorta, the umbilical arteries leading to the placenta are the dominant branches (Fig. 18-9). Small branches leading off from them (the *external iliac arteries*) supply the posterior appendage buds. When the placental circulation is stopped at birth, the umbilical arteries are reduced to small vessels nourishing local tissues, and the remainder of the umbilical vessels become fibrous cords that course along the wall of the bladder, which itself is the remains of the old allantoic stalk.

THE VEINS

The development of the venous system is anatomically complex, particularly for students who have not had previous training in gross or comparative anatomy. Therefore, this section will outline only the broad aspects of venous development, with emphasis on major rearrangements that take place as the adult venous pattern is established. The development of veins in general is characterized by the initial appearance of complex capillary plexuses and the preferential utilization and expansion of individualized channels within the plexus. Also common in venous development is the formation of new channels, which in time take over the function of preexisting main channels. Thus, several generations of precursor channels are often incorporated into the substance of a single large adult vein.

The simplest venous pattern, seen in early embryos (Fig. 18-14A), mirrors the arrangement of the arterial system. The systemic venous system consists of cardinal veins, the *anterior* and *posterior cardinal veins*, which empty into the common cardinal vein and ultimately into the sinus venosus. In the extraembryonic circulation, the *vitelline veins* drain the yolk sac; the *umbilical (allantoic) veins* drain the allantois and later return blood to the body from the placenta.

Within the cephalic part of the embryo the originally symmetrical anterior cardinal veins, which are the primary venous drainage channels from the head (Fig. 8-17), are transformed into the *internal jugular veins* (Fig. 18-11). Although initially symmetrical, the base of the left internal jugular vein diminishes greatly in size and develops into a new channel (*left brachiocephalic vein*; Fig. 18-11) that shunts the venous blood from the left side of the head into the base of the original right anterior cardinal vein. Blood from the anterior part

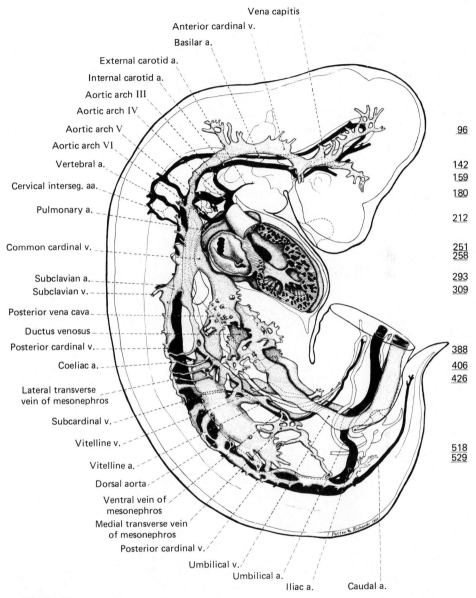

Vena capitis
Anterior cardinal v.
Basilar a.
External carotid a.
Internal carotid a.
Aortic arch III
Aortic arch IV
Aortic arch V
Aortic arch VI
Vertebral a.
Cervical interseg. aa.
Pulmonary a.
Common cardinal v.
Subclavian a.
Subclavian v.
Posterior vena cava
Ductus venosus
Posterior cardinal v.
Coeliac a.
Lateral transverse vein of mesonephros
Subcardinal v.
Vitelline v.
Vitelline a.
Dorsal aorta
Ventral vein of mesonephros
Medial transverse vein of mesonephros
Posterior cardinal v.
Umbilical v.
Umbilical a.
Iliac a.
Caudal a.

96
142
159
180
212
251
258
293
309
388
406
426
518
529

FIGURE 18-9
Reconstruction (×14) of the circulatory system of a 9.4-mm pig embryo. The numbered horizontal lines on the right indicate the levels of the cross sections drawn in Figures 12-16, 15-7, 17-10, 18-10, and 18-13. The use of a transparent straightedge placed at the level of the appropriate line will greatly help in the correlation of each of the cross sections with the lateral plan of the reconstruction.

of the body ultimately empties into the right atrium of the heart through the *superior vena cava* (Figs. 18-11 and 18-12E and F), which utilizes the channel of the original right common cardinal vein.

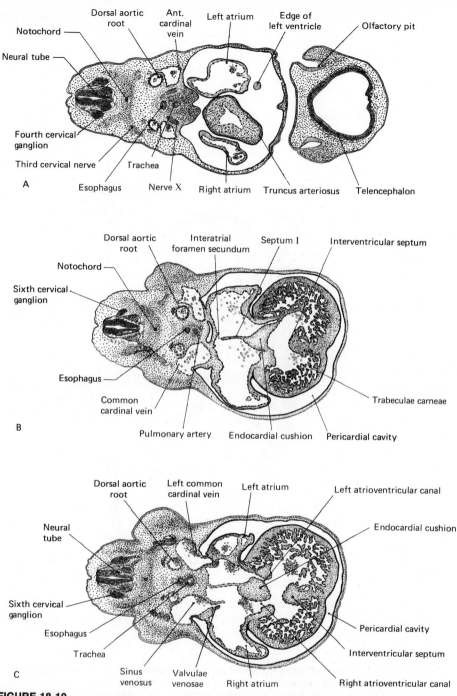

FIGURE 18-10

Three transverse sections from the series of a 9.4-mm pig embryo used in making the reconstructions illustrated in Figure 18-9. (Projection drawings, ×14.) By laying a straightedge across either of these reconstructions at the level of the marginal line bearing the serial number of a cross section, its relations within the body as a whole are precisely indicated. (A) Section 212 passing through the olfactory pits and the truncus arteriosus; (B) section 251 passing through the heart at the level of the interatrial foramen secundum; (C) section 258 passing through the heart at the level of the orifice of the sinus venosus and the atrioventricular canals.

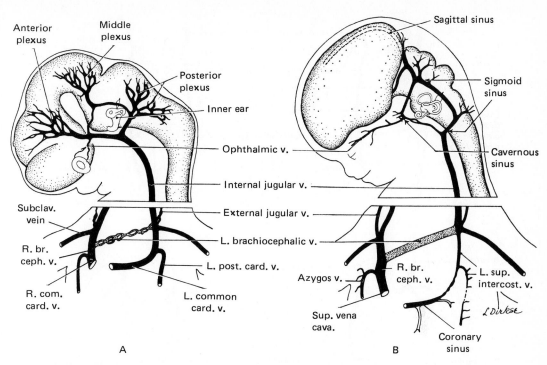

FIGURE 18-11
Schematic diagrams to show some of the major events in the formation of the superior vena cava. (A) Conditions late in the seventh week, when the left brachiocephalic vein is just beginning to establish a cross anastomosis between the right and left anterior cardinal (internal jugular) veins. (B) Conditions at about 10 weeks, when the left brachiocephalic anastomosis has radically reduced the blood flow to the left common cardinal. (*The scheme of rotating the head in reference to the trunk was borrowed from Hamilton, Boyd, and Mossman. The arrangement of vessels in the cephalic region was based on the work of Streeter and Padget.*) Abbreviations: *br. ceph.*, brachiocephalic; *com. card.*, common cardinal; *L*, left; *R*, right; *sup.*, superior; *v.*, vein.

The changes in the systemic veins of the posterior part of the body are much more radical than they are in the anterior part. They involve three pairs of cardinal veins and are closely connected with the rise and decline in prominence of the mesonephros. The *posterior cardinal veins* are the original drainage channels for the caudal half of the body (Fig. 18-12A and B). Somewhat later, a pair of *subcardinal veins* arises in association with the ventral margin of the mesonephros. Both pairs of veins receive numerous small side branches from the mesonephric kidneys. Accompanying the degeneration of the mesonephric kidneys, the *postcardinal* and *subcardinal* veins begin to break up as well (Fig. 17-16C). As the midparts of the posterior cardinal veins break up, blood coursing in the posterior segments goes through connecting channels in the mesonephros into the subcardinal veins. The last of the pairs of cardinal veins to appear are the *supracardinal veins* (Fig. 18-12D). These are located in the body wall dorsolateral to the aorta (Fig. 18-12I). The major systemic drainage channel from the posterior part of the adult body is the *inferior vena cava*, which, like the superior vena cava, drains into right atrium of the heart. It is formed by a complex sequence of takeovers of persisting segments of the three pairs of cardinal

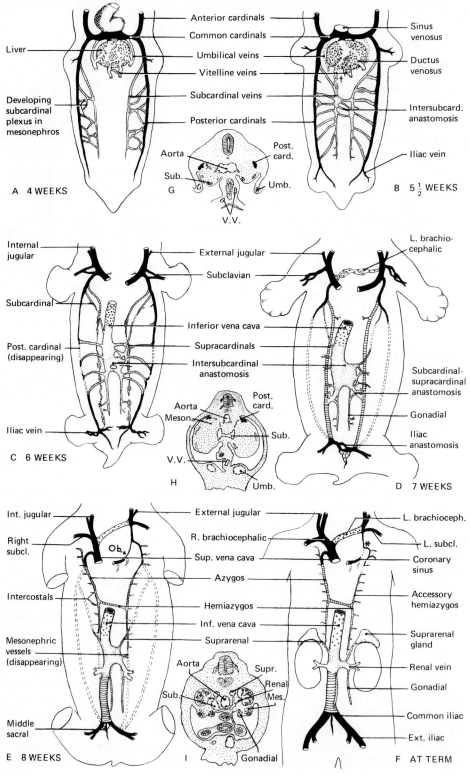

Liver

Developing subcardinal plexus in mesonephros

A 4 WEEKS

Anterior cardinals
Common cardinals
Umbilical veins
Vitelline veins
Subcardinal veins
Posterior cardinals

Aorta
Sub.
G
V.V.

Post. card.
Umb.

Sinus venosus
Ductus venosus
Intersubcard. anastomosis
Iliac vein

B 5 1/2 WEEKS

Internal jugular

Subcardinal

Post. cardinal (disappearing)

Iliac vein

C 6 WEEKS

External jugular
Subclavian
Inferior vena cava
Supracardinals
Intersubcardinal anastomosis

Aorta
Meson.
V.V.
H
Post. card.
Sub.
Umb.

L. brachio-cephalic

Subcardinal-supracardinal anastomosis
Gonadial
Iliac anastomosis

D 7 WEEKS

Int. jugular

Right subcl.

Intercostals

Mesonephric vessels (disappearing)

Middle sacral

E 8 WEEKS

External jugular
R. brachiocephalic
Sup. vena cava
Azygos
Hemiazygos
Inf. vena cava
Suprarenal

Aorta
Sub.
I
Supr.
Renal
Mes.
Gonadial

L. brachioceph.
L. subcl.
Coronary sinus
Accessory hemiazygos
Suprarenal gland
Renal vein
Gonadial
Common iliac
Ext. iliac

F AT TERM

602

veins. This is illustrated in Fig. 18-12 and will not be further described in the text.

The vitelline and umbilical veins follow an equally complex course, but the rearrangements of these vessels are intimately connected with the growth of the liver. Their bilaterally symmetrical course in the early embryo (Fig. 18-14A) is short-lived. With the growth of the liver, the vitelline veins break up into a large set of anastomosing channels which soon connect with one another and with the umbilical veins (Fig. 18-14B). Soon the cephalic segments of both umbilical veins degenerate (Fig. 18-14C), and the umbilical veins drain directly into the vitelline vascular plexus. One of these channels becomes favored and is transformed into a large channel, the *ductus venosus* (Fig. 18-14D), a temporary structure which represents one of the major adaptations for the embryonic circulation. As the right umbilical vein degenerates, the left umbilical vein serves as the sole channel for returning blood to the fetus from the placenta. The ductus venosus is large in order to accommodate the blood flow which passes through the ductus directly into the inferior vena cava.

Meanwhile, the original vitelline veins become the system of veins draining the gut, the *hepatic portal system* (Fig. 18-14B to 18-14D). The vascular plexus in the early liver is converted into a capillary bed and serves as an effective functional adaptation in the adult. The distal branches of the portal vein absorb digested food from the lining of the gut and transport it directly to the liver for metabolic processing. It is this second capillary network at the hepatic end of the portal vein that permits the broad and efficient distribution of digested foodstuffs to the liver.

Near the heart is another venous transformation of embryological significance. With the degeneration of the segment of both the left anterior cardinal vein where it joins the left common cardinal vein and the left posterior cardinal vein (Fig. 18-12E), the left common cardinal vein is left with almost no incoming blood from the body. However, it does not disappear. Instead, it becomes closely applied to the wall of the rapidly developing heart, and as the *coronary sinus* (Figs. 18-12B and 18-19E and F) it serves as the channel that drains the coronary veins and empties the blood into the right atrium of the heart.

The *pulmonary veins*, along with the lungs, are phylogenetically relatively new structures. It is not surprising therefore that we find the pulmonary veins arising independently and not by the conversion of old vascular channels. They originate as vessels which drain the lung buds and converge into a common trunk (Fig. 15-9F). As the heart grows, this trunk vessel is gradually absorbed

---

FIGURE 18-12
Schematic diagrams showing some of the steps in the development of the inferior vena cava. Cardinal veins are shown in black; subcardinals are stippled; supracardinals are horizontally hatched. Vessels arising independently on these three systems are indicated by small crosses. (*Based on the work of McClure and Butler.*) Abbreviations: *Ob.*, oblique vein of left atrium; *, left superior intercostal; †, mesenteric portion of inferior vena cava; *subcl.*, subclavian vein. (See colored insert)

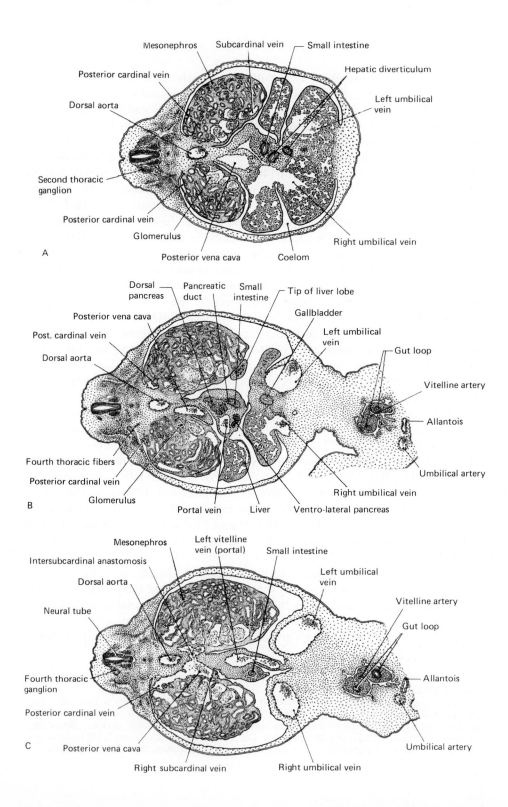

into the left atrial wall, and usually four of its original branches open directly into the left atrium as the main pulmonary veins of the adult (Fig. 18-15).

THE HEART

The heart arises from splanchnic mesoderm, but in both amphibians and birds prospective heart cells are localized in the early gastrula stage (Fig. 5-21). The evidence for an inductive effect by the endoderm on the precardiac mesoderm in the initiation of cardiac mesoderm in higher vertebrates is equivocal, but evidence of an inductive interaction between these two tissues exists in amphibians (Jacobson, 1960). In the cardiac lethal mutant in the axolotl an endodermal induction has been shown to be responsible for the formation of an adequate amount of myofibrillar material for beating of the heart (Lemanski et al., 1979).

The early establishment of the heart as two simple endocardial tubes on either side of the body and their subsequent fusion into a single cardiac tube was discussed in Chap. 8 (Figs. 8-15 and 8-16). The bilateral origin of the cardiac primordia is strikingly emphasized in a condition known as *cardia bifida* (Fig. 18-16). If the cardiac primordia are prevented from fusing by chemical or surgical means, separate independently beating hearts form on each side of the body (Gräper, 1907).

Formation of the Cardiac Loop and Establishment of Regional Divisions in the Heart

The primary factor that brings about the regional differentiation of the heart is the rapid elongation of the primitive cardiac tube, resulting in its bending into an S shape. The factors that cause the heart to bend are incompletely understood. It is now known that much of the impetus to the looping of the heart is intrinsic to the cardiac tube, for when the early heart is explanted and allowed to develop in vitro or as a free graft, the characteristic curving configuration appears (Orts Llorca and Gil, 1965). Other studies (Manasek et al., 1972) have shown that regional changes in cell shape along the early cardiac tube are to a great extent responsible for the looping.

FIGURE 18-13
Three transverse sections from the series of the 9.4-mm pig embryo used in making the reconstructions illustrated in Fig. 18-9. By laying a straightedge across either reconstruction at the level of the marginal line bearing the serial number of a cross section, its relations within the body as a whole are precisely indicated. The sections in this figure are especially helpful in working out the relations of the major veins of the abdominal region. (A) Section 388 passing through the body just cephalic to the attachment of the belly stalk. The belly stalk and the tip of the tail have not been included in the drawing. (B) Section 406 passing through the level of the pancreatic primordia. Tail and part of belly stalk not drawn. (C) Section 426 at the level of the intersubcardinal anastomosis. By cross-referring these sectional diagrams to the schematic plans of Fig. 18-12, it will be evident that the part of the posterior vena cava labeled in (A) and (B) is its *mesenteric portion* and that the part labeled in (C) is an early stage in the formation of the *interrenal portion* from the intersubcardinal anastomosis.

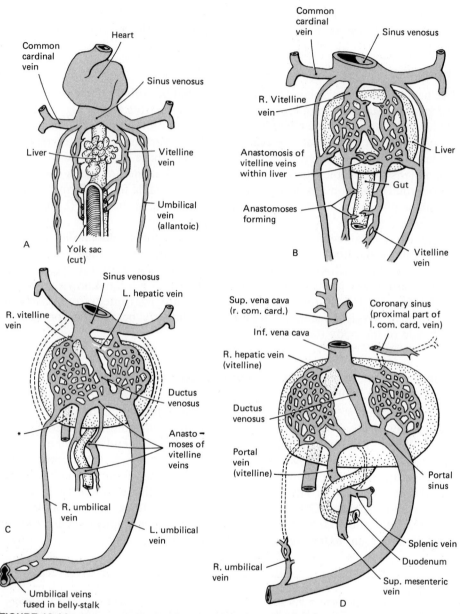

FIGURE 18-14

Diagrams showing the development of the hepatic portal circulation. (A) Based on conditions in pig embryos of 3 to 4 mm; (B) about 6 mm; (C) 8 to 9 mm; (C) 20 mm and older. The asterisk in (C) indicates the location of a hepatic portion of the inferior vena cava, which is formed by the enlarging and straightening of originally small and irregular hepatic sinusoids. When fully formed this portion of the cava lies in a notch on the dorsal side of the liver. (*Collaboration of Dr. Alexander Barry is gratefully acknowledged.*)

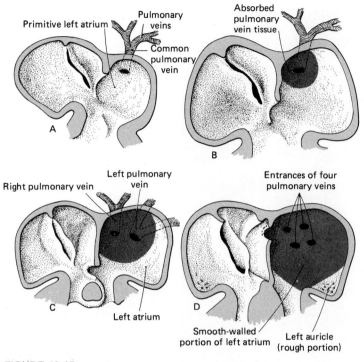

FIGURE 18-15
Absorption of the pulmonary vein into the wall of the left atrium of the human heart so that the four channels past the second branch point ultimately empty directly into the left atrium. (A) 5 weeks; (B) 5½ weeks; (C) 6 weeks; (D) 8–9 weeks.

During the period when the cardiac loop is being formed, the primary regional divisions of the heart become clearly differentiated. The *sinus venosus* is the thin-walled chamber in which the great veins become confluent, entering the heart at its primary caudal end (Fig. 18-19C). The *atrial region* is established by transverse dilation of the heart tube just cephalic to the sinus venosus (Figs. 18-18C to 18-18E and 18-19C to 18-19E).

The ventricle is formed by the bent midportion of the original cardiac tube. As this *ventricular loop* becomes progressively more extensive, it at first projects ventrally beneath the attached aortic and sinus ends of the heart (Fig. A-46F and A-46G). Later it is bent caudally so that the ventricle, formerly situated cephalic to the atrium, is brought into its characteristic adult position caudal to the atrium (Fig. A-46I to L). Between the atrium and ventricle the heart remains relatively undilated. This narrow connecting portion is the *atrioventricular canal* (Fig. 18-21A). The most cephalic part of the cardiac tube undergoes the least change in appearance, persisting as the *truncus arteriosus*, which connects the ventricle with the ventral aortic roots (Fig. 18-18). The

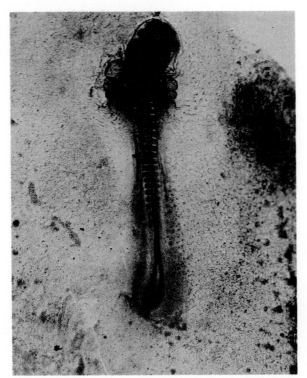

FIGURE 18-16
Chick embryo with surgically produced cardia bifida. This condition was produced by cutting the floor of the foregut at an early stage. The cardiac primordia are prevented from fusing. (*Courtesy of R. L. DeHaan.*)

transitional region where the ventricle narrows to join the truncus arteriosus is known as the *conus* (Fig. 18-18).

Almost from their earliest appearance the atrium and the ventricle show external indications of the impending division of the heart into right and left sides. A distinct median furrow appears at the apex of the ventricular loop (Figs. 18-18D and E and 18-19D and E). The atrium meanwhile has undergone rapid dilation and bulges out on either side of the midline (Fig. 18-19). These superficial features suggest that more important changes are going on internally. Among these internal factors are localized cell death (Pexieder, 1972) and relationships between the extracellular matrix and the early myocardial cells (Manasek, 1975).

Beginning of Cardiac Function

The metabolic requirements of the embryo require the heart to commence beating early in its development. In the chick, the earliest tentative beats begin while the paired cardiac primordia are still fusing (about 29 hours). In the human embryo, the heart begins to beat at the end of the third week. The strength and regularity of the heartbeat increase dramatically as fusion of the cardiac primordia continues and the major divisions of the heart are laid down.

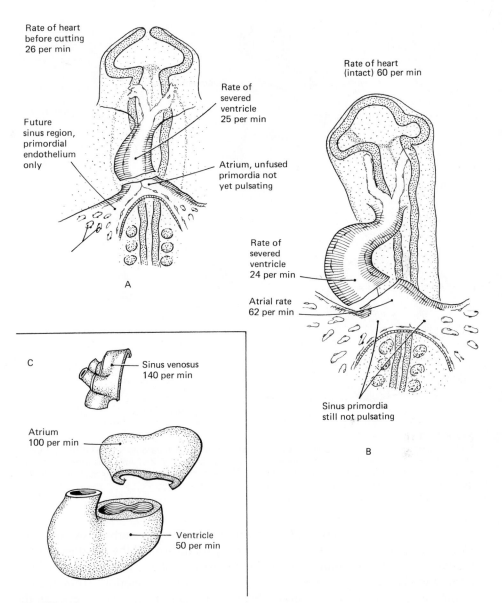

Rate of heart
before cutting
26 per min

Rate of
severed
ventricle
25 per min

Future
sinus region,
primordial
endothelium
only

Atrium, unfused
primordia not
yet pulsating

A

Rate of heart
(intact) 60 per min

Rate of
severed
ventricle
24 per min

Atrial rate
62 per min

Sinus primordia
still not pulsating

B

C

Sinus venosus
140 per min

Atrium
100 per min

Ventricle
50 per min

FIGURE 18-17
Diagrams showing the locations of cuts made in living hearts of young embryos. (A) Embryos of 10–12 somites. At this stage only the ventricle is pulsating, and cutting it away from the nonactive atrium has no appreciable effect on its rate. (B) Embryos of 13 to 15 somites. By this stage the atrium has begun to pulsate and, in the intact heart, acts as a pacemaker. Transection between atrium and ventricle shows the atrium maintaining essentially the rate of the intact heart, whereas the ventricle drops back to approximately the same slow rate it exhibited before the atrium became active and drove the ventricle at its own faster rate. (C) When the heart of a 4-day embryo is cut into three parts, each part beats at its characteristic rate. (*From Patten, 1956*, Univ. of Mich. Med. Bull. **22**:*1*.)

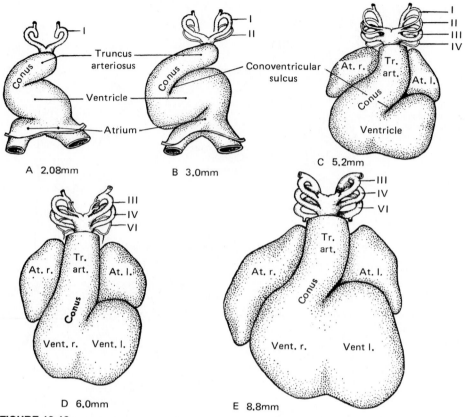

FIGURE 18-18
Ventral views of human embryonic hearts to show bending of the cardiac tube and the establish-ing of its regional divisions. (*After Kramer, 1942*, Am. J. Anat., *vol. 71*.) Abbreviations: *At. l., At. r.*, atrium left and right; *roman numerals I to VI*, aortic arches of corresponding numbers; *Tr. art.*, truncus arteriosus; *Vent. l., Vent. r.*, ventricle left and right.

Even before gross pulsations of the heart are detectable, spontaneous electrical excitatory waves can be detected parsing around the heart as the two cardiac primordia are joining in the midline (Hirota et al., 1983). The electrical signals appear to be transmitted through low-resistance junctions from one cell to the next, with only a temporary delay at the site of the fusion between the right and left cardiac primordia.

Pulsation of the myocardium appears in the major cardiac regions in the same sequence in which they are laid down. The first contractions appear in the ventricle before the atrium is fully established. The contraction rate of the ventricle is at first very slow (Fig. 18-17A). When the atrium begins to pulsate, the heart rate increases. Experiments involving transection of the early heart between the atrium and the ventricle have shown that the atrium has a faster inherent rate of pulsation (Fig. 18-17B) and takes control of cardiac rhythm as

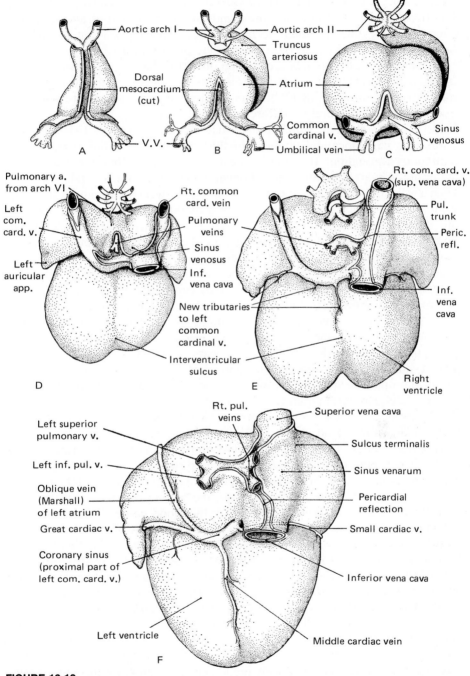

FIGURE 18-19
Six stages in the development of the heart, drawn in dorsal aspect to show the changing relations of the sinus venosus and great veins entering the heart. Abbreviations: *Peric. refl.*, pericardial reflections; *V.V.*, vitelline vein.

a pacemaker. In experimentally produced cardia bifida (DeHaan, 1959) the right heart begins to beat at a more rapid rate than the left at about the time when the sinus venosus would normally form. This suggests that the origin of the cardiac pacemaker is in the right cardiac primordium.

If transection experiments are carried out after the sinus venosus has been formed caudal to the atrium, one finds that its rate of contraction is higher than that of the atrium (Fig. 18-17C). This gradient in contraction rate within the myocardium is not merely a difference in rate from one chamber to another. There is also a gradient *within* the cardiac chambers (Barry, 1942). In other words, the cephalocaudal gradient in the rate of myocardial contraction is continuous throughout the tubular heart. This means that the pacemaking center of the heart is, in all stages of development, at its intake end. Therefore, when a wave of contraction starts at this area and sweeps throughout the length of the cardiac tube, it picks up the incoming blood and forces it out at the other end of the heart into the vessels by which it will be distributed to the body. One would have great difficulty postulating a more efficient type of pumping action in a simple tubular heart without valves.

There is an interesting relationship between the morphological development of the heart and its electrical activity. This can be demonstrated by making electrocardiographic recordings. In the chick the first traces of electrical activity are noted in embryos with 15 somites (somewhat more advanced than is shown in Fig. A-46C). A pattern of electrical activity resembling the ventricular pattern in the adult heart appears in embryos of 16 to 17 somites (Fig. A-46D), at which time the embryonic heart is practically all ventricle. As the atrium appears and shifts toward its normal position, the pattern of the embryonic electrocardiogram resembles more closely that of the adult; by the fourth day of development the electrocardiogram has practically assumed its adult configuration (Hoff et al., 1939). The blood pressure of an early (3-day) chick embryo is extremely low (0.61/0.43 mmHg), and the heart rate is about 140 beats per minute (Girard, 1973). The blood pressure at first rises exponentially, and later the rate of increase tapers off until just before hatching, when, with a pulse rate of 220 beats per minute, the blood pressure is 30/19 mmHg.

In the early embryo (3- to 4-day chick), the blood passes through the heart in two parallel streams which have less of a spinal character than was formerly thought (Arcilla, 1986). Although over the years a number of theories have attributed some of the major developmental changes in heart structure (e.g., formation of the interventricular system) to hemodynamic molding—the shaping of heart structure as a result of mechanical influences—the evidence still does not permit definitive statements to be made (Clark, 1984).

Partitioning of the Heart

Two conspicuous masses of loosely organized mesenchyme called *endocardial cushion tissue* develop in the walls of the narrowed portion of the heart between the atrium and the ventricle. The endocardial cushions, which act as primitive valves to aid in the forward propulsion of blood, consist of relatively large masses of an extracellular matrix rich in glycosaminoglycans, which has long been called *cardiac jelly*.

As the endocardial cushions take shape, the cardiac jelly is seeded by mesenchymal cells which arise by transformation of the endothelium that covers the endocardial cushions (Fig. 18-20). Markwald et al. (1984) have

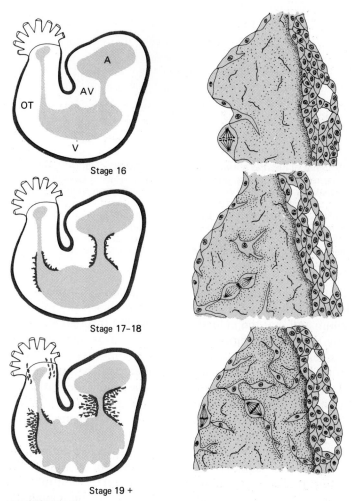

FIGURE 18-20
Seeding of the endocardial cushions by mesenchymal cells in the developing avian heart. The illustrations at the right are magnifications of the atrioventricular (AV) endocardial cushions. Some seeding also occurs in the outflow tract. Abbreviations: A, atrium; V, ventricle; AV, atrioventricular canal; OT, outflow tract. (*Adapted from R. Markwald et al., 1977, Am. J. Anat.* **148**:*85.*)

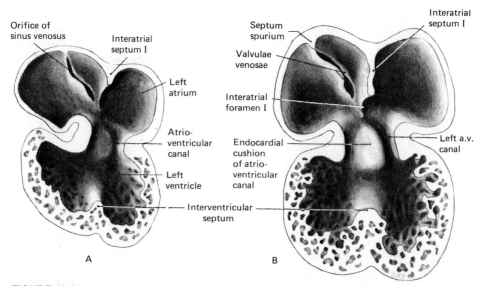

FIGURE 18-21
Semischematic drawings of interior of heart to show initial steps in its partitioning. (A) Cardiac septa are represented at stage reached in human embryos early in fifth week of development. Note especially the primary relations of interatrial septum primum. Based on original reconstructions of the heart of a 3.7-mm pig embryo and on Tandler's reconstructions of corresponding stages of the human heart. (B) Cardiac septa as they appear in human embryos of sixth week. Note restriction of interatrial foramen primum by growth of interatrial septum primum. Based on original reconstructions of the heart of 6-mm pig embryo, on Born's reconstructions of rabbit heart, and on Tandler's reconstructions of corresponding stages of the human heart.

provided evidence that the seeding of the cardiac jelly by transformed endothelial cells is the result of an inductive interaction between the myocardium and the endothelium and that the induction is mediated through extracellular matrix secreted by the myocardium (Fig. 18-20). This was demonstrated by explanatory endothelial monolayers onto collagen gels in vitro. When the endothelium was cocultured with myocardial cells, seeding of the intervening gel regularly occurred. In the absence of myocardium, seeding failed to occur even after 3 weeks in culture. Prominent constituents of the cardiac jelly are hyaluronic acid, a chondroitin sulfate proteoglycan, and a large number of glycoproteins.

Ultimately the endocardial cushions are transformed into dense connective tissue. One of these endocardial cushions is formed in the dorsal wall of the atrioventricular canal (Fig. 18-21B), and the other is formed on the ventral wall. When they meet, these two masses occlude the central part of the canal and separate it into right and left channels (Fig. 18-10C). Meanwhile, a muscular band (*the interventricular septum*) grows from the apex of the ventricle toward the atrium (Fig. 18-21).

At the same time a median partition appears in the cephalic wall of the atrium. Because another, closely related partition is destined to form here later, this one is called the *septum primum*. It is crescentic in shape, and the apices of the crescent extend all the way to the atrioventricular canal, where they merge with the endocardial cushions (Fig. 18-21). This leaves the atria separated from each other except for an opening called the *interatrial foramen primum*.

While these changes have been occurring, the sinus venosus has been shifted out of the midline so that it opens into the atrium to the right of the interatrial septum (Figs. 18-19D and 18-21). The heart is now in a critical stage of development. Its simple tubular form has been altered so that the four chambers characteristic of the adult heart are clearly recognizable. Partitioning of the heart into right and left sides is well under way, but there is as yet incomplete division of the bloodstream because there are still open communications from the right side to the left side in both atrium and ventricle. Further progress in the growth of the partitions, however, would leave the two sides of the heart completely separated. If this occurred now, the left side of the heart would become almost literally dry, for the sinus venosus, into which systemic, portal, and placental currents all enter, opens on the right of the interatrial septum, and not until much later do the lungs and their vessels develop sufficiently to return any considerable volume of blood to the left atrium. The partitions in the ventricle and in the atrioventricular canal progress rapidly to completion (Figs. 18-22 to 18-25), but an interesting series of events at the interatrial partition assures that an adequate supply of blood reaches the left atrium and thence the left ventricle.

Just when the septum primum is about to merge with the endocardial cushions of the atrioventricular canals, closing the interatrial foramen primum and isolating the left atrium, a new opening is established. This opening appears in the more cephalic part of the septum primum, first in the form of multiple small perforations resulting from the presence of dying cells in the area (Fig. 18-22). These rapidly expand and unite to form a single opening of considerable size, known as the *interatrial foramen secundum* (Fig. 18-23). By the time the interatrial foramen primum has been closed, the interatrial foramen secundum offers an effective alternative passageway. By this route enough of the blood entering the right atrium can make its way across the left atrium to equalize approximately the intake of the two sides of the heart.

Shortly after the formation of this secondary opening in the septum primum, a second interatrial partition makes its appearance just to the right of the first. Like the septum primum, the septum secundum is crescentic in form. However, the open end of its crescent is in a different place. Whereas the open part of the septum primum is directed toward the atrioventricular canal (Figs. 18-21B and 18-22), the open part of the septum secundum is directed toward the lower part of the sinus entrance (Fig. 18-23), which later becomes the opening of the inferior vena cava into the right atrium (Fig. 18-25). This is of vital functional significance, for it means that as the septum secundum grows, the opening remaining in it is carried out of line with the interatrial foramen secundum in the septum primum (Fig. 18-24). The opening in the septum

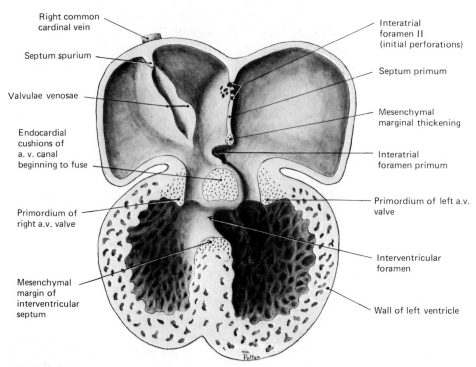

Right common cardinal vein

Septum spurium

Valvulae venosae

Endocardial cushions of a. v. canal beginning to fuse

Primordium of right a.v. valve

Mesenchymal margin of interventricular septum

Interatrial foramen II (initial perforations)

Septum primum

Mesenchymal marginal thickening

Interatrial foramen primum

Primordium of left a.v. valve

Interventricular foramen

Wall of left ventricle

FIGURE 18-22
Semischematic drawing showing the interior of the developing heart in the sixth week, at the stage when interatrial foramen primum is almost closed and interatrial foramen secundum is just being established by the appearance of multiple small perforations. (*From Patten, 1960, Am. J. Anat., vol. 107.*)

secundum, although it becomes smaller as development progresses, is not destined to be completely closed but will remain open as the *foramen ovale* (Fig. 18-25).

The flaplike remains of the septum primum overlying the persistent oval opening in the septum secundum constitute an efficient valvular mechanism between the two atria. When the atria are filling, some of the blood returning by way of the great veins can pass freely through the foramen ovale by merely pushing aside the flap of the septum primum. The inferior caval entrance lies adjacent to and is directed diagonally into the orifice of the foramen ovale (Figs. 18-24 and 18-25). Consequently, it is largely blood from the inferior vena cava which passes through the foramen ovale into the left atrium. When the atria start to contract, pressure of the blood within the left atrium forces the flap of the septum primum against the foramen ovale, effectively closing it against return flow into the right atrium. Without some such mechanism affording a supply of blood to its left side, the developing heart could not be partitioned in the midline, ready to assume its adult function of pumping two separate bloodstreams.

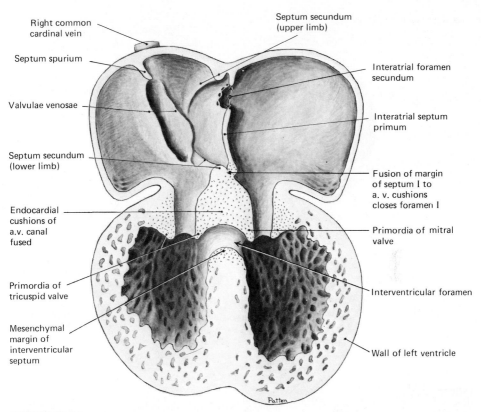

Right common cardinal vein

Septum spurium

Valvulae venosae

Septum secundum (lower limb)

Endocardial cushions of a.v. canal fused

Primordia of tricuspid valve

Mesenchymal margin of interventricular septum

Septum secundum (upper limb)

Interatrial foramen secundum

Interatrial septum primum

Fusion of margin of septum I to a. v. cushions closes foramen I

Primordia of mitral valve

Interventricular foramen

Wall of left ventricle

Patten

FIGURE 18-23
Semischematic drawing showing the interior of the developing heart, at the stage when interatrial foramen primum has been closed and interatrial foramen secundum has been well opened. Septum secundum is just being established to the right of septum primum. (*From Patten, 1960, Am. J. Anat., vol. 107.*)

While these changes are taking place in the main part of the heart, the truncus arteriosus is being divided into two separate channels. In a manner similar to that in the endocardial cushions, mesenchymal cells derived from local endocardium and also from the region of the aortic arches invade the cardiac jelly (Thompson and Fitzharris, 1979; Fig. 18-20). This process starts in the ventral aortic root between the fourth and sixth arches (Fig. 18-7). Continuing toward the ventricle, the division is effected by the formation of longitudinal ridges of readily molded young connective tissue of the same type as that making up the endocardial cushions of the atrioventricular canal. These ridges, called *truncoconal ridges*, bulge progressively farther into the lumen of the truncus arteriosus and finally meet to separate it into aortic and pulmonary channels (Fig. 18-26B). In the chick embryo at least, some of the cells that contribute to the distal part of the truncoconal ridges are derived from neural crest, possibly as a continuation of the neural crest contribution to the aortic

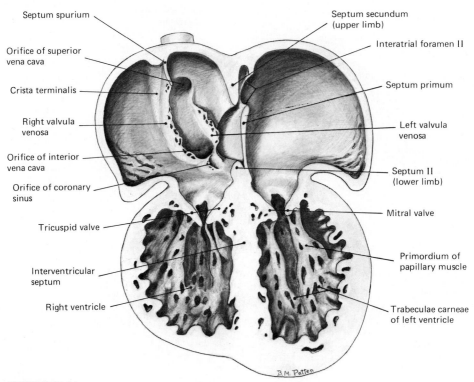

FIGURE 18-24
Schematic drawing to show the internal configuration of the heart in human embryos of the third month. At this stage resorption has begun to involve the valvulae venosae and septum spurium, as indicated by the many small perforations in their margins. Note the way the left venous valve is coming to lie against the septum secundum, with which it is already beginning to fuse. It usually leaves no recognizable traces in the adult, but occasionally delicate lacelike remains of it can be seen adherent to septum secundum and, more rarely, extending a way onto the valvula foraminis ovalis (cf. Fig. 18-25).

arch system (Kirby, 1987). The semilunar valves of the aorta and the main pulmonary trunk develop as local specializations of these truncus ridges. The truncoconal ridges follow a spiral course, and where they extend down into the ventricles they meet and become continuous with the margins of the interventricular septum. This reduces the size of the interventricular foramen but does not close it completely. Its final closure is brought about by a mass of endocardial cushion tissue from three sources: the interventricular septum, the endocardial cushions, and the truncoconal ridges. When the interventricular septum has been completed, the right ventricle leads into the pulmonary channel, and the left into the aorta. With this condition established, the heart is completely divided into right and left sides except for the interatrial valve at the foramen ovale, which must remain open throughout fetal life until, after birth, the lungs attain their full functional capacity and

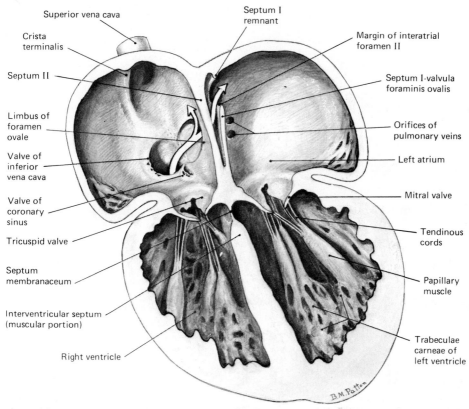

FIGURE 18-25
Schematic drawing to show the interrelations of septum primum and septum secundum during the latter part of fetal life. Note especially the way in which the lower part of the septum primum is situated so that it acts as a one-way valve at the oval foramen in the septum secundum. The split arrow indicates the way a considerable part of the blood from the inferior vena cava passes through the foramen ovale to the left atrium while the remainder eddies back into the right atrium to mingle with the blood being returned by way of the superior vena cava.

the full volume of the pulmonary stream passes through them to be returned to the left atrium.

Course and Balance of Blood Flow in the Fetal Heart

All the steps in the partitioning of the embryonic heart lead gradually toward the final adult condition in which the heart is completely divided into right and left sides. However, from the nature of its living conditions it is not possible for the fetus in utero to attain the adult type of circulation. The lungs, although fully formed and ready to function in the last part of fetal life, cannot actually begin their work until after birth. Yet in the first few minutes of its postnatal life a fetus must change from an existence submerged in the amniotic fluid to air breathing.

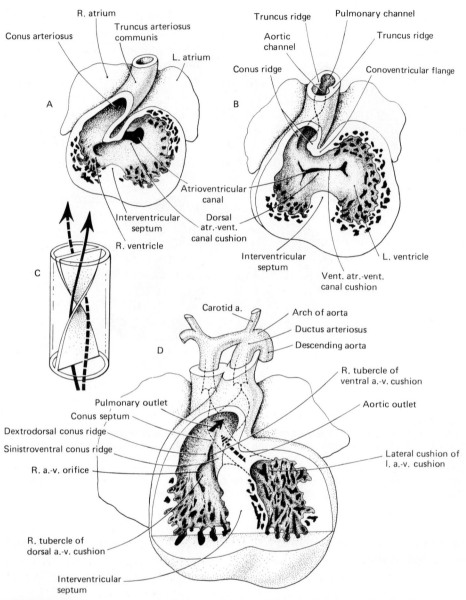

FIGURE 18-26
Semischematic dissections of the developing heart viewed in frontal aspect to show relations of importance in the subdivision of the truncus arteriosus and in establishing the aortic and pulmonary outlets. (C) Diagram illustrating the 180° spiraling of the septum that subdivides the truncus. (A, B, D after Kramer, 1942, Am. J. Anat. **71**:343.)

The rapid readjustment of the circulation at the time of birth is a dramatic and fascinating biological process. To understand how these changes are so promptly yet smoothly accomplished, it is necessary to have clearly in mind the manner in which the way for them has been prepared during intrauterine life. The very mechanisms which maintain cardiac balance during intrauterine life are perfectly adapted to keep the balance of the circulatory load on the new, postnatal basis. In the foregoing account of the development of the interatrial septal complex, emphasis was placed on the fact that at no time were the atria completely separated from each other. This permits the left atrium, throughout prenatal life, to receive a contribution of blood from the inferior vena cava and the right atrium by a transseptal flow, which compensates for the relatively small amount of blood entering the left atrium of the embryo directly by way of the pulmonary circuit and maintains an approximate balance of intake into the two sides of the heart.

The evidence from both anatomical and functional studies allows one to summarize the course followed by the blood in passing through the fetal heart as follows: The inferior caval entrance is so directed with reference to the foramen ovale that a considerable portion of its stream passes directly into the left atrium (Figs. 18-25 and 18-29). In fluctuating pressure conditions—for example, following uterine contractions which send a surge of placental blood through the umbilical vein—the placental flow may temporarily hold back any blood from entering the circuit by way of either the portal vein or the inferior caval tributaries (Fig. 18-29). In these conditions, the left atrium would be charged almost completely with fully oxygenated blood. Such conditions, however, would be only temporary and would be counterbalanced by periods when the portal and systemic veins poured enough blood into the common channels to load the heart for a time with mixed or depleted blood. The important thing physiologically is not the fluctuations but the maintenance of the average oxygen content of the blood at adequate levels. The interatrial communication in the heart of the fetus at term is considerably smaller than the inferior caval inlet. This would suggest that the portion of the inferior caval stream which could not pass through this opening into the left atrium would have to eddy back and mix with the rest of the blood in the right atrium. Angiocardiographic studies confirm this inference.

Compared with conditions in adult mammals, the mixing of oxygenated blood freshly returned from the placenta with depleted blood returning from a circuit of the body may seem inefficient, but this is a one-sided comparison. The fetus is an organism in transition. Starting with a simple ancestral plan of structure and living an aquatic life, it attains its full heritage slowly. It must be viewed in the light of both the primitive conditions from which it is emerging and the definitive conditions toward which it is progressing. Below the bird-mammal level, circulatory mechanisms with partially divided and undivided hearts and correspondingly unseparated bloodstreams meet all the needs of metabolism and growth. Maintenance of food, oxygen, and waste products at an average level which successfully supports life does not depend on "pure currents," although such separated currents undoubtedly make for higher efficiency in the rate of interchange of materials. From a comparative viewpoint, the fact that the mammalian fetus is supported by a mixed circulation seems quite natural.

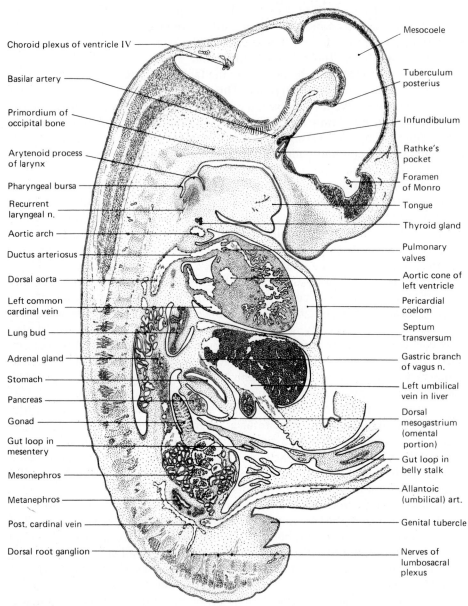

Choroid plexus of ventricle IV

Basilar artery

Primordium of occipital bone

Arytenoid process of larynx

Pharyngeal bursa

Recurrent laryngeal n.

Aortic arch

Ductus arteriosus

Dorsal aorta

Left common cardinal vein

Lung bud

Adrenal gland

Stomach

Pancreas

Gonad

Gut loop in mesentery

Mesonephros

Metanephros

Post. cardinal vein

Dorsal root ganglion

Mesocoele

Tuberculum posterius

Infundibulum

Rathke's pocket

Foramen of Monro

Tongue

Thyroid gland

Pulmonary valves

Aortic cone of left ventricle

Pericardial coelom

Septum transversum

Gastric branch of vagus n.

Left umbilical vein in liver

Dorsal mesogastrium (omental portion)

Gut loop in belly stalk

Allantoic (umbilical) art.

Genital tubercle

Nerves of lumbosacral plexus

FIGURE 18-27
Drawing (×10) of parasagittal section of 15-mm pig embryo. The section is in a plane slightly to the left of the midline and passes through the ductus arteriosus, lung bud, stomach, gonad, and metanephros.

From the standpoint of smooth postnatal circulatory readjustments, the larger the fetal pulmonary return becomes, the less will be the balancing transatrial flow and the less will be the change entailed by the assumption of

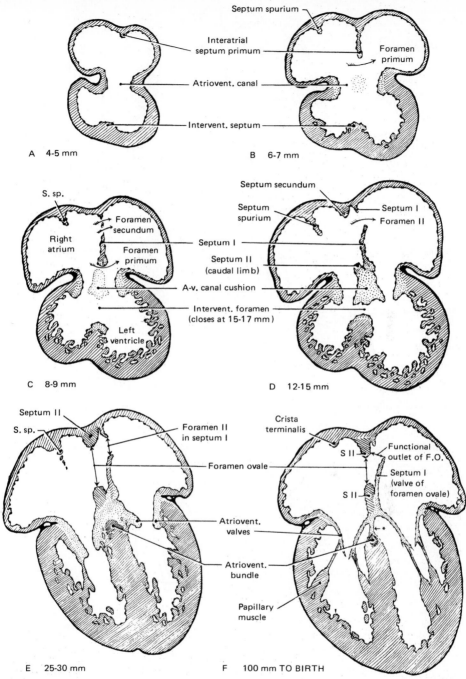

FIGURE 18-28
Sectional plans of embryonic heart in the frontal plane giving, specifically for the human embryo, a more precise picture of the rate of progress of partitioning than do the preceding schematic drawings. Stippled areas indicate endocardial cushion tissue, muscle is shown in diagonal hatching, and epicardium is shown in solid black. The lightly stippled areas in the atrioventricular canal in (B) and (C) indicate location of endocardial cushions of atrioventricular canal before they have grown sufficiently to fuse in the plane of the diagram. Abbreviations: *A-v.*, atrioventricular; *S. sp.*, septum spurium; asterisk and arrows in (F) indicate the membranous part of the interventricular septum.

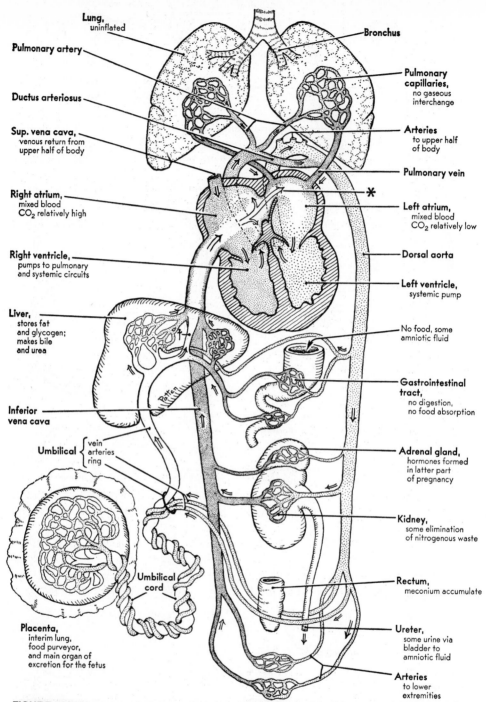

Lung,
uninflated

Bronchus

Pulmonary artery

Pulmonary
capillaries,
no gaseous
interchange

Ductus arteriosus

Sup. vena cava,
venous return from
upper half of body

Arteries
to upper half
of body

Pulmonary vein

Right atrium,
mixed blood
CO_2 relatively high

Left atrium,
mixed blood
CO_2 relatively low

Dorsal aorta

Right ventricle,
pumps to pulmonary
and systemic circuits

Left ventricle,
systemic pump

Liver,
stores fat
and glycogen;
makes bile
and urea

No food, some
amniotic fluid

Gastrointestinal
tract,
no digestion,
no food absorption

Inferior
vena cava

Umbilical { vein
arteries
ring

Adrenal gland,
hormones formed
in latter part
of pregnancy

Kidney,
some elimination
of nitrogenous waste

Umbilical
cord

Rectum,
meconium accumulate

Placenta,
interim lung,
food purveyor,
and main organ of
excretion for the fetus

Ureter,
some urine via
bladder to
amniotic fluid

Arteries
to lower
extremities

FIGURE 18-29

Plan of the fetal circulation at term. The ductus venosus in the liver is marked by †. The arrow in the heart marked by * indicates the passage of blood from the right atrium to the left as it occurs during atrial diastole. This flow pushes the valvula into the open positions represented here. When the atria contract, the valvula moves back against the septum, closing the foramen ovale against the return flow and thus forcing all the blood in the left atrium to enter the left ventricle. (See colored insert.) (*After Patten, in Fishbein, 1963, Birth Defects. Courtesy of the National Foundation and the J. B. Lippincott Company, Philadelphia.*)

lung breathing. Early in development, before the lungs have grown to any great extent, the pulmonary return is negligible and the flow from the right atrium through the interatrial ostium primum constitutes practically the entire intake of the left atrium. After the ostium primum is closed and while the lungs are only slightly developed, flow through the interatrial ostium secundum must still account for the major part of the blood entering the left atrium. During the latter part of fetal life the foramen ovale in the septum secundum becomes the gateway of the transseptal route. As the lungs grow and the pulmonary circulation increases in volume, a progressively smaller proportion of the left atrial intake comes by way of the foramen ovale and a progressively larger amount comes from the vessels of the growing lungs. But this circulation to the lungs, although increased compared with earlier stages, has been shown in angiocardiographic studies to be much less in volume than the caliber of the pulmonary vessels would lead one to expect. Until the lungs are inflated there are factors that restrict flow through the smaller pulmonary vessels. Therefore, even in the terminal months of pregnancy, a considerable right-left flow must still be maintained through the foramen ovale in order to keep the left atrial intake on a par with that of the right.

The balanced atrial intake thus maintained implies a balanced ventricular output. Although not in the heart itself, there is a mechanism in the closely associated great vessels which affords an adequate outlet from the right ventricle during the period when the pulmonary circuit is developing. When the pulmonary arteries are formed from the sixth pair of aortic arches, the right sixth arch soon loses its original connection with the dorsal aorta. On the left, however, a portion of the sixth arch persists as a large vessel connecting the pulmonary artery with the dorsal aorta (Figs. 18-27 and 18-29). This vessel, already familiar to us as the ductus arteriosus, remains open throughout fetal life and acts as a shunt, carrying over to the aorta whatever excess blood the pulmonary vessels at any particular phase of their development are not prepared to receive from the right ventricle. The ductus arteriosus functions as the "exercising channel" of the right ventricle because it makes it possible for the right ventricle to carry its full share of work throughout development and thus to be prepared for pumping all the blood through the lungs at the time of birth.

CHANGES IN THE CIRCULATION FOLLOWING BIRTH

The two most obvious changes that occur in the circulation at the time of birth are the abrupt cutting off of the placental bloodstream and the immediate assumption by the pulmonary circulation of the function of oxygenating the blood. One of the impressive things in embryology is the perfect preparedness for this event, which has been built into the very architecture of the circulatory system during its development. The shunt at the ductus arteriosus, which has been one of the factors in balancing ventricular loads throughout intrauterine development, and the valvular mechanism at the foramen ovale, which has at the same time been balancing atrial intakes, are perfectly adapted to effect the postnatal rebalancing of the circulation. The lessening of peripheral resistance in the small vessels of the lungs and the closure of the ductus arteriosus are the primary events, and the closure of the foramen ovale follows as a logical sequel (Fig. 18-1).

It has long been known that the lumen of the ductus arteriosus is gradually occluded postnatally by tissue overgrowth. Its earliest phases begin to be recognizable in the fetus as the time of birth approaches, and postnatally the process continues at an accelerated rate to terminate in complete anatomical occlusion of the lumen of the ductus about 6 to 8 weeks after birth. Barcroft, Barclay, and Barron have conducted an extensive series of experiments on sheep delivered by cesarean section which indicate that the ductus arteriosus closes functionally almost immediately. Following birth there appears to be a contraction of the circularly disposed smooth muscle in the wall of the ductus which promptly reduces the flow of blood through it. This reduction in the shunt from the pulmonary circuit to the aorta, acting together with the lowering of the resistance in the vascular bed of the lungs that accompanies their newly assumed respiratory activity, aids in raising the pulmonary circulation promptly to a full functional level. At the same time functional closure of the ductus paves the way for the ultimate anatomical obliteration of its lumen.

The effects of increased pulmonary circulation with the concomitant increase in the direct intake of the left atrium are manifested secondarily at the foramen ovale. Following birth, as the pulmonary return increases, compensatory blood flow from the right atrium to the left decreases correspondingly and shortly ceases altogether. In other words, when equalization of atrial intakes has occurred, the compensating one-way valve at the foramen ovale falls into disuse and the foramen may be regarded as functionally closed. Anatomical obliteration of the foramen ovale follows leisurely in the wake of its functional abandonment. There is a considerable interval following birth before the septum primum fuses with the septum secundum to seal the foramen ovale. This delay is, however, of no import because as long as the pulmonary circuit is normal and pressure in the left atrium equals or exceeds that in the right, the orifice between them is functionally inoperative.

With birth and the interruption of the placental circuit there follows the gradual fibrous involution of the umbilical vein and the umbilical arteries. The flow of blood in these vessels, of course, ceases immediately with the severing of the umbilical cord, but obliteration of the lumen is likely to take from 3 to 5 weeks, and isolated portions of these vessels may retain a vestigial lumen for much longer. Ultimately these vessels are reduced to fibrous cords. The old course of the umbilical vein is represented in the adult by the *round ligament* of the liver, extending from the umbilicus through the falciform ligament, and by the *ligamentum venosum* within the substance of the liver. The proximal portions of the umbilical arteries are retained in reduced relative size as the hypogastric or internal iliac arteries. The fibrous cords extending from these arteries on either side of the urachus toward the umbilicus represent the remains of the more distal portions of the old umbilical arteries. They are known in the adult as the *lateral umbilical ligaments*.

Much remains to be learned regarding the more precise physiology of the fetal circulation and the interaction of various factors during the transition from intrauterine to postnatal conditions. Nevertheless, with our present knowledge it is quite apparent that the changes in the circulation which occur following birth involve no revolutionary disturbances of the load carried by different parts of the heart. The fact that the pulmonary vessels are already so well developed before birth means that the changes which must occur following birth have been thoroughly prepared for, and the compensatory mechanisms at the foramen ovale and the ductus arteriosus which have been functioning all during fetal life are entirely competent to effect the final postnatal rebalancing of the circulation with a minimum of functional disturbance. It is still true that as individuals we crowd into a few crucial moments the change from water living to air living that in phylogeny must have been spread over eons of transitional amphibious existence. But as we learn more about this change in manner of living, it becomes apparent that we should marvel more at the completeness and perfection of the preparations for its smooth accomplishment and dwell less on the old theme of the revolutionary character of the changes involved.

DEVELOPMENT OF CHICK EMBRYOS FROM 18 HOURS TO 4 DAYS OF INCUBATION

STUDY OF SERIAL SECTIONS

To study embryos with any degree of thoroughness one must cut them into sections which are sufficiently thin to allow effective use of the microscope to ascertain cellular organization and detailed structural relationships. In the preparation of such material the entire embryo is cut into sections which are mounted on slides in the order in which they were cut. A sectional view of any region of the embryo is then available for study. In studying a section from a series, it is first necessary to determine the location in which it was cut through the embryo. The plane of the section under consideration, as well as the region of the embryo through which it passes, should be ascertained by comparing it with an entire embryo of the same age as that from which the section was cut. Only when the location of a section is known precisely can the structures appearing in it be correlated with the organization of the embryo as a whole. Probably nothing in the study of embryology causes students more difficulty than neglecting to locate sections accurately, with the consequent failure to appreciate the relationships of the structures seen in them. The importance of fitting the structures shown by sections properly into the general scheme of organization as it appears in whole mounts (Figs. A-1 and A-2) cannot be overemphasized.

TERMS OF LOCATION

In embryology it is necessary to designate the location of structures and the direction of growth processes by terms which are referable to the body of the embryo regardless of the position it occupies. Our ordinary terms of location refer to the direction of the action of gravity, such as *above*, *over*, and *under*.

629

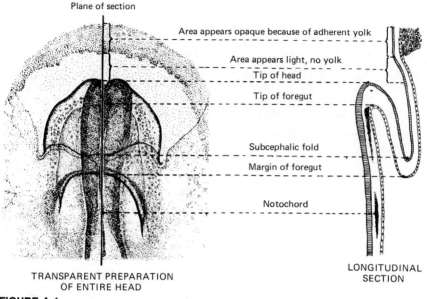

Plane of section

Area appears opaque because of adherent yolk

Area appears light, no yolk

Tip of head

Tip of foregut

Subcephalic fold

Margin of foregut

Notochord

TRANSPARENT PREPARATION
OF ENTIRE HEAD

LONGITUDINAL
SECTION

FIGURE A-1
Relation of longitudinal section of the embryonic head to the picture presented
by a head of the same age mounted entire as a transparent preparation.

These terms are not sufficiently accurate in this type of work because the embryo itself may lie in a great variety of positions. The correct adjectives of position are *dorsal*, pertaining to the back; *ventral*, the belly; *cephalic*, the head; *caudal*, the tail; *mesial*, the middle part; and *lateral*, the side of the embryonic body. In dealing with relations to the head, the adjective *cephalic* is inadequate and the term *rostral* (Latin, *beak of a bird*, *bow of a ship*) is used to designate the extreme anterior portion of the head or the relative location of intracephalic structures such as the various parts of the brain. Adverbs of fixed position are made in the usual way by adding *-ly* to the root of the adjective.

In addition to adverbs of position, corresponding adverbs of motion or direction are formed by adding the suffix *-ad* to the root of the adjective, as in *dorsad*, meaning *toward the back*; *cephalad*, meaning *toward the head*, etc. These adverbs should be applied only to the progress of processes or to the extension of structures toward the part indicated by their root. Thus, for example, we would say that the developing eye of an embryo was located in the later*al* wall of the forebrain or that the forebrain was in the cephal*ic* part of the embryo, but if we wished to express the idea that as the eye increases in size it moves farther to the side, we would say it grew later*ad*.

FROM THE PRIMITIVE-STREAK STAGE TO THE APPEARANCE OF SOMITES

Because the earliest stages of embryonic development of the chick occur within the reproductive tract of the hen, the laboratory study of chick embryos

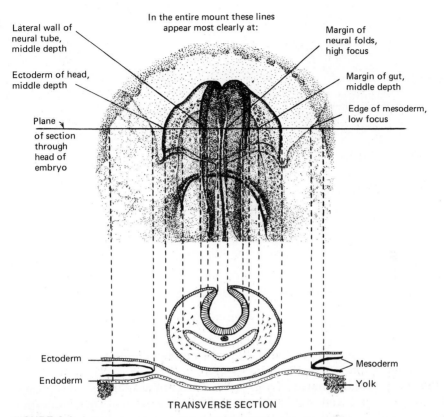

Lateral wall of
neural tube,
middle depth

In the entire mount these lines
appear most clearly at:

Margin of
neural folds,
high focus

Ectoderm of head,
middle depth

Margin of gut,
middle depth

Plane

Edge of mesoderm,
low focus

of section
through
head of
embryo

Ectoderm

Mesoderm

Endoderm

Yolk

TRANSVERSE SECTION

FIGURE A-2
Relation of transverse section of the embryonic head to the picture presented by an
entire head of the same age viewed as a transparent preparation.

commonly omits the early stages of cleavage and blastulation. Gastrulation and
the formation of the primitive streak were described in Chap. 5 (Figs. 5-12 to
5-16) and will not be repeated here. By 18 hours' incubation the primitive streak
is well established, and cells which have been invaginated by way of Hensen's
node push cephalad, initiating the formation of the notochord (Fig. 5-18B). In
chick embryos which have been incubated about 18 hours the notochord has
become markedly elongated to form a well-defined midline structure (Fig. A-3).
In older embryological treatises the young notochord was frequently called the
"head process," and chick embryos of about 18 hours' incubation have often
been spoken of as being in the "head-process stage."

Sections of embryos in the head-process stage (Fig. A-4) give much more
information about the structure and relations of the notochord than can be
gleaned from the study of whole mounts. Near its origin at Hensen's node, the
notochord is made up of diffusely arranged cells that merge caudally with the
recently turned under cells of the lateral portions of the mesoderm and the

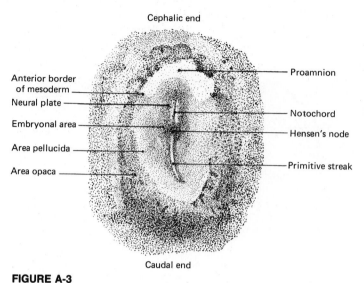

Cephalic end

Anterior border of mesoderm

Proamnion

Neural plate

Notochord

Embryonal area

Hensen's node

Area pellucida

Area opaca

Primitive streak

Caudal end

FIGURE A-3
Dorsal view (×14) of entire chick embryo of 18 hours' incubation.

endoderm. Its cephalic tip gradually develops a more definitely circumscribed, rodlike configuration.

In the sections diagrammed in Fig. A-4 a conventional scheme of representation has been employed to indicate each of the germ layers. The ectoderm is vertically hatched, the cells of the mesoderm are represented by heavy angular dots when they are isolated or by solid black lines when they lie arranged in the form of compact layers, and the endoderm is represented by stippling backed by a single line. A similar conventional representation of the different germ layers is observed in all diagrams of chick sections to facilitate following the way in which the organ systems of the embryo are constructed from the germ layers. The plane in which each of the sections diagrammed passes through the embryo is indicated by a line drawn on a small outline sketch of an embryo at a corresponding stage.

The Primitive Streak

Cross sections passing just caudal to Hensen's node are the most suitable for studying the relations of the primitive streak and the three germ layers (Fig. A-4C). In this region (Fig. A-4D) one can see the continuity between the cells moving into the primitive groove and the cells of the flattened early germ layers. For a discussion of cellular movements at the primitive streak and their role in formation of the mesoderm, see Figs. 5-17 and 5-18 and the associated text. In regions lateral to the primitive streak (Fig. A-4E) the three germ layers have a linear appearance in cross section. The space between any two germ

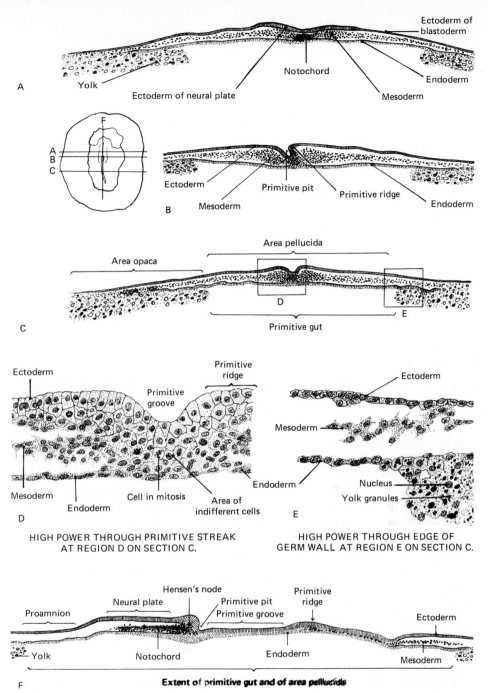

FIGURE A-4

Sections of 18-hour chick. The location of each section is indicated by a line drawn on a small outline sketch of an entire embryo of corresponding age. The letters affixed to the lines on the sketch indicating the location of the sections correspond with the letters designating the section diagrams. Each germ layer is represented by a different conventional scheme: ectoderm by vertical hatching; endoderm by fine stippling backed by a single line; the cells of the mesoderm, which at this stage do not form a coherent layer, by heavy angular dots. (A) Diagram of transverse section through neural plate and notochord. (B) Diagram of transverse section through primitive pit. (C) Diagram of transverse section through primitive streak. (D) Drawing showing cellular structure in primitive-streak region. (E) Drawing showing cellular structure at inner margin of germ wall. (F) Diagram of median longitudinal section passing through notochord and primitive streak.

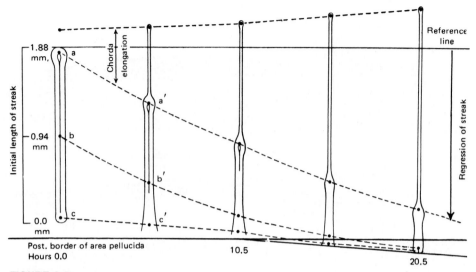

FIGURE A-5
Graphic summary of the regression of the primitive streak and the growth of the notochord. The scale is indicated by the length of the primitive streak given in millimeters at the left. The time is given in hours beyond the start of the experiments. (*From Spratt, 1947.*)

layers is an artifact that is due partly to shrinkage of the tissues and partly to dissolving of the delicate extracellular matrix during preparation of the slide.

The primitive streak soon begins to regress toward the caudal end of the embryo. As this occurs, cells still move into the retreating groove and Hensen's node, leaving in their wake the ever-elongating notochord (Fig. A-5). Emigration of cells from the caudal half of the streak contributes to the lateral and caudal expansion of both the intra- and extraembryonic tissues and also accounts for much of the shortening of the streak.

As the primitive streak thus undergoes a sort of dismemberment from its more caudal portion, Hensen's node moves farther and farther back toward the area opaca. By the end of the second day of incubation little more than the node and a very short portion of the streak persist. At this stage what remains of the once prominent node and the primitive streak is usually given a new name, the *end bud* or *tail bud* (Fig. A-31). Cells of the tail bud ultimately contribute to the neural tube, somites, mesenchyme, and caudal arteries of the tail but do not contribute to the notochord, ectoderm, or gut (Schoenwolf, 1977).

Growth of the Endoderm and Establishment of the Primitive Gut

Sections of embryos which have been incubated from 18 to 20 hours show how the endoderm has become organized into a coherent layer of cells merging peripherally with the yolk and overlapping it to a certain extent (Fig. A-4). This marginal area where the expanding germ layers merge with the underlying yolk

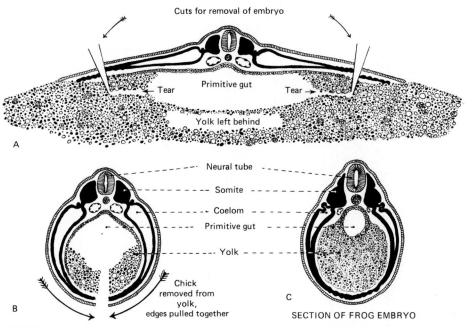

FIGURE A-6
(A) is a diagram showing how the usual method of removing chick embryos from the yolk in order to prepare them for microscopic study makes the sections appear as if the primitive gut had no ventral boundary. (B) and (C) show how removing a chick from the yolk and pulling its edges together ventrally facilitates comparisons with forms which do not develop with their growing bodies spread out on a large mass of yolk.

is known as the *germ wall* (Fig. A-4E). The cavity between the yolk and the endoderm is now termed the *primitive gut*. When embryos are removed for sectioning, the yolk floor of the primitive gut, not being adherent to the blastoderm, is left behind (Fig. A-6A). In contrast, the periphery of the blastoderm lies closely applied to the yolk. Some yolk adheres to this part of the blastoderm when it is removed (Fig. A-6A). Its presence clearly indicates why the *area opaca* appears less translucent than the *area pellucida* in surface views of entire embryos (Fig. A-3). One can obtain a better sense of perspective on the relations of the flat avian embryo by imagining how it would appear if the embryo were folded ventrally into a cylinder (Fig. A-6).

By about the twentieth hour of incubation we can see indications of a local differentiation of the region of the primitive gut that underlies the cephalic part of the embryo. The first part of the gut to acquire a floor, this endodermal pocket is called the *foregut* (Fig. A-7).

Growth and Early Differentiation of the Mesoderm

The mesoderm which arises from either side of the primitive streak spreads rapidly laterad, and at the same time each lateral wing of the mesoderm swings

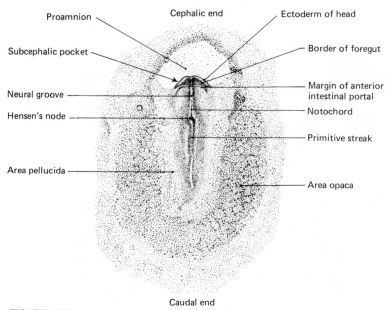

Proamnion

Subcephalic pocket

Neural groove

Hensen's node

Area pellucida

Cephalic end

Ectoderm of head

Border of foregut

Margin of anterior intestinal portal

Notochord

Primitive streak

Area opaca

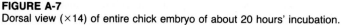

Caudal end

FIGURE A-7
Dorsal view (×14) of entire chick embryo of about 20 hours' incubation.

cephalad (Fig. A-8A to A-8C). The manner in which the mesoderm spreads out leaves a mesoderm-free area in the cephalic portion of the blastoderm. This region, known as the *proamnion*, is clearly recognizable in entire embryos by reason of its lesser density (Figs. A-7 and A-9). Despite the implications of its name, this region is not the precursor of the amnion.

The laterally spreading mesoderm tends to pull away from the notochord, leaving the notochord sharply outlined in the midline in a territory that for a time is relatively free of other mesodermal elements (Figs. A-8C and A-13C). Sections passing through the primitive streak of embryos of about 18 hours' incubation age show loosely aggregated masses of mesoderm extending to either side between the ectoderm and endoderm (Fig. A-4C to A-4E). Immediately adjacent to the midline the mesoderm is markedly thicker than it is farther laterad (Fig. A-4A to A-4C). These thickened zones of the mesoderm are the primordia of the dorsal mesodermal plates. Because of the way in which they are later divided into metamerically arranged cell masses or somites, they are frequently designated as the *segmental zones of the mesoderm*. The segmental zones are, in early stages, most clearly marked somewhat cephalic to Hensen's node, where the first somites will appear. In the embryo represented in Fig. A-9 there is just a suggestion of the lines of division separating the first somites. As the segmental zones of the mesoderm are followed from this region caudad on either side of the primitive streak, they gradually become less and less definite. Traced laterally, the mesoderm rapidly becomes less

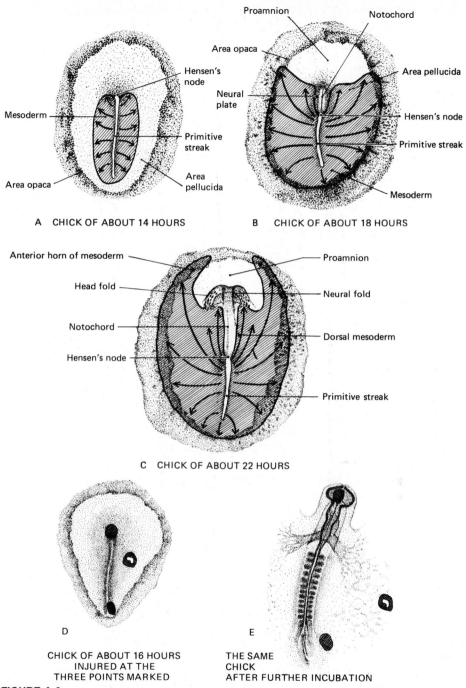

A CHICK OF ABOUT 14 HOURS

B CHICK OF ABOUT 18 HOURS

C CHICK OF ABOUT 22 HOURS

D CHICK OF ABOUT 16 HOURS
INJURED AT THE
THREE POINTS MARKED

E THE SAME
CHICK
AFTER FURTHER INCUBATION

FIGURE A-8
Diagrams outlining direction of growth from the primitive streak (A–C) show the progress of the mesoderm during the latter part of the first day of incubation. The areas into which the mesoderm has grown are indicated by diagonal hatching. (D) and (E) show the direction of growth as demonstrated by experimental methods. (*After Kopsch.*) (D) shows the location at which three injuries were made close to the primitive streak of a 16-hour embryo. (E) shows the position to which the injured areas are carried by growth of the same embryo subsequent to the operation.

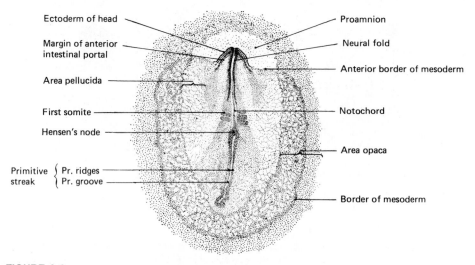

FIGURE A-9
Dorsal view (×16) of an entire chick embryo at the beginning of somite formation (about 22 to 23 hours of incubation).

thick. This thinner region along the sides of the embryo contains the primordial tissue from which the *intermediate zones* and *lateral plates of the mesoderm* are later differentiated.

The sheetlike layers of mesoderm that are characteristic of the midbody region do not extend to the cephalic part of the embryo. The mesoderm of the head is largely derived from cells which become detached from the more definitely organized layers of mesoderm lying farther caudally in the body and then migrate into the cephalic region.

Formation of the Neural Plate and Cephalic Region

In surface views of entire chicks incubated for about 18 hours (Fig. A-3) areas of greater density may be made out on either side of the notochord. These areas extend somewhat rostral to the cephalic end of the notochord, where they appear to blend with each other in the midline. Sections of this region (Fig. A-4A) show that the greater density seen in whole mounts is due to a thickening of the ectoderm, known as the *neural* (medullary) *plate*. In embryos of about 22 hours (Fig. A-9) the neural plate becomes longitudinally folded to establish a trough known as the *neural groove* (Figs. A-13 and A-14). The formation of the neural plate and its subsequent folding to form the neural groove are the first indications of the establishment of the central nervous system. The cephalic region of the embryonal area is thickened and protrudes above the general surface of the surrounding blastoderm as a rounded elevation. This prominence marks the region in which the head of the embryo will develop (Figs. A-7 and A-9). The crescentic fold which bounds it is termed the *head fold* and forms the first definite boundary of the body of the embryo. Throughout the course of development we shall find the cephalic region farther advanced in differentiation than are other parts of the body.

STRUCTURE OF 24-HOUR CHICKS

External Features

The cephalic region of the 24-hour chick undergoes rapid growth and extends anteriorly, overhanging the proamnion (Figs. A-10 and A-13E). The space between the head and the blastoderm is called the *subcephalic space* (Fig. A-13E). In the midline, the notochord, which is visible through the overlying ectoderm, can be readily traced into the cephalic region, where it will terminate somewhat short of the rostral end of the head (Fig. A-11).

The folding of the neural plate is much more clearly marked than it is at 22 hours (Figs. A-11 and A-12). Since the neural folds first appear in the cephalic region, the cephalic end of the neural groove is deeper and the neural folds are more prominent than they are caudally, where the neural folds have not yet formed from the neural plate (Fig. A-14).

Establishment of the Foregut

In the outgrowth of the head, the endoderm as well as the ectoderm has been involved. As a result the endoderm forms a pocket within the ectoderm, much

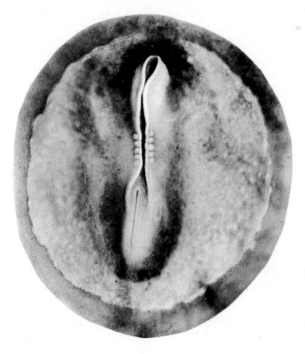

FIGURE A-10
Chick embryo of 25 to 26 hours photographed by reflected light to show its external configuration. Compare with Figs. A-11 and A-12.

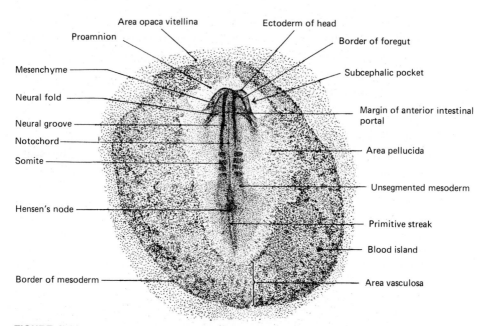

FIGURE A-11
Dorsal view (×16) of entire chick embryo having four pairs of somites (about 24 hours' incubation). Compare this figure of an embryo, which has been stained and cleared and then drawn by transmitted light, with Fig. A-10, which shows an embryo of about the same age photographed by reflected light.

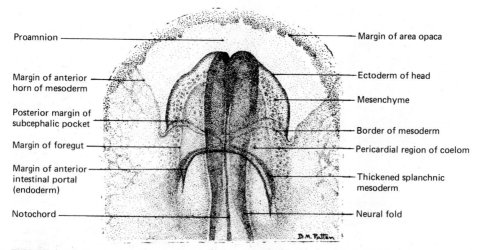

FIGURE A-12
Ventral view (×40) of cephalic region of chick embryo having five pairs of somites (about 25 to 26 hours of incubation).

like a small glove finger within a larger one. This endodermal pocket, or foregut, is the first part of the digestive tract to acquire a definite cellular floor. That part of the gut caudal to the foregut, where the yolk still constitutes the only floor, is termed the *midgut*. The opening from the midgut into the foregut is called the *anterior intestinal portal* (Figs. A-13E and A-14).

The topography of the foregut region at this stage can be made out very well by studying the ventral aspect of entire embryos. The margin of the anterior intestinal portal appears as a well-defined crescentic line. Comparison of Fig. A-12 with a sagittal section as diagrammed in Fig. A-13E will aid in making clear the relationships of the foregut to the head.

Regional Divisions of the Mesoderm

The first conspicuous metamerically arranged structures to appear in the chick are the *mesodermal somites*. The somites arise by division of the mesoderm of the dorsal or segmental zone to form blocklike cell masses. In the embryo shown in Fig. A-11, three pairs of somites are completely delimited and one can make out a fourth pair, which is not yet completely cut off from the dorsal mesoderm caudal to it.

Cross sections passing through the midbody region show the formation of the somites and the beginning of other changes in the mesoderm (Fig. A-13C). If one follows the mesoderm from the midline toward either side, one can make out (1) the *dorsal paraxial mesoderm*, which at this level has been organized into somites, (2) the *intermediate mesoderm*, a thin and narrow plate of cells connecting the dorsal and lateral mesoderm, and (3) the *lateral mesoderm*, which is distinguished from the intermediate mesoderm by being split into two sheetlike layers with a space between them. In 24-hour chick embryos, the intermediate mesoderm never becomes segmentally divided as does the dorsal paraxial mesoderm. The fact that it is potentially segmental in character is indicated, however, by the way in which it later gives rise to segmentally arranged nephric tubules.

The lateral mesoderm, like the intermediate mesoderm, shows no segmental division. In 24-hour embryos (Fig. A-13C) it is clearly differentiated from the intermediate mesoderm by being split horizontally into two layers with a space between them. The layer of lateral mesoderm lying next to the ectoderm is termed the *somatic mesoderm*, the layer next to the endoderm is termed the *splanchnic mesoderm*, and the cavity between the somatic and splanchnic mesoderms is the *coelom*. Because in development the somatic mesoderm and ectoderm are closely associated and undergo many foldings in common, it is convenient to designate the two layers together by the single term *somatopleure*. Similarly, the splanchnic mesoderm and the endoderm together are called the *splanchnopleure*.

The coelom, like the cell layers of the blastoderm, extends over the yolk peripherally beyond the embryonal area (Fig. A-13B and C). Later in development, foldings mark off the embryonic from the extraembryonic portion of the

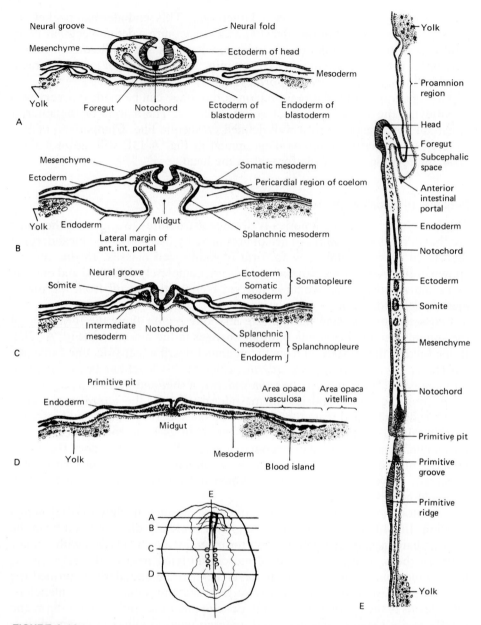

FIGURE A-13

Diagrams of sections of 24-hour chick. The sections are located on an outline sketch of the entire embryo. The conventional representation of the germ layers is the same as that employed in Fig. A-4 except that here, where its cells have become aggregated to form definite layers, the mesoderm is represented by solid black lines.

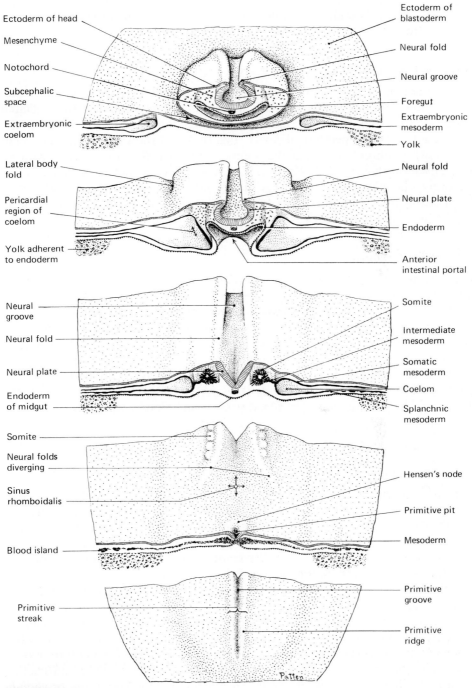

FIGURE A-14
Stereogram of 24-hour chick. Compare with Fig. A-13. (*This three-dimensional approach was suggested by Huettner's drawings in his* Fundamentals of Comparative Embryology of the Vertebrates, *Macmillan, New York.*)

germ layers. This same folding process divides the coelom into intraembryonic and extraembryonic regions.

The Pericardial Region

In the region of the anterior intestinal portal the coelomic chambers on either side show very marked local enlargement. Later in development these dilated regions are extended mesiad and break through into each other ventral to the foregut to form the pericardial cavity. Figure A-13B shows the great dilation of the coelom on either side of the anterior intestinal portal compared with its condition farther caudad (Fig. A-13C). Where the splanchnic mesoderm lies closely applied to the endoderm at the lateral margins of the portal it is noticeably thickened. It is from these areas of thickened splanchnic mesoderm that the paired primordia of the heart will later arise.

The Area Vasculosa

In a 24-hour chick the boundary between the area opaca and the area pellucida has the same appearance that it has in chicks of 18 to 20 hours. There is, however, a very marked difference between the proximal portion of the area opaca adjacent to the area pellucida and the more distal portions of the area opaca. The proximal region is much darker and has a somewhat mottled appearance (Fig. A-11) as a result of its invasion by mesoderm (Fig. A-13D). The mottled appearance of this region is due to the aggregation of mesoderm into cell clusters, or *blood islands* (Figs. A-11 and A-13D). These mark the initial step in the formation of blood vessels and blood corpuscles.

CHANGES BETWEEN 24 AND 33 HOURS OF INCUBATION

Early Formation of the Brain

In embryos of 27 to 28 hours, the head has elongated rapidly and projects free from the blastoderm for a considerable distance, with a corresponding increase in the depth of the subcephalic pocket and the length of the foregut. The neural folds in the cephalic region meet in the middorsal line, and their edges become fused. The process of fusion of the neural tube and its separation from the surface ectoderm can be followed in Fig. A-21A to A-21E.

By 27 hours of incubation the cephalic part of the neural tube, which will form the brain, is markedly enlarged compared with the more caudal part, which gives rise to the spinal cord. Three *primary brain vesicles* can be distinguished in the enlarged cephalic region of the neural tube (Fig. A-15). The most rostral vesicle is the *forebrain*, or *prosencephalon*. Marked off from it by a constriction is the *midbrain*, or *mesencephalon*. A very slight constriction marks the boundary between the mesecephalon and the *hindbrain*, or *rhombencephalon*, which is continuous with the future spinal cord without any definite point of transition. In somewhat older embryos (Fig. A-17) the lateral

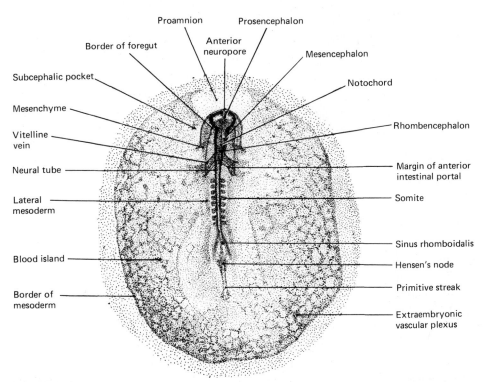

FIGURE A-15
Dorsal view (×14) of the entire chick embryo having 8 pairs of somites (about 27 to 28 hours' incubation).

walls of the prosencephalon become outpocketed to form a pair of rounded dilations known as the *primary optic vesicles*.

Ventral to the brain the notochord extends rostrally as far as a depression in the floor of the prosencephalon known as the *infundibulum* (Fig. A-16). As the result of an interaction with the ectoderm of the future oral cavity, the infundibulum develops into the neural hypophysis.

The closure of the neural folds takes place first near the rostral end of the neural groove and then progresses both cephalad and caudad. Closure of the extreme rostral end of the brain is delayed, and the persisting communication between the prosencephalon and the outside is called the *anterior neuropore* (Figs. A-15 and A-16). By 33 hours it is almost closed (Fig. A-18), and later it becomes entirely closed, leaving only a scarlike fissure in the anterior wall of the prosencephalon.

At the other end of the body, the neural tube has closed caudally as far as somite formation has progressed. Caudal to the most posterior somites, the neural groove is still open and the neural folds diverge to either side of Hensen's node (Fig. A-15). The lateral boundaries of the unclosed region at the posterior end of the neural tube delimit an open area known as the *sinus*

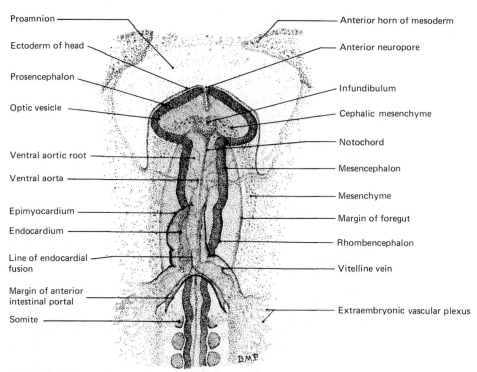

Proamnion

Ectoderm of head

Prosencephalon

Optic vesicle

Ventral aortic root

Ventral aorta

Epimyocardium

Endocardium

Line of endocardial fusion

Margin of anterior intestinal portal

Somite

Anterior horn of mesoderm

Anterior neuropore

Infundibulum

Cephalic mesenchyme

Notochord

Mesencephalon

Mesenchyme

Margin of foregut

Rhombencephalon

Vitelline vein

Extraembryonic vascular plexus

FIGURE A-16
Ventral view (×47) of cephalic and cardiac region of chick embryo of 9 somites (about 29 to 30 hours' incubation).

rhomboidalis (Fig. A-18). Hensen's node and the primitive pit lie in the floor of this still unclosed region of the neural groove; subsequently they are enclosed within it when the neural folds here finally fuse to complete the neural tube. This process in the chick is homologous with the enclosure of the blastopore by the neural folds in lower vertebrates (Fig. 5-4E). In forms where the blastopore does not become closed until after it is surrounded by the neural folds, it for a time constitutes an opening from the neural canal into the primitive gut known as the *neurenteric canal*. In the chick the fact that there is never an open blastopore precludes the establishment of an open neurenteric canal, but the primitive pit represents its homologue.

The Foregut and Heart

The anterior intestinal portal occupies progressively more caudal positions (Fig. A-22) as the margins on either side of the portal are constantly converging toward the midline. Such merging, along with elongation of the structures cephalic to the anterior intestinal portal, results in lengthening of the foregut. The resulting enlargement of coelomic space between the subcephalic pocket and the margin of the anterior intestinal portal is occupied by the developing heart.

In dorsal view the heart is largely concealed by the overlying rhomben-cephalon (Fig. A-15), but it may readily be made out by viewing the embryo from the ventral surface (Fig. A-16). At this stage the heart is a nearly straight tubular structure lying in the midline ventral to the foregut. Its dilated midregion has noticeably thickened walls. Cephalically the heart is continuous with a large median vessel, the *ventral aorta*; caudally it is continuous with the paired *vitelline veins*.

Organization in the Area Vasculosa

The mottled appearance of the area vasculosa gives way to a netted one as the blood islands are replaced by the *vitelline vascular plexus* on the yolk (Fig. A-18). The peripheral margin of the area vasculosa is marked by a dark band, the precursor of the *sinus terminalis* (*marginal sinus*). Through these vessels yolk is absorbed and transferred as food to the embryo. This vascular plexus becomes connected with the *vitelline (omphalomesenteric) veins*, the main venous return channels to the heart. Somewhat later the *vitelline arteries* extend peripherally and become connected with the yolk-sac plexus (cf. Figs. A-26, A-27 and A-28).

STRUCTURE OF CHICKS BETWEEN 33 AND 38 HOURS OF INCUBATION

Chicks which have been incubated from 33 to 38 hours are in a favorable stage to show some of the fundamental steps in the formation of the central nervous system and the circulatory system. In this section, therefore, attention has been concentrated on these two systems.

The Brain and Its Neuromeric Structure

The metameric arrangement of structures, which is so striking a feature in the body organization of all vertebrates, is masked in the head region of the adult by superimposed specializations. In the brain of young vertebrate embryos, however, the metamerism is still indicated. Dissections of the neural plate of chicks at the end of the first day of incubation show a series of 11 enlargements marked off from each other by constrictions (Fig. A-17).

With the closure of the neural tube and the establishment of the three primary brain vesicles, we can begin to trace the fate of the various neuromeric enlargements in the formation of the brain regions. The three most rostral neuromeres form the prosencephalon, neuromeres IV and V are incorporated into the mesencephalon, and neuromeres VI to XI become part of the rhombencephalon (Fig. A-17B). Rostrally, all but two of the interneuromeric constrictions soon disappear, namely, the one between the prosencephalon and the mesencephalon and the one between the mesencephalon and the rhomben-cephalon. The rhombencephalic neuromeres, however, remain clearly marked for a considerable period.

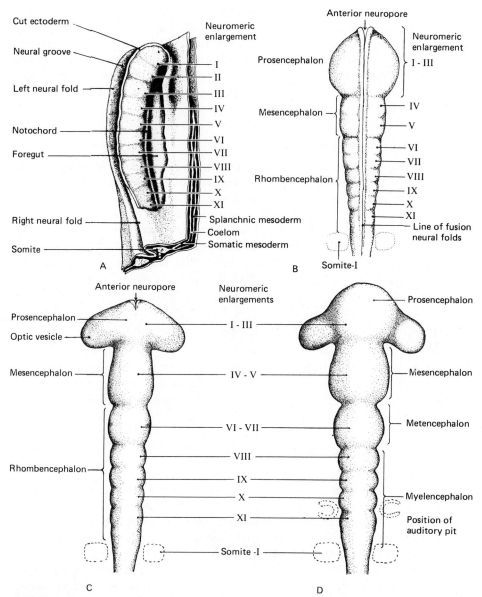

FIGURE A-17
Diagrams to show the neuromeric enlargements in the brain region of the neural tube. (A) Lateral view of neural plate from dissection of chick of 4 somites (24 hours). (B) Dorsal view of brain dissected out of 7-somite (26- to 27-hour) embryo. (C) Dorsal view of brain from 10-somite (30-hour) embryo. (D) Dorsal view of brain from 14-somite (36-hour) embryo. (*Based on figures of Hill.*)

By about 33 hours of incubation the *optic vesicles* are established as paired lateral outgrowths of the prosencephalon. They soon extend to occupy the full width of the head (Figs. A-17C and A-18). The distal portion of each of the

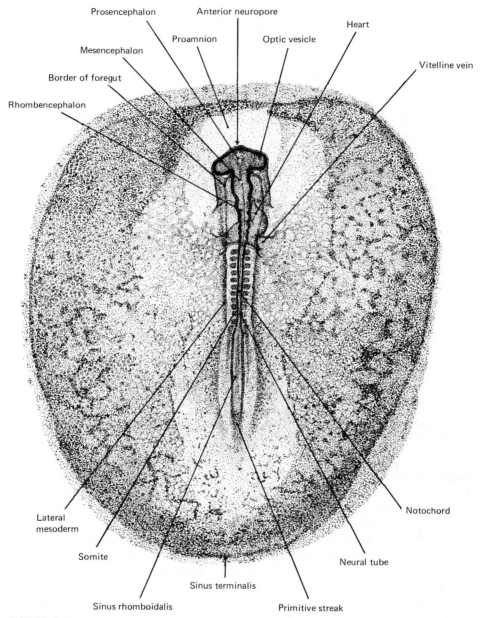

FIGURE A-18
Dorsal view (×17) of an entire chick embryo of 12 somites (about 33 hours' incubation).

vesicles thus comes to lie closely approximated to the superficial ectoderm, a
relationship of importance in the later development of the lens. At first the
cavities of the optic vesicles are broadly confluent with the cavity of the

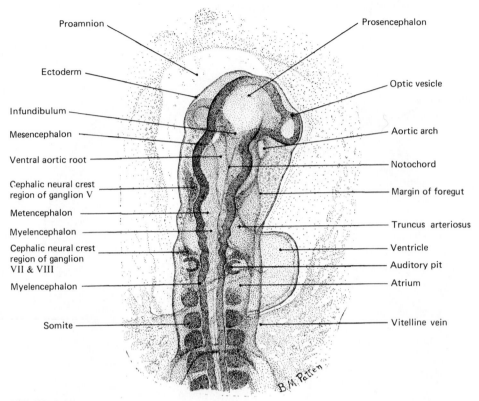

Proamnion

Ectoderm

Infundibulum

Mesencephalon

Ventral aortic root

Cephalic neural crest
region of ganglion V

Metencephalon

Myelencephalon

Cephalic neural crest
region of ganglion
VII & VIII

Myelencephalon

Somite

Prosencephalon

Optic vesicle

Aortic arch

Notochord

Margin of foregut

Truncus arteriosus

Ventricle

Auditory pit

Atrium

Vitelline vein

FIGURE A-19
Dorsal view (×50) of cephalic and cardiac regions of a chick embryo with the seventeenth somite just forming (about 38 hours' incubation).

prosencephalon. Somewhat later, constrictions mark more definitely the
boundaries between the optic vesicles and the prosencephalon (Figs. A-17D
and A-19). The infundibulum remains as a depression in the floor of the
prosencephalon (Figs. A-19 and A-20).

In chicks of about 38 hours, indications of the impending division of the
three primary vesicles to form the five regions characteristic of the adult brain
are already beginning to appear. In the establishment of the five-vesicle
condition of the brain, the prosencephalon is subdivided to form the *telencephalon* and *diencephalon*, the mesencephalon remains undivided, and the
rhombencephalon divides to form the *metencephalon* and *myelencephalon*
(Fig. A-19).

The ears also arise early in development. The first indication of the
formation of the sensory part of the ear becomes evident at about 35 hours of
incubation. At this age a pair of thickenings termed the *auditory placodes* arise
in the superficial ectoderm of the head. They are situated on the dorsolateral
surface opposite the most caudal interneuromeric constriction of the myelen-

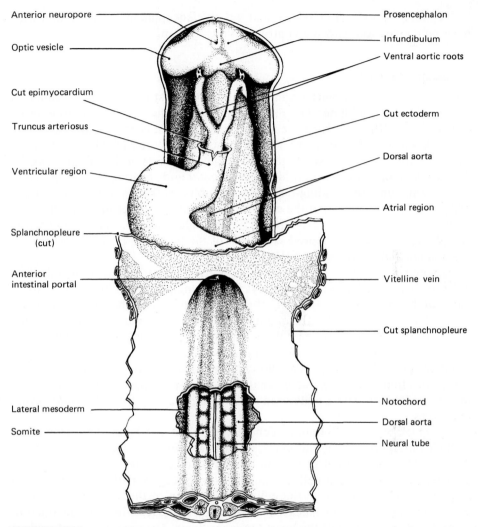

Anterior neuropore

Optic vesicle

Cut epimyocardium

Truncus arteriosus

Ventricular region

Splanchnopleure (cut)

Anterior intestinal portal

Lateral mesoderm

Somite

Prosencephalon

Infundibulum

Ventral aortic roots

Cut ectoderm

Dorsal aorta

Atrial region

Vitelline vein

Cut splanchnopleure

Notochord

Dorsal aorta

Neural tube

FIGURE A-20
Diagrammatic ventral view of dissection of a 35-hour chick embryo. The splanchnopleure of the yolk sac cephalic to the anterior intestinal portal, the ectoderm of the ventral surface of the head, and the mesoderm of the pericardial region have been removed to show the underlying structures. (*Modified from Prentiss.*)

cephalon. By 38 hours of incubation (Fig. A-19) the auditory placodes have become depressed below the general level of the ectoderm and form the walls of a pair of cavities, the *auditory pits*. When first formed, the walls of the auditory pits are directly continuous with the superficial ectoderm and their cavities are wide open to the outside. In later stages the openings into the pits become narrowed and finally closed, so that the pits become vesicles lying between the superficial ectoderm and the myelencephalon. At this point they have no connection with the central nervous system.

Formation of the Heart

The heart arises from paired primordia, which at first lie widely separated on either side of the midline. The paired condition of the heart at the time of its origin is due to the fact that the early embryo lies open ventrally, spread out on the yolk surface. The primordia of all ventral structures which appear at an early age are thus at first separated and lie on either side of the midline. As the embryo develops, a series of foldings undercut it and separate it from the yolk. At the same time, this folding-off process establishes the ventral wall of the gut and the ventral body wall of the embryo by bringing together in the midline the structures that were formerly spread out to right and left. As the embryo is completed ventrally the paired primordia of the heart are brought together in the midline and become fused (Figs. A-21 and A-22).

Although prospective cardiac tissue can be recognized much earlier, the first definite structural indications of heart formation appear in chicks of 25 to 26 hours in the region of the anterior intestinal portal. Where the splanchnopleure of either side bends toward the midline along the lateral margin of the intestinal portal, there is a marked regional thickening in the splanchnic mesoderm (Fig. A-21A). This pair of thickenings indicates where there has been rapid cell proliferation preliminary to the differentiation of the heart. Loosely associated cells can already be seen somewhat detached from the mesial face of the mesodermal layer. These cells soon become organized to form the *endocardial primordia*.

In a chick of about 26 hours, sections through a corresponding region show distinct differentiation of the endocardial and epimyocardial primordia (Fig. A-21B). The endocardial primordia are a pair of delicate tubular structures, with walls only a single cell in thickness, lying between the endoderm and mesoderm. They arise from the cells migrating out of the adjacent thickened mesoderm in the 25-hour chick. As their name indicates, they are destined to form the internal lining of the heart (*endocardium*). The greater part of each of the original mesodermal thickenings becomes applied to the lateral aspects of the endocardial tubes as the *epimyocardial primordium*, which will later differentiate into the external coat of the heart (*epicardium*) and the heavy muscular layers of the heart (*myocardium*). Careful study of the developing heart at this stage reveals the presence of *cardiac jelly* between the epimyocardium and the endocardium.

In chicks of 27 hours the lateral margins of the anterior intestinal portal have been undergoing concrescence, thus lengthening the foregut caudally and at the same time elongating the pericardial region. In this process the lateral margins of the portal swing in to meet each other and merge in the midline, and the endocardial tubes of the right side and the left side are brought toward each other beneath the newly completed floor of the foregut (Figs. A-21C and

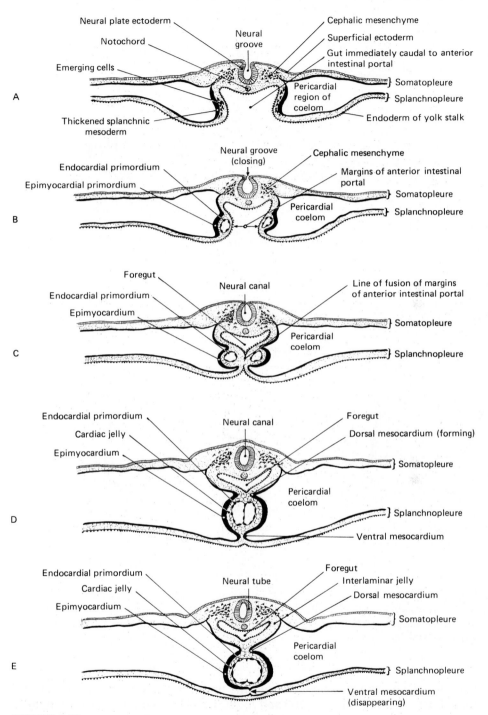

FIGURE A-21
Diagrams of transverse sections through the pericardial region of chicks at various stages to show the formation of the heart. For location of the sections, consult Fig. A-22. (A) At 25 hours; (B) at 26 hours; (C) at 27 hours; (D) at 28 hours; (E) at 29 hours.

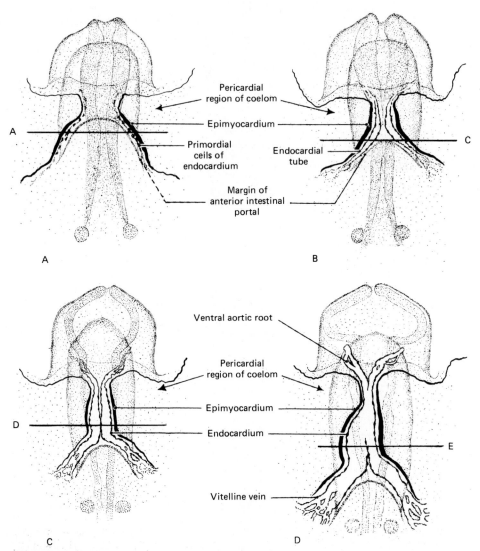

Pericardial region of coelom

Epimyocardium

Primordial cells of endocardium

Endocardial tube

Margin of anterior intestinal portal

Ventral aortic root

Pericardial region of coelom

Epimyocardium

Endocardium

Vitelline vein

FIGURE A-22
Ventral-view diagrams to show the origin and subsequent fusion of the paired primordia of the heart. The lines *A, C, D,* and *E* indicate the planes of the sections diagrammed in Fig. A-21A, C, D, and E, respectively. (A) Chick of 25 hours; (B) chick of 27 hours; (C) chick of 28 hours; (C) chick of 29 hours.

A-22B). In the 28-hour chick embryo the endocardial primordia are approximated to each other (Figs. A-21D and A-22C), and by 29 hours they fuse in their midregion to form a single tube (Figs. A-21E and A-22D).

At the same time the epimyocardial areas of the mesoderm are brought together, first ventrally (Fig. A-21D) and then dorsally to the endocardium (Fig.

A-21E). Where the splanchnic mesodermal layers of the opposite sides of the body become apposed to each other dorsal and ventral to the heart, they form double-layered supporting membranes called, respectively, the *dorsal mesocardium* and the *ventral mesocardium*. The ventral mesocardium is a transitory structure that disappears almost as soon as it is formed (Fig. A-21E). The dorsal mesocardium, although the greater part of it disappears in the next few hours of incubation, persists in embryos of the early part of the age range under consideration, suspending the heart in the pericardial region of the coelom.

The gross shape of the heart and its positional relations to other structures can be readily seen in entire embryos. The fusion of the paired cardiac primordia establishes the heart as a nearly straight tubular structure. It lies at the level of the rhombencephalon approximately in the midline, ventral to the foregut (Fig. A-16). By 33 hours of incubation the midregion of the heart is considerably dilated and bent to the right (Fig. A-18). At 38 hours the heart is bent so far to the right that it extends beyond the lateral body margin of the embryo (Fig. A-19). This bending process is correlated with the rupture of the dorsal mesocardium at the midregion of the heart.

Although there are not yet any sharply bounded subdivisions of the heart, its fundamental regions are beginning to take shape. From the heart's future intake end to its future discharging end, the regions are sinus venosus, atrium, ventricle, and truncus arteriosus. At the 36- and 38-hour stages the *sinus venosus* is only suggested. It is represented by the still paired primordia where the common cardinals enter the vitelline veins, and they in turn are becoming confluent with each other to enter the atrial portion of the tubular heart (Fig. A-25). The *atrium* is held close beneath the caudal part of the foregut by a persisting portion of the dorsal mesocardium. The *ventricle* is the part of the cardiac tube that makes a U-shaped bend to the right, bringing it into clear view at the side of the body in whole mounts (Fig. A-19). Swinging back to the midline, the ventricle is narrowed to form the discharging part of the heart known as the *truncus arteriosus* (Figs. A-20 and A-25).

From the way the paired cardiac primordia are at first located on either side of the anterior intestinal portal, it is evident that they can fuse with each other in a sequential process only as the "flooring in" of the foregut progresses (cf. Fig. A-21B to A-21D). The truncoventricular part of the heart is formed first (Fig. A-23A). Then the atrium is added caudal to the ventricle (Fig. A-23B and C). Finally, in stages older than those under discussion here, the sinus venosus is added caudal to the atrium (Fig. A-23D).

Formation of Intraembryonic Blood Vessels

Concurrently with the establishment of the heart, blood vessels arise within the body of the embryo. The large vessels connecting with the heart are the first of the intraembryonic channels to be established. The endothelial lining of the truncus is continued cephalad beneath the foregut as the ventral aorta (Fig. A-25). Almost immediately the ventral aorta bifurcates to form the paired

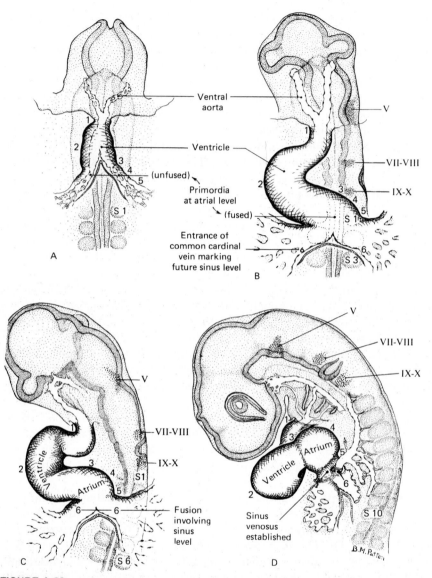

FIGURE A-23

The formation of the fundamental regions of the chick heart by progressive fusion of its paired primordia. (A) Ventral view at the 9-somite stage (about 28–29 hours), when the first contractions appear. The ventricular part of the heart is the only region where the fusion of the paired primordia has occurred and the myocardial investment has been formed. (B) Ventral view at the 16-somite stage (about 36–37 hours), when the blood first begins to circulate. The atrium and ventricle have been established, but the sinus venosus exists only as undifferentiated primordial channels, still paired and still lacking myocardial investment. (C) Ventro-sinistral view at the 19-somite stage (about 43 hours). Fusion of the paired primordia is just beginning to involve the sinus region. (D) Sinistral view at the 26-somite stage (about 51–53 hours). The sinus venosus is definitely established, and its investment with myocardium is well advanced. To facilitate following the progress of fusion, Arabic numerals have been placed against approximately corresponding locations. The 6 is located at the point of entrance of the common cardinal vein as determined from injected specimens. (*From Patten and Kramer, 1933.*)

ventral aortic roots. At the cephalic end of the foregut the ventral aortic roots turn dorsad, curve around the gut, and then extend caudad as the paired *dorsal aortae* (Figs. A-20, A-24B and C, and A-25). The curved vessels connecting the ventral aortic roots with the dorsal aortic roots are the *aortic arches*. At this age there is just a single pair, the first of the series of six which will appear as development progresses. Both the ventral aortic roots at the outlet end and the vitelline veins at the intake end are direct continuations of the paired endocardial primordia of the heart, which at this early stage consist of endothelium only. The epimyocardial coat is formed about the original endothelial tubes only where they are fused in the region destined to become the heart.

During early embryonic life the cardinal veins are the main afferent vessels of the intraembryonic circulation. The cardinal trunks are paired vessels that are symmetrically placed on either side of the midline. There are two pairs: the *anterior cardinal veins*, which return blood to the heart from the cephalic region of the embryo, and the *posterior cardinal veins*, which return blood from the caudal region. The anterior and posterior cardinal veins of the same side of the body become confluent dorsal to the level of the heart, the vessels formed by their junction being the *common cardinal veins* (ducts of Cuvier in the older literature). The right and left common cardinal veins turn ventrad, one on either side of the foregut, and enter the sinoatrial end of the heart along with the right and left vitelline veins (Fig. A-25). In chicks of 33 hours of incubation the anterior cardinal veins can usually be made out in sections (Fig. A-24B and C). By 38 hours the anterior cardinals and the common cardinals are readily recognized. The posterior cardinals appear somewhat later than do the anterior cardinals but are ordinarily discernible in the region of the common cardinals by 33 to 35 hours and are well established by 38 hours.

CHANGES BETWEEN 38 AND 50 HOURS OF INCUBATION

Starting at about 38 hours, the cephalic end of the embryo begins a series of bendings (*flexion*) and twistings (*torsion*) that change its overall shape from a straight line to a C-shaped configuration. The first flexion, occurring at the level of the midbrain as the *cranial flexure*, tips the forebrain ventrally toward the yolk. However, because of the presence of the large yolk mass, the anterior part of the embryo must twist toward the right in order to accommodate the cranial flexure (Fig. A-26). As time goes on, the torsion of the embryo extends caudad, with the result that an increasing proportion of the embryo lies on its left side. This allows the head to fold almost double upon itself (Fig. A-28).

The vitelline veins, which began to grow outward as extensions of the endocardial tubes of the heart in 33- to 36-hour embryos, soon connect with the capillary plexus on the yolk sac. Starting at about 40 hours, the vitelline arteries also begin to grow out as branches of the dorsal aorta. When the arterial extensions have become connected with the yolk sac vascular network, the extraembryonic vitelline circulatory arc is complete.

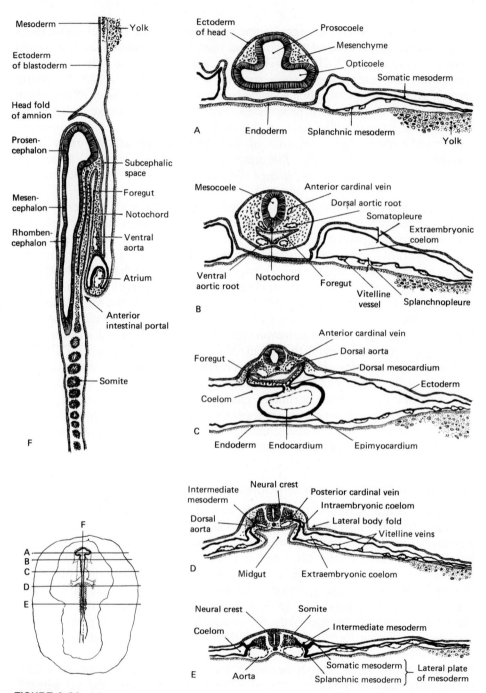

FIGURE A-24
Diagrams of sections of 33-hour chick. The location of each section is indicated on a small outline sketch of the entire embryo.

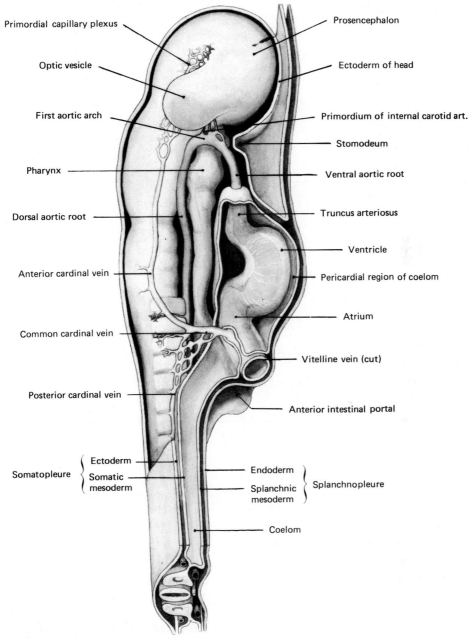

Primordial capillary plexus

Optic vesicle

First aortic arch

Pharynx

Dorsal aortic root

Anterior cardinal vein

Common cardinal vein

Posterior cardinal vein

Somatopleure
{ Ectoderm
{ Somatic mesoderm

Prosencephalon

Ectoderm of head

Primordium of internal carotid art.

Stomodeum

Ventral aortic root

Truncus arteriosus

Ventricle

Pericardial region of coelom

Atrium

Vitelline vein (cut)

Anterior intestinal portal

Endoderm
Splanchnic mesoderm
} Splanchnopleure

Coelom

FIGURE A-25
Diagrammatic lateral view of dissection of a 38-hour chick. The lateral body wall of the right side has been removed to show the internal structures. Note especially the relations of the pericardial region to that part of the coelom which lies farther caudally and the small anastomosing channels of the developing posterior cardinal vein from which a single main vessel is later derived. (See colored insert.)

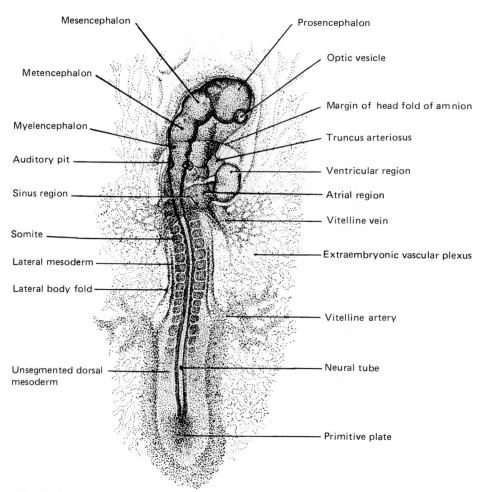

FIGURE A-26
Dorsal view (×20) of entire chick embryo having 19 pairs of somites (about 43 hours' incubation). Because of torsion the cephalic region appears in dextrodorsal view.

Beginning of the Circulation of Blood

As one might suspect, a mechanism as elaborate as the circulatory system does not go into full-scale operations all at once. Long before the actual circulation of blood commences, the heart has begun to pulsate tentatively and feebly. Its first contractions can usually be seen at the 9-somite stage (about 29 hours of incubation). At this age, the fusion of the paired cardiac primordia has not progressed beyond the ventricular region (Fig. A-23A). These first beats therefore take place in the ventricular myocardium. Their rhythmic recurrence is at first very slow, and the amplitude of the contractions is small. The pulsations, watched in living embryos, are obviously far short of the power needed to set the blood in motion.

Within 3 or 4 hours of the time of the appearance of the first beats the rate and the amplitude of the pulsations are both strikingly increased. Examination of the structure of the heart at this stage shows that the fusion of the paired primordia has now established the atrium behind the ventricle (Fig. A-23B).

While the heart has been building up an efficient beat, blood corpuscles have been forming in the blood islands of the yolk sac. At the same time, the vessels from the area vasculosa to the heart have been formed so that there is an open path for the corpuscles to enter the heart. Just before the actual circulation of blood begins, some of these corpuscles can be seen in the afferent vessels floating in fluid and shuttling back and forth with each heartbeat. The last links in the chain to be forged are the arterial channels from the dorsal aortae out to the yolk sac. At about the 16- to 17-somite stage (38 to 40 hours of incubation), these vessels open all the way out to the meshwork of small channels on the yolk sac, which is dotted with blood islands. If one is watching a living embryo when the final channels open, one sees the shuttling of corpuscles in the afferent vessels near the heart give way to a jerky progression (Patten and Kramer, 1933).

The routes followed by the blood that has begun circulating should make apparent the functional significance of the entire system. The heart is the logical point at which to begin tracing the course of either the embryonic ciculation or the vitelline circulation. From the heart the blood of the extraembryonic vitelline circulation passes through the ventral aortic roots, then through the aortic arches and along the dorsal aortae, and out through the vitelline arteries to the plexus of vessels on the yolk (Fig. A-27). In the small vessels which ramify in the membranes enveloping the yolk, the blood absorbs food material which has been made soluble by the digestive action of the endodermal cells lining the yolk sac. In young embryos, before the allantoic circulation has appeared, the vitelline circulation is also involved in the oxygenation of the blood. The great surface exposure presented by the multitude of small vessels on the yolk makes it possible for the blood to take up oxygen which penetrates the porous shell and the albumen. After acquiring food material and oxygen, the blood is collected by the sinus terminalis and the vitelline veins, which return the blood to the heart (Fig. A-27).

The blood of the intraembryonic circulation, leaving the heart, enters the ventral aortic roots, passes by way of the aortic arches into the dorsal aortae, and is distributed through branches from the dorsal aortae to the body of the embryo. It is returned from the cephalic part of the body by the anterior cardinal veins, and from the caudal part of the body by the posterior cardinals. The anterior and posterior cardinals discharge together through the common cardinal veins into the sinoatrial region of the heart (Fig. A-27).

In the heart the blood of the extraembryonic circulation and the intraembryonic circulation is mixed. The mixed blood in the heart is not as rich in oxygen and food material as the blood that comes to the heart from the vitelline circulation, nor is it as low in food and oxygen content as the blood returned to the heart from the intraembryonic circulation, where these materials are drawn

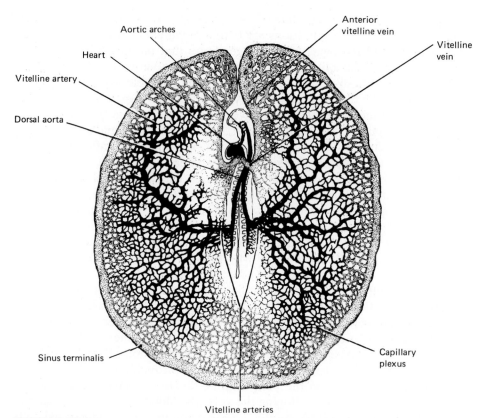

FIGURE A-27
The vitelline circulation of a chick of about 44 hours' incubation. Diagrammatic ventral view based on Popoff's figures of injected embryos. The arteries are shown in solid black; the veins are stippled. Note the rich plexus of small, freely anastomosing vessels in the splanchnopleure of the yolk sac.

upon by the growing tissues of the embryo. Nevertheless, it carries a sufficient proportion of food and oxygen to supply the growing tissues as it is distributed to the body of the embryo.

STRUCTURE OF CHICKS FROM 50 TO 60 HOURS OF INCUBATION

In chicks incubated for 55 hours (Fig. A-28) the entire head has been freed from the yolk by the caudal progression of the subcephalic fold, and through torsion, the entire anterior half of the embryo lies on its left side. At the extreme posterior end, the beginning of the caudal fold marks off the tail region of the embryo from the extraembryonic membranes. The head fold of the amnion has progressed caudad, inpocketing the embryo nearly to the level of the vitelline arteries.

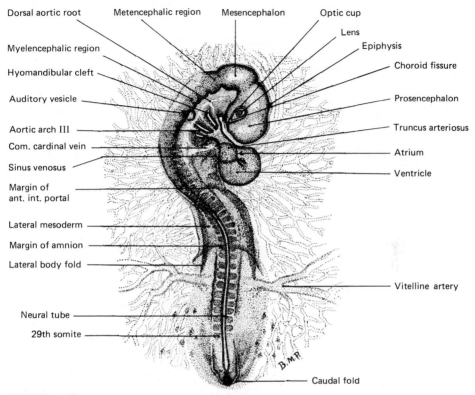

FIGURE A-28
Dextrodorsal view (×17) of entire embryo of 29 somites (about 55 hours' incubation).

The cranial flexure has by now caused the brain nearly to double back upon itself (Fig. A-28). At the level of the heart, where the myelencephalon passes into the spinal cord, another major flexure, the *cervical flexure*, begins to be apparent.

The Nervous System

The rostral part of the brain has undergone rapid enlargement, and a slight constriction indicates the impending division of the prosencephalon into the telencephalon and diencephalon (Fig. A-32). A small middorsal evagination from the diencephalon, the *epiphysis* (Fig. A-29), will later become the pineal body of the adult. Meanwhile, in the floor of the diencephalon the infundibular process enlarges and lies close to a newly formed ectodermal invagination known as *Rathke's pocket* (Fig. A-29). The epithelium of Rathke's pocket becomes separated from the superficial ectoderm and, with the infundibulum, will form the *hypophysis* or *pituitary gland*.

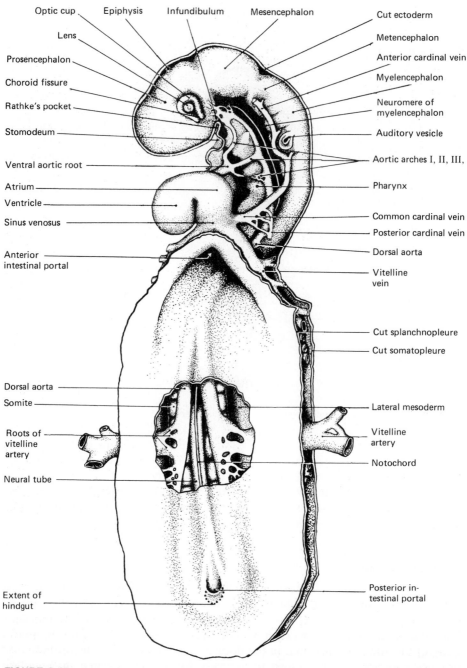

FIGURE A-29
Diagram of dissection of chick of about 50 hours. The splanchnopleure of the yolk sac cephalic to the anterior intestinal portal, the ectoderm of the left side of the head, and the mesoderm in the pericardial region have been dissected away. A window has been cut in the splanchnopleure of the dorsal wall of the midgut to show the origin of the vitelline arteries. (*Modified from Prentiss.*)

The optic vesicles have changed their form from the broad outpocketings to double-walled *optic cups* (Fig. A-30B) connected to ventrolateral walls of the diencephalon by narrow *optic stalks*. The cuplike arrangement results from the invagination of the distal ends of the optic vesicles. The invaginated layer is termed the sensory layer because it is destined to give rise to the sensory layer of the retina. The outer layer of the optic cup will become the pigmented layer of the retina. A break in the continuity of the rim of the optic cup, the *choroid fissure* (Fig. A-29), permits the passage of the central artery to the retina (Fig. 13-3). The lens of the eye first appears as a thickened ectodermal placode, which sinks beneath the surface to form a lens vesicle that ultimately detaches from the surface ectoderm (Fig. A-30B).

Caudal to the diencephalon the brain shows relatively little change in complexity, although indications of a subdivision of the hindbrain into the metencephalon and myelencephalon can be seen (Figs. A-28 and A-29). As a result of continued invagination, the auditory pits can now be called *auditory vesicles* (Figs. A-29 and A-30A).

In the spinal cord region of the neural tube the lateral walls have become thickened at the expense of the lumen, so that the neural canal appears slitlike in sections of embryos of this age (Fig. A-30E). The neural tube is complete, with the closure of both the anterior neuropore and the sinus rhomboidalis.

At the extreme caudal end of the embryo, the future spinal cord in the tail is being formed in a different manner from the rest of the cord. At the area of the tail bud (Fig. A-31), the formation of the neural tube by closure of a neural groove ceases. To complete the remainder of the neural tube, the tail bud undergoes a process of cavitation that is sometimes called *secondary neurulation* (Criley, 1969).

As the neural tube closes and separates from the overlying ectoderm, groups of ectodermal cells, the *neural crest*, form first aggregates along the junction between neural tube and superficial ectoderm (Fig. 6-13A). Later (36 hours; Fig. 6-13B), the aggregations temporarily form a confluent mass located between the dorsal wall of the neural tube and the overlying ectoderm. As development progresses (55 hours), the cells of the neural crest migrate ventrolaterally on either side of the neural tube (Fig. 6-13C).

The Digestive Tract

As the first region of the digestive tract to be established (Fig. 7-2), the foregut is the most advanced in its development. In younger embryos of this age range, there is still no mouth opening. The mouth is still represented by an ectodermal depression, the *stomodeum* (Fig. A-29). Deep in the stomodeum a region of stomodeal ectoderm abuts directly upon pharyngeal endoderm to form the *oral plate*, which will soon break down, bringing the stomodeum and pharynx into open communication. The pharynx, a prominent feature of the gut at this stage, is somewhat flattened dorsoventrally and has a larger lumen than does the esophagus (cf. Fig. A-30B and C). The relation of the pharynx to the branchial

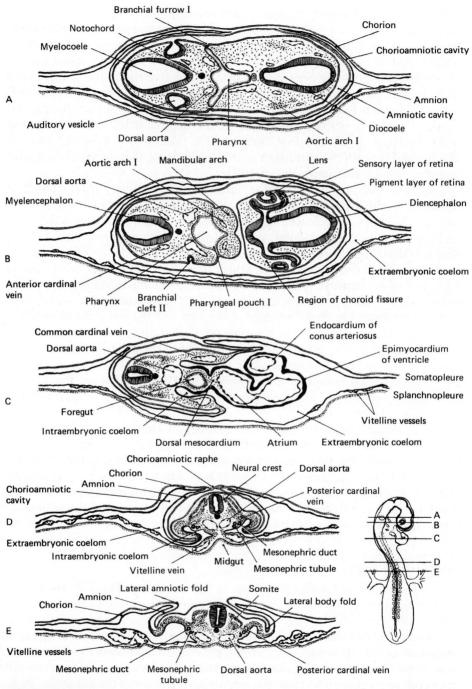

FIGURE A-30
Diagrams of transverse sections of 55-hour (30-somite) chick. The location of the sections is indicated on an outline sketch of the entire embryo.

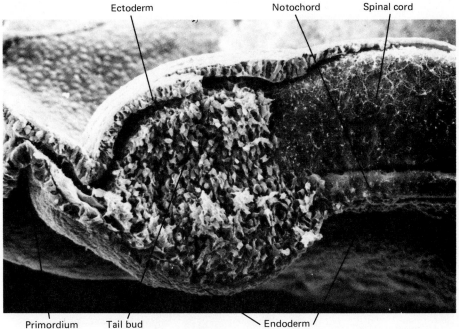

Ectoderm Notochord Spinal cord

Primordium Tail bud Endoderm
of
allantois

FIGURE A-31
Scanning electron micrograph of the tail bud region of a Hamburger-Hamilton stage 15 (50–55 hours) chick embryo. The spinal cord and notochord merge with the tail bud. (*From Schoenwolf, G. C., 1978, in Scanning Electron Microscopy, vol. II. SEM Inc. Courtesy of the author.*)

arches and clefts will be covered in the next section. Starting at the caudal end of the pharynx, the *esophagus* remains a straight tube which leads into the *midgut*. Opening directly into the yolk duct, the midgut still lacks a floor (Figs. A-29 and A-32). As the body walls of the embryo fold up into a cylinder, a larger portion of the gut is added to the foregut and hindgut at the expense of the midgut (Fig. 7-2). The *hindgut* first takes shape at about 50 hours in a manner similar to the folding off of the foregut. Not until later does the hindgut show any local specialization in development.

The Branchial Clefts and Arches

At this stage the chick embryo has unmistakable *branchial* (*gill*) *arches* and *branchial clefts*. Although only transitory, they are morphologically of great importance because of their significance as structures exemplifying recapitulation and also because of their participation in the formation of some of the ductless glands, the eustachian tubes, and the face and jaws. Moreover, the location of the aortic arches within the correspondingly numbered branchial

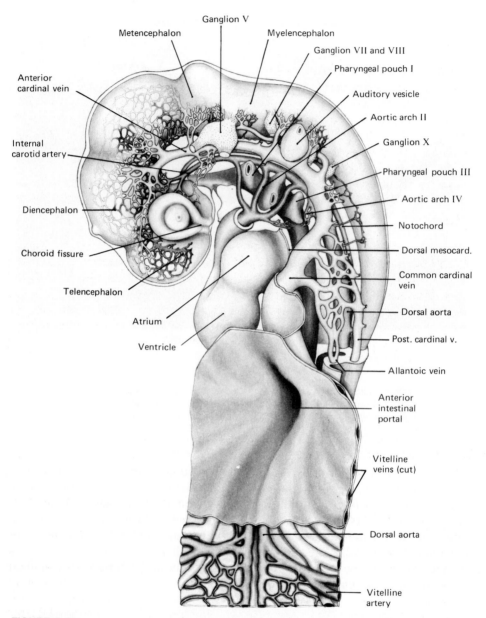

FIGURE A-32
Drawing to show the deeper structures of the cephalothoracic region of a 60-hour chick, exposed from the left. The basis of the illustration was a wax-plate reconstruction made from serial sections; the smaller vessels and the primordial capillary plexuses were added from injected specimens. Note especially the relations of the aortic arches to the pharyngeal pouches and the way in which the large veins enter the sinus venosus. Although the vitelline veins are not fully exposed by this dissection, the bulges they cause in the endoderm on either side of the anterior intestinal portal clearly suggest the way the main right and left veins become confluent with each other to enter the sinus venosus as a short median trunk. (See colored insert.)

arches and their relations to the pharyngeal pouches are among the most characteristic and important structural features of young embryos.

The *branchial clefts* were formed by the meeting of ectodermal depressions, the *branchial furrows*, with diverticula from the lateral walls of the pharynx, the *pharyngeal pouches*. In sections the branchial clefts may be seen to be closed by a thin layer of tissue composed of the ectoderm of the floor of the branchial furrow and the endoderm at the distal extremity of the pharyngeal pouch (Fig. A-30A). Although some of the clefts never open and others open only for a short time, the term *cleft* is usually used to designate these structures which are potentially clefts, whether open or not.

Between adjacent branchial clefts, the lateral body walls about the pharynx are thickened. Each of these lateral thickenings meets and merges in the midventral line with the corresponding thickening of the opposite side of the body. Thus the pharynx is encompassed laterally and ventrally by a series of archlike thickenings, the branchial arches. The branchial arches, like the branchial clefts, are designated by number, beginning at the anterior end of the series. Branchial arch I lies cephalic to the first branchial cleft, between it and the mouth. Because of the part it plays in the formation of the mandible it is also designated as the *mandibular arch*. Branchial arch II is frequently termed the *hyoid arch*, and branchial cleft I, because of its position between the mandibular and hyoid arches, is known as the *hyomandibular cleft*. Posterior to the hyoid arch the branchial arches and clefts are ordinarily designated by numbers only.

The Circulatory System

During the period between 30 and 55 hours of incubation the heart is growing more rapidly than is the body of the embryo in the region where it lies. The unattached midregion of the heart first becomes U shaped and then is twisted on itself to form a loop. The atrial region of the heart is forced somewhat to the left, and the truncus is thrown across the atrium by being twisted to the right dorsally. The ventricular region constitutes the loop proper (Fig. A-23). This twisting process reverses the original cephalocaudal relations of the atrial and ventricular regions.

The bending and subsequent twisting of the heart lead toward its division into separate chambers. However, no indication of the actual partitioning of the heart into right and left sides is yet apparent. It is still essentially a tubular organ through which the blood passes directly, without any division into separate channels.

During the period from 50 to 80 hours of incubation, interesting changes become increasingly apparent in the ventricular portions of the cardiac wall. Beginning in embryos of 50 to 55 hours and rapidly becoming more marked, the myocardium shows irregular projections extending into the cardiac jelly. These projections are the start of the *trabeculae carneae*, which are such conspicuous features of the interior of the adult ventricles. The spaces among the trabeculae carnae bring the blood in close relationship to the growing cardiac muscle during the period before coronary circulation to the myocardium has been formed.

In 33- to 38-hour chicks the ventral aortae communicate with the dorsal aortae over a single pair of aortic arches which bend around the anterior end of the pharynx (Figs. A-20 and A-25). Even embryos as old as 43 to 45 hours may still show only the first pair of aortic arches (Fig. A-33), but soon thereafter, as the branchial arches caudal to the mandibular arch differentiate, new aortic arches appear within them. These aortic arches are designated by the postoral number of the branchial arch in which they lie. Thus the original aortic arches lying in the first (mandibular) arch are spoken of as the first aortic arches, those developing in the second (hyoid) arch are the second aortic arches, and so on. In chicks of about 50 hours' incubation, the second pair of aortic arches has usually made its appearance (Fig. A-34). The third arch at this stage is usually represented by some capillary sprouts (Fig. A-34). By 60 hours, the third aortic arch has become quite sizable and sprouts indicating the start of the fourth arch are usually identifiable (Fig. A-32).

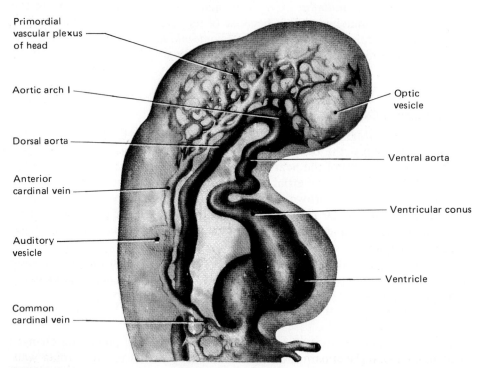

FIGURE A-33
Dextral view of cephalic and cardiac regions of chick of about 45 hours' incubation. The blood vessels have been injected to show the capillary plexus of the cephalic region. In the drawing the heart and arteries are differentiated from the veins and capillaries by darker shading. (*From Minot after Evans.*)

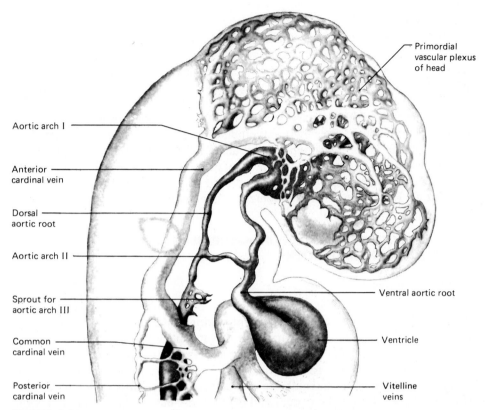

Aortic arch I

Anterior
cardinal vein

Dorsal
aortic root

Aortic arch II

Sprout for
aortic arch III

Common
cardinal vein

Posterior
cardinal vein

Primordial
vascular plexus
of head

Ventral aortic root

Ventricle

Vitelline
veins

FIGURE A-34
Dextral view of cephalic and cardiac regions of injected chick of about 50 hours' incubation. This
figure shows a later stage in the development of the anterior cardinal vein from the primary capillary plexus of the cephalic region. A second aortic arch has been completed, and plexiform outgrowth of vascular endothelium from the dorsal aortic root toward the ventral aorta indicates the
impending formation of the third aortic arch. (*From Minot after Evans.*)

The dorsal aortae arise as vessels paired throughout their entire length (Fig.
A-20). As development progresses they fuse in the midline to form the unpaired
dorsal aorta familiar in adult anatomy. This fusion first takes place at the
cardiac level and then progresses cephalad and caudad. Cephalically it never
extends to the pharyngeal region. By 50 hours the fusion has progressed caudad
well beyond the anterior intestinal portal (Fig. A-29).

The Somites

By 55 to 60 hours, the embryo possesses about 30 pairs of somites. At this time
the cells of the ventromedial part of the somite (the sclerotome) have become
mesenchymal in character and are extending toward the notochord. Ultimately
they aggregate about the notochord and participate in the formation of the body
of the vertebra.

The Urinary System

From the intermediate mesoderm, a segmentally arranged system of urine-secreting units gradually differentiates. The structural details of the early urinary system are dealt with in Chap. 17. Starting at around 36 hours, solid buds of mesodermal cells, representing pronephric tubules, appear as pairs in sequence from the fifth to the sixteenth somites. They are connected to a longitudinal duct (the primary nephric duct) that makes its way toward the cloaca. By 55 hours, the newly developing nephric tubules are of a different morphological type: mesonephric tubules. In 55-hour chick embryos, the *mesonephros*, as the complex of mesonephric tubules and the mesonephric duct is called, begins to bulge into the dorsal part of the coelomic cavity (Fig. A-30D and E). There are no recognizable gonads yet.

DEVELOPMENT OF THE CHICK DURING THE THIRD AND FOURTH DAYS OF INCUBATION

External Features

In 3-day chicks torsion has progressed to the point where the embryo lies with its left side on the yolk to a level just posterior to the heart (Fig. A-35). By 4 days the entire body has been turned so that its entire left side lies on the yolk (Fig. A-36). Meanwhile, the cranial and cervical flexures have increased to somewhat greater than right angles in 3-day chicks. Because of the broad attachment to the yolk, the midbody of 3-day chicks is slightly concave dorsally (Fig. A-36). By 4 days the caudal end of the body has also undergone flexion, leaving the embryo C shaped (Fig. A-36).

The face and oral region are similar in their development to those of a 5-week human embryo (cf. Fig. 14-1B), with *nasal pits*, each surrounded by a U-shaped elevation consisting of a *nasolateral* and a *nasomedial process*. The two nasomedial processes merge with each other in the midline and fuse laterally with the *maxillary processes*, as in mammals, to form the upper jaw (Fig. 14-1B to 14-1E). The lower jaw arises by merging in the midline of the right and left *mandibular processes* of the first branchial arch. Formation of the beak does not begin for several more days. The branchial-arch system continues to develop, and by 3 days a fourth branchial cleft has appeared (Fig. A-35).

Both the anterior and posterior appendage buds are grossly evident at 3 days. At this stage they consist of homogeneous mesoderm. The outer covering of ectoderm is marked by the conspicuously thickened *apical ectodermal ridge* (Figs. 11-5B and A-39E).

The *allantois* in the 3-day chick is still small and is concealed by the posterior appendage buds. By 4 days it has undergone rapid enlargement and projects beyond the confines of the body as a stalked vesicle of considerable size (Fig. A-36). In subsequent days (Fig. 7-5) the allantois becomes very large.

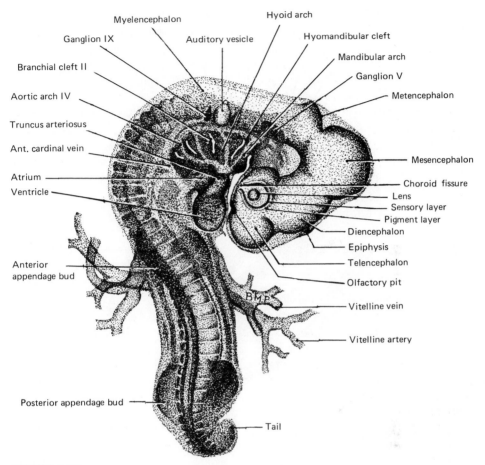

FIGURE A-35
Dextrodorsal view (×16) of entire chick embryo of 36 somites (about 3 days' incubation).

The Nervous System

The five major divisions of the brain are well delineated. As early as the third day but more prominently in the fourth, the *telencephalic vesicles* have invaginated from the lateral walls of the forebrain (Fig. A-36). Their cavities are continuous with the lumen of the median portion of the brain through the *foramina of Monro* (Fig. 12-13E). The telencephalic vesicles eventually become the cerebral hemispheres.

The lateral walls of the *diencephalon* at this stage show little differentiation except ventrally, where the optic stalks merge into the walls of the brain. The *epiphysis* continues as a median evagination from the roof of the diencephalon. The infundibular depression in the floor of the diencephalon has become appreciably deepened and lies close to Rathke's pocket (Fig. A-37) with which it is destined to fuse in the formation of the *hypophysis*. Changes in the walls

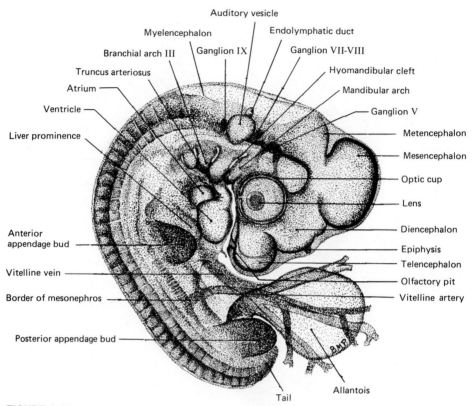

FIGURE A-36
Dextral view (×16) of entire chick embryo of 41 somites (about 4 days' incubation). Stained and cleared preparation drawn by transmitted light.

leading to the formation of the *thalami* and the *hypothalamus* have not yet occurred.

The *mesencephalon* does not show any specializations beyond a thickening of its walls. Later its dorsal walls will become thicker to form the *corpora quadrigemina*, four symmetrically placed elevations in the adult brain. The superior pair (*superior colliculi*) constitute the brain center for visual reflexes and are very prominent in birds. The *inferior colliculi* are the center for auditory reflexes.

The *metencephalon* remains well delineated from the mesencephalon by a constriction, but it remains poorly delimited from the myelencephalon. The metencephalon shows practically no differentiation in 4-day chicks. Later, its roof undergoes extensive enlargement to form the *cerebellum*. Its base, like that of the mesencephalon, contains the main nerve fiber tracts that connect the brain with the spinal cord.

The *myelencephalon* is recognizable by its thin dorsal wall, which is characteristic of its adult derivative, the medulla. Its ventral and lateral walls

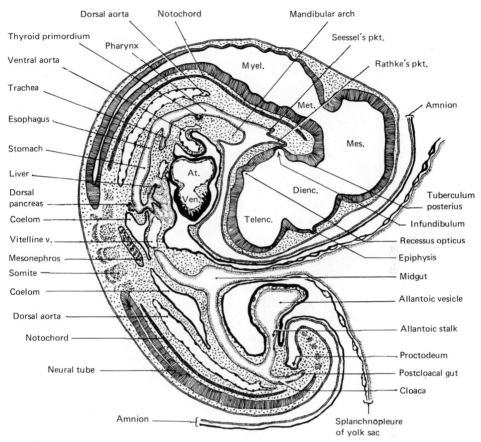

FIGURE A-37
Diagram of median longitudinal section of 4-day chick. Because of a slight bend in the embryo the section is parasagittal in the middorsal region, but for the most part it passes through the embryo in the sagittal plane.

will serve as a conduction path between the brain and spinal cord and as a reflex center for involuntary activities such as breathing.

Prominent in 4-day embryos are ganglia of the sensory components of the cranial nerves. The most prominent is the *semilunar (gasserian) ganglion* of the fifth (*trigeminal*) nerve (Fig. A-38), which lies opposite the most anterior neuromere of the myelencephalon. Just cephalic to the auditory vesicle is a mass of neural crest cells which constitute the primordium of the ganglia of the seventh and eighth nerves. Caudal to the auditory vesicle one can make out the superior and inferior sensory ganglia of the ninth and tenth cranial nerves (Fig. A-38).

As development progresses, the lateral walls of the spinal cord become greatly thickened in contrast to the dorsal and ventral walls, which remain thin.

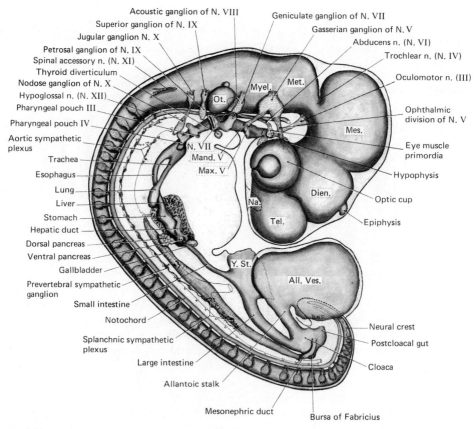

FIGURE A-38
Reconstruction of nervous, digestive, and urinary systems of 4-day chick (original ×51, reproduced ×16). Abbreviations: *All. Ves.*, allantoic vesicle; *Dien.*, diencephalon; *Mand. V*, mandibular division of cranial nerve V (trigeminal); *Max. V*, maxillary division of cranial nerve V (trigeminal); *Mes.*, mesencephalon; *Met.*, metencephalon; *Myel.*, myelencephalon; *Na.*, location of nasal pit; *N. VII*, seventh cranial nerve (facial); *Ot.*, otic vesicle; *Tel.*, lateral telencephalic vesicle; *Y. St.*, yolk stalk.

In this process the lumen (central canal) becomes compressed laterally until it appears in cross section as little more than a vertical slit (Fig. A-41).

Prominent along the spinal cord are regular spinal (dorsal root) ganglia which contain the cell bodies of the neural crest-derived sensory neurons.

When first formed from the neural crest cells, the spinal ganglion has no connection with the cord. The dorsal root is established by the growth of nerve fibers from cells of the spinal ganglion into the cord. At the same time fibers grow distad from these cells to form the peripheral part of the nerve. The fibers arising from cells of the dorsal root ganglion conduct sensory impulses toward the cord. Coincident with the establishment of the dorsal root, the ventral root is formed by fibers which grow out from cells located in the ventral part of the lateral plate of the cord. Most of the fibers which thus arise from cells in the cord and pass out through the ventral root conduct motor impulses from the brain and cord to the muscles with which they are associated.

The Digestive and Respiratory Systems

During the third day the *oral plate*, the thin membrane of ectoderm and endoderm that separates the stomodeal depression from the foregut, breaks down, allowing the foregut to communicate with the outside. Following the rupture of the oral plate, the originally shallow stomodeal depression deepens as a result of growth of the surrounding structures. The original site of the oral plate becomes the region of transition from the oral cavity to the pharynx.

Caudal to the oral opening the foregut has become flattened dorsoventrally and widened laterally to form the pharynx. On either side the pharyngeal lumen shows a series of extensions or bays known as the *pharyngeal pouches* (Fig. A-38). Each pharyngeal pouch lies opposite an external *branchial groove* (Fig. A-41B). This leaves in these areas only a thin layer of tissue (the *branchial plate*) separating the pharyngeal lumen from the outside. This layer is composed internally of endoderm and externally of the ectoderm of the bottom of the branchial groove (Fig. A-41B). Between adjacent branchial grooves or clefts the lateral walls are greatly thickened and filled with closely packed mesenchymal cells (Fig. A-39B). These thickened areas are known as the *branchial arches*. One of the most important relationships to grasp in the study of young embryos is the manner in which the *aortic arches* lie embedded in the tissue of the branchial arch of corresponding number. The stereogram appearing in Fig. A-40 emphasizes these relationships.

Several important glandular structures arise from portions of the pharyngeal endoderm. The *thyroid gland* starts out as a median diverticulum from the floor of the pharynx between the first and second pairs of pharyngeal pouches (Figs. A-37 and A-38). Sometime after the fourth day, two pairs of parathyroid primordia, as well as primordia of the thymus gland, arise from the endodermal lining of the third and fourth pharyngeal pouches.

The respiratory tract appears in 3-day chicks as a midventral groove in the pharynx. Beginning just posterior to the level of the fourth pharyngeal pouches and extending caudad, the laryngotracheal groove (Fig. A-39B) deepens rapidly and closes off to form the *tracheal tube*, which continues to grow caudad and bifurcates to form a pair of lung buds.

At 4 days the esophagus remains a thin tube, but the segment of the gut that will become stomach is already slightly dilated (Fig. A-38). Just caudal to the stomach one finds the primordia of the liver and pancreas. After arising as a simple diverticulum from the ventral wall of the gut at the end of the second day, the original hepatic evagination has grown as a mass of branching cords of endodermal cells, and by 4 days it has become quite massive (Fig. A-38). The part of the invagination closest to the gut remains open and will eventually

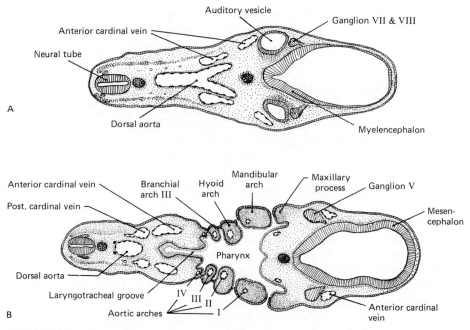

FIGURE A-39
Diagrams of five representative transverse sections of a 3-day chick. Locations of the sections are indicated on the small outline sketch of the entire embryo.

serve as the drainage duct of the liver and gallbladder. Regional differentiation of the drainage system occurs later in development.

In birds, the pancreas arises as a single median dorsal and a pair of ventral pancreatic buds. The dorsal evagination appears at about 72 hours, and the ventral evaginations appear toward the end of the fourth day. Later in development, the masses of cellular cords derived from the three pancreatic primordia grow together and become fused into a single glandular mass.

In chicks of 4 days, the midgut has been practically replaced by the extension of the foregut and hindgut. It consists only of the restricted region where the yolk stalk leads from the gut tract to the yolk sac (Fig. A-37).

The beginning of the formation of the *cloaca* is indicated in chicks of 4 days' incubation by a dilation of the posterior portion of the hindgut (Fig. A-38). Although extensive differentiation in the cloacal region does not appear until later in development, certain of its fundamental relationships are established at this stage. The cloaca of an adult bird is the common chamber into which the intestinal contents, the urine, and the products of the reproductive organs are received for discharge.

Arising from the dorsal wall of the cloaca in birds is an endodermal outgrowth known as the *bursa of Fabricius*. Later in development it becomes heavily infiltrated with lymphoid cells. Its function long remained obscure. It is

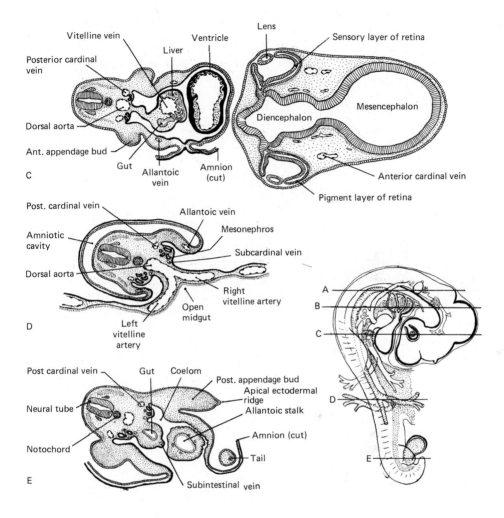

now recognized that the bursa of Fabricius is one of the major components of the immune defense system of birds and that if it is extirpated, the ability of the bird to produce humoral antibodies is sharply reduced (Glick and Chang, 1957).

Although the urinary system is not yet developed to conditions which resemble those in the adult, the parts of it which have been established are already definitely associated with the cloaca. The proximal portion of the allantoic stalk, which is the homologue of the urinary bladder of mammals, opens directly into the cloaca (Fig. A-38). The ducts which drain the developing excretory organs also open into the cloacal region on either side of the allantoic stalk. There is at this stage little indication of the formation of the gonads. The relation of the sexual ducts to the cloaca can be discerned only by the study of older embryos.

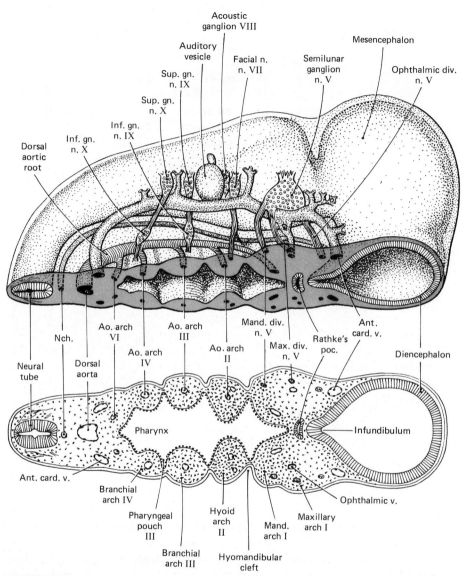

FIGURE A-40

The lower drawing is a slightly schematized representation of a transverse section through the pharyngeal region of a 3½-day chick. This section is keyed to a stereogram showing cephalic structures above the level of the section. The illustration was planned primarily to show the relations of the aortic arches to the branchial arches. (*Method of presentation suggested by some of the drawings in* Huettner's Fundamentals of Comparative Embryology of the Vertebrates, *Macmillan, New York.*) Abbreviations: *Mand. div. N. V*, mandibular division of fifth (trigeminal) cranial nerve; *Max. div. N. V*, maxillary division fo fifth nerve; *Nch.*, notochord; *Rath. poc.*, Rathke's pocket.

The Circulatory System

The 4-day chick embryo possesses two extraembryonic circulatory arcs—the vitelline and allantoic arcs—in addition to the intraembryonic system of vessels. The heart is already well developed.

The pattern of the *vitelline circulation* in chicks of 4 days is shown in Fig. A-43. Blood from the dorsal aorta leaves the body via the paired *vitelline arteries*, which branch repeatedly and end up as a large capillary network lying on the yolk sac. After picking up nutritive materials from the yolk, the blood flows into venous collecting channels and finally makes its way either into the *marginal sinus* or directly into one of the larger vitelline veins. The proximal portions of the main *vitelline veins* have fused to form an unpaired median vessel within the body of the embryo. Through this vessel, the blood returning from the vitelline circuit eventually reaches the heart.

The *allantoic circulatory arc* is not yet highly developed in 4-day embryos because of the small size of the allantois. The *allantoic arteries* arise by means of the prolongation and enlargement of a pair of ventral segmental vessels arising from the aorta at the level of the allantoic stalk. Their size increases rapidly as the allantois increases in extent. From them the blood is distributed in a rich plexus of vessels which spread over the mesoderm of the allantois (Fig. A-45). Later, when the allantois has expanded to meet the chorion as the *chorioallantoic membrane*, the allantoic vascular plexus serves as the main site of gas exchange for the embryo. The blood from the allantois is collected and returned to the heart by way of the *allantoic veins*, which enter the body of the embryo through the allantoic stalk (Fig. A-42), course through the lateral body walls (Fig. A-41E to A-41G), and ultimately empty into the sinus venous (Fig. A-44).

The *intraembryonic circulation* of the 4-day chick begins with the *ventral aorta*, which leads into the series of aortic arches. Aortic arches I and often II have disappeared as main channels, leaving only the third, fourth, and sixth pairs of arches. The right and left arches are still symmetrical at this time. The regression of the first two aortic arches plays a prominent role in the formation of important arteries supplying the head—the *external* and *internal carotid arteries*, which are extensions of the ventral and dorsal aortic roots, respectively.

The *dorsal aortae*, originally paired throughout much of their length, are by 4 days fused as far cephalically as the posterior pharynx (Fig. A-41B and C). As the dorsal aorta extends caudally, it gives off the paired vitelline arteries. Occasionally, early traces of the unpaired *coeliac artery* can be seen in 4-day chicks (Fig. A-44). Farther caudally, the paired allantoic arteries branch out toward the allantois (Fig. A-44). Throughout the length of the aorta, paired segmental arteries supply the somites and their derivatives.

The intraembryonic venous system is represented principally by the *anterior* and *posterior cardinal veins*, which drain the head and trunk and converge into the *common cardinal veins* as they turn medially to enter the sinus venosus (Fig. A-44). The posterior cardinal veins lie just dorsal to the mesonephric kidneys

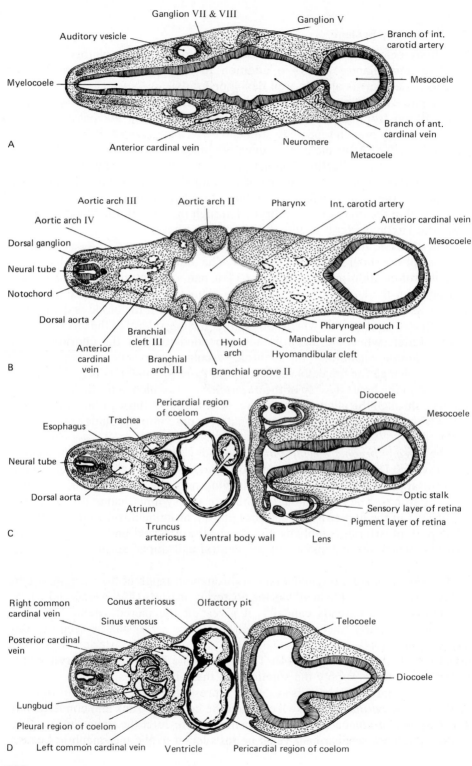

A

Myelocoele

Auditory vesicle

Ganglion VII & VIII

Ganglion V

Branch of int. carotid artery

Mesocoele

Anterior cardinal vein

Neuromere

Branch of ant. cardinal vein

Metacoele

B

Aortic arch III

Aortic arch IV

Dorsal ganglion

Neural tube

Notochord

Dorsal aorta

Aortic arch II

Pharynx

Int. carotid artery

Anterior cardinal vein

Mesocoele

Anterior cardinal vein

Branchial cleft III

Branchial arch III

Hyoid arch

Branchial groove II

Pharyngeal pouch I

Mandibular arch

Hyomandibular cleft

C

Esophagus

Trachea

Pericardial region of coelom

Neural tube

Dorsal aorta

Atrium

Truncus arteriosus

Ventral body wall

Diocoele

Mesocoele

Optic stalk

Sensory layer of retina

Pigment layer of retina

Lens

D

Right common cardinal vein

Conus arteriosus

Sinus venosus

Olfactory pit

Telocoele

Posterior cardinal vein

Lungbud

Pleural region of coelom

Left common cardinal vein

Ventricle

Pericardial region of coelom

Diocoele

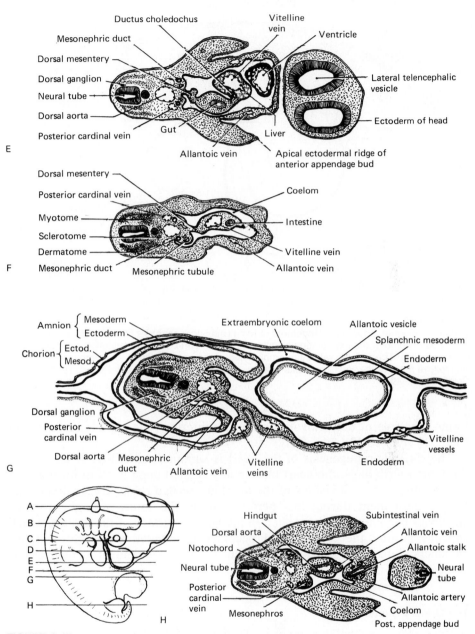

FIGURE A-41
Diagrams of transverse sections of a 4-day chick. The location of the sections is indicated on a small outline sketch of the entire embryo.

throughout their length (Figs. A-39D and E and A-41E to A-41H). Situated ventrally in the mesonephroi are the small, irregular *subcardinal veins* (Fig. A-44), which are a relic of the renal portal system of more primitive ancestral

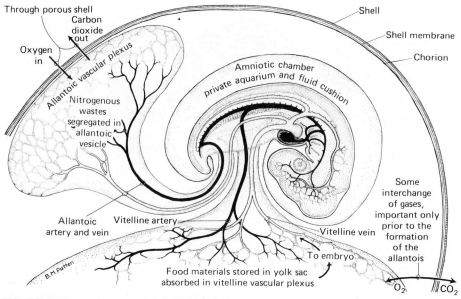

FIGURE A-42
Schematic diagram showing arrangement of main circulatory channels in young chick embryo. The sites of some of the extraembryonic interchanges important in bioeconomics are indicated by the labeling. The vessels within the embryo carry food and oxygen to all its growing tissues and relieve them of the waste products incident to their metabolism. (*From Patten, 1951*, Am. Scientist, *vol. 39.*)

forms. The role of the veins in forming the inferior vena cava is discussed in Chap. 17 (Fig. 18-12).

The early morphogenesis of the heart is summarized in Fig. A-46. By the third and fourth days, the heart has already twisted upon itself to form a loop and the major gross divisions of the heart are recognizable. Expansion of the atria is prominent during this period. During the fourth day the truncus arteriosus becomes closely applied to the ventral surface of the atrium, which expands around the truncus and becomes subdivided into right and left chambers. Separation of the common ventricle into right and left ventricles has just begun.

The Urinary System

It is possible to recognize pronephric tubules in chick embryos of 38 to 40 hours (Fig. 17-2A). In chicks of 50 to 55 hours the mesonephric ducts and primordial tubules are clearly identifiable just lateral to the somites (Fig. A-30D and E). By the fourth day mesonephric tubules are well under way in their development. The metanephric tubules, which constitute the excretory units of the permanent kidneys, do not appear until considerably later in development, and the genital organs which become intimately interrelated with the urinary organs are

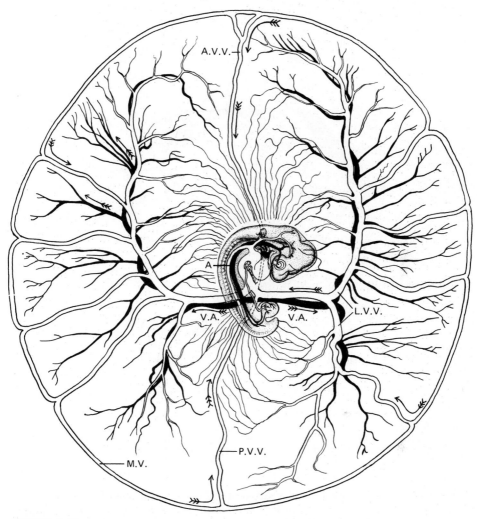

FIGURE A-43
Diagram to show course of vitelline circulation in chick of about 4 days. (*After Lillie.*) Abbreviations: A, dorsal aorta; *A. V. V.*, anterior vitelline vein; *L. V. V.*, lateral vitelline vein; *M. V.*, marginal vein (sinus terminalis); *P. V. V.*, posterior vitelline vein; *V. A.*, vitelline artery. The direction of blood flow is indicated by arrows.

also relatively late in making their appearance. The later stages of the development of the urinary system as well as the entire sequence of stages in the formation of the genital organs are covered in Chap. 17.

The Coelom and Mesenteries

In adult birds and mammals the body cavity consists of three regions: pericardial, pleural, and peritoneal. The *pleural cavities* are paired, each of the

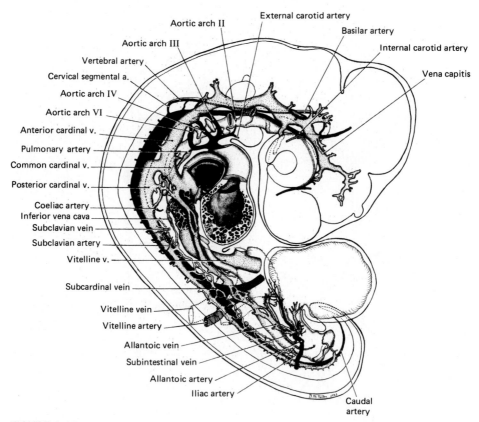

FIGURE A-44
Reconstruction of circulatory system of 4-day chick (original ×51, reproduced ×18). From the same embryo as that represented in Fig. A-38. These illustrations should be studied together.

pleural chambers being a laterally situated sac containing one of the lungs. The *pericardial chamber* containing the heart and the *peritoneal chamber* containing the viscera other than the lungs and heart are unpaired. These regions of the adult body cavity are formed by the reshaping and partitioning of the primary body cavity, or coelom, of the embryo.

In the chick the coelom arises by a splitting of the lateral mesoderm of either side of the body (Fig. A-47A and B). It is therefore at first a paired cavity. Unlike the coelom of some of the more primitive vertebrates, the coelom of the chick never shows any indications of segmental pouches corresponding in arrangement with the somites. Instead, the right and left coelomic chambers extend cephalocaudally without interruption through the entire lateral plates of mesoderm.

The coelomic chambers are not limited to the region in which the body of the embryo is developing. They extend on either side into the mesoderm, which, in common with the other germ layers, spreads out over the yolk surface. Large

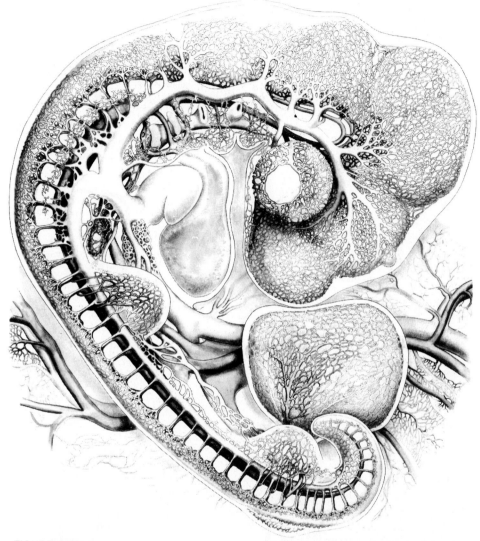

FIGURE A-45
The blood vessels of a 4-day chick. The basis of the illustration was the same wax-plate reconstruction from which Fig. A-44 was drawn. The smaller vessels and the primordial capillary plexuses were added from injected specimens. The labeled diagram of Fig. A-44 will serve as a
means of identifying the main vessels. (See colored insert.)

parts of the primitive coelomic chambers thus come to be extraembryonic in
their associations (Chap. 7; Figs. 7-1 and 7-4). The portion of the coelom that
gives rise to the embryonic body cavities is first marked off by the series of
folds which separate the body of the embryo from the yolk (Fig. A-47C and D).
As the closure of the ventral body wall progresses (Fig. A-47E and F), the

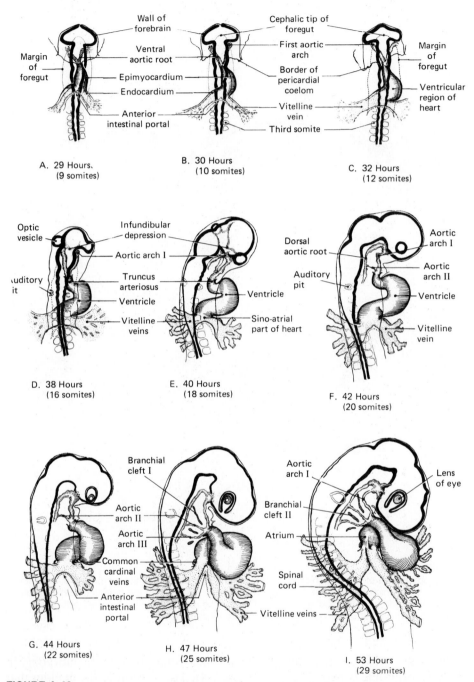

FIGURE A-46
Projection drawings of the heart and great vessels of chick embryos of various ages. Some of the major topographical features of the embryos have been outlined to emphasize the relations of the heart within the body. Abbreviations: *roman numerals I–VI*, aortic arches of the designated numbers; *Sin. ven.*, sinus venosus; *Vent.*, ventricle.

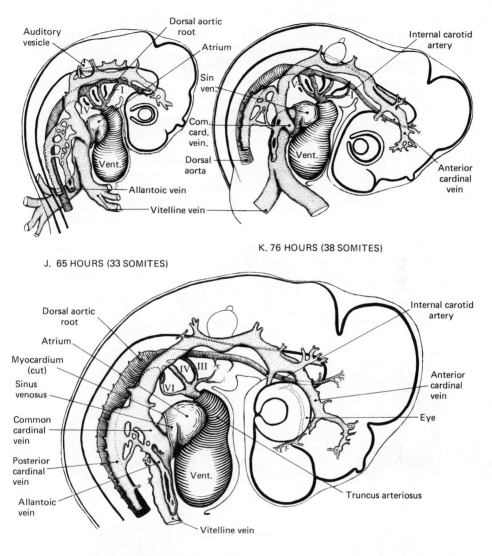

J. 65 HOURS (33 SOMITES)

K. 76 HOURS (38 SOMITES)

L. 100 HOURS (45 SOMITES)

embryonic coelom becomes completely separated from the extraembryonic. The delayed closure of the ventral body wall in the yolk-stalk region results in the embryonic coelom and extraembryonic coelom retaining an open communication at this point for a considerable time after they have been completely separated elsewhere.

The same folding process that establishes the ventral body wall completes the gut ventrally (Fig. A-47C to A-47F). Meanwhile the right and left coelomic chambers are expanded mesiad. As a result the newly closed gut comes to lie

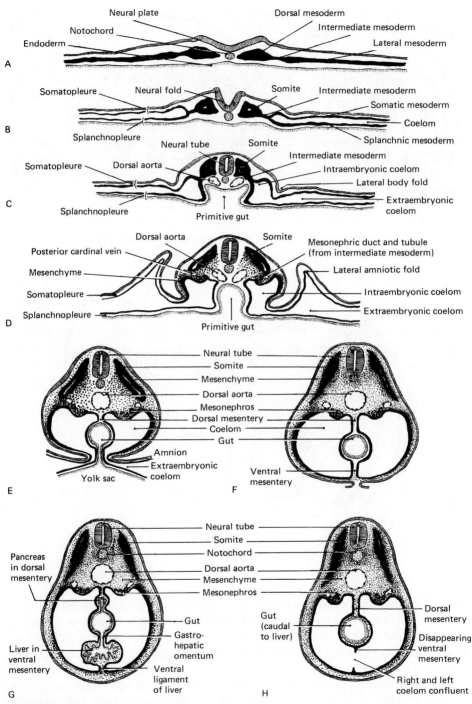

FIGURE A-47
Schematic diagrams of cross sections at various stages to show the establishment of the coelom and mesenteries. For explanation, see text.

suspended between the two layers of splanchnic mesoderm which constitute the mesial walls of the right and left coelomic chambers. The double layer of splanchnic mesoderm which thus becomes apposed to the gut and supports it in the body cavity is known as the *primary mesentery*. The part of the mesentery dorsal to the gut, suspending it from the dorsal body wall, is the *dorsal mesentery*, and the part ventral to the gut, attaching it to the ventral body wall, is the *ventral mesentery*.

When the dorsal and ventral mesenteries are intact, they constitute a complete membranous partition dividing the body cavity into right and left halves. The primary dorsal mesentery persists in large part, but the ventral mesentery soon is extensively resorbed (Fig. A-47H), bringing the right and left coelomic chambers into confluence ventral to the gut and establishing the unpaired condition of the body cavity characteristic of the adult.

In their relation to the other mesenteries of the body, the mesocardia may be regarded as special regions of the ventral mesentery. In the most cephalic part of the body cavity, the gut lies embedded in the dorsal body wall instead of being suspended by a dorsal mesentery as it is farther caudally (cf. Figs. A-21E and A-47F). A ventral mesentery is, however, developed in the same manner anteriorly as it is posteriorly, and when the heart is formed it is suspended in the most anterior part of this ventral mesentery. The dorsal and ventral mesocardia may therefore be thought of as the parts of the primary ventral mesentery lying dorsal to the heart and ventral to the heart, respectively (Fig. A-21D). When the ventral mesocardium, and a little later the dorsal mesocardium, break through, the primary right and left coelomic chambers become confluent to form the pericardial region of the body cavity (Fig. A-25).

Summary of Figures

Students using this text as the basis for laboratory study of chick or pig development have sometimes experienced difficulty in locating the appropriate illustrative material because of its incorporation into areas of the text concerned with a particular organ system. To increase the usefulness of this text in the laboratory, this Appendix summarizes the figures illustrating key stages in the development of both the chick and the pig. The chick material is subdivided into groups of both whole mounts and sectioned preparations, whereas the pig material is largely based upon serial sections as well as reconstructions derived from them.

CHICK WHOLE MOUNTS

Figure number	Incubation age
5-12A	3–4 hours
5-12B	5–6 hours
5-12C	7–8 hours
5-12D	10–12 hours
5-14	16 hours
A-3	18 hours
A-7	20 hours
A-9	22–23 hours
A-11	24 hours

692

Figure number	Incubation age
A-10	25–26 hours
A-12	25–26 hours—ventral view, cephalic region
A-15	27–28 hours
A-16	29–30 hours—ventral view, cephalic and cardiac region
A-18	33 hours
A-19	38 hours—dorsal view, cephalic and cardiac region
A-25	38 hours—lateral view, dissection
7-3	40 hours
A-26	43 hours
A-27	44 hours—vitelline circulation
A-33	45 hours—lateral view, cephalic and cardiac region
A-29	50 hours—ventral view, dissection
A-34	50 hours—lateral view, cephalic and cardiac region
A-28	55 hours
A-22	60 hours—ventrolateral view, dissection
A-35	72 hours
A-36	96 hours
A-38	96 hours—lateral view, reconstruction of nervous, digestive, and urinary system
A-44, A-45	96 hours—lateral view, reconstruction of circulatory system

CHICK SECTIONS

18 Hours

Figure A-4—longitudinal and transverse sections

24 Hours

Figure A-13—longitudinal and transverse sections
Figure A-14—stereogram

33 Hours

Figure A-24—longitudinal and transverse sections

55 Hours

Figure A-30—transverse sections

72 Hours

Figure A-39—transverse sections

96 Hours

Figure A-37—longitudinal section
Figure A-41—transverse sections

PIG

Figure number	Length
	5 mm
8-9	whole mount
8-13	sagittal section
8-18	parasagittal section
8-19	parasagittal section
8-14	transverse sections through 8 levels
	10 mm
18-9	reconstruction of circulatory system and key to transverse sections
12-10	sagittal section
12-16	transverse sections through head and pharynx—3 levels
15-7	transverse sections through thorax and upper abdomen—3 levels
18-10	transverse sections through cardiac area—3 levels
18-13	transverse sections through abdominal area—3 levels
19-10	transverse sections through pelvic area—2 levels
	15 mm
18-27	parasagittal section
12-25	parasagittal section through head
12-26	parasagittal section through head
16-7	reconstruction of pharynx
16-8	transverse sections through pharyngeal region—4 levels
17-11	transverse sections through pelvic region—3 levels

REFERENCES FOR
COLLATERAL READING

**EMBRYOLOGY - ITS SCOPE, HISTORY, AND SPECIAL FIELDS -
CHAPTER 1**

Adelmann, H. B., 1942. *The Embryological Treatises of Hieronymus Fabricus of Aquapendente*. Cornell University Press, Ithaca, N.Y., 376 pp.

———, 1966. *Marcello Malpighi and the Evolution of Embryology*. Cornell University Press, Ithaca, N.Y., 5 vols., 2475 pp.

Baserga, R., and D. Malamud, 1969. *Autoradiography: Techniques and Application*. Hoeber-Harper, New York, 281 pp.

Berns, M. W., and C. Salet, 1972. Laser microbeams for partial cell irradiation. *Int. Rev. Cytol.*, **33**:131–156.

Briggs, R., and T. J. King, 1952. Transplantation of living nuclei from blastula cells into enucleated frogs' eggs. *Proc. Nat. Acad. Sci.*, **38**:455–463.

Brothers, A. J., 1976. Stable nuclear activation dependent on a protein synthesized during oogenesis. *Nature*, **260**:112–115.

Bullough, W. S., E. B. Laurence, O. H. Iverson, and K. Elgjo, 1967. The vertebrate epidermal chalone. *Nature*, **214**:578–580.

Burnet, F. M., 1969. *Cellular Immunology*. Cambridge University Press, London, 725 pp.

Carpenter, G., and S. Cohen, 1978. Biological and molecular studies of the mitogenic effects of human epidermal growth factor. In Papaconstantinou and Rutter, *Molecular Control of Proliferation and Differentiation*. Academic Press, Inc., New York, pp. 13–31.

Ebert, J. D., 1966. *Interacting Systems in Development*. Holt, Rinehart and Winston, Inc., New York, 227 pp.

Edelman, G. M., 1983. Cell adhesion molecules. *Science*, **219**:450–457.

Etzler, M. E., 1985. Plant lectins: Molecular and biological aspects. *Ann. Rev. Plant Physiol.*, **36**:209–234.

Fraser, F. C., H. Kalter, B. E. Walker, and T. D. Fainstat, 1954. The experimental production of cleft palate with cortisone and other hormones. *J. Cell. Comp. Physiol.*, **43**, suppl. 1:237–259.

————, B. E. Walker, and D. C. Taylor, 1957. Experimental production of congenital cleft palate: Genetic and environmental factors. *Pediatrics*, **19**:782–787.

Glücksmann, A., 1951. Cell death in normal vertebrate ontogeny. *Biol. Revs. Cambridge Phil. Soc.*, **26**:59–86.

Goeddel, D. V., D. G. Kleid, F. Bolivar et al., 1979. Expression in *Escherichia coli* of chemically synthesized genes for human insulin. *Proc. Natl. Acad. Sci. USA*, **76**:106–110.

Gospodarowicz, D., J. S. Moran, and A. L. Mescher, 1978. Cellular specificities of fibroblast growth factor and epidermal growth factor. In Papaconstantinou and Rutter, *Molecular Control of Proliferation and Differentiation*. Academic Press, Inc., New York, pp. 33–63.

Gould, G. M., and W. L. Pyle, 1937. *Anomalies and Curiosities of Medicine*. Sydenham Publishers, New York, 968 pp.

Grobstein, C., 1956. Transfilter induction of tubules in mouse metanephrogenic mesenchyme. *Exp. Cell. Res.*, **10**:424–440.

Gross, P. R., and G. H. Cousineau, 1964. Macromolecule synthesis and the influence of actinomycin on early development. *Exp. Cell Res.*, **33**:368–395.

Gurdon, J. B., 1962. The developmental capacity of nuclei taken from intestinal epithelium cells of feeding tadpoles. *J. Embryol. Exp. Morphol.*, **10**:622–640.

Gurwitsch, A. G., 1944. *The Theory of Biological Fields* (Russian). Izdatel. Sovietskaya Nauka, Moscow, 155 pp.

Gustafson, T., and L. Wolpert, 1963. The cellular basis of morphogenesis and sea urchin development. *Int. Rev. Cytol.*, **15**:139–214.

Hakomori, S., 1986. Glycosphingolipids. *Sci. Amer.*, **254**(5):44–53.

Harrison, R. G., 1907. Observations on the living developing nerve fiber. *Anat. Rec.*, **1**:116–118.

Hay, E. D., 1977. Embryonic induction and tissue interaction during morphogenesis. In Littlefield and deGrouchy, *Birth Defects*. Int. Congr. Series No. 432. Excerpta Medica, Amsterdam, pp. 126–140.

Hay, E. D. (ed.), 1981. *Cell Biology of the Extracellular Matrix*. Plenum Press, New York, 417 pp.

Illmensee, K., and P. Hoppe, 1981. Nuclear transplantation in *Mus musculus*. *Cell*, **23**:9–18.

Konigsberg, I. R., 1963. Clonal analysis of myogenesis. *Science*, **140**:1273–1284.

Larsen, W. J., 1983. Biological implications of gap junction structure, distribution and composition: A review. *Tissue and Cell*, **15**:645–671.

Lehtonen, E., 1976. Transmission of signals in embryonic induction. *Med. Biol.*, **54**:108–128.

Lowenstein, W. R., 1970. Intercellular communication. *Sci. Am.*, May, pp. 79–86.

Malacinski, G. M., and S. V. Bryant (eds.), 1984. *Pattern Formation*. The Macmillan Company, New York, 626 pp.

Markert, C. L. (ed.), 1975. *Isoenzymes*. Vols. I-IV, Academic Press, Inc., New York, vols. 1–4.

Meyer, A. W., 1936. *An Analysis of the De Generatione Animalium of William Harvey*. Stanford University Press, Stanford, Calif., 167 pp.

————, 1939. *The Rise of Embryology*. Stanford University Press, Stanford, Calif., 367 pp.

Milstein, C., 1980. Monoclonal antibodies. *Sci. Am.*, **243**(4):66–74.

Mintz, B., 1971. Clonal basis of mammalian differentiation. In Control Mechanisms of Growth and Differentiation. *Symp. Soc. Exp. Biol.*, **25**:345–368.

Moscona, A., and H. Moscona, 1952. The dissociation and aggregation of cells from organ rudiments of the early chick embryo. *J. Anat.*, **6**:287–301.

Needham, J., 1931. *Chemical Embryology*. Cambridge University Press, Cambridge, vols. 1–3, 2021 pp.

————, 1959. *A History of Embryology*. Cambridge University Press, Cambridge, 2d ed., 303 pp.

Oppenheimer, J. M., 1967. *Essays on the History of Embryology and Biology*. M.I.T., Cambridge, Mass., 374 pp.

Papaconstantinou, J., and W. J. Rutter (eds.), 1978. *Molecular Control of Proliferation and Differentiation*. Academic Press, Inc., New York, 264 pp.

Pierschbacher, M. D., and E. Rouslahti, 1984. Cell attachment activity of fibronectin can be duplicated by small synthetic fragments of the molecule. *Nature* (London), **309**:30–33.

Roux, W., 1888. Beiträge zur Entwicklungsmechanik des Embryo. V. Über die künstliche Hervorbringung "halber" Embryonen durch Zerstörung einer der beiden ersten Furchungszellen, sowie über die Nachentwicklung (Postgeneration) der fehlenden Körperhalfte. *Virchows Arch.*, **114**:419–521.

Roux, W., 1888. Contributions to the developmental mechanisms of the embryo. On the artificial production of half-embryos by destruction of one of the first two blastomeres, and the later development (postgeneration of the missing half of the body) (German). *Virchows Arch. Path. Anat. u. Physiol. u. Klin Med.*, **114**:113–153. (English translation in Willier, B. H., and J. M. Oppenheimer, 1974, *Foundations of Experimental Embryology*, 2d ed., Hafner Press, New York, pp. 2–37.)

Saunders, J. W., M. T. Gasseling, and L. C. Saunders, 1962. Cellular death in morphogenesis of the avian wing. *Dev. Biol.*, **5**:147–178.

Saxén, L., and S. Toivonen, 1962. *Primary Embryonic Induction*. Prentice-Hall, Inc., Englewood Cliffs, N.J., 271 pp.

Spemann, H., 1901. Ueber Korrelationen in der Entwicklung des Auges. *Verh. Anat. Ges. Jena Verslag.* (Bonn), **15**:61–79.

————, 1912. Zur Entwicklung des Wirbeltierauges. *Zool. Jahrb., Abt. allgem. Zool.*, **32**:1–98.

————, and H. Mangold, 1924. Ueber Induktion von Embryonalanlagen durch Implantation ortfremder Organisatoren. *Arch. mikroskop. Anat. Entwmech.*, **100**:599–638.

Thompson, D. W., 1959. *On Growth and Form*. Cambridge University Press, Cambridge, 2d ed., 1116 pp.

Toivonen, S., D. Tarin, L. Saxén, P. J. Tarin, and J. Wartiovaara, 1975. Transfilter studies on neural induction in the newt. *Differentiation*, **4**:1–7.

Toole, B. P., 1982. Developmental role of hyaluronate. *Conn. Tissue Res.*, **10**:93–100.

Townes, P. L., and J. Holtfreter, 1955. Directed movements and selective adhesion of embryonic amphibian cells. *J. Exp. Zool.*, **128**:53–120.

Trinkaus, J. P., 1969. *Cells into Organs: The Forces that Shape the Embryo*. Prentice-Hall, Englewood Cliffs, N.J., 237 pp.

Waddington, C. H., 1956. *Principles of Embryology*. The Macmillan Company, New York, 3d printing, 1960, 510 pp.

Weiss, P., 1939. *Principles of Development*. H. Holt and Co., New York, 601 pp.

Weiss, P., and A. C. Taylor, 1960. Reconstruction of complete organs from single-cell suspensions of chick embryos in advanced stages of differentiation. *Proc. Nat. Acad. Sci.*, **46**:1177–1185.

Wilson, H. V., 1907. On some phenomena of coalescence and regeneration in sponges. *J. Exp. Zool.*, **5**:245–258.

Wolpert, L., 1969. Positional information and the spatial pattern of cellular differentiation. *J. Theor. Biol.*, **25**:1–47.

Wolpert, L., 1971. Positional information and pattern formation. *Curr. Top. Devel. Biol.*, **6**:183–224.

SEXUAL CYCLE, GAMETOGENESIS, AND FERTILIZATION (CHAPTERS 2 AND 3)

Ancel, P., and P. Vintemberger, 1948. Recherches sur le déterminisme de la symétrie bilaterale dans l'oeuf des amphibiens. *Bull. Biol. Suppl.* **31**:1–182.

Austin, C. R., 1961. *The Mammalian Egg*. Blackwell Scientific Publications, Ltd., Oxford, 183 pp.

———, 1978. Patterns in metazoan fertilization. *Curr. Top. Dev. Biol.*, **12**:1–9.

Baccetti, B. (ed.), 1970. *Comparative Spermatology*. Academic Press, Inc., New York, 573 pp.

Barr, M. L., and E. G. Bertram, 1949. A morphological distinction between neurones of the male and female and the behavior of the nucleolar satellite during accelerated nucleoprotein synthesis. *Nature* (London), **16**:676.

Bartelmez, G. W., 1937. Menstruation. *Physiol. Rev.*, **17**:28–72.

Bascom, K. F., and H. L. Osterud, 1925. Quantitative studies of the testicle. II. Pattern and total tubule length in the testicles of certain common mammals. *Anat. Rec.*, **31**:159–169.

Bedford, J. M., 1970. Sperm capacitation and fertilization in mammals. *Biol. of Reproduction*, vol. 2, suppl. 2, pp. 128–158.

Bellairs, R., 1964. Biological aspects of the yolk of the hen's egg. *Adv. Morphogen.*, **4**:217–272.

———, 1965. The relationship between oocyte and follicle in the hen's ovary as shown by electron microscopy. *J. Embryol. Exp. Morph.*, **13**:215–233.

Biggers, J. D., and A. W. Schuetz (eds.), 1972. *Oogenesis*. University Park Press, Baltimore, 543 pp.

Blandau, R. J., and R. Hayashi. Ovulation and egg transport in mammals. Film, University of Washington Press, Seattle.

Bliel, J. D., and P. M. Wassarman, 1980a. Synthesis of zona pellucida proteins by denuded and follicle-enclosed mouse oocytes during culture in vitro. *Proc. Natl. Acad. Sci.*, USA, **77**:1029–1033.

Bliel, J. D., and P. M. Wassarman, 1980b. Structure and function of the zona pellucida: Identification and characterization of the proteins of the mouse oocytes' zona pellucida. *Devel. Biol.*, **76**:185–202.

Board, R. G., and R. Fuller, 1974. Non specific antimicrobial defenses of the avian egg, embryo and neonate. *Biol. Rev.*, **49**:15–49.

Brachet, J., 1977. An old enigma: The gray crescent of amphibian eggs. *Curr. Top. Dev. Biol.*, **11**:133–186.

Browder, L. W. (ed.), 1985. "Developmental Biology. Vol. 1, Oogenesis." Plenum Press, New York, 632 pp.

Brown, D. D., and I. G. Dawid, 1968. Specific gene amplification in oocytes. *Science*, **160**:272–280.

Brown, D. D., and K. Sugimoto, 1973. 5S DNAs of *Xenopus laevis* and *Xenopus mulleri*: Evolution of a gene family. *J. Mol. Biol.*, **78**:397–415.

Chang, M. C., 1955. The maturation of rabbit oocytes in culture and their maturation, activation, fertilization and subsequent development in the fallopian tubes. *J. Exp. Zool.*, **128**:379–406.

Clermont, Y., 1972. Kinetics of spermatogenesis in mammals: Seminiferous epithelium cycle and spermatogonial renewal. *Physiol. Rev.*, **52**:198–236.

Conrad, R. M., and H. B. Scott, 1938. The formation of the egg in the domestic fowl. *Physiol. Rev.*, **18**:481–494.

Corner, G. W., 1928–1929. Physiology of the corpus luteum. I. *Am. J. Physiol.*, **86**:74–81; II. *Am. J. Physiol.*, **88**:326–339; III. *Am. J. Physiol.*, **88**:340–346.

Crosignani, P. G., and D. R. Mishell (eds.), 1976. *Ovulation in the Human*. Academic Press, Inc., New York, 317 pp.

Cross, N. L., and R. P. Elinson, 1980. A fast block to polyspermy in frogs mediated by changes in the membrane potential. *Dev. Biol.*, **75**:187–198.

Curtis, A. S. G., 1960. Cortical grafting in *Xenopus laevis*. *J. Embryol. Exp. Morphol.*, **8**:163–173.

———, 1962. Morphogenetic interactions before gastrulation in the amphibian, *Xenopus laevis*—the cortical field. *J. Embryol. Exp. Morph.*, **10**:410–422.

Dubois, R., 1969. Le mécanisme d'entrée des cellules germinales primordiales dans le réseau vasculaire, chez l'embryon de poulet. *J. Emb. Exp. Morphol.*, **21**:255–270.

Eddy, E. M., 1975. Germ plasm and differentiation of the germ cell line. *Int. Rev. Cytol.*, **43**:229–281.

Eddy, E. M., J. M. Clark, D. Gong, and B. A. Fenderson, 1981. Origin and migration of primordial germ cells in mammals. *Gamete Res.*, **4**:333–362.

Edwards, R. G., 1980. *Conception in the Human Female*. Academic Press, London, 1087 pp.

Enders, A. C. (ed.), 1963. *Delayed Implantation*. University of Chicago Press, Chicago, 318 pp.

Epel, D., 1980. Fertilisation. *Endeavour, N.S.* **4**:26–31.

Erickson, G. F., D. A. Magoffin, C. A. Dyer, and C. Hofeditz, 1985. The ovarian androgen producing cells: A review of structure/function relationships. *Endocrine Revs.*, **6**:371–399.

Eyal-Giladi, H., M. Ginsburg, A. Farbarov, 1981. Avian primordial germ cells are of epiblastic origin. *J. Embryol. Exp. Morph.*, **65**:139–147.

———, H., S. Kochav, and M. K. Menashi, 1976. On the origin of primordial germ cells in the chick embryo. *Differentiation*, **6**:13–16.

Fawcett, D. W., 1975. The mammalian spermatozoon. *Dev. Biol.*, **44**:394–436.

———, 1979. The cell biology of gametogenesis in the male. *Perspect. Biol. Med.*, **22**:S56-S73.

———, 1986. *A Textbook of Histology*. W. B. Saunders Co., Philadelphia, 1017 pp.

Flickinger, R., and D. E. Rounds, 1956. The maternal synthesis of egg yolk proteins as demonstrated by isotopic and serological means. *Biochem. Biophys. Acta*, **22**:38–42.

Fox, C. A., S. J. Meldrum, and B. W. Watson, 1973. Continuous measurement by radio-telemetry of vaginal pH during human coitus. *J. Reprod. Fert.*, **33:**69–75.

Gall, J. G., and H. G. Callan, 1962. ^{3}H-Uridine incorporation in lampbrush chromosomes. *Proc. Nat. Acad. Sci.*, **48:**562–570.

Gardner, R. L., 1978. Developmental potency of normal and neoplastic cells of the early mouse embryo. In "Birth Defects," J. W. Littlefield and J. Grouchy (eds.), *Excerpta Medica International Congress Series* No. 432, pp. 154–166.

Gartler, S. M., and A. D. Riggs, 1983. Mammalian x-chromosome inactivation. *Am. Rev. Genet.*, **17:**155–190.

Gerhart, J., M. Danilchik, J. Roberts, B. Rowning, and J.-P. Vincent, 1986. Primary and secondary polarity of the amphibian oocyte and egg. In *Gametogenesis and the Early Embryo*, J. G. Gall (ed.). Alan R. Liss, Inc., New York, pp. 305–319.

Gilbert, S. F., 1985. *Developmental Biology*. Sinauer Associates, Sunderland, Mass., 726 pp.

Gimlich, R. L., and J. C. Gerhart, 1984. Early cellular interactions promote embryonic axis formation in *Xenopus laevis*. *Devel. Biol.*, **104:**117–130.

Gwatkin, R. B. L., 1977. *Fertilization Mechanisms in Man and Mammals*. Plenum, New York, 161 pp.

Hafez, E. S. E., and T. N. Evans, 1973. *Human Reproduction*. Harper & Row, Hagerstown, Md., 778 pp.

Halbert, S. A., P. Y. Tam, and R. J. Blandau, 1976. Egg transport in the rabbit oviduct: The roles of cilia and muscle. *Science*, **191:**1052–1053.

Hamerton, J. L., 1961. Sex chromatin and human chromosomes. *Int. Rev. Cytol.*, **12:**1–68.

Hansbrough, J. R., and D. L. Garbers, 1981. Speract-purification and characterization of a peptide associated with eggs that activates spermatozoa. *J. Biol. Chem.*, **256:**1447–1452.

Hansel, W., and E. M. Convey, 1983. Physiology of the estrous cycle. *J. Animal Sci.* **57:**(Suppl. 2):404–424.

Hartman, C. G., 1936. *Time of Ovulation in Women*. The Williams & Wilkins Company, Baltimore, 226 pp.

Heller, C. G., and Y. Clermont, 1963. Spermatogenesis in man: An estimate of its duration. *Science*, **140:**184–186.

Henking, H. von, 1891. Untersuchungen über die ersten Entwicklungsvörgange in den Eiern der Insekten. *Zeitsch. f. wiss. Zool.*, **51:**685–763.

Hill, R. T., E. Allen, and T. C. Kramer, 1935. Cinemicrographic studies of rabbit ovulation. *Anat. Rec.*, **63:**239–245.

Hunter, G. L., G. P. Bishop, C. E. Adams, and L. E. Rowson, 1962. Successful long-distance aerial transport of fertilized sheep ova. *J. Reprod. Fert.*, **3:**33–40.

Jaffe, L. A., M. Gould-Somero, and L. Z. Holland, 1982. Studies of the mechanism of the electrical polyspermy block using voltage clamp during cross-species fertilization. *J. Cell Biol.*, **92:**616–621.

Jatrou, K., and G. H. Dixon, 1978. Protamine messenger RNA: Its life history during spermatogenesis in rainbow trout. *Fed. Proc.*, **37:**2526–2533.

Kemp, N. E., 1956. Electron microscopy of growing oocytes of *Rana pipiens*. *J. Biophys. Biochem. Cytol.*, **2:**281–292.

Kessel, R. G., 1985. Annulate lamellae (porous cytomembranes): with particular emphasis on their possible role in differentiation of the female gamete. In "Developmental Biology, vol. 1. Oogenesis," L. W. Browder (ed.). Plenum Press, New York, pp. 179–234.

Kirschner, M. W., J. L. Gerhart, K. Hara, and G. A. Ubbels, 1980. Initiation of the cell cycle and establishment of bilateral symmetry in *Xenopus* eggs. *38th Growth Symposium*, pp. 187–216.

Kochav, S., and H. Eyal-Giladi, 1971. Bilateral symmetry in chick embryo determination by gravity. *Science*, **171**:1027–1029.

Lillie, F. R., 1919. *Problems of Fertilization*. University of Chicago Press, Chicago, 278 pp.

Longo, F. J., 1984. Pronuclear events. In "Biology of Fertilization," vol. 3, C. B. Metz and A. Monroy (eds.). Academic Press, New York, pp. 252–298.

Lynn, J. W., and E. L. Chambers, 1984. Voltage clamp studies of fertilization in sea urchin eggs. I. Effect of clamped membrane potential on sperm entry, activation and development. *Devel. Biol.*, **102**:98–109.

Lyon, M. F., 1961. Gene action in the X chromosome of the mouse (*Mus musculus L.*). *Nature*, **190**:372–373.

MacGregor, H. C., 1972. The nucleolus and its genes in amphibian oogenesis. *Biol. Rev.*, **47**:177–210.

Maller, J. L., 1985. Oocyte maturation in amphibians. In "Developmental Biology, vol. 1, Oogenesis," L. W. Browder (ed.), Plenum Press, New York, pp. 289–311.

Mann, T., 1953. Biochemical aspects of semen. In Wolstenholm, *Ciba Symposium on Mammalian Germ Cells*. Little, Brown and Company, Boston, pp. 1–8.

Masui, Y., and C. L. Markert, 1971. Cytoplasmic control of nuclear behavior during meiotic maturation of frog oocytes. *J. Exp. Zool.*, **177**:129–146.

McClung, C. E., 1902. The accessory chromosome—sex determinant? *Biol. Bull.*, **3**:43–84.

Metz, C. B., and A. Monroy, 1967–1969. *Fertilization: Comparative Morphology, Biochemistry and Immunology*. Academic Press, Inc., New York, vols. I and II, 489 and 553 pp.

Metz, C. B., and A. Monroy (eds.), 1985. *Biology of Fertilization*. Vols. 1–3. Academic Press, New York, pp. 391, 475 and 489.

Meyer, D. B., 1964. The migration of primordial germ cells in the chick embryo. *Dev. Biol.*, **10**:154–190.

Meyerhof, P. G., and Y. Masui, 1979. Properties of a cytostatic factor from *Xenopus laevis* eggs. *Devel. Biol.*, **72**:182–187.

Midgley, A. R., V. L. Gray, P. L. Keyes, and J. S. Hunter, 1973. Human reproductive endocrinology. In Hafez and Evans, *Human Reproduction*. Harper & Row, Hagerstown, Md., pp. 201–236.

Mintz, B., and K. Illmensee, 1975. Normal genetically mosaic mice produced from malignant teratocarcinoma cells. *Proc. Natl. Acad. Sci. USA*, **72**:3585–3589.

———, and E. S. Russell, 1957. Gene-induced embryological modifications of primordial germ cells in the mouse. *J. Exp. Zool.*, **134**:207–237.

Moses, M. J., 1968. Synaptinemal complex. *Annu. Rev. Genet.*, **2**:363–412.

Nieuwkoop, P., 1977. Origin and establishment of embryonic polar axes in amphibian development. *Curr. Top. Dev. Biol.*, **11**:115–132.

———, and L. A. Sutasurya, 1976. Embryological evidence for a possible polyphyletic origin of the recent amphibians. *J. Embryol. Exp. Morphol.*, **35**:159–167.

———, and ———, 1979. *Primordial Germ Cells in the Chordates*. Cambridge University Press, Cambridge, 187 pp.

Old, R. W., H. G. Callan, and K. W. Gross, 1977. Localization of histone gene transcripts in newt lampbrush chromosomes by *in situ* hybridization. *J. Cell Sci.*, **27**:57–80.

O'Malley, B. W., W. L. McGuire, P. O. Kohler, and S. G. Korenman, 1969. Studies on the mechanism of steroid hormone regulation of synthesis of specific proteins. *Recent Prog. Horm. Res.*, **25**:105–160.

Palmiter, R. D., T. M. Wilkie, H. Y. Chen, and R. L. Brinster, 1984. Transmission distortion and mosaicism in an unusual transgenic mouse pedigree. *Cell*, **36**:869–877.

Papanicolaou, G. N., 1933. The sexual cycle in the human female as revealed by vaginal smears. *Am. J. Anat.*, supp. to vol. 52, pp. 519–637.

Parker, G. H., 1931. Passage of sperms and of eggs through oviducts in terrestrial vertebrates. *Phil. Trans. Roy. Soc. London*, ser. B, **219**:381–419.

Pincus, G., 1936. *The Eggs of Mammals*. The Macmillan Company, New York, 160 pp.

Richards, J. S., 1979. Hormonal control of ovarian follicular development: A 1978 perspective. *Recent Prog. Horm. Res.*, **35**:343–373.

———, and A. R. Midgley, 1976. Protein hormone action: A key to understanding ovarian follicular and luteal cell development. *Biol. Reprod.*, **14**:82–94.

Romanoff, A. L., and A. J. Romanoff, 1949. *The Avian Egg*. John Wiley & Sons, Inc., New York, 918 pp.

Roosen-Runge, E. C., 1962. The process of spermatogenesis in mammals. *Biol. Rev.*, **37**:343–377.

Schatten, G., 1982. Motility during fertilization. *Internat. Rev. Cytol.*, **79**:59–163.

Schimke, R. T., G. S. McKnight, D. J. Shapiro, D. Sullivan, and R. Palacios, 1975. Hormonal regulation of ovalbumin synthesis in the chick oviduct. *Recent Prog. Horm. Res.*, **31**:175–208.

Schroeder, P. C., and P. Talbot, 1985. Ovulation in the animal kingdom: A review with an emphasis on the role of contractile processes. *Gamete Res.*, **11**:191–221.

Schuel, H., 1984. The prevention of polyspermic fertilization in sea urchins. *Biol. Bull.*, **167**:271–309.

Schuetz, A. W., 1974. Role of hormones in oocyte maturation. *Biol. Reprod.*, **10**:150–178.

———, and N. Dubin, 1981. Progesterone and prostaglandin secretion by ovulated rat cumulus cell-oocyte complexes. *Endocrinology*, **108**:457–463.

Shapiro, B. M., R. W. Schackmann, and C. A. Gabel, 1981. Molecular approaches to the study of fertilization. *Am. Rev. Biochem.*, **50**:815–843.

———, ———, ———, C. A. Foerder, M. L. Farance, E. M. Eddy, and S. J. Klebanoff, 1980. Molecular alterations in gamete surfaces during fertilization and early development. In "The Cell Surface: Mediator of Developmental Processes," S. Subtelny and N. K. Wessels (eds.), Academic Press, New York, pp. 127–149.

Shettles, L. B., 1957. The living human ovum. *Obstet. Gynecol.*, **10**:359–365.

Smith, L. D., 1966. The role of a "germinal plasm" in the formation of primordial germ cells in *Rana pipiens*. *Dev. Biol.*, **14**:330–347.

———, 1975. Molecular events during oocyte maturation. In Weller, *The Biochemistry of Animal Development*. Academic Press, Inc., New York, vol. 3, pp. 1–46.

———, and M. A. Williams, 1975. Germinal plasm and primordial germ cells. *33d Symp. Soc. Devel. Biol.*, pp. 3–24.

Socher, S. H., and B. O'Malley, 1973. Estrogen mediated cell proliferation during chick oviduct development and its modulation by progesterone. *Dev. Biol.*, **30**:411–417.

Spirin, A. S., 1966. On "masked" forms of messenger RNA in early embryogenesis and in other differentiating systems. *Curr. Top. Devel. Biol.*, **1**:1–38.

Stambaugh, R., 1978. Enzymatic and morphological events in mammalian fertilization. *Gamete Res.*, **1**:65–86.

Suzuki, N., K. Nomura, H. Ohtake, and S. Isaka, 1981. Purification and the primary structure of sperm activating peptides from the jelly coat of sea urchin eggs. *Biochem. Biophys. Res. Commun.*, **99**:1238–1244.

Taylor, T. G., 1970. How an eggshell is made. *Sci. Am.*, March, pp. 89–95.

Tindall, D. J., D. R. Rowley, L. Murthy, L. I. Lipshultz, and C. H. Chang, 1985. Structure and biochemistry of the Sertoli cell. *Int. Rev. Cytol.*, **94**:127–149.

Tsai, S. Y., S. E. Harris, M. J. Tsai, and B. W. O'Malley, 1976. Effects of estrogen on gene expression in chick oviduct. *J. Biol. Chem.*, **251**:4713–4721.

Vintemberger, P., and J. Clavert, 1960. Sur le déterminisme de la symétrie bilatérale chez les oiseaux. XIII. *C. R. Soc. Biol.,* Paris, **154**:1072–1076.

von Baer, K. E., 1828. *Entwicklungsgeschichte des Hünchens im Eie*. Bornträger, Koningsberg, p. 315.

Wallace, R. A., 1985. Vitellogenesis and oocyte growth in nonmammalian vertebrates. In "Developmental Biology, vol. 1. Oogenesis," L. W. Browder (ed.). Plenum Press, New York, pp. 127–177.

Wallace, R. A., and E. W. Bergink, 1974. Amphibian vitellogenin: Properties, hormonal regulation of hepatic synthesis and ovarian uptake, and conversion to yolk proteins. *Am. Zool.*, **14**:1159–1175.

Wassarman, P. R., 1987. The biology and chemistry of fertilization. *Science*, **235**:553–560.

Willier, B. H., 1937. Experimentally produced sterile gonads and the problem of the origin of germ cells in the chick embryo. *Anat. Rec.*, **70**:78–112.

Wilson, E. G., 1905. The chromosomes in relation to the determination of sex in insects. *Science*, **22**:500–502.

Wischnitzer, S., 1966. The ultrastructure of the cytoplasm of the developing amphibian egg. *Adv. Morphogen.*, **5**:131–179.

―――, 1967. The ultrastructure of the nucleus of the developing amphibian egg. *Adv. Morphogen.*, **6**:173–198.

Witschi, E., 1948. Migration of the germ cells of human embryos from the yolk sac to the primitive gonadal folds. *Carnegie Cont. to Emb.*, **32**:67–80.

Wylie, C. C., J. Heasman, A. Snape, M. O'Driscoll, and S. Holwill, 1985. Primordial germ cells of *Xenopus laevis* are not irreversibly determined early in development. *Devel. Biol.*, **112**:66–72.

Wynn, R. M., 1977. *Biology of the Uterus*. Plenum Press, New York, 748 pp.

Young, W. C. (ed.), 1961. *Sex and Internal Secretions*, vols. I and II. Williams & Wilkins Co., Baltimore, 1609 pp.

Zamboni, L., 1971. *Fine Morphology of Mammalian Fertilization*. Harper & Row, New York, 223 pp.

CLEAVAGE AND THE FORMATION OF THE GERM LAYERS (CHAPTERS 4 AND 5)

Azar, Y., and H. E. Eyal-Giladi, 1981. Interaction of epiblast and hypoblast in the formation of the primitive streak and the embryonic axis, as revealed by hypoblast-rotation experiments. *J. Embryol. Exp. Morphol.*, **61**:133–144.

Beier, H. M., and R. R. Maurer, 1975. Uteroglobin and other proteins in rabbit blastocyst fluid after development *in vivo* and *in vitro*. *Cell Tissue Res.*, **159**:1–10.

Bellairs, R., 1986. The primitive streak. *Anat. Embryol.*, **174**:1–14.

———, F. W. Lorenz, and T. Dunlap, 1978. Cleavage in the chick embryo. *J. Embryol. Exp. Morphol.*, **43**:55–69.

Bennett, M. V. L., 1973. Function of electrotonic junctions in embryonic and adult tissues. *Fed. Proc.*, **32**:65–75.

Bluemink, J. G., and S. W. deLaat, 1973. New membrane formation during cytokinesis in normal and cytochalasin B-treated eggs of *Xenopus laevis*. I. Electron microscopic observations. *J. Cell Biol.*, **59**:89–108.

Borland, R. M., J. D. Biggers, and C. P. Lechene, 1977. Studies in the composition and formation of mouse blastocoele fluid using electron probe microanalysis. *Dev. Biol.*, **55**:9–32.

Boucaut, J.-C., and T. Darribere, 1983. Fibronectin in early amphibian embryos. Migrating mesodermal cells contact fibronectin established prior to gastrulation. *Cell Tissue Res.*, **234**:135–145.

Boycott, A. E., C. Diver, S. L. Garstang, and F. M. Turner, 1930. The inheritance of sinistrality in *Linnea peregra* (Mollusca, Pulmonata). *Phil. Trans. Roy. Soc. Lond.* (Biol.), **219**:51–131.

Brachet, J., 1977. An old enigma: The gray crescent of amphibian eggs. *Curr. Top. Dev. Biol.*, **11**:133–186.

Brick, I., and C. Weinberger (eds.), 1984. Symposium on gastrulation. *Am. Zool.*, **24**:537–688.

Briggs, R., and J. T. Justus, 1968. Partial characterization of the component from normal eggs which corrects the maternal effect of gene *o* in the Mexican axolotl (*Ambystoma mexicanum*). *J. Exp. Zool.*, **167**:105–116.

———, and T. J. King, 1952. Transplantation of living nuclei from blastula cells into enucleated frogs' eggs. *Proc. Nat. Acad. Sci.*, **38**:455–463.

Brown, D. D., and I. B. Dawid, 1968. Specific gene amplification in oocytes. *Science*, **160**:272–280.

Curtis, A. S. G., 1962. Morphogenetic interactions before gastrulation in the amphibian *Xenopus laevis*—the cortical field. *J. Embryol. Exp. Morphol.*, **10**:410–422.

Czihak, G. (ed.), 1975. *The Sea Urchin Embryo*. Springer-Verlag, Berlin, 700 pp.

Davidson, E. H., 1976. *Gene Activity in Early Development*, 2d ed., Academic Press, Inc., New York, 452 pp.

Dawid, I. B., and T. D. Sargent, 1986. Molecular embryology in amphibians: New approaches to old questions. *Trends in Genet.*, **2**:47–50.

Denker, H.-W., 1983. Cell lineage, determination and differentiation in earliest developmental stages in mammals. *Bibliotheca Anat.*, **24**:22–58.

Ducibella, T., and E. Anderson, 1975. Cell shape and membrane changes in the eight-cell mouse embryo: Prerequisites for morphogenesis of the blastocyst. *Dev. Biol.*, **47**:45–58.

Edelman, G. M., W. J. Gallin, A. Delowvee, B. A. Cunningham, and J.-P. Thiery, 1983. Early epochal maps of two different cell adhesion molecules. *Proc. Natl. Acad. Sci USA*, **80**:4384–4388.

Evans, T., E. Rosenthal, J. Youngblom, D. Distel, and T. Hunt, 1983. Cyclin: A protein specified by maternal nRNA in sea urchin eggs that is destroyed at each cleavage division. *Cell*, **33**:389–396.

Eyal-Giladi, H., 1984. The gradual establishment of cell commitments during the early stages of chick development. *Cell Differen.*, **14**:245–255.

————, and S. Kochav, 1976. From cleavage to primitive streak formation: A complementary normal table and a new look at the first stages of the development of the chick. I. General morphology. *Dev. Biol.* **49**:321–337.

————, and M. Wolk, 1970. The inducing capacities of the primary hypoblast as revealed by transfilter induction studies. *Wilhelm Roux' Arch.*, **165**:226–241.

Fehilly, C. B., S. M. Willadsen, and E. M. Tucker, 1984. Interspecific chimaerism between sheep and goat. *Nature*, **307**:634–636.

Fink, R. D., and D. R. McClay, 1985. Three cell recognition changes accompanying the ingression of sea urchin primary mesenchyme cells. *Devel. Biol.*, **107**:66–74.

Gerhart, J. C., 1980. Mechanisms regulating pattern formation in the amphibian egg and early embryo. In "Biological Regulation and Development. vol. 2. Molecular Organization and Cell Function," R. F. Goldberger (ed.). Plenum Press, New York, pp. 133–316.

Gross, P. R., and G. H. Cousineau, 1964. Macromolecule synthesis and the influence of actinomycin on early development. *Exp. Cell Res.*, **33**:368–395.

Gulyas, B. J., 1975. A reexamination of cleavage patterns in eutherian mammalian eggs: Rotation of blastomere pairs during second cleavage in the rabbit. *J. Exp. Zool.*, **193**:235–248.

Gurdon, J. B., 1974. *The Control of Gene Expression in Animal Development.* Harvard University Press, Cambridge, Mass., 160 pp.

Gustafson, T., and L. Wolpert, 1967. Cellular movement and contact in sea urchin morphogenesis. *Biol. Rev.*, **42**:442–498.

Hara, K., 1977. The cleavage pattern of the axolotl egg studied by cinematography and cell counting. *Wilhelm Roux' Arch*, **181**:73–87.

Hardin, J., and L. Y. Cheng, 1986. The mechanisms and mechanics of archenteron elongation during sea urchin gastrulation. *Devel. Biol.*, **115**:490–501.

Harvey, E. B., 1936. Parthenogenetic merogony or cleavage without nuclei in *Arbacia punctulata. Biol. Bull.*, **71**:101–121.

Hay, E. D., 1968. Organization and fine structure of epithelium and mesenchyme in the developing chick embryo. In Fleischmajer and Billingham, *Epithelial-Mesenchymal Interactions.* The Williams and Wilkins Company, Baltimore, Md., pp. 31–55.

Holtfreter, J., 1943–1944. A study of the mechanics of gastrulation. I. *J. Exp. Zool.*, **94**:261–318; II. *J. Exp. Zool.*, **95**:171–212.

Hörstadius, S., 1939. The mechanics of sea urchin development studied by operative methods. *Biol. Rev.*, **14**:132–179.

Kalt, M. R., 1971. The relationship between cleavage and blastocoel formation in *Xenopus laevis.* I. Light microscopic observations. *J. Embryol. Exp. Morphol.*, **26**:37–49.

Katow, H., and M. Solursh, 1980. Ultrastructure of primary mesenchyme cell ingression in the sea urchin *Lytechinus pictus. J. Exp. Zool.*, **213**:231–246.

Keller, R. E., 1981. An experimental analysis of the role of bottle cells and the deep marginal zone in gastrulation of *Xenopus laevis. J. Exp. Zool.*, **216**:81–101.

Leikola, A., 1976. Hensen's node—the "organizer" of the amniotic embryo. *Experientia*, **32**:269–277.

Lewis, W. H., and C. G. Hartmann, 1933. Early cleavage stages of the egg of the monkey (*Macacus rhesus*). *Carnegie Cont. to Emb.*, **24**:187–201.

Luckett, W. P., 1978. Origin and differentiation of the yolk sac and extraembryonic mesoderm in presomite human and rhesus monkey embryos. *Am. J. Anat.*, **152**:59–98.

McClay, D. R., and C. A. Ettensohn, 1987. Cell recognition during sea urchin gastrulation. In "Genetic Regulation of Development," W. F. Loomis (ed.). A. R. Liss, New York, pp. 111–128.

———, and G. M. Wessel, 1985. The surface of the sea urchin embryo at gastrulation: A molecular mosaic. *Trends in Genetics*, **1**:12–16.

McLaren, A., 1976. *Mammalian Chimaeras*. Cambridge, University Press, Cambridge, 154 pp.

Meinecke-Tillmann, S., and B. Meinecke, 1984. Experimental chimaeras—removal of reproductive barrier between sheep and goat. *Nature*, **307**:637–638.

Minor, R. R., P. S. Hoch, T. R. Koszalka, R. L. Brent, and N. A. Kefalides, 1976. Organ cultures of the embryonic rat parietal yolk sac. I. Morphologic and autoradiographic studies of the deposition of the collagen and noncollagen glycoprotein components of basement membrane. *Devel. Biol.*, **48**:344–364.

Mintz, B., 1964. Formation of genetically mosaic mouse embryos and early development of "lethal" (T^{12}/T^{12})-normal" mosaics. *J. Exp. Zool.*, **157**:273–292.

Nakamura, O., and S. Toivonen (eds.), 1978. *Organizer-A Milestone of a Half Century from Spemann*. Elsevier/North-Holland, Amsterdam, 379 pp.

Nicolet, G., 1971. Avian gastrulation. *Adv. Morphogen.*, **9**:231–262.

Nieuwkoop, P. D., 1973. The "organization center" of the amphibian embryo: Its origin, spatial organization and morphogenetic action. *Adv. Morphogen.*, **10**:1–39.

———, 1977. Origin and establishment of embryonic polar axes in amphibian development. *Curr. Top. Dev. Biol.*, **11**:115–132.

Pasteels, J., 1945. On the formation of the primary entoderm of the duck (*Anas domestica*) and on the significance of the bilaminar embryo in birds. *Anat. Rec.*, **93**:5–21.

Patterson, J. T., 1913. Polyembryonic development in *Tatusia novemcincta*. *J. Morphol.*, **24**:559–683.

Rappaport, R., 1971. Cytokinesis in animal cells. *Int. Rev. Cytol.*, **31**:169–213.

———, 1974. Cleavage. In Lash and Whittaker, *Concepts in Development*. Sinauer Associates, Stamford, Conn., pp. 76–98.

Rosenquist, G. C., 1966. A radioautographic study of labeled grafts in the chick blastoderm. Development from primitive streak stages to stage 12. *Carnegie Cont. to Emb.*, **38**:71–110.

Rudnick, D., 1944. Early history and mechanics of the chick blastoderm. *Quart. Rev. Biol.*, **19**:187–212.

Sanders, E. J., 1986. Mesoderm migration in the early chick embryo. In L. Browder (ed.). *Developmental Biology*, vol. 2. Plenum Press, New York, pp. 449–480.

Schuetz, A. W., and N. Dubin, 1981. Progesterone and prostaglandin secretion by ovulated rat cumulus cell-oocyte complexes. *Endocrinology*, **108**:457–463.

Schultz, G. A., 1986. Molecular biology of the early mouse embryo. *Biol. Bull.*, **171**:291–309.

Selman, G. G., and M. M. Perry, 1970. Ultrastructural changes in the surface layers of the newt's egg in relation to the mechanism of its cleavage. *J. Cell Sci.*, **6**:207–227.

Slack, J. M. W., 1984. The early amphibian embryo—A hierarchy of developmental decisions. In Malacinski and Bryant, *Pattern Formation. A Primer in Developmental Biology*. The Macmillan Company, New York, pp. 457–480.

Slack, J. M. W., 1984. Cell lineage labels in the early amphibian embryo. *BioEssays*, **1**:5–8.

Slack, C., A. E. Warner, and R. L. Warren, 1973. The distribution of sodium and potassium in amphibian embryos during early development. *J. Physiol.*, **232**:297–312.

Snell, G. D., and L. C. Stevens, 1966. Early embryology. In Green, *Biology of the Laboratory Mouse*. McGraw-Hill Book Co., New York, pp. 205–245.

Solursh, M., 1986. Migration of sea urchin primary mesenchyme cells. In Browder, L., *Developmental Biology*, vol. 2. Plenum Press, New York, pp. 391–431.

Spemann, H., 1928. Die Entwicklung seitlicher und dorso-ventraler Keimhälften bei verzögerter Kernversorgung. *Z. Wiss. Zool.*, **132**:104–134.

——, 1938. *Embryonic Development and Induction*. Reprinted 1962 by Hafner Publishing Company, New York, 401 pp.

——, and H. Mangold, 1924. Ueber Induktion von Embryonalanlagen durch Implantation ortfremder Organistoren. *Arch. Mikrosk. Anat. Entwmech.*, **100**:599–638.

Spratt, N. T., Jr., 1946. Formation of the primitive streak in the explanted chick blastoderm marked with carbon particles. *J. Exp. Zool.*, **103**:259–304.

——, and H. Haas, 1965. Germ layer formation and the role of the primitive streak in the chick. I. Basic architecture and morphogenetic tissue movements. *J. Exp. Zool.*, **158**:9–38.

Tarkowski, A. K., 1961. Mouse chimeras developed from fused eggs. *Nature* (London), **190**:857–860.

——, and J. Wroblewska, 1967. Development of blastomeres of mouse eggs isolated at the 4- and 8-cell stage. *J. Embryol. Exp. Morphol.*, **18**:155–180.

Theiler, K., 1972. *The House Mouse. Development and Normal Stages from Fertilization to 4 Weeks of Age*. Springer-Verlag, Berlin, 168 pp.

Townes, P. L., and J. Holtfreter, 1955. Directed movements and selective adhesion of embryonic amphibian cells. *J. Exp. Zool.*, **128**:53–120.

Trinkhaus, J. P., 1969. *Cells into Organs*. Prentice-Hall, Inc., Englewood Cliffs, N.J., 237 pp.

Vakaet, L., 1970. Cinephotomicrographic investigations of gastrulation in the chick blastoderm. *Arch. Biol.*, **81**:387–426.

Vogt, W., 1929. Gestaltungsanalyse am Amphibienkeim mit örtlicher Vitalfarbung. II. Gastrulation and Mesodermbildung bei Urodelen und Anuren. *Wilhelm Roux' Arch.*, **120**:385–706.

Waddington, C. H., 1933. Induction by the primitive streak and its derivatives in the chick. *J. Exp. Biol.*, **10**:38–46.

Warner, A. E., 1985. The role of gap junctions in amphibian development. *J. Embryol. Exp. Morph.*, **89**, (Suppl.):365–380.

NEURULATION AND SOMITE FORMATION (CHAPTER 6)

Barth, L. G., and L. J. Barth, 1974. Ionic regulation of embryonic induction and cell differentiation in *Rana pipiens*. *Dev. Biol.*, **39**:1–22.

Bellairs, R., and P. A. Portch, 1977. Somite formation in the chick embryo. In Ede, Hinchliffe, and Balls, *Vertebrate Limb and Somite Morphogenesis*. Cambridge University Press, Cambridge, pp. 449–463.

Bellairs, R., D. A. Ede, and J. W. Lash (eds.), 1986. *Somites in Developing Embryos*. Plenum Press, New York, 320 pp.

Black, I. B., 1982. Stages of neurotransmitter development in autonomic neurons. *Science*, **215**:1198–1204.

Bunge, R., M. Johnson, and C. D. Ross, 1978. Nature and nurture in development of the autonomic neuron. *Science*, **199**:1409–1416.

Burnside, B., 1971. Microtubules and microfilaments in newt neurulation. *Dev. Biol.*, **26**:419–441.

———, 1973. Microtubules and microfilaments in amphibian neurulation. *Am. Zool.*, **13**:989–1006.

———, and A. G. Jacobson, 1968. Analysis of morphogenetic movements in the neural plate of the newt, *Taricha torosa*. *Dev. Biol.*, **18**:537–553.

Chevallier, A., M. Kieny, A. Mauger, and P. Sengel, 1977. Developmental fate of the somitic mesoderm in the chick embryo. In Ede, Hinchliffe, and Balls, *Vertebrate Limb and Somite Morphogenesis*. Cambridge University Press, Cambridge, pp. 421–432.

Cooke, J., 1977. The control of somite number during amphibian development: Models and experiments. In Ede, Hinchliffe, and Balls, *Vertebrate Limb and Somite Morphogenesis*. Cambridge University Press, Cambridge, pp. 433–448.

Crossin, K. L., C.-M. Chuong, and G. M. Edelman, 1985. Expression sequences of cell adhesion molecules. *Proc. Natl. Acad. Sci. USA*, **82**:6942–6946.

Gallera, J., 1971. Primary induction in birds. *Adv. Morphogen.*, **9**:149–180.

Gearhart, J. D., and B. Mintz, 1972. Clonal origins of somites and their muscle derivates: Evidence from allophenic mice. *Dev. Biol.*, **29**:27–37.

Gordon, R., 1985. A review of the theories of vertebrate neurulation and their relationship to the mechanics of neural tube birth defects. *J. Embryol. Exp. Morph.*, **89**, Suppl.:229–255.

Hay, E. D., 1968. Organization and fine structure of epithelium and mesenchyme in the developing chick embryos. In Fleischmajer, *Epithelial-Mesenchymal Interactions*. The Williams & Wilkins Company, Baltimore, Md., pp. 31–55.

———, and S. Meier, 1974. Glycosaminoglycan synthesis by embryonic inductors: Neural tube, notochord and lens. *J. Cell Biol.*, **62**:889–898.

Holtfreter, J., 1947. Changes of structure and the kinetics of differentiating embryonic cells. *J. Morphol.*, **80**:57–92.

Holtfreter, J., 1947. Observations on the migration, aggregation and phagocytosis of embryonic cells. *J. Morphol.*, **80**:25–55.

———, 1968. Mesenchyme and epithelia in inductive and morphogenetic processes. In Fleishmajer and Billingham, *Epithelial-Mesenchymal Interactions*. The Williams & Wilkins Company, Baltimore, Md., pp. 1–30.

Holtzer, H., and S. R. Detwiler, 1953. An experimental analysis of the development of the spinal column. III. Induction of skeletogenous cells. *J. Exp. Zool.*, **123**:335–369.

———, and R. Mayne, 1973. Experimental morphogenesis: The induction of somitic chondrogenesis by embryonic spinal cord and notochord. *Pathobiology of Development*, Am. Soc. Microbiol. and Pathol., pp. 52–64.

Horstadius, S. O., 1950. *The Neural Crest*. Oxford University Press, London, 111 pp.

Jacobson, A. G., 1966. Inductive processes in embryonic development. *Science*, **152**:25–34.

———, and R. Gordon, 1976. Changes in the shape of the developing vertebrate nervous system analyzed experimentally, mathematically and by computer simulation. *J. Exp. Zool.*, **197**:191–246.

Johnson, K. E., 1985. Frog gastrula cells adhere to fibronectin-sepharose beads. In Edelman, *Molecular Determinants of Animal Form*. Alan R. Liss, Inc., New York, pp. 271–292.

Johnston, M. C., A. Bhakdinaronk, and Y. C. Reid, 1973. An expanded role of the neural crest in oral and pharyngeal development. In *Oral Sensation and Perception-Development in the Fetus and Infant*, J. F. Basma (ed.). U.S. Government Printing Office, Washington, D.C., pp. 37–52.

Karfunkel, P., 1973. The mechanisms of neural tube formation. *Int. Rev. Cytol.*, **38**:245–272.

Langman, J., and G. R. Nelson, 1968. A radioautographic study of the development of the somite in the chick embryo. *J. Embryol. Exp. Morphol.*, **19**:217–226.

Lash, J. W., 1968. Somitic mesenchyme and its response to cartilage induction. In Fleischmajer and Billingham, *Epithelial-Mesenchymal Interactions*. The Williams & Wilkins Company, Baltimore, Md., pp. 165–172.

———, and D. Ostrovsky, 1986. On the formation of somites. In L. Browder, *Developmental Biology*, vol. 2. Plenum Press, New York, pp. 547–563.

LeDourin, N., 1982. *The Neural Crest*. Cambridge University Press, Cambridge, 259 pp.

———, and M. A. Teillet, 1974. Experimental analysis of the migration and differentiation of neuroblasts of the autonomic nervous system and of neuroectodermal mesenchymal derivatives, using a biological cell marking technique. *Dev. Biol.*, **41**:162–184.

Lehtonen, E., 1976. Transmission of signals in embryonic induction. *Med. Biol.*, **54**:108–128.

Leussink, J. A., 1970. The spatial distribution of inductive capacities in the neural plate and archenteron roof of urodeles. *Neth. J. Zool.*, **20**:1–79.

Lipton, B. H., and A. G. Jacobson, 1974. Experimental analysis of the mechanisms of somite morphogenesis. *Dev. Biol.*, **38**:91–103.

Malacinski, G. M., and B. W. Youn, 1982. The structure of the anuran amphibian notochord and a re-evaluation of its presumed role in early embryogenesis. *Differentiation*, **21**:13–21.

Meier, S., 1984. Somite formation and its relationship to metameric patterning of the mesoderm. *Cell Differen.*, **14**:235–243.

Newgreen, D. F., and C. A. Erickson, 1986. The migration of neural crest cells. *Int. Rev. Cytol.*, **103**:89–143.

Nieuwkoop, P. D., 1966. Induction and pattern formation as primary mechanisms in early embryonic differentiation. In *Cell Differentiation and Morphogenesis, International Lecture Course*. North-Holland Publishing Company, Amsterdam, pp. 120–143.

Noden, D., 1975. An analysis of the migratory behavior of avian cephalic neural crest cells. *Devel. Biol.*, **42**:106–130.

———, 1978. The control of avian cephalic neural crest cytodifferentiation. I. Skeletal and connective tissues. II. Neural tissues. *Dev. Biol.*, **67**:296–312; **67**:313–329.

———, 1983. The role of the neural crest in patterning of avian cranial skeletal, connective, and muscle tissues. *Devel. Biol.*, **96**:144–165.

———, 1984. Craniofacial development: New views on old problems. *Anat. Rec.*, **208**:1–13.

Rawles, M. E., 1948. Origin of melanophores and their role in development of color patterns in vertebrates. *Physiol. Rev.*, **28**:383–408.

Revel, J.-P., P. Yip, and L. L. Chang, 1973. Cell junctions in the early chick embryo— a freeze etch study. *Dev. Biol.*, **35**:302–317.

Saxén, L., 1978. Two-gradient hypothesis of primary embryonic induction. *Med. Biol.*, **56**:293–298.

———, and S. Toivonen, 1962. *Primary Embryonic Induction*. Logos Press, London, 271 pp.

Schroeder, T. E., 1973. Cell constriction: Contractile role of microfilaments in division and development. *Am. Zool.*, **13**:949–960.

Sheridan, J. D., 1966. Electrophysiological study of special connections between cells in the early chick embryo. *J. Cell Biol.*, **31**:C1-C5.

Spemann, H., 1938. *Embryonic Development and Induction*. Reprinted by Hafner Publishing Company, New York, 1962, 401 pp.

Toivonen, S., D. Tarin, and L. Saxén, 1976. The transmission of morphogenetic signals from amphibian mesoderm to ectoderm in primary induction. *Differentiation*, **5**:49–55.

Tosney, K. W., 1982. The segregation and early migration of cranial neural crest cells in the avian embryo. *Devel. Biol.*, **89**:13–24.

Trelstad, R. L., J.-P. Revel, and E. D. Hay, 1966. Tight junctions between cells in the early chick embryo as visualized with the electron microscope. *J. Cell Biol.*, **31**:C6-C10.

Weston, J. A., 1970. The migration and differentiation of neural crest cells. *Adv. Morphogen.*, **8**:41–114.

Weston, J. A., 1986. Phenotypic diversification in neural crest-derived cells: The time and stability of commitment during early development. *Curr. Top. Devel. Biol.*, **20**:195–210.

FETAL MEMBRANES AND PLACENTATION (CHAPTER 7)

Adamstone, F. B., 1948. Experiments on the development of the amnion in the chick. *J. Morphol.*, **83**:359–371.

Beaconsfield, P., 1980. The placenta. *Sci. Am.*, **243**:95–102.

Blandau, R. J., 1949. Embryo-endometrial interrelationship in the rat and guinea pig. *Anat. Rec.*, **104**:331–359.

Böving, B. G., 1971. Biomechanics of implantation. In *The Biology of the Blastocyst*, R. J. Blandau (ed.). University of Chicago Press, Chicago, pp. 423–442.

Boyd, J. D., and W. J. Hamilton, 1970. *The Human Placenta*. W. Heffer & Sons, Ltd., Cambridge, England, 365 pp.

Dhouailly, D., 1978. Feather-forming capacities of the avian extra-embryonic somato-pleure. *J. Embryol. Exp. Morphol.*, **43**:279–287.

Dunn, B. E., J. S. Graves, and T. P. Fitzharris, 1981. Active calcium transport in the chick chorioallantoic membrane requires interaction with the shell membranes and/or shell calcium. *Devel. Biol.*, **88**:259–268.

Enders, A. C., 1965. A comparative study of the fine structure of the trophoblast in several hemochorial placentae. *Am. J. Anat.*, **116**:29–67.

Harris, J. W. S., and E. M. Ramsey, 1966. The morphology of human uteroplacental vasculature. *Carnegie Cont. to Emb.*, **38**:43–58.

Hertig, A. T., and J. Rock, 1945. Two human ova of the previllous stage, having a developmental age of about 7 and 9 days respectively. *Carnegie Cont. to Emb.*, **31**:65–84.

Heuser, C. H., 1927. A study of the implantation of the ovum of the pig from the stage of the bilaminar blastocyst to the completion of the fetal membranes. *Carnegie Cont. to Emb.*, **19**:229–243.

Luckett, W. P., 1974. Comparative development and evolution of the placenta in primates. In Luckett, *Reproductive Biology of the Primates*. Contrib. to Primatology, vol. 3. S. Karger, Basel, pp. 142–234.

———, 1975. The development of primordial and definitive amniotic cavities in early rhesus monkey and human embryos. *Am. J. Anat.*, **144**:149–168.

———, 1978. Origin and differentiation of the yolk sac and extraembryonic mesoderm in presomite human and rhesus monkey embryos. *Am. J. Anat.*, **152**:59–98.

Kearns, M., and P. K. Lala, 1982. Bone marrow origin in decidual cell precursors in the pseudopregnant mouse uterus. *J. Exp. Med.*, **155**:1537–1554.

Markee, J. E., R. A. Pasqualetti, and J. C. Hinsey, 1936. Growth of intraocular endometrial and transplants in spinal rabbits. *Anat. Rec.*, **64**:247–253.

Mossman, H. W., 1937. Comparative morphogenesis of the fetal membranes and accessory uterine structures. *Carnegie Cont. to Emb.*, **26**:129–246.

Plentl, A., 1966. Formation and circulation of amniotic fluid. *Clin. Obstet. Gynecol.*, **9**:427–439.

Ramsey, E. M., 1962. Circulation in the intervillous space of the primate placenta. *Am. J. Obstet. Gynecol.*, **84**:1649–1663.

———, 1965. The placenta and fetal membranes. In Greenhill, *Obstetrics*, 13th ed., W. B. Saunders Company, Philadelphia, pp. 101–136.

Schuetz, A. W., and N. Dubin, 1981. Progesterone and prostaglandin secretion by ovulated rat cumulus cell-oocyte complexes. *Endocrinology*, **108**:457–463.

Tao, T. W., and A. T. Hertig, 1965. Viability and differentiation of human trophoblast in organ culture. *Am. J. Anat.*, **116**:315–327.

Wangensteen, O. D., 1972. Gas exchange by a bird's embryo. *Resp. Physiol.*, **14**:64–74.

Wilkin, P., 1965. Organogenesis of the human placenta. In DeHaan and Ursprung, *Organogenesis*. Holt, Rinehart and Winston, Inc., New York, pp. 743–769.

Wislocki, G. B., 1929. On the placentation of primates, with a consideration of the phylogeny of the placenta. *Carnegie Cont. to Emb.*, **20**:51–80.

Young, M. F., P. P. Minghetti, and N. W. Klein, 1980. Yolk sac endoderm: Exclusive site of serum protein synthesis in the early chick embryo. *Devel. Biol.*, **75**:239–245.

YOUNG MAMMALIAN EMBRYOS (CHAPTER 8)

Arey, L. B., 1938. The history of the first somite in human embryos. *Carnegie Cont. to Emb.*, **27**:233–269.

Boyden, E. A., 1933. A laboratory atlas of the 13-mm pig embryo. The Wistar Institute Press, Philadelphia, 100 pp.

Corner, G. W., 1929. A well-preserved human embryo of 10 somites. *Carnegie Cont. to Emb.*, **20**:81–102.

Davis, C. L., 1923. Description of a human embryo having twenty paired somites. *Carnegie Cont. to Emb.*, **15**:1–51.

Fallon, J. F., and B. K. Simandl, 1978. Evidence for a role for cell death in the disappearance of the embryonic human tail. *Am. J. Anat.*, **152**:111–130.

Hertig, A. T., J. Rock, E. C. Adams, and W. J. Mulligan, 1954. On the preimplantation stages of the human ovum: A description of four normal and four abnormal specimens ranging from the second to the fifth day of development. *Carnegie Cont. to Emb.*, **35**:199–220.

Heuser, C. H., and G. L. Streeter, 1929. Early stages in the development of pig embryos, from the period of initial cleavage to the time of the appearance of limb buds. *Carnegie Cont. to Emb.*, **20**:1–29.

Lewis, F. T., 1902. The gross anatomy of a 12-mm pig. *Am. J. Anat.*, **2**:211–226.

Murphy, M. E., and E. C. Carlson, 1978. Ultrastructural study of developing extracellular matrix of vitelline blood vessels of the early chick embryo. *Am. J. Anat.*, **151**:345–375.

Streeter, G. L., 1945. Developmental horizons in human embryos. Description of age group XIII, embryos about 4 or 5 mm long, and age group XIV, period of indentation of the lens vesicle. *Carnegie Cont. to Emb.*, **31**:27–63.

———, 1948. Developmental horizons in human embryos. Description of age groups XV, XVI, XVII, and XVIII, being the third issue of a survey of the Carnegie collection. *Carnegie Cont. to Emb.*, **32**:133–204.

———, 1951. Developmental horizons in human embryos. Description of age groups XIX, XX, XXI, XXII, and XXIII, being the fifth issue of a survey of the Carnegie collection. Prepared for publication by C. H. Heuser and G. W. Corner. *Carnegie Cont. to Emb.*, **34**:165–196.

Thyng, F. W., 1911. The anatomy of a 7.8-mm pig embryo. *Anat. Rec.*, **5**:17–45.

CELLULAR DIFFERENTIATION AND THE SKELETAL AND MUSCULAR SYSTEMS (CHAPTER 9)

Barnett, T., C. Pachl, J. P. Gergen, and P. C. Wensink, 1980. The isolation and characterization of yolk protein genes. *Cell*, **21**:729–738.

Bassett, C. A. L., 1971. Biophysical principles affecting bone structure. In Bourne, *The Biochemistry and Physiology of Bone*, Academic Press, Inc., New York, vol. 3, pp. 1–76.

Beermann, W., 1952. Chromomerenkonstanz und specifische Modifikationen der Chromosomenstruktur in der Entwicklung und Organdifferenzierung von *Chironomus tentans*. *Chromosoma*, **5**:139–198.

Blau, H. M., G. K. Pavlath, E. C. Hardeman, C.-P. Chiu, L. Silberstein, S. G. Webster, S. C. Miller, and C. Webster, 1985. Plasticity of the differentiated state. *Science*, **230**:758–766.

Bonner, P. H., and T. R. Adams, 1982. The involvement of nerves in chick myoblast differentiation. In Kelley, Goetinck, and MacCabe, *Limb Development and Regeneration*, Part B. Alan R. Liss, Inc., New York, pp. 349–358.

Bosma, J. F. (ed.), 1976. *Symposium on Development of the Basicranium*. DHEW Publication No. (NIH) 76–989, U.S. Government Printing Office, Washington, D.C., 700 pp.

Caplan, A. I., M. Y. Fiszman, and H. M. E. Eppenberger, 1983. Molecular and cell isoforms during development. *Science*, **221**:921–927.

Carlson, B. M. 1973. The regeneration of skeletal muscle—a review. *Am. J. Anat.*, **137**:119–150.

Christ, B., and R. Čihák (eds.), 1986. *Development and Regeneration of Skeletal Muscles*. S. Karger, Basel, 221 pp.

Čihák, R., 1972. Ontogenesis of the skeleton and intrinsic muscles of the human hand and foot. *Ergebnisse des Anatomie und Entwicklungsgeschichte*, **46**(1):1–194.

Conklin, E. G., 1905. The organization and cell-lineage of the ascidian egg. *J. Acad. Nat. Sci., Philadelphia*, Ser. 2, **13**:1–119.

Davidson, E. H., and R. J. Britten, 1979. Regulation of gene expression: Possible role of repetitive sequences. *Science*, **204**:1052–1059.

deBeer, G. R., 1937. *The Development of the Vertebrate Skull*. Oxford University Press, London, 552 pp.

Drachman, D. B. (ed.), 1974. Trophic functions of the neuron. *Ann. N.Y. Acad. Sci.*, **228**:1–423.

Hall, B. K., 1974. Chondrogenesis of the somitic mesoderm. *Adv. Anat. Embryol. Cell Biol.*, **53**(4):1–50.

———, 1978. *Developmental and Cellular Skeletal Biology*. Academic Press, Inc., New York, 304 pp.

Holtzer, H., 1970. Proliferative and quantal cell cycles in the differentiation of muscle, cartilage and red blood cells. In Padykula, *Gene Expression in Somitic Cells*. Academic Press, Inc., New York, pp. 69–88.

———, J. Biehl, and S. Holtzer, 1985. Induction-dependent and lineage-dependent models for cell diversification are mutually exclusive. In Evans et al., *Advances in Neuroblastoma Research*. Alan R. Liss, Inc., New York, pp. 3–11.

———, and S. R. Detwiler, 1953. An experimental analysis of the development of the spinal column. III. Induction of skeletogenous cells. *J. Exp. Zool.*, **123**:335–370.

———, J. Sasse, A. Horwitz, P. Antin, and M. Pacifici, 1986. Myogenic lineages and myofibrillogenesis. In Christ and Čihák, *Development and Regeneration of Skeletal Muscles*. S. Karger, Basel, pp. 109–125.

Hozumi, N., and S. Tonegawa, 1976. Evidence for somatic rearrangement of immunoglobulin genes coding for variable and constant regions. *Proc. Natl. Acad. Sci. USA*, **73**:3628–3632.

Illmensee, K., and A. P. Mahowald, 1974. Transplantation of posterior pole plasm in *Drosophila*. Induction of germ cells at the anterior pole of the egg. *Proc. Natl. Acad. Sci. USA*, **71**:1016–1020.

Jacob, H. J., B. Christ, and B. Brand, 1986. On the development of trunk and limb muscles in avian embryos. In Christ and Čihák, *Development and Regeneration of Skeletal Muscles*. S. Karger, Basel, pp. 1–23.

Jacob, H. J., B. Christ, and M. Grim, 1982. Problems of muscle pattern formation and of neuromuscular relations in avian limb development. In Kelley, Goetinck, and MacCabe, *Limb Development and Regeneration*, Part B. Alan R. Liss, New York, pp. 333–342.

Jotereau, F. V., and N. M. LeDouarin, 1978. The developmental relationship between osteocytes and osteoclasts: A study using the quail-chick nuclear marker in endochondral ossification. *Dev. Biol.*, **63**:253–265.

Kahn, A. J., and D. J. Simmons, 1975. Investigation of cell lineage in bone using a chimera of chick and quail embryonic tissue. *Nature*, **258**:325–326.

Lash, J. W., 1968. Somitic mesoderm and its response to cartilage induction. In Fleischmajer and Billingham, *Epithelial-Mesenchymal Interactions*. The Williams & Wilkins Company, Baltimore, Md., pp. 165–172.

Liboff, A. R., and R. A. Rinaldi (eds.), 1974. *Electrically Mediated Growth Mechanisms in Living Systems*. Ann. N.Y. Acad. Sci., **238**:1–593.

Manasek, F. J., 1968. Embryonic development of the heart. I. A light and electron microscopic study of myocardial development in the early chick embryo. *J. Morphol.*, **125**:329–366.

——, 1968. Mitosis in developing cardiac muscle. *J. Cell Biol.*, **37**:191–196.

Mauro, A. (ed.), 1979. *Muscle Regeneration*. Raven Press, New York, 560 pp.

Merrifield, P. A., R. S. Compton, and I. R. Konigsberg, 1984. Cell cycle dependence of differentiation in synchronous skeletal muscle myocytes. *Exp. Biol. Med.*, **9**:1–9.

Mintz, B., and W. B. Baker, 1967. Normal mammalian muscle differentiation and gene control of isocitrate dehydrogenase synthesis. *Proc. Nat. Acad. Sci.*, **58**:592–598.

Moss, M. L., C. R. Noback, and G. G. Robertson, 1955. Critical developmental horizons in human fetal long bones: Correlated quantitative and histological criteria. *Am. J. Anat.*, **97**:155–175.

Murray, P. D. F., 1936. *Bones. A Study of the Development and Structure of the Vertebrate Skeleton*. Cambridge University Press, New York, 203 pp.

Noden, D. M., 1983. The embryonic origins of avian cephalic and cervical muscles and associated connective tissues. *Am. J. Anat.*, **168**:257–276.

Old, R. W., H. G. Callan, and K. W. Gross, 1977. Localization of histone gene transcripts in newt lampbrush chromosomes by in situ hybridization. *J. Cell Sci.*, **27**:57–80.

O'Rahilly, R., 1961. The developmental anatomy of the extensor assembly. *Acta Anat.*, **47**:363–375.

Pearson, M. L., and H. F. Epstein (eds.), 1982. *Muscle Development. Molecular and Cellular Control*. Cold Spring Harbor Lab., Cold Spring Harbor, N.Y., 581 pp.

Rumyantsev, P. P., 1967. Electron microscopic analysis of cell elements in the differentiation and proliferation processes in the developing myocardium (Russian). *Arkh. Anat. Gistol. Embryol.*, **52**:67–77.

——, 1982. *Cardiomyocytes in Processes of Reproduction, Differentiation and Regeneration* (Russian). Nauka, Leningrad, 288 pp.

Seed, J., and S. D. Hauschka, 1984. Temporal separation of the migration of distinct myogenic precursor populations into the developing chick wing bud. *Dev. Biol.*, **106**:389–393.

Sensenig, E. C., 1949. The early development of the human vertebral column. *Carnegie Cont. to Emb.*, **33**:21–41.

Snow, M. H., 1977. Myogenic cell formation in regenerating rat skeletal muscle injured by mincing I and II. *Anat. Rec.*, **188**:181–218.

Straus, W. L., and M. E. Rawles, 1953. An experimental study of the origin of the trunk musculature and ribs in the chick. *Am. J. Anat.*, **92**:471–509.

Wachtler, F., and M. Jacob, 1986. Origin and development of the cranial skeletal muscles. In Christ and Čihák, *Development and Regeneration of Skeletal Muscles*. S. Karger, Basel, pp. 24–46.

Wakelam, M. J. O., 1985. The fusion of myoblasts. *Biochem. J.*, **228**:1–12.

Weismann, A., 1892. *Das Keimplasma*. G. Fischer, Jena, 628 pp.

Whalen, R. G., L. S. Bugaisky, G. S. Butler-Browne, S. M. Sell, K. Schwartz, and I. Pinset-Härström, 1982. Characterization of myosin isozymes appearing during rat muscle development. In Pearson and Epstein (eds.), *Muscle Development*. Cold Spring Harbor, N.Y., pp. 25–33.

——, S. M. Sell, A. Erickson, and L. E. Thornell, 1982. Myosin subunit types in skeletal and cardiac tissues and their developmental distribution. *Devel. Biol.*, **91**:478–484.

White, N. K., P. H. Bonner, D. R. Nelson, and S. D. Hauschka, 1975. Clonal analysis of vertebrate myogenesis IV. Medium-dependent classification of colony-forming cells. *Devel. Biol.*, **44**:346–361.

Whittaker, J. R., G. Ortolani, and N. Farinella-Ferruzza, 1977. Autonomy of acetylcholinesterase differentiation in muscle-lineage cells of ascidian embryos. *Devel. Biol.*, **55**:196–200.

Wilson, E. B., 1904. Experimental studies on germinal localization. I. The germ regions in the egg of *Dentalium*. II. Experiments on the cleavage-mosaic in *Patella* and *Dentalium*. *J. Exp. Zool.*, **1**:1–72.

Zelená, J., 1957. The morphogenetic influence of innervation on the ontogenetic development of muscle spindles. *J. Embryol. Exp. Morph.*, **5**:283–292.

THE SKIN AND ITS DERIVATIVES (CHAPTER 10)

Assheton, R., 1896. Notes on the ciliation of the ectoderm of the amphibian embryo. ·*Quart. J. Micros. Sci.*, **38**:465–484.

Bullough, W. S., 1972. The control of epidermal thickness. *Brit. J. Dermatol.*, **87**:187–199.

Chuang, H.-H., and R.-X. Dai, 1961. Concerning the conductivity of the embryonic epithelium in the amphibian (in Chinese). *Kexue Tongbao*, **12**:41–43.

Chuang-Tseng, M. P., H. H. Chuang, C. Sandri, and K. Akert, 1982. Gap junctions and impulse propagation in embryonic epithelium of amphibia. *Cell Tissue Res.*, **225**:249–258.

Cohen, S., and G. A. Elliott, 1963. The stimulation of epidermal keratinization by a protein isolated from the submaxillary gland of the mouse. *J. Investig. Dermatol.*, **40**:1–6.

Coulombre, J. L., and A. J. Coulombre, 1971. Metaplastic induction of scales and feathers in the corneal anterior epithelium of the chick embryo. *Develop. Biol.*, **25**:464–478.

Dale, B. A., K. A. Holbrook, J. R. Kimball, M. Hoff, and T.-T. Sun, 1985. Expression of epidermal keratins and filaggrin during human fetal skin development. *J. Cell Biol.*, **101**:1257–1269.

Dürnberger, H., and K. Kratochwil, 1980. Specificity of time interaction and origin of mesenchymal cells in the androgen response of the embryonic mammary gland. *Cell*, **19**:465–471.

Fuchs, E., and H. Green, 1980. Changes in keratin gene expression during terminal differentiation of the keratinocyte. *Cell*, **19**:1033–1042.

Goetinck, P. F., and M. J. Sekellick, 1972. Observation on collagen synthesis, lattice formation, and morphology of scaleless and normal embryonic skin. *Devel. Biol.*, **28**:636–648.

Hogg, N. A. S., C. J. Harrison, and C. Tickle, 1983. Lumen formation in the developing mouse mammary gland. *J. Embryol. Exp. Morphol.*, **73**:39–57.

Holbrook, K. A., 1983. Structure and function of the developing human skin. In Goldsmith, *Biochemistry and Physiology of the Skin*, vol. 1. Oxford University Press, New York, pp. 64–101.

———, and G. F. Odland, 1975. The fine structure of developing human epidermis: Light, scanning and transmission electron microscopy of the periderm. *J. Investig. Dermatol.*, **65**:16–38.

Kratochwil, K., 1971. In vitro analysis of the hormonal basis for the sexual dimorphism in the embryonic development of the mouse mammary gland. *J. Embryol. Exp. Morphol.*, **25**:141–153.

———, and P. Schwartz, 1976. Tissue interaction in androgen response of embryonic mammary rudiment of mouse: Identification of target time for testosterone. *Proc. Natl. Acad. Sci. USA*, **73**:4041–4044.

Landström, U., 1977. On the differentiation of prospective ectoderm to a ciliated cell pattern in embryo of *Ambystoma mexicanum*. *J. Embryol. Exp. Morphol.*, **41**:23–32.

Marks, R., S. Barton, and R. Marshall, 1983. Aspects of the physiology and pathophysiology of desquamation. In Seiji and Bernstein (eds.), *Normal and Abnormal Epidermal Differentiation*. University of Tokyo Press, Tokyo, pp. 195–205.

Mauger, A., M. Démarchez, and P. Sengel, 1984. Role of extracellular matrix and dermal-epidermal junction architecture in skin development. In Kemp and Hinchliffe (eds.), *Matrices and Cell Differentiation*. Alan R. Liss, Inc., New York, pp. 115–128.

McAleese, S. R., and R. H. Sawyer, 1982. Avian scale development. IX. Scale formation by scaleless (sc/sc) epidermis under the influence of normal scale dermis. *Devel. Biol.*, **89**:493–502.

Nickerson, M., 1944. An experimental analysis of barred pattern formation in feathers. *J. Exp. Zool.*, **95**:361–397.

Ohsugi, K., and H. Ide, 1983. Melanophore differentiation in *Xenopus laevis* with special reference to dorsoventral pigment pattern formation. *J. Embryol. Exp. Morphol.*, **75**:141–150.

Potten, C. S., 1974. The epidermal proliferative unit: The possible role of the central basal cell. *Cell Tissue Kinet.*, **7**:77–88.

Rawles, M. E., 1948. Origin of melanophores and their role in development of color patterns in vertebrates. *Physiol. Rev.*, **28**:383–408.

Sakakura, T., Y. Nishizuka, and C. J. Dawe, 1976. Mesenchyme-dependent morphogenesis and epithelium-specific cytodifferentiation in mouse mammary gland. *Science*, **194**:1439–1441.

Sakakura, T., Y. Sakagami, and Y. Nishizuka, 1982. Dual origin of mesenchymal tissues participating in mouse mammary gland morphogenesis. *Develop. Biol.*, **91**:202–207.

Sawyer, R. H., 1972. Avian scale development I and II. *J. Exp. Zool.*, **181**:365–384, 385–408.

———, 1983. The role of epithelial-mesenchymal interactions in regulating gene expression during avian scale morphogenesis. In Sawyer and Fallon, *Epithelial-Mesenchymal Interactions in Development*. Praeger Publishers, New York, pp. 115–146.

Sengel, P., 1958. Recherches expérimentales sur la différenciation des germes plumaires et du pigment de la peau de l'embryon de poulet en culture *in vitro*. *Ann. Sci. Nat.* (*Zool.*), **11**:430–514.

———, 1976. *Morphogenesis of Skin*. Cambridge University Press, Cambridge, 277 pp.

———, 1983. Epidermal-dermal interactions during formation of skin and cutaneous appendages. In Goldsmith, *Biochemistry and Physiology of the Skin*, vol. 1. Oxford University Press, New York, pp. 102–131.

Spemann, H., and O. Schotté, 1932. Ueber xenoplastische Transplantation als Mittel zur Analyse der embryonalen Induktion. *Naturwissenschaften*, **20**:463–467.

Topper, Y. J., and C. S. Freeman, 1980. Multiple hormone interactions in the developmental biology of the mammary gland. *Physiol. Rev.*, **60**:1049–1106.

Twitty, V., 1949. Developmental analysis of amphibian pigmentation. *Growth Symposium*, **9**:133–161.

Willier, B. H., and M. E. Rawles, 1944. Genotypic control of feather color pattern as demonstrated by the effects of a sex-linked gene upon the melanophores. *Genetics*, **29**:309–330.

Wolff, K., and G. Stingl, 1983. The Langerhans cell. *J. Investig. Dermatol.*, **80** (Suppl. 6): 17S-21S.

LIMB DEVELOPMENT (CHAPTER 11)

Amprino, R., 1965. Aspects of limb morphogenesis in the chicken. In DeHaan and Ursprung, *Organogenesis*. Holt, Rinehart and Winston, Inc., New York, pp. 255–281.

Cairns, J. W., 1965. Development of grafts from mouse embryos to the wing bud of the chick embryo. *Dev. Biol.*, **12**:36–52.

Cameron, J., and J. F. Fallon, 1977. The absence of cell death during development of free digits in amphibians. *Dev. Biol.*, **55**:331–338.

Caplan, A. I., and S. Kautroupas, 1973. The control of muscle and cartilage development in the chick limb: The role of differential vascularization. *J. Embryol. Exp. Morphol.*, **29**:571–583.

Chevallier, A., M. Kieny, A. Mauger, and P. Sengel, 1977. Developmental fate of the somitic mesoderm in the chick embryo. In Ede, Hinchliffe, and Balls, *Vertebrate Limb and Somite Morphogenesis*. Cambridge University Press, Cambridge, pp. 421–432.

Christ, B., H. J. Jacob, and M. Jacob, 1977. Experimental analysis of the origin of the wing musculature in avian embryos. *Anat. Embryol.*, **150**:171–186.

Čihák, R., 1972. Ontogenesis of the skeleton and intrinsic muscles of the human hand and foot. *Adv. Anat. Embryol. Cell Biol.*, **46**:1–194.

Ede, D. A., J. R. Hinchliffe, and M. Balls (eds.), 1977. *Vertebrate Limb and Somite Morphogenesis*. Cambridge University Press, Cambridge, 498 pp.

Fallon, J. F., and A. I. Caplan (eds.), 1983. *Limb Development and Regeneration*. Parts A and B. Alan R. Liss, Inc., New York, 639 and 434 pp.

———, and G. M. Crosby, 1975. Normal development of the chick wing following removal of the polarizing zone. *J. Exp. Zool.*, **193**:449–455.

———, and R. O. Kelley, 1977. Ultrastructural analysis of the apical ectodermal ridge during vertebrate limb morphogenesis. *J. Embryol. Exp. Morphol.*, **41**:223–232.

Forsthoefel, P. F., 1963. Observations on the sequence of blastemal condensations in the limbs of the mouse embryo. *Anat. Rec.*, **147**:129–138.

Grim, M., 1972. Ultrastructure of the ulnar portion of the contrahent muscle layer in the embryonic human hand. *Folia Morphol. (Praha)*, **20**:113–115.

Harrison, R. G., 1918. Experiments on the development of the forelimb of *Ambystoma*, a self-differentiating equipotential system. *J. Exp. Zool.*, **25**:413–461.

———, 1921. On relations of symmetry in transplanted limbs. *J. Exp. Zool.*, **32**:1–136.

Hinchliffe, J. R., and D. R. Johnson, 1980. *The Development of the Vertebrate Limb*. Clarendon Press, Oxford, 268 pp.

Iten, L. E., 1982. Pattern specification and pattern regulation in the embryonic chick limb bud. *Am. Zool.*, **22**:117–129.

Kelley, R. O., and J. F. Fallon, 1976. Ultrastructural analysis of the apical ectodermal ridge during morphogenesis. I. The human forelimb with special reference to gap junctions. *Dev. Biol.*, **51**:241–256.

MacCabe, J. A., A. J. Calandra, and B. W. Parker, 1977. *In vitro* analysis of the distribution and nature of a morphogenetic factor in the developing chick wing. In Ede, Hinchliffe, and Balls, *Vertebrate Limb and Somite Morphogenesis*. Cambridge University Press, Cambridge, pp. 25–39.

Milaire, J., 1969. Some histochemical considerations of limb development. In Swinyard, *Limb Development and Deformity: Problems of Evaluation and Rehabilitation*. Charles C Thomas, Springfield, Ill., pp. 70–77.

Pautau, M. P., 1977. Dorso-ventral axis determination of chick limb bud development. In Ede, Hinchcliffe, and Balls, *Vertebrate Limb and Somite Morphogenesis*. Cambridge University Press, Cambridge, pp. 257–266.

Piatt, J., 1956. Studies on the problem of nerve pattern. I. Transplantation of the forelimb primordium to ectopic sites in *Ambystoma*. *J. Exp. Zool.*, **131**:173–202.

————, 1957. Studies on the problem of nerve pattern. II. Innervation of the intact forelimb by different parts of the central nervous system in *Ambystoma*. *J. Exp. Zool.*, **134**:103–126.

Rubin, L., and J. W. Saunders, 1972. Ectodermal-mesodermal interactions in the growth of limb buds in the chick embryo: Constancy and temporary limits of the ectodermal induction. *Dev. Biol.*, **28**:94–112.

Saunders, J. W., 1948. The proximodistal sequence of origin on the parts of the chick wing and the role of the ectoderm. *J. Exp. Zool.*, **108**:363–403.

————, 1969. The interplay of morphogenetic factors. In Swinyard, *Limb Development and Deformity: Problems of Evaluation and Rehabilitation*. Charles C Thomas, Springfield, Ill., pp. 89–100.

————, J. M. Cairns, and M. T. Gasseling, 1957. The role of the apical ridge of ectoderm in the differentiation of the morphological structure and inductive specificity of limb parts in the chick. *J. Morphol.*, **101**:57–87.

————, M. T. Gasseling, and L. C. Saunders, 1962. Cellular death in morphogenesis of the avian wing. *Devel. Biol.*, **5**:147–178.

————, and M. T. Gasseling, 1968. Ectodermal-mesenchymal interactions in the origin of limb symmetry. In Fleischmajer and Billingham, *Epithelial-Mesenchymal Interactions*. The Williams and Wilkins Company, Baltimore, Md., pp. 78–97.

Seichert, V., and Z. Rychter, 1971. Vascularization of the developing anterior limb of the chick embryo. *Folia Morphol. (Praha)*, **19**:367–377.

Shellswell, G. B., and L. Wolpert, 1977. The pattern of muscle and tendon development in the chick wing. In Ede, Hinchliffe, and Balls, *Vertebrate Limb and Somite Morphogenesis*. Cambridge University Press, Cambridge, pp. 71–86.

Stark, R. J., and R. L. Searls, 1973. A description of chick wing bud development and a model of limb morphogenesis. *Dev. Biol.*, **33**:138–153.

Sullivan, G. E., 1962. Anatomy and embryology of the wing musculature of the domestic fowl (*Gallus*). *Australian J. Zool.*, **10**:458–518.

Summerbell, D., and L. S. Honig, 1982. The control of pattern across the antero-posterior axis of the chick limb bud by a unique signalling region. *Am. Zool.*, **22**:105–116.

————, J. H. Lewis, and L. Wolpert, 1973. Positional information in chick limb morphogenesis. *Nature*, **244**:492–496.

Swett, F. H., 1937. Determination of limb-axes. *Quart. Rev. Biol.*, **12**:322–339.

Tickle, C., B. Alberts, L. Wolpert, and J. Lee, 1982. Local application of retinoic acid to the limb bud mimics the action of the polarizing region. *Nature* (London), **298:**564–565.

Wilson, D. J., and A. Orr-Urtereger, 1986. Aspects of vascular differentiation in the developing chick wing. *Acta Histochem.*, Suppl., **32:**151–157.

Wolpert, L., J. Lewis, and D. Summerbell, 1975. Morphogenesis of the vertebrate limb. In *Cell Patterning.* Ciba Foundation Symposium 29 (new series), pp. 95–130.

Zwilling, E., 1949. The role of epithelial components in the developmental origin of the "wingless" syndrome of chick embryos. *J. Exp. Zool.*, **111:**175–187.

———, 1956. Interaction between limb bud ectoderm and mesoderm in the chick embryo. I. *J. Exp. Zool.*, **132:**157–172; II. *J. Exp. Zool.*, **132:**173–188; III. *J. Exp. Zool.*, **132:**219–240; IV. *J. Exp. Zool.*, **132:**241–254.

———, 1961. Limb morphogenesis. *Adv. Morphogen.*, **1:**301–330.

NERVOUS SYSTEM AND SENSE ORGANS
(CHAPTERS 12 AND 13)

Angevine, J. B., and R. L. Sidman, 1961. Autoradiographic study of cell migration during histogenesis of cerebral cortex in the mouse. *Nature*, **192:**766–768.

Anniko, M., 1983. Embryonic development of vestibular sense organs and their innervation. In Romand, *Development of Auditory and Vestibular Systems.* Academic Press, New York, pp. 375–423.

Balinsky, B. I., 1925. Transplantation des Ohrbläschens bei *Triton. Roux Arch.*, **105:**718–731.

Bartelmez, G. W., 1922. The origin of the otic and optic primordia in man. *J. Comp. Neurol.*, **34:**201–232.

Bennett, M. R., 1983. Development of neuromuscular synapses. *Physiol. Revs.*, **63:**915–1048.

Black, I. B., 1982. Stages of neurotransmitter development in autonomic neurons. *Science*, **215:**1198–1204.

Boulder Committee, The, 1970. Embryonic vertebrate central nervous system: Revised terminology. *Anat. Rec.*, **166:**257–262.

Bradley, R. M., and C. M. Mistretta, 1975. Fetal sensory receptors. *Physiol. Rev.*, **55:**352–382.

Bunge, R., M. Johnson, and C. D. Ross, 1978. Nature and nurture in development of the autonomic neuron. *Science*, **199:**1409–1416.

Clarke, P. G. H., 1985. Neuronal death in the development of the vertebrate nervous system. *Trends in Neurol. Sci.*, August, pp. 345–349.

Clayton, R. M., 1970. Problems of differentiation in the vertebrate lens. *Curr. Top. Dev. Biol.*, **5:**115–180.

Coulombre, A. J., 1956. The role of intraocular pressure in the development of the chick eye. I. Control of eye size. *J. Exp. Zool.*, **133:**211–226.

———, 1965. The Eye. In DeHaan and Ursprung, *Organogenesis.* Holt, Rinehart and Winston, Inc., New York, pp. 219–251.

———, and J. L. Coulombre, 1957. The role of intraocular pressure in the development of the chick eye: III. Ciliary body. *Am. J. Ophthal.*, **44**(4), part 2:85–92.

———, and ———, 1958. Corneal development. I. Corneal transparency. *J. Cell. and Comp. Physiol.*, **51:**1–11.

Coulombre, J. L., and A. J. Coulombre, 1963. Lens development. Fiber elongation and lens orientation. *Science*, **142**:1489–1490.

Detwiler, S. R., 1920. On the hyperplasia of nerve centers resulting from excessive peripheral loading. *Proc. Nat. Acad. Sci.*, **6**:96–101.

———, 1936. *Neuroembryology: An Experimental Study*. The Macmillan Company, New York, 218 pp.

Gans, C., and R. G. Northcutt, 1983. Neural crest and the origin of vertebrates: A new head. *Science*, **220**:268–274.

Geren, B. B., 1954. The formation from the Schwann cell surface of myelin in the peripheral nerves of chick embryos. *Exp. Cell Res.*, **7**:558–562.

Gimlich, R. L., and J. Cooke, 1983. Cell lineage and the induction of second nervous systems in amphibian development. *Nature* (London), **306**:471–473.

Goodman, C., and M. J. Bastiani, 1984. How embryonic nerve cells recognize one another. *Sci. Am.*, **251**(6):58–66.

Gorski, R. A., R. E. Harlan, C. D. Jacobson, J. E. Shryne, and A. M. Southam, 1980. Evidence for the existence of a sexually dimorphic nucleus in the preoptic area of the rat. *J. Comp. Neurol.*, **193**:529–539.

Gottlieb, G., 1976. Conceptions of prenatal development: Behavioral embryology. *Psych. Rev.*, **83**:215–234.

Goy, R. W., and B. S. McEwen, 1980. *Sexual Differentiation of the Brain*. MIT Press, Cambridge, Mass., 223 pp.

Hamburger, V., 1958. Regression versus peripheral control of differentiation in motor hypoplasia. *Am. J. Anat.*, **102**:365–410.

———, 1975. Changing concepts in developmental neurobiology. *Perspect. Biol. Med.*, **18**:162–178.

Harris, M. J., and M. J. McLeod, 1982. Eyelid growth and fusion in fetal mice. *Anat. Embryol.*, **164**:207–220.

Harrison, R. G., 1908. Embryonic transplantation and the development of the nervous system. *Anat. Rec.*, **2**:385–410.

———, 1910. The outgrowth of the nerve fiber as a mode of protoplasmic movement. *J. Exp. Zool.*, **9**:787–848.

Hay, E. D., 1980. Development of the vertebrate cornea. *Internat. Rev. Cytol.*, **63**:263–322.

———, and J. W. Dodson, 1973. Secretion of collagen by corneal epithelium. *J. Cell Biol.*, **57**:190–213.

———, and J.-P. Revel, 1969. *Fine Structure of the Developing Avian Cornea*. Monogr. Dev. Biol., S. Karger, Basel, vol. 1, 144 pp.

Hirose, G., and M. Jacobson, 1979. Clonal organization of the central nervous system of the frog. I. Clones stemming from individual blastomeres of the 16-cell and earlier stages. *Dev. Biol.*, **71**:191–202.

Hooker, D., 1952. *The Prenatal Origin of Behavior*. Porter Lectures, series 18, University of Kansas Press, Lawrence, 143 pp.

Hughes, A., 1961. Cell degeneration in the larval ventral horn of *Xenopus laevis* (Daudin). *J. Embyol. Exp. Morphol.*, **9**:269–284.

———, 1968. *Aspects of Neural Ontogeny*. Logos Press, London, 249 pp.

Jacobson, M., 1978. *Developmental Neurobiology*. Plenum Press, New York, 2d ed., 562 pp.

———, 1985. Clonal analysis and cell lineages of the vertebrate central nervous system. *An. Rev. Neurosci.*, **8**:71–102.

Källén, B., 1953. On the significance of the neuromeres and similar structures in vertebrate embryos. *J. Embryol. Exp. Morphol.*, **1**:387–392.

Keynes, R. J., and C. D. Stern, 1984. Segmentation in the vertebrate nervous system. *Nature* (London), **310**:786–789.

Konyukhov, B. V., and M. P. Vakhrusheva, 1969. Abnormal development of eyes in mice homozygous for the fidget gene. *Teratology*, **2**:147–158.

Landmesser, L., 1984. The development of specific motor pathways in the chick embryo. *Trends in Neurol. Sci.*, **7**:336–339.

Langman, J., R. L. Guerrant, and B. A. Freeman, 1966. Behavior of neuroepithelial cells during closure of the neural tube. *J. Comp. Neurol.*, **127**:399–412.

Le Douarin, N. M., 1986. Cell line segregation during peripheral nervous system ontogeny. *Science*, **231**:1515–1522.

———, and M.-A. M. Teillet, 1974. Experimental analysis of the migration and differentiation of neuroblasts of the autonomic nervous system and of neuroecto-dermal mesenchymal derivatives using a biological cell marking technique. *Devel. Biol.*, **41**:162–184.

Letourneau, P. C., 1982. Nerve fiber growth and its regulation by extrinsic factors. In Spitzer (ed.), *Neuronal Development*. Plenum Press, New York, pp. 213–254.

Levi-Montalcini, R., 1958. Chemical stimulation of nerve growth. In McElroy and Glass, *The Chemical Basis of Development*. The Johns Hopkins Press, Baltimore, Md., pp. 646–664.

———, 1976. The nerve growth factor: Its role in growth, differentiation and function of the sympathetic adrenergic neuron. *Prog. Brain Res.*, **45**:235–258.

———, and P. U. Angeletti, 1961. Growth control of the sympathetic system by a specific protein factor. *Quart. Rev. Biol.*, **36**:99–108.

Lewis, W. H., 1904. Experimental studies on the development of the eye in amphibia. I. On the origin of the lens. *Am. J. Anat.*, **3**:505–536.

———, 1905. Experimental studies on the development of the eye in amphibia. II. On the cornea. *J. Exp. Zool.*, **2**:431–446.

Lewis, J., A. Chevallier, M. Kieny, and L. Wolpert, 1981. Muscle nerves do not develop in chick wing devoid of muscle. *J. Embryol. Exp. Morph.*, **64**:211–232.

Lim, D. J., and M. Anniko, 1985. Developmental morphology of the mouse inner ear. *Acta Oto-Laryngol.*, Suppl., **422**:1–69.

Lopashov, G. V., and O. G. Stroeva, 1964. *Development of the Eye*. Israel Program for Scientific Translations, Jerusalem, 177 pp.

Mann, I., 1964. *The Development of the Human Eye*. Grune & Stratton, Inc., New York, 3d ed., 316 pp.

Meier, S., and E. D. Hay, 1974. Control of corneal differentiation by extracellular materials. Collagen as a promoter and stabilizer of epithelial stroma production. *Dev. Biol.*, **38**:249–270.

Noden, D. M., and T. Van De Water, 1986. The developing ear: Tissue origins and interactions. In Ruben et al. (eds)., *The Biology of Change in Otolaryngology*. Elsevier, Amsterdam, pp. 15–46.

Nottebohm, F., 1980. Testosterone triggers growth of brain vocal control nuclei in adult female canaries. *Brain Res.*, **189**:429–436.

O'Rahilly, R., and E. Gardner, 1977. The developmental anatomy and histology of the human central nervous system. In Vinken and Bruyn, *Handbook of Clinical Neurology, vol. 30: Congenital Malformations of the Brain and Skull*. North-Holland Publishing Company, Amsterdam, pp. 15–40.

Papaconstantinou, J., 1967. Molecular aspects of lens differentiation. *Science*, **156**:338–346.

Patterson, P. H., 1978. Environmental determination of autonomic neurotransmitter functions. *Ann. Rev. Neurosci*, **1**:1–17.

Piatigorsky, J., 1981. Lens differentiation in vertebrates. *Differentiation*, **19**:134–153.

Purves, D., and J. W. Lichtman, 1985. *Principles of Neural Development*. Sinauer, Sunderland, Mass., 433 pp.

Rakic, P., 1975. Cell migration and neuronal ectopias in the brain. In Bergsma, *Morphogenesis and Malformation of Face and Brain*. Birth Defects: Original Article Series, II (7):95–129.

———, and R. L. Sidman, 1973. Weaver mutant mouse cerebellum: Defective neuronal migration secondary to specific abnormality of Bergmann glia. *Proc. Natl. Acad. Sci. USA*, **70**:240–244.

Reeder, R., and E. Bell, 1965. Short- and long-lived messenger RNA in embryonic chick lens. *Science*, **150**:71–72.

Reyer, R. W., 1977. The amphibian eye: Development and regeneration. In Crescitelli, *Handbook of Sensory Physiology, vol. VII, 15: The Visual System in Vertebrates*. Springer-Verlag, Berlin, pp. 309–330.

Ruben, R. J., and T. R. Van De Water, 1983. Recent advances in the developmental biology of the inner ear. In Gerber and Mencher, *Development of Auditory Behavior*. Grune and Stratton, New York, p. 3.

Sauer, F. C., 1935. The cellular structure of the neural tube. *J. Comp. Neurol.*, **63**:13–23.

———, and B. E. Walker, 1959. Radioautoradiographic study of interkinetic nuclear migration in the neural tube. *Proc. Soc. Exp. Biol. Med.*, **101**:557–560.

Schmechel, D. E., and P. Rakic, 1979. A Golgi study of radial cells in developing monkey telencephalon: Morphogenesis and transformation into astrocytes. *Anat. Embryol.*, **156**:115–152.

Spemann, H., 1912. Zur Entwicklung des Wirbeltierauges. *Zool. Jahrb., Abt. allg. Zool. Physiol. Tiere*, **32**:1–98.

———, 1938. *Embryonic Development and Induction*. Yale University Press, New Haven, Conn., 401 pp.

Streeter, G. L., 1906. On the development of the membranous labyrinth and the acoustic and facial nerves in the human embryo. *Am. J. Anat.*, **6**:139–166.

———, 1922. Development of the auricle in the human embryo. *Carnegie Cont. to Emb.*, **14**:111–138.

Toole, B. P., and R. L. Trelstad, 1971. Hyaluronate production and removal during corneal development in the chick. *Dev. Biol.*, **26**:28–35.

Tosney, K. W., and L. T. Landmesser, 1985. Development of the major pathways for neurite outgrowth in the chick hindlimb. *Devel. Biol.*, **109**:193–214.

Weiss, P., 1934. *In vitro* experiments on the factors determining the course of the outgrowing nerve fiber. *J. Exp. Zool.*, **68**:393–448.

———, and H. B. Hiscoe, 1948. Experiments on the mechanism of nerve growth. *J. Exp. Zool.*, **107**:315–396.

Windle, W. F., and D. W. Orr, 1934. The development of behavior in chick embryos: Spinal cord structure correlated with early somatic motility. *J. Comp. Neurol.*, **60**:287–307.

Zelená, J. 1964. Development, degeneration and regeneration of receptor organs. In Singer and Schadé, *Mechanisms of Neural Regeneration*. Progess in Brain Research, vol. 13, pp. 175–213.

Zwann, J., and R. W. Hendrix, 1973. Changes in cell and organ shape during early development of the ocular lens. *Am. Zool.*, **13**:1039–1049.

———, and A. Ikeda, 1968. Macromolecular events during differentiation of the chicken eye lens. *Exp. Eye Res.*, **7**:301–311.

THE FACE, ORAL REGION, AND TEETH (CHAPTER 14)

Bernfield, M., S. D. Banerjee, J. E. Koda, and A. C. Rapraeger, 1984. Remodelling of the basement membrane as a mechanism of morphogenetic tissue interaction. In Trelstad (ed.), *The Role of Extracellular Matrix in Development*. Alan R. Liss, New York, pp. 545–572.

———, R., R. H. Cohn, and S. D. Banerjee, 1973. Glycosaminoglycans and epithelial organ formation. *Am. Zool.*, **13**:1067–1083.

Bevelander, G., 1941. The development and structure of the fiber system of dentin. *Anat. Rec.*, **81**:79–97.

Bradley, R. B., and C. M. Mistretta, 1975. Fetal sensory receptors. *Physiol. Rev.*, **55**:352–382.

Dahlberg, A. A. (ed.), 1971. *Dental Morphology and Evolution*. University of Chicago Press, Chicago, 350 pp.

Ferguson, M. W. J., 1978. Palatal shelf elevation in the Wistar rat fetus. *J. Anat.*, **125**:555–577.

———, and L. S. Honig, 1984. Epithelial-mesenchymal interactions during vertebrate palotogenesis. *Curr. Top. Devel. Biol.*, **19**:137–164.

Green, R. M., and R. M. Pratt, 1976. Developmental aspects of secondary palate formation. *J. Embryol. Exp. Morphol.*, **36**:225–245.

Grobstein, C., 1953. Epithelio-mesenchymal specificity in the morphogenesis of mouse submandibular rudiments in vitro. *J. Exp. Zool.*, **124**:383–413.

Hall, B. K., 1987. Tissue interactions in the development and evolution of the vertebrate head. In Maderson (ed.), *Development and Evolution of the Neural Crest*. John Wiley & Sons, in press.

Koch, W. E., 1967. In vitro differentiation of tooth rudiments of embryonic mice. I. Transfilter interaction of embryonic incisor tissues. *J. Exp. Zool.*, **165**:155–170.

Kollar, E. J., 1981. Tooth development and dental patterning. In Connelly, Brinkley, and Carlson, *Morphogenesis and Pattern Formation*, Raven Press, New York, pp. 87–102.

———, and G. R. Baird, 1969. The influence of the dental papilla on the development of tooth shape in the embryonic mouse tooth germs. *J. Embryol. Exp. Morphol.*, **21**:131–148.

———, and C. Fisher, 1980. Tooth inductions in chick epithelium: Expression of quiescent genes for enamel synthesis. *Science*, **207**:993–995.

Kraus, B. S., H. Kitamura, and R. A. Latham, 1966. *Atlas of Developmental Anatomy of the Face*. Hoeber Harper, New York, 378 pp.

Lawson, K. A., 1974. Mesenchyme specificity of rodent salivary gland development: The response of salivary epithelium to lung mesenchyme in vitro. *J. Embryol. Exp. Morphol.*, **32**:469–493.

Mistretta, C. M., 1972. Topographical and histological study of the developing rat tongue, palate and taste buds. In Bosma, *Third Symposium on Oral Sensation and Perception: The Mouth of the Infant*. Charles C Thomas, Springfield, Ill., pp. 163–187.

Noden, D. M., 1984. Craniofacial development: New views on old problems. *Anat. Rec.*, **208**:1–13.

Northcutt, R. G., and C. Gans, 1983. The genesis of neural crest and epidermal placodes: A reinterpretation of vertebrate origins. *Quart. Rev. Biol.*, **58**:1–28.

Patten, B. M., 1961. The normal development of the facial region. In Pruzansky, *Congenital Anomalies of the Face and Associated Structures*. Charles C Thomas, Springfield, Ill., pp. 11–45.

Ross, R. B., and M. C. Johnston, 1972. *Cleft Lip and Palate*. The Williams & Wilkins Company, Baltimore, Md., Chapter 2.

Ruch, J.-V., 1985. Epithelial-mesenchymal interactions in formation of mineralized tissues. In Butler (ed.), *The Chemistry and Biology of Mineralized Tissues*. EBSCO Media, Birmingham, Ala., pp. 54–61.

Slavkin, H. C., M. L. Snead, M. Zeichner-David, P. Bringas, and G. L. Greenberg, 1984. Amelogenin gene expression during epithelial-mesenchymal interactions. In Trelstad, *The Role of Extracellular Matrix in Development*. Alan R. Liss, New York, pp. 221–253.

Snead, M. L., M. Zeichner-David, T. Chandra, K. J. H. Rolson, S. L. C. Woo, and H. C. Slavkin, 1983. Construction and identification of mouse amelogenin cDNA clones. *Proc. Natl. Acad. Sci. USA*, **80**:7254–7258.

Thesleff, I., and K. Hurmerinta, 1981. Tissue interactions in tooth development. *Differentiation*, **18**:75–88.

DIGESTIVE AND RESPIRATORY SYSTEMS AND THE BODY CAVITIES AND MESENTERIES (CHAPTER 15)

Alescio, T., and A. Cassini, 1962. Induction in vitro of tracheal buds by pulmonary mesenchyme grafted on tracheal epithelium. *J. Exp. Zool.*, **150**:83–94.

Avery, M. E., N.-S. Wang, and H. W. Taeusch, 1973. The lung of the newborn infant. *Sci. Am.*, **228**:74–85.

Bernfield, M. R., and S. H. Banerjee, 1972. Acid mucopolysaccharide (glycosamino-glycan) at the epithelial-mesenchymal interface of mouse embryo salivary glands. *J. Cell Biol.*, **52**:664–673.

Colony, P. C., 1983. Successive phases of human fetal intestinal development. In Kretchmer and Minkowski, *Nutritional Adoption of the Gastrointestinal Tract of the Newborn*. Raven Press, New York, pp. 3–28.

Cullen, T. S., 1916. *Embryology, Anatomy, and Diseases of the Umbilicus, together with Diseases of the Urachus*. W. B. Saunders Company, Philadelphia, 680 pp.

Elliott, R., 1933. A contribution to the development of the pericardium. *Am. J. Anat.*, **48**:355–390.

Emery, J. (ed.), 1969. *The Anatomy of the Developing Lung*. William Heinemann, Ltd., London, 223 pp.

Jackson, C. M., 1909. On the developmental topography of the thoracic and abdominal viscera. *Anat. Rec.*, **3**:361–396.

Johnson, L. R., 1985. Functional development of the stomach. *Ann. Rev. Physiol.*, **47**:199–215.

Jost, A., 1962. Hormonal factors controlling the storage of glycogen in the fetal liver. In Cori, Foglia, Leloir, and Ochoa, *Perspectives in Biology*. Elsevier, Amsterdam, pp. 174–178.

Lawson, K. A., 1974. Mesenchyme specificity of rodent salivary gland development: The response of salivary epithelium to lung mesenchyme in vitro. *J. Embryol. Exp. Morphol.*, **32**:469–493.

Le Douarin, N. M., 1975. An experimental analysis of liver development. *Med. Biol.*, **53**:427–455.

Lewis, F. T., 1912. The form of the stomach in human embryos with notes upon the nomenclature of the stomach. *Am. J. Anat.*, **13**:477–503.

Lim, S.-S., and F. N. Low, 1977. Scanning electron microscopy of the developing alimentary canal in the chick. *Am. J. Anat.*, **150**:149–174.

Mall, F. P., 1891. Development of the lesser peritoneal cavity in birds and mammals. *J. Morph.*, **5**:165–179.

———, 1897. Development of the human coelom. *J. Morphol.*, **12**:395–453.

Mathan, M., P. C. Moxey, and J. S. Trier, 1976. Morphogenesis of fetal rat duodenal villi. *Am. J. Anat.*, **146**:73–92.

Moog, F., 1951. The functional differentiation of the small intestine. II. The differentiation of alkaline phosphomonoesterase in the duodenum of the mouse. *J. Exp. Zool.*, **118**:187–207.

O'Rahilly, R., 1978. The timing and sequence of events in the development of the human digestive system and associated structures during the embryonic period proper. *Anat. Embryol.*, **153**:123–136.

———, and E. A. Boyden, 1973. The timing and sequence of events in the development of the human respiratory system during the embryonic period proper. *Z. Anat. Entwickl.-Gesch.*, **141**:237–250.

Pictet, R., and W. J. Rutter, 1972. Development of the embryonic endocrine pancreas. In Greep and Astwood, *Handbook of Physiology*, Section 7: Endocrinology, vol. 1. Endocrine Pancreas. American Physiological Society, Washington, D.C., pp. 25–66.

Ponder, B. A. J., G. H. Schmidt, M. M. Wilkinson, M. J. Wood, M. Monk, and A. Reid, 1985. Derivations of mouse intestinal crypts from single progenitor cells. *Nature*, **313**:689–691.

de Reuck, A. V. S., and R. Porter (eds.), 1967. *Development of the Lung*. J. & A. Churchill, Ltd., London, 408 pp.

Rudnick, D., 1933. Developmental capacities of the chick lung in chorioallantoic grafts. *J. Exp. Zool.*, **66**:125–154.

———, 1952. Development of the digestive tube and its derivatives. *Ann. N.Y. Acad. Sci.*, **55**:109–116.

Spooner, B. S., and N. K. Wessels, 1970. Mammalian lung development: Interactions in primordium formation and bronchial morphogenesis. *J. Exp. Zool.*, **175**:445–454.

———, and ———, 1972. An analysis of salivary gland morphogenesis: Role of cytoplasmic microfilaments and microtubules. *Dev. Biol.*, **27**:38–54.

Thompson, A. B. R., and M. Keelan, 1986. The development of the small intestine. *Canad. J. Physiol. Pharmacol.*, **64**:13–29.

Wells, L. J., 1954. Development of the human diaphragm and pleural sacs. *Carnegie Cont. to Emb.*, **35**:107–134.

———, and E. A. Boyden, 1954. The development of the bronchopulmonary segments in human embryos of horizons XVII to XIX. *Am. J. Anat.*, **95**:163–201.

Wessels, N. K., 1970. Mammalian lung development: Interactions in formation and morphogenesis of tracheal buds. *J. Exp. Zool.*, **175**:455–466.

———, 1977. *Tissue Interactions and Development*. W. A. Benjamin, Inc., Menlo Park, Calif., 276 pp.

DUCTLESS GLANDS, PHARYNGEAL DERIVATIVES, AND LYMPHOID SYSTEM (CHAPTER 16)

Atwell, W. J., 1926. The development of the hypophysis cerebri in man, with special reference to the pars tuberalis. *Am. J. Anat.*, **37**:159–193.

Auerbach, R., 1966. Embryogenesis of immune systems. In Wolstenholme and Porter, *The Thymus*. Ciba Foundation Symposium. J. & A. Churchill, Ltd., London, pp. 39–49.

Burnet, M., 1969. *Cellular Immunology*. Cambridge University Press, Cambridge, Books 1 and 2, 726 pp.

Challis, J. R. G., C. T. Jones, J. S. Robinson, and G. O. Thorburn, 1977. Development of fetal pituitary-adrenal functions. *J. Steroid Biochem.*, **8**:471–478.

Gruenwald, P., 1946. Embryonic and postnatal development of the adrenal cortex, particularly the zona glomerulosa and accessory nodules. *Anat. Rec.*, **95**:391–421.

Hamburgh, M., and E. J. W. Barrington (eds.), 1971. *Hormones in Development*. Appleton-Century-Crofts, Inc., New York, 854 pp.

Hamilton, H. L., and G. W. Hinsch, 1957. The fate of the second visceral pouch in the chick. *Anat. Rec.*, **129**:357–369.

Hillemann, H. H., 1943. An experimental study of the development of the pituitary gland in chick embryos. *J. Exp. Zool.*, **93**:347–373.

Jost, A., 1961. The role of fetal hormones in prenatal development. *Harvey Lect. Series*, **55**:201–226.

Kingsbury, B. F., 1915. The development of the human pharynx. I. The pharyngeal derivatives. *Am. J. Anat.*, **18**:329–397.

Marx, J. L., 1985. The immune system "belongs to the body." *Science*, **227**:1190–1192.

Mitskevich, M. S., 1959. *Glands of Internal Secretion in the Embryonic Development of Birds and Mammals*. National Science Foundation and Israel Program for Scientific Translations, Washington, D.C., 304 pp.

Moore, M. A. S., and J. J. T. Owen, 1966. Experimental studies on the development of the bursa of Fabricius. *Dev. Biol.*, **14**:40–51.

Norris, E. H., 1937. The parathyroid glands and the lateral thyroid in man: Their morphogenesis, histogenesis, topographic anatomy and prenatal growth. *Carnegie Cont. to Emb.*, **26**:247–294.

———, 1938. The morphogenesis and histogenesis of the thymus gland in man: In which the origin of the Hassall's corpuscles of the human thymus is discovered. *Carnegie Cont. to Emb.*, **27**:191–208.

Owen, J. J. T., and E. J. Jenkinson, 1981. Embryology of the lymphoid system. *Prog. Allergy*, **29**:1–34.

Politzer, G., and F. Hann, 1935. Über die Entwicklung der branchiogenen Organe beim Menschen. *Zeitschr. f. Anat. u. Entw.*, **104**:670–708.

Rothenberg, E., and J. P. Lugo, 1985. Differentiation and cell division in the mammalian thymus. *Devel. Biol.*, **112**:1–17.

Shain, W. G., S. R. Hilfer, and V. G. Fonte, 1972. Early organogenesis of the embryonic chick thyroid. I. Morphology and biochemistry. *Dev. Biol.*, **28**:202–218.

Silverstein, A. M., 1964. Ontogeny of the immune response. *Science*, **144**:1423–1428.

Smith, R. T., R. A. Good, and P. A. Miescher (eds.), 1967. *Ontogeny of Immunity*. University of Florida Press, Gainesville, 208 pp.

Šterzl, J., and I. Říha, 1970. *Developmental Aspects of Antibody Formation and Structure*. Academia Publishing House, Prague, vols. 1 and 2, 1054 pp.

Sucheston, M. E., and M. S. Cannon, 1968. Development of zonular patterns in the human adrenal gland. *J. Morphol.*, **126**:477–492.

Weller, G. L., Jr., 1933. Development of the thyroid, parathyroid and thymus glands in man. *Carnegie Cont. to Emb.*, **24**:93–140.

Woods, J. E., G. W. De Vries, and R. C. Thommes, 1971. Ontogenesis of the pituitary-adrenal axis in the chick embryo. *Gen. Comp. Endocrinol.*, **17**:407–415.

THE UROGENITAL SYSTEM (CHAPTER 17)

Abdel-Malek, E. T., 1950. Early development of the urinogenital system in the chick. *J. Morphol.*, **86**:599–626.

Bardin, C. W., and J. F. Catterall, 1981. Testosterone: A major determinant of extragenital sexual dimorphism. *Science*, **211**:1285–1294.

Bok, G., and U. Drews, 1983. The role of the Wolffian ducts in the formation of the sinus vagina: An organ culture study. *J. Embryol. Exp. Morph.*, **73**:275–295.

Burns, R. K., 1961. Role of hormones in the differentiation of sex. In Young, *Sex and Internal Secretions*. The Williams & Wilkins Company, Baltimore, Md., 3d ed., vol. 1, pp. 16–158.

Byskov, A. G., 1978. The meiosis inducing interaction between germ cells and rete cells in the fetal mouse gonad. *Ann. Biol. Anim. Biochem. Biophys.*, **18**(2B):327–334.

———, and L. Saxén, 1976. Induction of meiosis in fetal mouse testis *in vitro*. *Dev. Biol.*, **52**:193–200.

Cunha, G. R., H. Fujii, B. L. Newbauer, J. M. Shannon, L. Sawyer, and B. A. Reese, 1983. Epithelial-mesenchymal interactions in prostatic development. I. *J. Cell Biol.*, **96**:1662–1670.

De Vries, G. J., J. P. C. DeBruin, H. B. M. Uylings, and M. A. Corner (eds.), 1984. *Sex Differences in the Brain*, *Prog. in Brain Res.*, **61**:1–516.

Ekblom, P., 1984. Basement membrane proteins and growth factors in kidney differentiation. In Trelstad (ed.), *The Role of Extracellular Matrix in Development*. Alan R. Liss, New York, pp. 173–206.

Erickson, R. S., 1968. Inductive interactions in the development of the mouse metanephros. *J. Exp. Zool.*, **169**:33–42.

Fox, H., 1963. The amphibian pronephros. *Quart. Rev. Biol.*, **38**:1–25.

Fraser, E. A., 1950. The development of the vertebrate excretory system. *Biol. Rev.*, **25**:159–187.

Friebová, Z., 1975. Formation of the chick mesonephros. I. General outline of development. *Folia Morphol.*, **23**:19–28.

Gluecksohn-Schoenheimer, S., 1943. The morphological manifestations of a dominant mutation in mice affecting tail and urogenital system. *Genetics*, **28**:341–348.

Grobstein, C., 1955. Inductive interaction in the development of the mouse metanephros. *J. Exp. Zool.*, **130**:319–340.

Gruenwald, P., 1939. The mechanism of kidney development in human embryos as revealed by an early stage in the agenesis of the ureteric buds. *Anat. Rec.*, **75**:237–247.

Hamilton, W. J., J. D. Boyd, and H. W. Mossman, 1972. *Human Embryology*. The Williams & Wilkins Company, Baltimore, Md., 4th ed, 646 pp.

Holtfreter, J., 1943. Experimental studies on the development of the pronephros. *Rev. Canad. Biol.*, **3**:220–250.

Huber, G. C., 1905. On the development and shape of uriniferous tubules of certain of the higher mammals. *Am. J. Anat.*, **4**(suppl):1–98.

Jokelainen, P., 1963. An electron microscope study of the early development of the rat metanephric nephron. *Acta Anat.*, **52**(suppl. 47):1–73.

Josso, N., and J.-Y. Picard, 1986. Anti-Müllerian hormone. *Physiol. Revs.*, **66**:1038–1090.

Jost, A., 1972. A new look at the mechanisms controlling sex differentiation in mammals. *Johns Hopkins Med. J.*, **130**:38–53.

Koff, A. K., 1933. Development of the vagina in the human fetus. *Carnegie Cont. to Emb.*, **24**:59–90.

Lehtonen, E., 1976. Transmission of signals in embryonic induction. *Med. Biol.*, **54**:108–128.

Lillie, F. R., 1917. The freemartin: A study of the action of sex hormones in the foetal life of cattle. *J. Exp. Zool.*, **23**:371–452.

Naftolin, F., and other authors, 1981. Understanding the bases of sex differences (and other articles). *Science*, **211**:1263–1324.

Ohno, W., 1978. The role of H-Y antigen in primary sex determination. *J. Amer. Med. Assoc.*, **239**:217–220.

Ohno, S., 1985. The Y-linked testis determining gene and H-Y plasma membrane antigen gene: Are they one and the same? *Endocrine Revs.*, **6**:421–431.

Oliver, J., 1968. *Nephrons and Kidneys*. Harper & Row, New York, 116 pp.

O'Rahilly, R., 1977. The development of the vagina in the human. In Blandau and Bergsma, *Morphogenesis and Malformation of the Genital System*. Birth Defects: Original Article Series, vol. 13(2). Alan R. Liss, Inc., New York, pp. 123–136.

Osathanondh, V., and E. L. Potter, 1963. Development of the human kidney as shown by microdissection. I. *Arch. Path.*, **76**:271–276; II. *Arch. Path.*, **76**:277–289; I. *Arch. Path.*, **76**:290–302.

———, and ———, 1966. Development of the human kidney as shown by microdissection. IV and V. *Arch. Path.*, **82**:391–411.

Poole, T. J., and M. S. Steinberg, 1981. Amphibian pronephric duct morphogenesis: Segregation, cell rearrangement and directed migration of the *Ambystoma* duct rudiment. *J. Embryol. Exp. Morph.*, **63**:1–16.

———, and ———, 1984. Different modes of pronephric duct origin among vertebrates. *Scanning Electron Microscopy*, 1984 (I):475–482.

Potter, E. L., 1965. Development of the human glomerulus. *Arch. Path.*, **80**:241–255.

Price, D., 1936. Normal development of the prostate and seminal vesicles of the rat with a study of experimental postnatal modifications. *Am. J. Anat.*, **60**:79–127.

———, 1947. An analysis of the factors influencing growth and development of the mammalian reproductive tract. *Physiol. Zool.*, **20**:213–247.

Sariola, H., 1985. Interspecies chimeras: An experimental approach for studies on embryonic angiogenesis. *Med. Biol.*, **63**:43–65.

Saxén, L., P. Ekblom, and E. Lehtonen, 1981. The kidney as a model system for determination and differentiation. In Ritzén et al., *The Biology of Normal Human Growth*, Raven Press, New York, pp. 117–127.

———, and E. Lehtonen, 1978. Transfilter induction of kidney tubules as a function of the extent and duration of intercellular contacts. *J. Embryol. Exp. Morphol.*, **47**:97–109.

———, H. Sariola, and E. Lehtonen, 1986. Sequential cell and tissue interactions governing organogenesis of the kidney. *Anat. Embryol.*, **175**:1–6.

Segal, S. J., 1985. Sexual differentiation in vertebrates. In Halvorson and Monroy (eds.), *The Origin and Evolution of Sex*. Alan R. Liss, New York, pp. 263–270.

Spaulding, M. H., 1921. The development of the external genitalia in the human embryo. *Carnegie Cont. to Emb.*, **13:**67–88.

Thesleff, I., and P. Ekblom, 1984. Role of transferrin in branching morphogenesis, growth and differentiation of the embryonic kidney. *J. Embryol. Exp. Morphol.*, **82:**147–161.

Torrey, T. W., 1965. Morphogenesis of the vertebrate kidney. In DeHaan and Ursprung, *Organogenesis*. Holt, Rinehart and Winston, Inc., New York, pp. 559–579.

———, 1971. *Morphogenesis of the Vertebrates*. John Wiley & Sons, Inc., New York, 3d ed., 529 pp.

Van Wagenen, G., 1965. *Embryology of the Ovary and Testis, Homo sapiens and Macaca mulatta*. Yale University Press, New Haven, Conn., 256 pp.

Vernier, R. L., and A. Birch-Anderson, 1962. Studies of the human fetal kidney. I. Development of the glomerulus. *J. Pediat.*, **60:**754–768.

Wachtel, S. S., S. Ohno, G. C. Koo, and E. A. Boyse, 1975. Possible role of H-Y antigen in primary determination of sex. *Nature*, **257:**235–236.

Waddington, C. H., 1938. The morphogenetic function of a vestigial organ in the chick. *J. Exp. Biol.*, **15:**371–376.

Witschi, E., 1939. Modification of the development of sex in lower vertebrates and in mammals. In Allen, Danforth, and Doisy, *Sex and Internal Secretions*. The Williams & Wilkins Company, Baltimore, 2d ed., pp. 145–226.

———, 1948. Migration of the germ cells of human embryos from the yolk sac to the primitive gonadal folds. *Carnegie Cont. to Emb.*, **32:**67–80.

Zackson, S. L., and M. S. Steinberg, 1986. Cranial neural crest cells exhibit directed migration on the pronephric duct pathway. Further evidence for an *in vitro* adhesion gradient. *Devel. Biol.*, **117:**342–353.

THE CIRCULATORY SYSTEM (CHAPTER 18)

Arcilla, R. A., 1986. Role of intracardiac streaming upon early cardiac development. Colloque INSERM (*Cardiovascular and Respiratory Physiology in the Fetus and Neonate*), **133:**33–45.

Barclay, A. E., K. J. Franklin, and M. M. L. Prichard, 1944. *The Foetal Circulation and Cardiovascular System, and the Changes that They Undergo at Birth*. Blackwell Scientific Publications, Ltd., Oxford, 275 pp.

Barcroft, J., 1946. *Researches on Prenatal Life*. Blackwell Scientific Publications, Ltd., Oxford, vol. 1, 292 pp.

Barron, D. H., 1944. The changes in the fetal circulation at birth. *Physiol. Rev.*, **24:**277–295.

Barry, A., 1942. The intrinsic pulsation rates of fragment of the embryonic chick heart. *J. Exp. Zool.*, **91:**119–130.

Bloom, W., and G. W. Bartelmez, 1940. Hematopoiesis in young human embryos. *Am. J. Anat.*, **67:**21–53.

Boucek, R. J., W. P. Murphy, and G. H. Paff, 1959. Electrical and mechanical properties of chick embryo heart chambers. *Circulation Res.*, **7:**787–793.

Clark, E. B., 1984. Functional aspects of cardiac development. In Zak, R. (ed.), *Growth of the Heart in Health and Disease*. Raven Press, New York, pp. 81–103.

Comline, R. S., K. W. Cross, G. S. Dawes, and P. W. Nathanielsz (eds.), 1973. *Foetal and Neonatal Physiology*. Cambridge University Press, Cambridge, 641 pp.

Congdon, E. D., 1922. Transformation of the aortic-arch system during the development of the human embryo. *Carnegie Cont. to Emb.*, **14**:47–110.

Dawes, G. S., J. C. Mott, and J. G. Widdicombe, 1955. The patency of the ductus arteriosus in newborn lambs and its physiological consequences. *J. Physiol.*, **128**:361–383.

———, ———, and ———, 1955. Closure of the foramen ovale in newborn lambs. *J. Physiol.*, **128**:384–395.

DeHaan, R. L., 1959. Cardia bifida and the development of pacemaker function of the early chick heart. *Develop. Biol.*, **1**:586–602.

Dieterlen-Lievre, R., 1978. Yolk sac erythropoiesis. *Experientia*, **34**:284–289.

Dieterlen-Lievre, F., 1984. Blood in chimeras. In Le Douarin and McLaren (eds.). *Chimeras in Developmental Biology*. Academic Press, New York, pp. 133–163.

Evans, H. M., 1909. On the development of the aortae, cardinal and umbilical veins and other blood vessels of vertebrate embryos from capillaries. *Anat. Rec.*, **3**:498–518.

Everett, N. B., and R. J. Johnson, 1951. A physiological and anatomical study of the closure of the ductus arteriosus in the dog. *Anat. Rec.*, **110**:103–111.

Girard, H., 1973. Arterial pressure in the chick embryo. *Am. J. Physiol.*, **224**:454–560.

Goss, C. M., 1942. The physiology of the embryonic mammalian heart before circulation. *Am. J. Physiol.*, **137**:146–152.

———, 1952. Development of the median coordinated ventricle from the lateral hearts in rat embryos with 3 to 6 somites. *Anat. Rec.*, **112**:761–796.

Graeper, L., 1907. Untersuchungen über die Herzbildung der Vögel. *Wilhelm Roux', Arch.*, **24**:375–410.

Heuser, C. H., 1923. The branchial vessels and their derivatives in the pig. *Carnegie Cont. to Emb.*, **15**:121–139.

Hirota, A., T. Sakai, S. Fujii, and K. Kamino, 1983. Initial development of conduction pattern of spontaneous action potential in early embryonic precontractile chick heart. *Devel. Biol.*, **99**:517–523.

Hoff, E. C., T. C. Kramer, D. DuBois, and B. M. Patten, 1939. The development of the electrocardiogram of the embryonic heart. *Am. Heart J.*, **17**:470–488.

Jacobson, A. G., 1960. Influences of ectoderm and endoderm on heart differentiation in the newt. *Dev. Biol.*, **2**:138–154.

Johnstone, P. N., 1971. *Studies on the Physiological Anatomy of the Embryonic Heart*. Charles C Thomas, Publisher, Springfield, Ill., 139 pp.

Kirby, M. L., 1987. Cardiac morphogenesis. Recent research advances. *Pediat. Res.*, **21**:219–224.

———, T. F. Gale, and D. E. Stewart, 1983. Neural crest cells contribute to normal aorticopulmonary septation. *Science*, **220**:1059–1061.

Knezevic, A., N. Petrovic, and D. Radivoyevic, 1971. The effect of the humoral erythropoietic stimulation factor of different origins on chick embryo hematopoiesis. *Jugoslav. Physiol. Pharmacol. Acta*, **7**:421–429.

Kramer, T. C., 1942. The partitioning of the truncus and conus and the formation of the membranous portion of the interventricular septum in the human heart. *Am. J. Anat.*, **71**:343–370.

Lemanski, L. F., D. J. Paulson, and C. S. Hill, 1979. Normal anterior endoderm corrects the heart defect in cardiac mutant salamanders (*Ambystoma mexicanum*). *Science*, **204**:860–862.

Lemez, L., 1964. The blood of chick embryos. Quantitative embryology at a cellular level. *Adv. Morphogen.*, **3:**197–245.

Lind, J., and C. Wegelius, 1949. Angiocardiographic studies on the human foetal circulation. *Pediatrics*, **4:**391–400.

Manasek, F. J., 1975. The extracellular matrix in the early embryonic heart. In Lieberman and Sano, *Developmental Aspects of Cardiac Cellular Physiology*. Raven Press, New York, pp. 1–20.

———, M. B. Burnside, and R. E. Waterman, 1972. Myocardial cell shape as a mechanism of embryonic heart looping. *Dev. Biol.*, **29:**349–371.

Markwald, R. R., T. P. Fitzharris, H. Bank, and D. H. Bernanke, 1978. Structural analysis on the matrical organization of glycosaminoglycans in developing endocardial cushions. *Dev. Biol.*, **62:**292–316.

———, ———, and F. J. Manasek, 1977. Structural development of endocardial cushions. *Am. J. Anat.*, **148:**85–120.

———, R. B. Runyan, G. T. Kitten, F. M. Funderburg, D. H. Bernanke, and P. R. Brauer, 1984. Use of collagen gel cultures to study heart development: Proteoglycan and glycoprotein interactions during the formation of endocardial cushion tissue. In Trelstad (ed.), *The Role of Extracellular Matrix in Development*. Alan R. Liss, Inc., New York, pp. 323–350.

McClure, C. F. W., and E. G. Butler, 1925. The development of the vena cava inferior in man. *Am. J. Anat.*, **35:**331–383.

Nigon, V., and J. Godet, 1976. Genetic and morphogenetic factors in hemoglobin synthesis during higher vertebrate development: An approach to cell differentiation mechanisms. *Int. Rev. Cytol.*, **46:**79–176.

Padget, D. H., 1948. The development of the cranial arteries in the human embryo. *Carnegie Cont. to Emb.*, **32:**205–261.

———, 1957. The development of the cranial venous system in man, from the viewpoint of comparative anatomy. *Carnegie Cont. to Emb.*, **36:**79–140.

Patten, B. M., 1931. The closure of the foramen ovale. *Am. J. Anat.*, **48:**19–44.

———, 1949. Initiation and early changes in the character of the heart beat in vertebrate embryos. *Physiol. Rev.*, **29:**31–47.

———, 1951. The first heart beats and the beginning of the embryonic circulation. *Am. Scientist*, **39:**224–243.

———, 1956. The development of the sinoventricular conduction system. *Univ. Mich. Med. Bull.*, **22:**1–21.

———, 1960. The development of the heart. In Gould, *The Pathology of the Heart*. Charles C Thomas, Publisher, Springfield, Ill., 2d ed., pp. 24–92.

———, and T. C. Kramer, 1933. The initiation of contraction in the embryonic chick heart. *Am. J. Anat.*, **53:**349–375.

———, ———, and A. Barry, 1948. Valvular action in the embryonic chick heart by localized apposition of endocardial masses. *Anat. Rec.*, **102:**299–311.

Pexieder, T., 1972. The tissue dynamics of heart morphogenesis. I. The phenomena of cell death. *Z. Anat. Entwickl.-Gesch.*, **138:**241–253.

Reagan, F. P., 1929. A century of study upon the development of the eutherian vena cava inferior. *Quart. Rev. Biol.*, **4:**179–212.

Rifkind, R. A., 1974. Erythroid cell differentiation. In Lash and Whittaker, *Concepts of Development*. Sinauer Associates, Stamford, Conn., pp. 149–162.

Rychter, Z., 1962. Experimental morphology of the aortic arches and the heart loop in chick embryos. *Adv. Morphogen.*, **2:**333–371.

Sabin, F. R., 1916. The origin and development of the lymphatic system. *Johns Hopkins Hosp. Rept.*, **17**:347–440.

Thompson, R. P., and T. P. Fitzharris, 1979. Morphogenesis of the truncus arteriosus of the chick embryo heart: The formation and migration of mesenchymal tissue. *Am. J. Anat.*, **154**:545–556.

Walls, E. W., 1947. The development of the specialized conducting tissue of the human heart. *J. Anat.*, **81**:93–110.

Wilt, F. W., 1965. Erythropoiesis in the chick embryo: The role of endoderm. *Science*, **147**:1588–1590.

APPENDIX 1

Boyden, E. A., 1918. Vestigial gill filaments in chick embryos with a note on similar structures in reptiles. *Am. J. Anat.*, **23**:205–235.

———, 1924. An experimental study of the development of the avian cloaca, with special reference to a mechanical factor in the growth of the allantois. *J. Exp. Zool.*, **40**:437–471.

Coulombre, A. J., 1955. Correlations of structural and biochemical changes in the developing retina of the chick. *Am. J. Anat.*, **96**:153–189.

Criley, B. B., 1969. Analysis of the embryonic sources and mechanisms of development of posterior levels of chick neural tubes. *J. Morph.*, **128**:465–501.

Davis, C. L., 1924. The cardiac jelly of the chick embryo. *Anat. Rec.*, **27**:201–202.

DeHaan, R. L., 1959. Cardia bifida and the development of pacemaker function in the early chick heart. *Dev. Biol.*, **1**:586–602.

de la Cruz, M. V., S. Muñoz-Armas, and L. Muñoz-Castellanos, 1972. *Development of the Chick Heart*. The Johns Hopkins Press, Baltimore, 80 pp.

Fraser, R. C., 1954. Studies on the hypoblast of the young chick embryo. *J. Exp. Zool.*, **126**:340–400.

———, 1960. Somite genesis in the chick. III. The role of induction. *J. Exp. Zool.*, **145**:151–167.

Gaertner, R. A., 1949. Development of the posterior trunk and tail of the chick embryo. *J. Exp. Zool.*, **111**:157–174.

Glick, B., 1964. The bursa of Fabricius and the development of immunologic competence. In Good and Gabrielson, *The Thymus in Immunobiology*. Hoeber, New York, pp. 343–358.

Grabowski, C. T., 1956. The effects of the excision of Hensen's node on the early development of the chick embryo. *J. Exp. Zool.*, **133**:301–344.

———, 1962. Neural induction and notochord formation by mesoderm from the node area of the early chick blastoderm. *J. Exp. Zool.*, **150**:233–246.

Hamburger, V., 1948. The mitotic patterns in the spinal cord of the chick embryo and their relation to histogenic processes. *J. Comp. Neurol.*, **88**:221–283.

———, and H. L. Hamilton, 1951. A series of normal stages in the development of the chick embryo. *J. Morphol.*, **88**:49–92.

Hammond, W. S., 1954. Origin of thymus in the chicken embryo. *J. Morphol.*, **95**:501–522.

Hillemann, H. H., 1943. An experimental study of the development of the pituitary gland in chick embryos. *J. Exp. Zool.*, **93**:347–373.

Hinsch, G. W., and H. L. Hamilton, 1956. The developmental fate of the first somite of the chick. *Anat. Rec.*, **125**:225–246.

Hughes, A. F. W., 1934. On the development of the blood vessels in the head of the chick. *Phil. Trans. Roy. Soc. London*, ser. B, **224**:75–161.

Hunt, T. E., 1931. An experimental study of the independent differentiation of the isolated Hensen's node and its relation to the formation of axial and nonaxial parts in the chick embryo. *J. Exp. Zool.*, **59**:395–427.

———, 1937. The development of gut and its derivatives from the mesectoderm and mesentoderm of early chick blastoderms. *Anat. Rec.*, **68**:349–369.

Ivey, W. D., and S. A. Edgar, 1952. The histogenesis of the esophagus and crop of the chicken, turkey, guinea fowl and pigeon, with special reference to ciliated epithelium. *Anat. Rec.*, **114**:189–212.

Locy, W. A., and O. Larsell, 1916. The embryology of the bird's lung based on observations of the domestic fowl. *Am. J. Anat.*, **19**:447–504; **20**:1–44.

Patten, B. M., 1922. The formation of the cardiac loop in the chick. *Am. J. Anat.*, **30**:373–397.

———, 1925. The interatrial septum of the chick heart. *Anat. Rec.*, **30**:53–60.

———, and T. C. Kramer, 1933. The initiation of contraction in the embryonic chick heart. *Am. J. Anat.*, **53**:349–375.

Romanoff, A. L., 1931. Cultivation of the chick embryo in an opened egg. *Anat. Rec.*, **48**:185–189.

Rudnick, D., 1935. Regional restriction of potencies in the chick during embryogenesis. *J. Exp. Zool.*, **71**:83–99.

Sabin, F. R., 1917. Origin and development of the primitive vessels of the chick and the pig. *Carnegie Cont. to Emb.*, **6**:61–124.

Schoenwolf, G. C., 1977. Tail (end) bud contributions to the posterior region of the chick embryo. *J. Exp. Zool.*, **201**:227–246.

———, 1979. Histological and ultrastructural observations of tail bud formation in the chick embryo. *Anat. Rec.*, **193**:131–148.

Spratt, N. T., Jr., 1948. Development of the early chick blastoderm on synthetic media. *J. Exp. Zool.*, **107**:39–64.

———, 1955. Analysis of the organizer center in the early chick embryo. I. Localization of prospective notochord and somite cells. *J. Exp. Zool.*, **128**:121–164.

———, 1957. Analysis of the organizer center in the early chick embryo. II. Studies of the mechanics of notochord elongation and somite formation. *J. Exp. Zool.*, **134**:577–612.

———, and H. Haas, 1960. Integrative mechanisms in development of the early chick blastoderm. I. Regulative potentiality of separated parts. *J. Exp. Zool.*, **145**:97–137.

———, and ———, 1961. Integrative mechanisms in development of the early chick blastoderm. II. Role of morphogenetic movements and regenerative growth in synthetic and topographically disarranged blastoderms. *J. Exp. Zool.*, **147**:57–93.

Vaage, S., 1969. The segmentation of the primitive neural tube in chick embryos (*Gallus domesticus*). *Ergeb. Anat. Entwicklungsges.*, **41**(3):1–88.

Waddington, C. H., 1935. Induction by the primitive streak and its derivatives in the chick. *J. Exp. Biol.*, **10**:38–46.

———, 1934. Experiments on embryonic induction. I. The competence of the extraembryonic ectoderm in the chick. II. Experiments on coagulated organizers in the chick. III. A note on inductions by chick primitive streak transplanted to the rabbit embryo. *J. Exptl. Biol.*, **11**:211–227.

———, and G. A. Schmidt, 1933. Induction by heteroplastic grafts of the primitive streak in birds. *Arch. f. Entw.-mech. d. Organ.*, **128**:521–563.

——, ——, and B. J. Fowler, 1954. The role of the neural tube and notochord in development of the axial skeleton of the chick. *Am. J. Anat.*, **95**:337–399.

——, C. R. Goodheart, G. Goodheart, and G. Lindberg, 1955. The influence of adjacent structures upon the shape of the neural tube and neural plate of chick embryos. *Anat. Rec.*, **122**:539–559.

Weiss, P., and G. Andres, 1952. Experiments on the fate of embryonic cells (chick) disseminated by the vascular route. *J. Exp. Zool.*, **121**:449–488.

Willier, B. H., 1927. The specificity of sex, of organization, and of differentiation of embryonic chick gonads as shown by grafting experiments. *J. Exp. Zool.*, **46**:409–465.

——, 1954. Phases in embryonic development. *J. Cell. and Comp. Physiol.*, **43**, suppl., **1**:307–318.

Yntema, C. L., and W. S. Hammond, 1955. Experiments on the origin and development of the sacral autonomic nerves in the chick embryo. *J. Exp. Zool.*, **129**:375–413.

INDEX

Ablation, 47
Acetylcholine, 241
Acetylcholinesterase, *322, 323*
Acid:
  hyaluronic, 20, 240, 342
  retinoic, 406
  Acrosin, 135
  Acrosomal process, 121
  Acrosomal reaction, 121, *122, 134; 136*
  Acrosome, 89
  Actin, 18, 121, 152, 331
  Actinomycin D, 52, 181, 184
  Adenyl cyclase, 12
  Adrenal:
  cortex, *543*
  medulla, 542, *543*
Afterbirth, 264, {289}
Air chamber, in egg, *116,* 117, *253*
Albinism, 54
Albumen, *105,* 115, *116,* 256
Alizarin red, 352
Alkaline phosphatase, 75, 343, 401
Allantois, 251
  avian, *254, 260-261, 262*
  mammalian, {*288,*} *289*
Allophenic mice (*See* Tetraparental mice)
Alpha-amanitin, 184
Ameloblast, 501, *505*
Amelogenin, 503
Amnion, 251
  avian, *254, 257-259*
  head fold of, *259, 291*
  lateral fold, *253*
  mammalian, *272,* {*298*}
  tail fold, *259, 291*
Amniote, 251

Amphibian embryo:
  cleavage and formation of blastula in, 158-163
  establishment of polarity in, 141-147
  gastrulation in, 191-199
  limb development, 392-397
  neurulation in, 229-235
  primary induction in, 223-229
Ampulla:
  of inner ear, 481
  of uterine tube, *57*
Anamniote, 251
Anaphase, *78, 81, 82*
Anastomosis:
  intersubcardinal, *602*
  subcardinal-supra cardinal, *502*
Androgen-binding protein, 72
Anterior pituitary (*See* Hypophysis, pars distalis)
Antibody, 542, 546
  maternal, 546
  monoclonal, 46
Antigen, 46
Antrum, 107
Anus, *57, 231*
Aorta:
  branches of, 596-598
  dorsal, *305, 309*
Appendage bud (*See* Limb bud)
Appendix:
  of epididymis, *60,* 573
  of testis, *60*
Apteria, 373
Aqueduct, cerebral of (of Sylvius), 439, *440*
Arc, vitelline, 256
Arch:
  aortic, 310, *680*
  branchial, 667, *680*

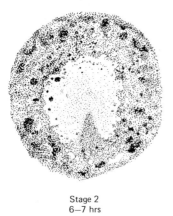

Stage 2
6—7 hrs

Stage 5
19-22 hrs

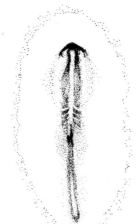

Stage 8
26—29 hrs

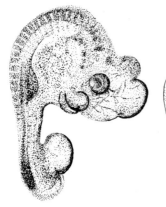

Stage 20
70—72 hrs

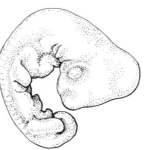

Stage 24
4 days

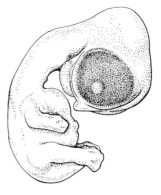

Stage 30
6½ days